Proceedings of the 6th International Conference on Photonics and Imaging in Biology and Medicine (PIBM 2007)

Advances in Biomedical Photonics and Imaging

Advances in Biomedical Photonics and Imaging

Proceedings of the 6th International Conference on Photonics and Imaging in Biology and Medicine (PIBM 2007)

Wuhan, P R China 4 – 6 November 2007

Editors

Qingming Luo
Britton Chance Center for Biomedical Photonics, China

Lihong V Wang
Washington University in St Louis, USA

Valery V Tuchin
Saratov State University, Russia

World Scientific

NEW JERSEY • LONDON • SINGAPORE • BEIJING • SHANGHAI • HONG KONG • TAIPEI • CHENNAI

Published by

World Scientific Publishing Co. Pte. Ltd.
5 Toh Tuck Link, Singapore 596224
USA office: 27 Warren Street, Suite 401-402, Hackensack, NJ 07601
UK office: 57 Shelton Street, Covent Garden, London WC2H 9HE

British Library Cataloguing-in-Publication Data
A catalogue record for this book is available from the British Library.

ADVANCES IN BIOMEDICAL PHOTONIC AND IMAGING
Proceedings of the 6th International Conference on Photonics and Imaging in Biology and Medicine (PIBM 2007)

ISBN-13 978-981-283-233-7
ISBN-10 981-283-233-5

Printed in Singapore by World Scientific Printers

PREFACE

The 6th International Conference on Photonics and Imaging in Biology and Medicine (PIBM 2007) was held during November 4-6, 2007 at Wuhan National Laboratory for Optoelectronics (WNLO) in the campus of the Huazhong University of Science and Technology (HUST), Wuhan, P.R. China. The present volume contains the texts of a selection from the invited talks delivered at the conference and the poster presentations.

PIBM is the largest international biomedical photonics conference in Asia. It was initially held at HUST biyearly since 1999. After three times held in Wuhan (1999, 2001, and 2003), it was hosted once in Tianjin (2005), and then, back to Wuhan yearly since 2006. PIBM is designed to bring together scientists, engineers and clinical researchers from a variety of disciplines engaged in applying optical science, photonics and imaging technologies to problems in biology and medicine. The scope of this conference ranges from basic research to instrumentation engineering, to biological and clinical studies. It is recognized as one of the largest and most comprehensive international conferences in China, and represents the highest level of the worldwide research in this field. In the past eight years, 5 volumes of proceedings with 510 papers were published by SPIE (International Society for Optical Engineering). All the published papers were indexed by Ei Compendex. More and more young researchers presented and exchanged their innovative ideas on this friendly and professional platform, which made PIBM an unforgettable annual meeting in Wuhan.

This year, PIBM attracted distinguished scholars in biomedical photonics and imaging field from all over the world, such as United States, United Kingdom, Russia, Canada, Israel, France, Japan, Sudan, and China. The conference gave attendees the opportunity to interact one-on-one with presenters. Multiple poster sessions allowed lively discussions on the latest research. During the three days, PIBM focused on 5 topics including 24 invited lectures and 89 poster presentations. This volume includes a selection of 6 invited papers, and 69 excellent subscriptions, some of which are recommended to a new journal, Journal of Innovation in Optical Health Science (JIOHS).

The major topics covered at the conference and presented in this volume include:

- ✧ Tissue Optics and Diffuse Optical Imaging
- ✧ Optical Molecular Imaging
- ✧ Multiphoton Microscopy in Biomedical Sciences
- ✧ Photonic Therapeutics, Diagnostics and Instrumentations
- ✧ Multimodal and Hybrid Biomedical Imaging

The Secretariat and Local Organizing Committee deserve great appreciation in creating a smoothly run and productive conference with comprehensive instructive lectures and innovation work presented by posters. The faculties and students from Britton Chance Center for Biomedical Photonics were dedicated in the reception and service work during the conference days. It is a pleasure to thank all of them for their hard work. We also gratefully thank the financial support by National Natural Science Foundation of China (30710303094, 90508003), 111 Project (B07038), 863 Project (2006AA02Z343, 2006AA020801), Program for Changjiang Scholars and Innovative Research Team in University (IRT0623), and the organizing and coordinating help from WNLO and HUST.

Finally, we would like to thank all of the authors for their contributions to PIBM 2007 and all the members of the Program Committee, listed on the following page, for their cooperation and time spent reviewing submissions.

Wuhan, P. R. China
December, 2007

Qingming Luo
Lihong V. Wang
Valery V. Tuchin
Conference Chairs

CONFERENCE HONORARY CHAIRS

Britton Chance, University of Pennsylvania (USA)
Bingkun Zhou, Tsinghua University (China)

CONFERENCE HONORARY CHAIRS

Qingming Luo, Huazhong University of Science and Technology (China)
Lihong V. Wang, Washington University in St Louis (USA)
Valery V. Tuchin, Saratov State University (Russia)

PROGRAM COMMITTEE

Jing Cheng, Tsinghua University (China)
Wei R. Chen, University of Central Oklahoma (USA)
Wei R. Chen, Yale University (USA)
Zhongping Chen, University of California/Irvine (USA)
Arthur Chiou, National Yang-Ming University (China)
Zhihua Ding, Zhejiang University (China)
Yuwen Qin, National Natural Science Foundation of China (China)
Steven L. Jacques, Oregon Medical Laser Center (USA)
Huabei Jiang, University of Florida (USA)
Xingde Li, University of Washington at Seattle (USA)
Hong Liu, University of Oklahoma (USA)
Zuhong Lu, Southeast University (China)
Hui Ma, Tsinghua University (China)
Atsushi Maki, Hitachi, Ltd. (Japan)
Avraham Mayevsky, Bar Ilan University (Israel)
Igor V Meglinski, Cranfield University (UK)
Stephen P Morgan, University of Nottingham (UK)
Shoko Nioka, University of Pennsylvania (USA)
Vasilis Ntziachristos, Harvard University (USA)
Yingtian Pan, State University of New York at Stony Brook (USA)
Alexander V. Priezzhev, M.V. Lomonosov Moscow State University (Russia)
Ammasi Periasamy, University of Virginia (USA)
Alexander Savitsky, A.N. Bach Institute of Biochemistry, RAS (Russia)
Peter T. C. So, Massachusetts Institute of Technology (USA)
Q. Charles Su, Illinois State University (USA)
Mamoru Tamura, Hokkaido University (Japan)
Bruce Tromberg, University of California/Irvine (USA)
Andong Xia, Institute of Chemistry, Chinese Academy of Sciences (China)
Shusen Xie, Fujian Normal University (China)

ORGANIZING COMMITTEE

LOCAL SECRETARIAT

6th International Conference on Photonics and Imaging in Biology and Medicine (PIBM 2007)

CONTENTS

Invited Papers

SHEDDING LIGHT ON LIFE: OPTICAL ASSESSMENT OF MITOCHONDRIAL FUNCTION AND TISSUE VITALITY IN BIOLOGY AND MEDICINE

MAYEVSKY AVRAHAM

The Mina & Everard Goodman Faculty of Life-Sciences and The Gonda Multidisciplinary Brain Research Center Bar-Ilan University, Ramat-Gan, 52900, Israel
Email: mayevsa@mail.biu.ac.il

The involvement of mitochondrial dysfunction in various pathophysiological conditions, developed in experimental and clinical situations, is widely documented. Nevertheless, real time monitoring of mitochondrial function *in-vivo* is very rare. The pressing question is how the mitochondria of intact tissues behave under *in-vivo* conditions as compared to isolated mitochondria that had been described by Chance and Williams over 50 years ago. This subject has been recently discussed in detail (Mayevsky and Rogatsky 2007). We reviewed the subject of evaluating mitochondrial function by monitoring NADH fluorescence together with microcirculatory blood flow, Hemoglobin oxygenation and tissue reflectance. These 4 parameters represent the vitality of the tissue and could be monitored in vivo, using optical spectroscopy, in animal models as well as in clinical practice. It is a well known physiological hypothesis that, under emergency conditions, the sympathetic nervous system will give preference to the most vital organs in the body, namely the brain, heart and adrenal glands. The less vital organs, such as the skin, GI-tract, and Urethral wall, will become hypoperfused and their mitochondrial activity will be inhibited. The monitoring of the less vital organs may reveal critical tissue conditions that may manifest an early phase of body deterioration. The aim of the current presentation is to review the experimental and preliminary clinical results accumulated using a new integrated medical device – the "CritiView" which enabled, for the first time, monitoring 4 parameters from the tissue using a single optical probe. The CritiView is a computerized optical device that integrates hardware and software in order to provide real time information on tissue vitality. In preliminary clinical testing, we used a 3-way Foley catheter that includes a bundle of optical fibers enabling the monitoring of the 4 parameters, representing the vitality of the urethral wall (a less vital organ).We found that the exposure of patients to metabolic imbalances in the operation room led to changes in tissue blood flow and inhibition of mitochondrial function in the urethral wall. In conclusion, the new device "CritiView" could provide reliable, real time data on mitochondrial function and tissue vitality in experimental animals as well as in patients.

1. Introduction

Normal physiological function of any organ is dependent on its capacity to produce biological energy from substrates available to it. A main source of energy is glucose which is oxidized through a complex biochemical chain, primarily through a series of enzymatic reactions known as the Krebs cycle. The end-product of these reactions is the production of ATP by oxidative phosphorylation occurred in the mitochondria. Thus, production of biological energy is dependent upon the functional integrity of the Krebs cycle, the normal mitochondrial function and availability of oxygen. Disruption of oxygen supply may produce disease which is manifested by physiological dysfunction. It is therefore of clinical importance to develop integrative experimental approaches which would make possible elucidation of the interaction between biochemical events, such as mitochondrial function, and their physiological consequences.

In 1914 Barcroft described in details the role of oxygen in the function of cells and tissues and concluded that "there is no instance in which it can be proven that an organ increases its activity under physiological conditions, without also increasing in its call for oxygen", and "in no organ excited by any form of stimulation can it be shown that positive work is done without the blood supply having to respond to a call for oxygen"[3]. Less then ten years earlier, the involvement of adenine nucleotides in yeast fermentation was described [18]. This discovery opened a period of 100 years of intensive research activities on mitochondrial function. Table 1 describes the history of NADH monitoring in studying mitochondrial function. As can be seen in the table, the contribution of Prof. Britton Chance to this important field is obvious. Also, the shift from a single parameter to the multiparametric monitoring approach including NADH redox state is presented.

Table 1. Milestones in Monitoring of Mitochondrial activity using NADH fluorescence and other physiological parameters, from animal studies to clinical use.

Year	Discovery	Author(s)
1906	Involvement of adenine containing nucleotides in yeast fermentation	Harden & Young [20]
1935	Complete structure of "Hydrogen transferring Coenzyme" in erythrocytes	Warburg et al [56]
1936	Definition of the two cofactors DPN and TPN	Warburg O. [55]
1951	A shift in the absorption spectrum of DPNH with Alcohol dehydrogenase	Theorell& Bonnichsen [53]
1951	Development of a rapid and sensitive Spectrophotometer	Chance& Legallias [10]
1952	Monitoring of pyridine nucleotide enzymes	Chance [5]
1954	Development of the double beam spectrophotometer	Chance [4]
1957	The first detailed study of NADH using Fluorescence spectrophotometry	Duysens & Amesz [15]
1958	Measurement of NADH fluorescence in isolated mitochondria	Chance & Baltscheffsky [6]
1959	Measurement of muscle NADH fluorescence *in vitro*	Chance & Jobsis [9]
1962	*In vivo* monitoring of NADH fluorescence from the brain and kidney	Chance et al [7]
1965	Comparison between NADH fluorescence in vivo and enzymatic analysis	Chance et al. [8]
1968	Monitoring tissue reflectance in addition to NADH fluorescence	Jöbsis & Stansby [24]
1971	The first attempt to monitor the human brain in neurosurgry	Jöbsis et al. [23]
1973	The first fiber optic NADH fluorometer used in the brain of an awake animal	Chance et al. [11]; Mayevsky & Chance [37]
1982	Simultaneous monitoring of NADH in vivo in four different organs	Mayevsky & Chance [39]
1982	Monitoring of NADH, ECoG, DC potential and extracellular ions *in vivo*	Friedli et al [17]
1985	Monitoring of brain NADH together with 31P NMR Spectroscopy	Mayevsky et al. [45]
1991	Monitoring of NADH , CBF, ECoG, and extracellular ions in animals and neurosurgical patients	Mayevsky et al. [42]
1996	Detection of cortical spreading depression in a comatose patient	Mayevsky et al. [41]
2000	"Tissue Spectroscope" a new device for monitoring of NADH and TBF	Mayevsky et al. [44]
2006	Monitoring of tissue vitality (NADH, TBF and HbO2) by the "CritiView"	Mayevsky et al. [35]

During the last decades a large number of publications had shown the very wide involvement of mitochondrial dysfunction in many pathological conditions as shown in Figure 1. It is clear that normal mitochondrial function is a critical factor and need in order to keep the physiological and biochemical activities of all cells, tissues and organs in the body [36,45].

In this article, the brain will be used as a target organ but the same principles of mitochondrial function could be applied to other organs and systems in the body as well.

The mechanisms behind the development of brain damage during pathological states such as ischemia, hypoglycemia and epilepsia were investigated during the past two decades using various experimental

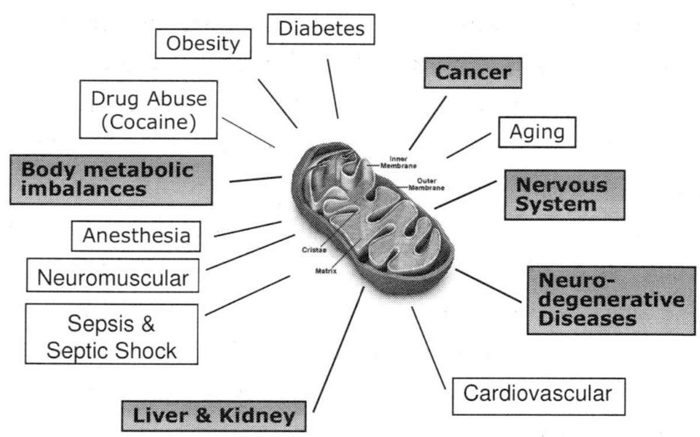

Figure 1. The involvement of mitochondrial dysfunction in various pathophysiological conditions. Mitochondrial dysfunction was described as a critical step in the development of many states of disease in patients.

approaches. Also, the mechanisms behind other human brain related diseases such as Alzheimer or alcohol intoxcication and other drug addiction, are poorly understood due to the lack of suitable experimental models and monitoring capabilities.

1.1. *Energy Metabolism and Brain Functions*

Normal brain activity depends upon the continuous supply of oxygen (carried by the blood) due to the high O_2 consumption as well as a very limited reserve of dissolved O_2 in the tissue. The brain consumes approximately 20% of the total O_2 used by the body per unit time. Over 50% of the energy consumed by the brain is utilized for active transport mechanisms (such as Na^+K^+ATPase) responsible for the normal distribution of ions around the cell membranes [15]. ATP is the sole form of energy available for cellular functions and is produced by the oxidative phosphorylation process which takes place in the mitochondria. Direct coupling exists between the extracellular levels of various ions (such as K^+, Ca^{2+}, Na^+ or H^+) and the metabolic activity of the mitochondria, reflected in the oxidation reduction state of the respiratory chain enzymes (such as NADH - Nicotine Adenine Dinucleotides).

A decrease in O_2 supply to the brain will result in a decrease in ATP levels and in the inhibition of active transport mechanisms (such as Na^+ K^+ ATPase). Extracellular K^+ levels will then increase until restoration of normal energy supply. The redox state of the mitochondria is a sensitive indicator of the intracellular metabolic state and can be used for the evaluation of the cellular energy status [12, 30]. Any change in the electrical activity of the brain will result in an activation of the ion pumps in an effort to restore normal ion distribution. In the pioneering work described by Chance and Williams [12], they describe several metabolic states for the isolated mitochondria, which depend upon the availability of O_2, substrate and ADP. The "resting state", state 4, exhibited high O_2 and substrate levels, with the limiting factor being the ADP level. Any increase in the energy demand will result in the acceleration of electron flow through the respiratory chain and the oxidation of the various carriers. The active state, during which O_2 consumption is increased, was defined as state 3. In state 3, cerebral blood flow will increase in order to compensate for the increased O_2 consumption [47], while in state 4, 99% of the NADH will be in the reduced form. The "resting" brain in vivo is probably somewhere between state 4 and state 3 [31,32]. In order to evaluate the functional state of the brain in vivo it is necessary to monitor numerous parameters relevant to various brain functions. A multiparametric monitoring device was developed for this purpose and applied to an experimental animal model of stroke [33] and to other pathological states [34].

As shown in Fig. 2, the supply of O_2 to all tissues is based on the same principles namely, that the

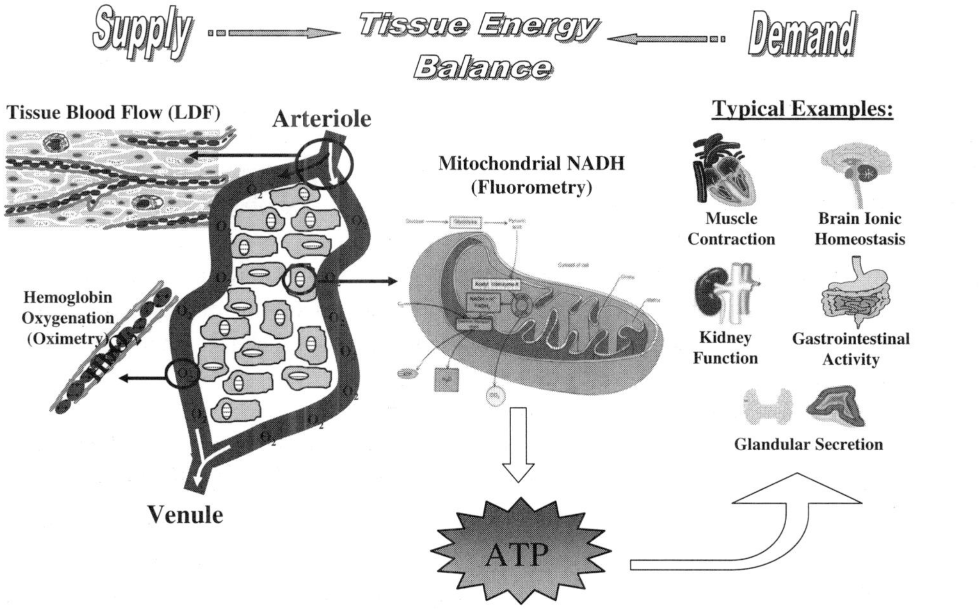

Figure 2. The relationship between oxygen supply and demand in the various tissues in the body. The redox state of the mitochondria could be used as an indicator to oxygen balance in the tissues.

oxygenated blood flows in the tissue is delivering the O_2 to the cells. Oxygen's demand on the other hand is dependent on the energy needs of every organ. In other words in order to measure O_2 supply to various tissues one can use the same technology while for O_2 demand, different techniques need to be applied to the different organs. In general terms, when the balance between O_2 supply and O_2 demand is negative (oxygen debt is created) the function of the tissue or organ will be affected and a pathological state is developed. Typical clinical examples of conditions characterized by a severe negative O_2 balance are myocardial infarction or stroke where due to occlusion of blood vessel in the heart or the brain the supply of O_2 to a specific region in the vital organ is limited and the function is then inhibited or severely compromised. These two examples represent dramatic and acute pathological events that create very specific manifestations and the diagnosis of such conditions are relatively straight forward.

Conversely, subclinical O_2 delivery insufficiency is a condition that may result from a number of several pathological states in patients with acute critical Illness. The underlying characteristic here can be defined as a circulatory failure and this may result from different conditions associated with Shock such as sepsis, cardiac insufficiency, hemorrhage, neurogenic shock etc.

In order to assess the functioning brain *in vivo*, we have developed a unique approach by which various parameters representing the several functions of the brain are simultaneously monitored from the same area in real-time mode. The key parameter that was developed was the monitoring of the intactness of mitochondrial function. The energy state and metabolism are evaluated by monitoring the cerebral blood flow and volume (Laser Doppler flowmetry) as well as intramitochondrial redox state (surface NADH fluorometry reflectometry). The ion homeostasis is determined by measuring the extracellular K^+, Na^+ Ca^{2+} activities (ion selective surface mini-electrodes). The electrical activities are assessed by the spontaneous ECoG activity (bipolar electrodes) as well as by the extracellular direct current steady potential (Ag/AgCl-electrodes).

The various aforementioned transducer probes are located on the surface of the cerebral cortex by using the **Multiprobe Assembly-MPA** which has been adapted for routine usage in experimental animals as well as under operating room conditions [29,41,45]. The MPA is connected to various specific detectors and amplifiers and the data are displayed on a multi-channel computerized recording system. We record up to 16 channels at the same time and only the most advanced computer technology can cope with the need for data acquisition and storage as well as on-line and off-line analyses.

Using the MPA in conjunction with the computerized monitoring and acquisition system will help the neuroscientist and the neurosurgeon to evaluate the status of the brain during various procedures. It is believed that such a system will contribute to the greater success in drug development and usage as well during brain surgery and thus to benefit the patient.

Figure 2 also shows the parameters that can be monitored on-line even in the awake state *in vivo*. Relative CBF is monitored by Laser Doppler Flowmetry [54]. Mitochondrial redox state is measured by monitoring NADH fluorescence [30]. Surface minielectrodes are used for monitoring extracellular levels of K^+, Na^+, Ca^{2+} and H^+. The functional state of the brain is evaluated by monitoring electrical activities (ECoG and DC steady potential).

Our approach is to monitor the brain at the "tissue level" rather than the "cellular level". We believe that the various elements of the brain: neurons, glial cell and capillaries, act as an integrated system. Therefore, we monitor the brain functions using "mini"-probes rather than the "micro"-probes commonly used by other investigators for monitoring ionic homeostasis [17]. The evaluations of other parameters, such as CBF and NADH redox state, were also adapted to the "tissue" level rather than to the "cellular" level. For all these reasons, our strategy was to develop a multiparametric monitoring assembly in which all the probes have the same type of contact with the sampled tissue volume [13,16,29,40,43]. The probes do not penetrate the tissue itself, thus avoiding severe damage to the brain or formation of an artificial environment around the sensor, as is created around a penetrating microelectrode. Also, the surface monitoring approach is more acceptable for clinical applications.

1.2. *Spreading Depression and Spreading Depression-like Depolarization*

Spreading depression (SD), discovered by Leao (1944)[28], is a phenomenon initiated by a rapid depolarization of neuronal tissue in the gray mater of the cerebral cortex (Figure 3 left side), with a massive redistribution of ions between the intracellular and extracellular spaces, a decrease in the DC steady potential and a depression of the spontaneous electrical activity (EEG or ECoG). No deleterious effects on neuronal tissue function was reported under conditions of increased oxygen supply compensating for the extra oxygen needed by the SD process [49,51]. The SD has a wave (Figure 3 right side) shape propagating from the point of its initiation to the entire ipsilateral cerebral hemisphere. Under hypoxic or ischemic conditions, depolarization events similar to Spreading Depression may develop, mostly in the ischemic penumbra [19]. These responses are termed, hypoxic or ischemic SD-like depolarization (HSD, ISD) which may start in a small focus or develop simultaneously in few sites depending on the metabolic state of the tissue. The HSD or the ISD may spreads at the same velocity as SD in a normal tissue It was suggested that ischemic depolarization waves developing after focal cerebral ischemia, cause an expansion of the core-infracted tissue into the penumbra region [19]. Following SD, CBF usually increases above its basal level, while under HSD or ISD, CBF show no hyperemia but rather a clear decrease [52]. Another major difference between SD and HSD or ISD is the state of mitochondrial enzymes which become oxidized under SD, but reduced (increased) after the onset of HSD or ISD [37,39,48,50]. Oxygen is a limiting factor in HSD and the ISD, therefore the time needed for the recovery from depolarization in the ischemic tissue is longer as compared to the normal SD (Figure 2). It was also hypothesized that normoxic SD may protect the brain from ischemic injury [23,24]. Considering all these aspects, we believe that increasing knowledge in this field, to be accumulated in the proposed project, may contribute to a better understanding of brain ischemia and/or stroke phenomena. Also, we are proposing to use the SD response as a standard tool for the evaluation of the ischemic damage or the reversibility of the insult which is a new approach.

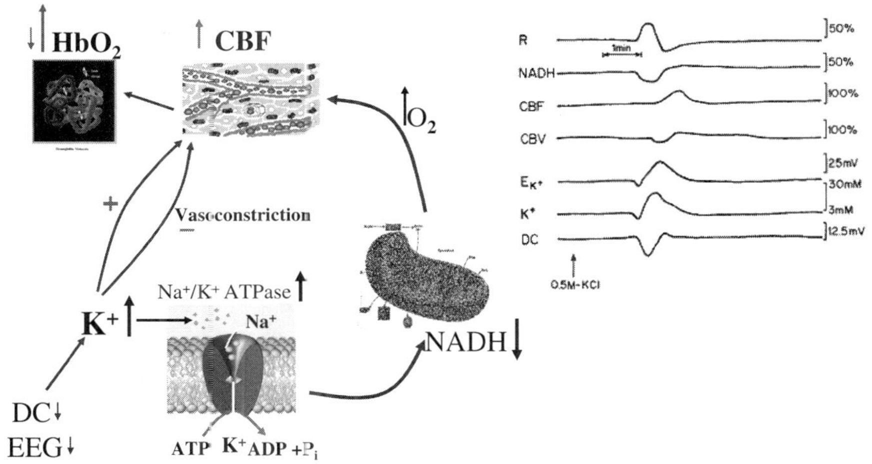

Figure 3. Hemodynamic, metabolic, ionic and electrical activities of the brain recorded after initiation of a wave of Spreading Depression (right side). A possible mechanism showing the cascade of events is presented in the left side.

1.3. *Body Homeostatic Compensatory Mechanisms*

It is well known and documented that the autonomic nervous system (ANS) and mainly its sympathetic branch, including the adrenal gland, play a major role in mounting the compensatory mechanisms of the body to O_2 deficiency. The rapid compensatory reaction to a decrease in blood volume (hypovolemia), for example, includes redistribution of blood flow to various organs and giving preference to the most vital organs in the body, namely to the, brain, heart and adrenal glands [2]. These compensatory mechanisms are overcome if the insult is severe enough or the underlying conditions are not withdrawn within a reasonable time. As a central protection mechanism, blood flow redistribution will occur and the three protected organs (brain, heart and adrenal gland) will receive more blood and O_2, while the peripheral organs or areas (skin and muscles), as well as others non vital visceral organs, will undergo vasoconstriction and subsequent decrease in blood flow and O_2 supply. This mechanism is presented in Figure 4.

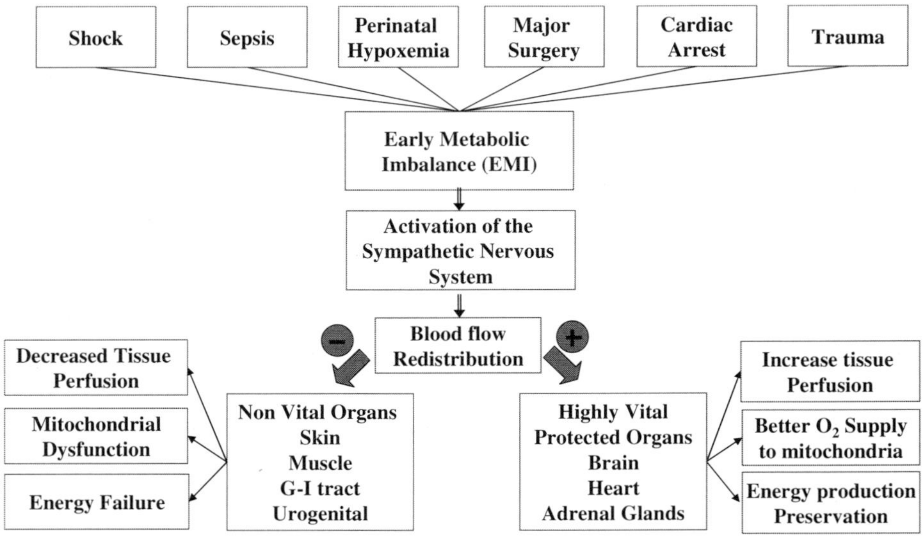

Figure 4. Schematic presentation of various pathological states developed under various clinical situations, which lead to the development of early emergency imbalance (EMI). As a result, blood flow redistribution will lead to an increase in blood flow to the most vital organs and a decrease in blood flow to the less vital organs.

1.4. *Pathophysiology of Critically Ill Patients*

The pattern of the physiopathological cascade of events that may occur in many clinical conditions associated with an imbalance between oxygen delivery and oxygen consumption is shown in Fig 2. Tissue hypoxia is the common denominator. Various pathological states as well as major surgery may lead to metabolic disturbances and may end up in cellular energy derangement. Consequently, compromised global O_2 balance can be associated with significantly high morbidity and mortality in large number of patients. The most common example of such a pathological state is sepsis, which is a major cause of death in critically ill patients [1,20].

Sepsis represents a large group of patient's illnesses in which early identification and accurate quantification of impaired O_2 balance is difficult to accomplish and the consequences of the delayed recognition of subclinical shock are associated with increased mortality. Other conditions such as hypotension, systemic hypoxemia, or early acute respiratory distress syndrome (ARDS) may benefit from "precocious" interventions if adequate monitoring techniques capable of identifying reversible critical thresholds of metabolic or energetic failure of the mitochondria could be identified before the onset of irreversible cellular damage has occurred. The same problem of O_2 imbalance may develop in preterm or full term infants hospitalized in the neonatal ICU or even in the newborn during delivery.

The six conditions shown in the upper part of Fig. 4 are some of the most common events associated with impaired O_2 balance in clinical practice. Under all those situations the metabolic state of the body will deteriorate and energy failure will develop if the appropriate corrective measures are not implemented. These clinical conditions could occur in any area of the hospital or in the prehospital field. Current monitoring mechanisms designed to identify manifestations of early metabolic imbalance (EMI) can only be applied in sophisticated ICU environments. Also, patients that undergo major surgery such as cardiac bypass, neurosurgical or organ transplantation as well as newborns during delivery or fragile elderly patients admitted to general internal medicine wards are at an increased risk of EMI.

1.5. *Monitoring of Critically Ill Patients in Medical Practice*

There is a strong evidence in the literature that the severity of impaired tissue perfusion associated with cardiovascular failure and other states of shock is a major determinant of outcome. The lack of standard endpoints or targets of complete resuscitation after severe hypoperfusion may account for increased incidence of multiple organ dysfunction syndrome (MODS) and poor outcome in critically ill trauma patients. A vast amount of resources in critical care research are being directed to identify and analyze generic markers of incomplete resuscitation such as splanchnic hypoperfusion, tissue acidosis, and impaired systemic oxygen delivery. Early identification of such factors and prompt interventions to correct them may result in reduced incidence of MODS, reduced length of stay and overall improvement of ICU outcomes.

If specific markers of incomplete resuscitation at the cellular level can be identified by means of applying new and less invasive sensors, then perhaps thresholds of irreversible cellular damage can be delineated and interventions be designed to prevent or ameliorate cellular injury and consequently MODS. The search for the perfect indicator as well as the most representative organ or tissue in the body to be monitored is an ongoing process. Ince and Sinaasappel (1999) [21] concluded that "to evaluate the severity of microcirculatory distress and the effectiveness of resuscitation strategies, new clinical technologies aimed at the microcirculation will need to be developed. It is anticipated that optical spectroscopy will play a major role in the development of such tools". In a recent published paper Kruse summarized the effort done by various investigators regarding the perfect indicator of dysoxia, which could be defined as a state of supply-dependent oxygen consumption [27].

Figure 5 present's results of a typical experiment showing the responses of a vital organ (brain) and less vital organ (skin of the scalp) to two perturbations. When the 2 common carotid arteries were occluded (left side), blood supply to the 2 organs decreased and therefore TBF decreased to the brain and skin. As a result, the NADH was elevated in the 2 organs indicating a low intramitochondrial oxygen levels. During the reperfusion stage the TBF in the brain showed a typical hyperemic response that was missing in the skin.

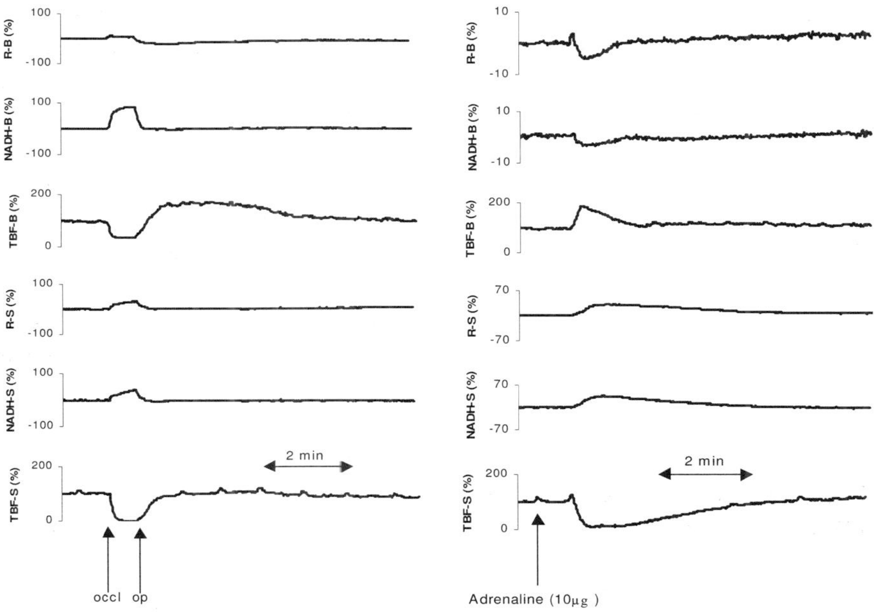

Figure 5. The responses of the rat brain and scalp skin to ischemia (left) and IV injection of Adrenaline (right). Occl ,Op- occlusion and opening of the two common carotid arteries. R-B, NADH-B and TBF-B: reflectance, NADH fluorescence and blood flow in the brain. R-S, NADH-S and TBF-S: reflectance, NADH fluorescence and blood flow in the skin.

When adrenaline was injected IV, a clear differentiation between the responses of the 2 organs was recorded. In the brain a large increase in TBF was followed by an oxidation of NADH. In the skin ischemic responses were recorded namely, a large decrease in TBF was recorded together with elevation in NADH. Those results were in accordance with the previously published material that show the difference in the responses of vital and less vital organs [25,26].

References

1. Angus DC, Linde-Zwirble WT, Lidicker J, Clermont G, Carcillo J and Pinsky MR.

"Epidemiology of severe sepsis in the United States: analysis of incidence, outcome, and associated costs of care". Crit Care Med 2001 Jul; 29(7):1303-10 29: 1303-1310.

2. Barber A. Shock. In: Principles of surgery, edited by S.I. Schwartz. New-York: McGraw-Hill, p.101-122, 1994.
3. Barcroft J. The respiratory function of the blood. Cambridge Univ. Press, 1914.
4. Chance B. "Spectrophotometry of intracellular respiratory pigments". Science 1954, 120: 767-775.
5. Chance B. "Spectra and reaction kinetics of respiratory pigments of homogenized and intact cells". Nature 1952, 169: 215-221.
6. Chance B and Baltscheffsky H. "Respiratory enzymes in oxidative phosphorylation". J Biol Chem 1958, 233: 736-739.
7. Chance B, Cohen P, Jobsis F and Schoener B. "Intracellular oxidation-reduction states in vivo". Science 1962, 137: 499-508.
8. Chance B, Jamieson D and Coles H. "Energy-linked pyridine nucleotide reduction: inhibitory effects of hyperbaric oxygen in vitro and in vivo". Nature 1965, 206: 257-263.
9. Chance B and Jobsis F. "Changes in fluorescence in a frog sartorius muscle following a twitch". Nature 1959, 184: 195-196.
10. Chance B and Legallias V. "Rapid and sensitive spectrophotometry. A stopped-flow attachment for a stabilized quartz spectrophotometer". The Review of Scientific Instruments 1951, 22: 627-638.
11. Chance B, Oshino N, Sugano T and Mayevsky A. Basic principles of tissue oxygen determination from mitochondrial signals. Internat. Symposium on Oxygen Transport to Tissue. In: Adv. Exp. Med. Biol., N.Y.: Plenum Pub. Corp., p. 239-244, 1973.
12. Chance B and Williams GR. "Respiratory enzymes in oxidative phosphorylation. I. Kinetics of oxygen utilization". J Biol Chem 1955, 217: 383-393.
13. Crowe W, Mayevsky A and Mela L. "Application of a solid membrane ion selective electrode to in vivo measurements". Am J Physiol 1977, 233: C56-C60.
14. Duysens LNM and Amesz J. "Fluorescence spectrophotometry of reduced phosphopyridine nucleotide in intact cells in the near-ultraviolet and visible region". Biochim Biophys Acta 1957, 24: 19-26.
15. Erecinska M and Silver IA. "ATP and brain function". J Cereb Blood Flow Metab 1989, 9: 2-19.
16. Friedli CM, Sclarsky DS and Mayevsky A. "Multiprobe monitoring of ionic, metabolic and electrical activities in the awake brain". Am J Physiol 1982, 243: R462-R469.
17. Hansen AJ. "Effect of anoxia on ion distribution in the brain". Physiol Rev 1985, 65: 101-148.
18. Harden A and Young W. "The Alcoholic Ferment of Yeast-Juice". Proc Roy Soc 1906, 77: 405-120.
19. Hartings JA, Rolli ML, Lu XC and Tortella FC. "Delayed secondary phase of peri-infarct depolarizations after focal cerebral ischemia: relation to infarct growth and neuroprotection". J Neurosci 2003, 23: 11602-11610.
20. Ince C and Sinaasappel M. "Microcirculatory oxygenation and shunting in sepsis and shock". Crit Care Med 1999, 27: 1369-1377.
21. Jobsis FF, O'Conner MJ, Rosenthal M and VanBuren JM. Fluorometric monitoring of metabolic activity in the intact cerebral cortex. In: Neurophysiology studied in man, Amsterdam: Excerpta Medica, p. 18-26, 1971.
22. Jobsis FF and Stainsby WN. "Oxidation of NADH during contractions of circulated mammalian skeletal muscle". Respir Physiol 1968, 4: 292-300.
23. Kawahara N, Ruetzler CA and Klatzo I. "Protective Effect of Spreading Depression Against Neuronal Damage Following Cardiac-Arrest Cerebral-Ischemia". Neurol Res 1995, 17: 9-16.
24. Kobayashi S, Harris VA and Welsh FA. "Spreading Depression Induces Tolerance of Cortical-Neurons to Ischemia in Rat-Brain". J CBF Metab 1995, 15: 721-727.
25. Kraut A, Barbiro-Michaely E and Mayevsky A. "Differential effects of norepinephrine on brain and other less vital organs detected by a multisite multiparametric monitoring system". Med Sci Monit 2004, 10: BR215-BR220.
26. Kraut, A., Zurovsky, Y., and Mayevsky, A. Simultaneous hemodynamic and metabolic responses to oxygen deficiency in the brain and other organs of the rat. Neuroscience Letters Supplement 51, S24. 1998.
27. Kruse JA. "Searching for the perfect indicator of dysoxia". Crit Care Med 1999, 27: 469-471.
28. Leao AAP. "Spreading depression of activity in cerebral cortex". J Neurophysiol 1944, 7: 359-390.

29. Mayevsky A. "Multiparameter monitoring of the awake brain under hyperbaric oxygenation". J Appl Physiol 1983, 54: 740-748.
30. Mayevsky A. "Brain NADH redox state monitored in vivo by fiber optic surface fluorometry". Brain Res Rev 1984, 7: 49-68.
31. Mayevsky A. "Brain energy metabolism of the conscious rat exposed to various physiological and pathological situations". Brain Res 1976, 113: 327-338.
32. Mayevsky A. Shedding light on the awake brain. In: Frontiers in Bienergetics: From Electrons to Tissues, edited by Dutton PL, Leigh J and Scarpa A. New York: Academic Press, p. 1467-1476, 1978.
33. Mayevsky A. "Level of ischemia and brain functions in the Mongolian gerbil in vivo". Brain Res 1990, 524: 1-9.
34. Mayevsky A. Cerebral blood flow and brain mitochondrial redox state responses to various perturbations in gerbils. In: Oxygen Transport to Tissue XIV, edited by Erdmann W. New York - London: Plenum Press, p. 707-716, 1992.
35. Mayevsky A, Blum Y, Dekel N, Deutsch A, Halfon R, Kremer S, pevzner E, Sherman E and Barnea O. "The CritiView - A new fiber optic based optical device for the assessment of tissue vitality". Proc SPIE 2006, 6083: 60830Z-1-60830Z-9.
36. Mayevsky A and Chance B. "Oxidation-reduction states of NADH in vivo: From animals to clinical use". Mitochondrion 2007, 7: 330-339.
37. Mayevsky A and Chance B. "Metabolic responses of the awake cerebral cortex to anoxia hypoxia spreading depression and epileptiform activity". Brain Res 1975, 98: 149-165.
38. Mayevsky A and Chance B. A new long-term method for the measurement of NADH fluorescence in intact rat brain with implanted cannula. In: Int. Symposium on Oxygen Transport to Tissue. Adv. Exp. Med. Biol., New York: Plenum Press, p. 239-244, 1973.
39. Mayevsky A and Chance B. "Repetitive patterns of metabolic changes during cortical spreading depression of the awake rat". Brain Res 1974, 65: 529-533.
40. Mayevsky A and Chance B. "Intracellular oxidation reduction state measured in situ by a multichannel fiber-optic-surface fluorometer". Science 1982, 217: 537-540.
41. Mayevsky A, Doron A, Manor T, Meilin S, Zarchin N and Ouaknine GE. "Cortical spreading depression recorded from the human brain using a multiparametric monitoring system". Brain Res 1996, 740: 268-274.
42. Mayevsky A, Flamm ES, Pennie W and Chance B. "A fiber optic based multiprobe system for intraoperative monitoring of brain functions". SPIE Proc 1991, 1431: 303-313.
43. Mayevsky A, Friedli CM and Reivich M. "Metabolic, ionic and electrical responses of the gerbil brain to ischemia". Am J Physiol 1985, 248: R99-R107.
44. Mayevsky A, Manor T, Pewzner E, Deutsch A, Etziony R and Dekel N. "Real time optical monitoring of tissue vitality in vivo". SPIE 2002, 4616: 30-39.
45. Mayevsky A and Rogatsky GG. "Mitochondrial function in vivo evaluated by NADH fluorescence: from animal models to human studies". Am J Physiol Cell Physiol 2007, 292: C615-C640.
46. Mayevsky A, Subramanian VH, Nioka S, Barlow C, Haselgrove J and Chance B. "Brain energy metabolism evaluated simultaneously in the newborn dog by 31P NMR spectroscopy and NADH fluorometry/reflectometry in vivo". J CBF Metab 1985, Supplement: 400-401.
47. Mayevsky A and Weiss HR. "Cerebral blood flow and oxygen consumption in cortical spreading depression". J CBF Metab 1991, 11: 829-836.
48. Mayevsky A, Zarchin N and Sonn J. Brain redox state and O2 balance in experimental spreading depression and ischemia. In: Migraine - Basic Mechanisms and Treatment, edited by Lehmenkuhler A, Grotemeyer K-H and Tegtmeier F. Munchen-Wier, Baltimore: Urban & Schwarzenberg, p. 379-393, 1993.
49. Nedergaard M and Hansen AJ. "Spreading depression is not associated with neuronal injury in the normal brain". Brain Res 1988, 449: 395-398.
50. Raffin CN, Harrison M, Sick TJ and Rosenthal M. "EEG suppression and anoxic depolarization: influences on cerebral oxygenation during ischemia". J Cereb Blood Flow Metab 1991, 11: 407-415.
51. Somjen GG. "Mechanisms of spreading depression and hypoxic spreading depression-like depolarization". Physiol Rev 2001, 81: 1065-1096.

52. Sonn J and Mayevsky A. "Effects of brain oxygenation on metabolic, hemodynamic, ionic and electrical responses to spreading depression in the rat". Brain Res 2000, 882: 212-216.
53. Theorell H and bonnichsen R. "Studies on liver alcohol dehydrogenase. I. Equilibria and initial reaction velocities". Acta Chem Scand 1951, 5: 1105-1126.
54. Wadhwani KC and Rapoport SI. Blood flow in the central and peripheral nervous systems. In: Laser Doppler Blood Flowmetry, edited by Shephard AP and Oberg PA. Boston: Kluwer Academic Pub., p. 265-304, 1990.
55. warburg O. Heavy metal prosthetic groups and enzyme action. Clarendon Press Oxford, 1949.
56. warburg O, Christian W and Griese A. "Hydrogen-transferring coenzyme;its composition and mode of action". J Biochem 1935, 282: 157-205.

INTERSTITIAL PHOTODYNAMIC THERAPY (PDT) OF SOLID TUMOR

ZHENG HUANG,[1] YANG K. CHEN[2], HEPING XU[3], FRED W. HETZEL[1]
[1]Department of Radiation Oncology and [2]Department of Gastroenterology
University of Colorado at Denver and Health Sciences Center, Aurora, CO, USA
[3]Regional Cancer Care, Thunder Bay Regional Health Sciences Centre, Thunder Bay, ON, Canada
Correspondence to: zheng_huang@msn.com

Photodynamic therapy (PDT) involves the administration of photosensitizer followed by local illumination with visible light of specific wavelength(s). In the presence of oxygen molecules, the light illumination of photosensitizer can lead to a series of photochemical reactions and consequently the generation of cytotoxic species. The quantity and location of PDT-induced cytotoxic species determine the nature and consequence of PDT. Much progress has been seen in both basic research and clinical application in recent years. Although the majority of approved PDT clinical protocols have primarily been used for the treatment of superficial lesions of both malignant and non-malignant diseases, interstitial PDT approaches for the ablation of deep-seated solid tumors are now being investigated worldwide. The complexity of geometry and inhomogeneity of solid tumor pose a great technical challenge on the implementation of minimally invasive interstitial PDT. This presentation will discuss the recent progress and challenges of various forms of interstitial PDT.

1. Introduction

Solid tumor is an abnormal mass of body tissue other than blood, bone marrow, or the lymphatic system. It may be non-malignant (benign or not cancerous) or malignant (cancerous). Different types of solid tumors are named by the cell origin, for instance sarcomas, carcinomas, and lymphomas. Treatment of solid tumor varies based on the type, location and stage of tumor. The term of solid tumor used in this article refers to the malignant solid tumors.

Photodynamic therapy (PDT) is a site-specific treatment modality. It involves the local or systemic administration of a photosensitizer (i.e. photosensitizing or photoactive drug) followed by irradiating the targeted lesion site with non-thermal visible light of appropriate wavelength(s) (e.g. 500 – 800 nm). In the presence of molecular oxygen, the light irradiation of photosensitizer and energy transfer can lead to a series of photochemical reactions (i.e. Type I and Type II reaction) and generation of various cytotoxic species (e.g. singlet oxygen and other reactive oxygen species, ROS), and consequently induce apoptosis and/or necrosis of targeted lesion. PDT can be performed in various forms in a non-invasive or minimally invasive fashion.[1] While the majority of approved PDT protocols treat superficial lesions of skin and luminal organs, such as actinic keratosis and Barrett's esophagus, interstitial and intra-operative approaches have been tested for the ablation of a broad range of superficial or bulky solid tumors (e.g. head and neck, neuroblastoma, breast, lung, gastrointestinal, and genitourinary, etc.).[2-5] The microenvironment and physiology of solid tumors are very different from those of normal tissue. The complexity of geometry and the inhomogeneity nature of solid tumor still pose great technical challenges on the implementation of minimally invasive interstitial PDT. This article will discuss recent progress in technical development in interstitial PDT.

2. Photosensitizer

The photosensitizer is considered to be a critical element. Although a major downside of PDT is the prolonged skin photosensitization and the patient needs to stay away from sunlight exposure for a period of time after the admission of photosensitizer (e.g. 1 to 6 weeks), for the majority of tumor patients the light avoidance is tolerable.

The majority of photosensitizers possess a hetero-cyclic ring structure similar to that of chlorophyll or hematoporphyrin. Traditionally, the porphyrins and those photosensitizers developed in the 1970s and early 1980s are called the 1st generation photosensitizer (e.g. Photofrin). Porphyrin derivatives or synthetics made since the late 1980s are called the 2nd generation photosensitizer (e.g. Foscan, Verteporfin). The 3rd generation photosensitizers generally refer to the modifications such as biologic conjugates (e.g. antibody conjugate, liposome conjugate) and built-in photo quenching or bleaching capability.[6] In general, for solid tumor PDT an ideal photosensitizer should meet at least some of the following criteria:

- a commercially available pure chemical,
- low dark toxicity but strong photocytotoxicity,
- good selectivity towards tumor cells,
- longer wavelength allowing deeper penetration,
- rapid removal from the body, and
- multiple administration routes (oral, intravenous, topical, intratumor, etc.).

These criteria provide a general guideline for comparison. Although some photosensitizers satisfy all of or some of these criteria, there are currently only a few PDT photo-sensitizers that have received official approval around the world. These include, but are not limited to:

- Photofrin (630 nm, Axcan Pharma, Inc.)
- Levulan (630 nm, DUSA Pharmaceuticals, Inc.)
- Metvix (630 nm, PhotoCure ASA.)
- Foscan (652 nm, Biolitec AG)
- Laserphyrin (664 nm, Meiji Seika Kaisha, Ltd.)
- Visudyne (693 nm, Novartis Pharmaceuticals)

Several 2nd generation photosensitizers (e.g. HPPH - 665 nm, SnET2 - 665 nm, Motexafin Lutetium - 732 nm) have been investigated in several preclinical and clinical trials for various solid tumors.[7-9] Although these photosensitizers show the selectivity towards tumor cells and they are ideal for cellular-targeting PDT, vascular-targeting PDT is of growing interest.

Vascular-targeting PDT is characterized by a very short drug to light interval (DLI), typically 0 – 30 minutes after the completion of IV injection of photosensitizer. The photosensitizer used in this type of PDT should have a fast clearance and therefore might have minimal selectivity toward tumor cells. Under this unique approach, light irradiation takes place while the photosensitizers are still circulating in the vascular compartment and therefore cause vascular damages through low-density lipoprotein receptor-mediated endocytosis pathways and lead to thrombosis and micro-vessel occlusion. Vascular-targeting PDT has been used primarily for the management of the neovascularization lesion (e.g. wet age-related macular degeneration, AMD) and capillary malformations (e.g. port wine stains, PWS).[10,11] Recently, vascular-targeting PDT-mediated by Tookad (also known as WST09, 763 nm) has been investigated for the curative or palliative treatment of solid tumors (e.g. prostate cancer) by targeting the tumor vasculature.[12,13] It has been hypothesized that the massive shutdown of pathological and normal vessels can deprive the supply of oxygen and nutrients and subsequently achieve tumor ablation. It can be expected that most likely the antivascular PDT can also bypass the drug resistant nature of solid tumors.[14] Although the vascular-targeting PDT might change the tradition criteria of photosensitizer selection, longer wavelength and rapid clearance might be the key for designing photosensitizer for antivascular PDT.

For ablating of a bulky solid tumor, a combination strategy, for instance, cellular-targeting plus vascular-targeting or systemic plus local drug delivery, might be more effective. (see details in Section 5.)

3. Light Source and Light Delivery

The broad spectra of non-coherent and coherent light sources can be chosen to match the optimal absorption peak of a specific photosensitizer, although the wavelength is a determining factor of the tissue penetration depth of the light used. The choice of a specific light delivery mode (superficial or interstitial) in clinical settings is usually based on the nature and location of the disease. The optimal light dose can be obtained by adjusting the fluence rate and fluence.

The first light sources used in PDT were non-coherent light (e.g. conventional arc lamps). Non-coherent light sources are safer, easy to use and less expensive. They can produce spectra of wavelengths to accommodate various photosensitizers although they can be used in conjunction with optical filers to output selective wavelength(s). The non-coherent light sources are mainly used for superficial lesion. The disadvantages of conventional lamps include significant thermal effect, low light intensity and difficulty in controlling light dose.

The most commonly used PDT light sources are lasers. They produce high energy monochromatic light of a specific wavelength with a narrow bandwidth. The laser light can be focused, passed down an optical fiber and directly delivered to the target site through a specially designed illuminator tip (e.g. microlens, cylindrical or spherical diffuser). In some PDT procedures, balloon catheters and light diffusing media have been used in conjunction with laser and optic fiber to provide uniform light distribution and better positioning for treating a luminal lesion. Tapered diffuser and side-fire tip might be suitable for intraluminal and intratumor irradiation, although they are not readily available.[15] For complicated anatomical structure and high risk location, specially designed light applicator is needed.[16] Argon dye, potassium-titanyl-phosphate (KTP) dye, metal vapor lasers and most recently diode lasers have been used for clinical PDT around the world. One advantage of the diode laser is that it can be engineered into a multi-channel unit to meet a highly sophisticated interstitial PDT procedure, which may require multi-channel diode lasers and each independent light output channel to simultaneously provide the light sources of variable power.[17]

Light emitting diode (LED) is another emerging PDT light applicator. LED can generate high energy light of desired wavelengths and can be assembled in a range of geometries and sizes. Large LED array may be more suitable for flat surface illumination of wide-area superficial lesion.[18,19] For intra-operative PDT of brain tumor, LED-probe may be arranged in a cylinder tip to fit into a balloon catheter,[20] whereas, for interstitial PDT of liver cancer, a thin light delivery LED catheter or bar can be implanted into the tumor percutaneously.[21,22]

Typically, in order to achieve the total ablation of massive solid tumor, it is often involved the implantation of multiple diffuser fibers or LED bars. It should be fully recognized that the characterization of light penetration and distribution in solid tumor is important due to the significant inter- and intra-patient differences in the tissue optical properties. Several recent studies suggest that a real-time drug/light dosimetry measurement and feedback system for monitoring and adjusting drug concentration and light distribution during interstitial PDT should be considered.[17,23] Furthermore, when using purely vascularly targeting photosensitizer (e.g. Tookad), the interstitial PDT has a limitation of "treatment time window" since photo-sentitizer administration to light irradiation has to be carried out as a single session in a short period of time, which requires a complicated treatment planning.

4. Tissue Oxygenation

The effectiveness of PDT is determined by many factors. Besides intrinsic target tissue sensitivity to ROS, the other important factors include concentration of the sensitizer, absorption of light energy, and availability of molecular oxygen in the target tissue during light irradiation. The availability of molecular oxygen during irradiation has a profound effect on the treatment outcome. It can be summarized as without oxygen, PDT will have literally no antitumor effect. It is known that hypoxic tumor cells of solid tumor are generally resistant to PDT. There are two types of hypoxia: (i) pre-existing hypoxic cells, which are a result of tumor physiological development, exist in many solid tumors; (ii) the PDT itself can also induce acute hypoxia due to fast depletion of local oxygen supply. Thus, it is noted that the generation of ROS and PDT efficacy can be affected by intra-tumor oxygen tension (pO_2 in cellular-targeting PDT), hemoglobin oxygen saturation (SO_2 in vascular-targeting PDT) and microenvironment of solid tumors.[24-26]

Tumor pO_2 can be measured by oxygen sensitive electrodes, fluorescent quenching-based optical probes, hypoxia markers or non-invasive phosphorescence real time imaging techniques. Hypoxia inducible factor (e.g. HIF-1a and HIF-2a) and expression of vascular endothelial growth factor (VEGF) have also been used as indicators of hypoxia. It has also been reported that positron emission tomography (PET) and magnetic resonance spectroscopy could also detect hypoxic cells in solid tumors.

Several techniques have been proposed to deal with tissue oxygen depletion during PDT. Fractionating light irradiation into controlled light/dark periods or reducing the fluence rate has been extensively studied.[27-29] These techniques are to promote tissue oxygen re-perfusion to compensate for the oxygen depletion caused by the photochemical reactions. Although limited improvement of tumor response has been reported, there are some disadvantages associated with these techniques. Reducing fluence rate or fractionating a dose only affects hypoxic cells. The pre-existing hypoxic cells are not affected by these techniques. Furthermore, reducing irradiation rate or fractionation also significantly increases the time required for delivering of a specific light dose.

Clinical trials of supplementing hyper-oxygenation during radiation therapy have shown statistically significant enhancement of tumor response when combined with hyperbaric oxygen (HBO) and this has inspired an attempt of PDT-hyperoxygenation combination. For instance, one group performed PDT inside a hyperbaric chamber for various indications.[30-33] Although the results are not conclusive, they have shown that hyperoxygenation has extended the survival period in patients with esophageal carcinomas. In comparison to combination therapy of radiotherapy and HBO, it is true that PDT combined with HBO poses far less technical difficulty. Yet, it still needs a pressurized chamber which is not widely available.

Preclinical studies have shown that normobaric hyper-oxygenation is as effective as, if not more than, hyperbaric oxygenation.[24,25] Hyperoxygenation can oxygenate pre-existing hypoxic cells to make them more susceptible to PDT treatment and compensate for oxygen depletion due to PDT photochemical reactions by boosting the oxygen availability. It has been shown that above certain levels of oxygen availability there is little gain in the PDT cell killing effect. However, the gain in the technical simplicity and statistically equivalency in the biological improvement clearly indicates that normobaric hyper-oxygenation is a feasible modality to be combined with PDT than HBO. Although intra-operative PDT procedures are performed under normobaric hyper-oxygenation (up to 100% oxygen breathing), the added effect of hyper-oxygenation on the clinical outcomes needs further investigation.

5. Biological Targets

The choice of treatment depends on the clinical goal and the prognostic profile of tumor. The latter consists of stage of disease, tumor size, location and grade, and lymph node status. Tailored molecular chemotherapy might interact on various gene and protein markers of tumor. Antitumor PDT (either curative or palliative) used for the ablation of tumor mass is a site-specific treatment modality which generally acts on biological targets non-specifically.

5.1. *Direct Cytotoxicity*

The PDT induced cytotoxic species have a short lifetime and only act within a limited distance. Therefore, the subcellular localization of photosensitizer is critical in PDT-mediated cytotoxicities. The subcellular localization is determined by the chemical properties, formulation, concentration and delivery route, the microenvironment of the lesion, and to certain degrees the phenotype of the target cells. The plasma membrane, intracellular membranes and organelles such as the endoplasmic reticulum, Golgi apparatus, lysosomes, mitochondria and nucleus, have been identified as subcellular PDT targets for many photosensitizers. Some photosensitizers might distribute very broadly in these membranes and organelles, some specifically in lysosomes or mitochondria and some might redistribute during light irradiation.

The cell genotype, photosensitizer subcellular location and concentration, light dose and pO2 might determine whether cell death occurs through apoptosis or necrosis. In general, the mode of cell death switches from apoptosis to necrosis with the increase of the intensity of the insult. High-dose PDT induced apoptosis might be a stochastic rather than a threshold effect. Interestingly, some researchers show that subthreshold low-dose ALA-PDT can induce selective apoptosis of tumor cells, without detectable (apoptotic or necrotic) damage to normal cells. This might offer the possibility of treating patients with low-dose ALA-PDT to minimize PDT damages to the surrounding normal tissue.[34]

In recent years there has been a steady increase in knowledge about the roles of the intracellular signaling machinery in cell response to PDT and, particularly, apoptotic pathways in PDT. Understanding PDT-induced unique changes and molecular events in signaling, transcription factors and regulating mechanisms may provide a means to modulate or enhance the cellular PDT effects at the molecular level. Two major apoptotic pathways have been characterized: (i) the death receptor-mediated, or extrinsic pathway, and (ii) the mitochondria-mediated apoptosis, or intrinsic pathway. The links between PDT effects and several apoptotic pathways have been clearly identified in numerous photosensitizers and cell lines in both in vitro and in vivo studies. In general, experimental evidences show a large heterogeneity in the mechanisms leading to cell death in cellular-targeted PDT. However, in addition to the activation of the molecular machinery leading to direct cell death, PDT may also initiate some metabolic reactions that could protect cells from oxidative damage. Therefore, it has been suggested that the control of these protective mechanisms is likely to enhance the cytotoxicity of cellular PDT on target cells.[35]

There is no doubt that advances in cell and molecular biology allow a better understanding of subcellular pathways and consequences of PDT-induced direct cytotoxicity. Although such advances help the rational design and choice of photosensitizers, the gold standard of evaluation of the efficacy of clinical PDT still relies on the gross response of the lesion or disease to PDT. For instance, in tumor ablation, the typical clinically relevant response or biological endpoint would be acute tumor necrosis, the duration and rate of this response, and the length of the tumor-free period. These might be affected not only by the direct cytotoxicity but other PDT effects as well as the tumor sensitivity to various PDT effects.

5.2. *Vascular Effect*

The growth of solid tumors is dependent on their capacity to acquire blood supply. Therefore, much effort has been directed towards the development of anti-angiogenic agents which inhibit the process of neovascularization in solid tumors. More recently, it has become apparent that the destruction of the established tumor vasculature represents a complementary avenue for developing new therapeutic opportunities.

It has long been recognized since the early 1980s that an additional indirect mechanism, coexisting with the direct cytotoxicity of PDT, is the vascular effect in which vascular damage causes ischemic death and therefore provides another approach to treat solid tumors.[36] Later, preclinical studies demonstrate that the mechanisms underlying the vascular effects differ greatly with photosensitizers, PDT approaches, and target tissues.

Irradiation of photosensitizers, either confined in the blood circulation or accumulated in endothelial cells or bound to the vessel walls, results in collateral damage to endothelial cells, which can be characterized by the loss of tight junctions between cells and exposure of vascular basement membranes. This primary damages within the vessel lumen lead to the formation of thrombogenic sites and initiation of a physiological cascade of reactions such as platelet aggregation, the release of vasoactive molecules, leukocyte adhesion, increase in vascular permeability and vessel constriction. Microvascular collapse, blood flow stasis and tissue hemorrhages can lead to persistent post-PDT tumor hypoxia and long-term tumor control.[37-39]

Photosensitizers bound to carrier molecules, such as albumin, HDL or LDL, appear to have an active affinity to endothelial cells and tumor microvascular endothelium because of the existence of specific receptors in high numbers in these structure. This might be an additive factor in photosensitizer uptake and retention in the tumor tissue. Although one can expect that the localization of photosensitizer in both vascular and target cell compartment might produce a stronger combination effect, several investigations have also shown that tumor response to vascular-targeted PDT does not correlate with cellular concentration of photosensitizer in the treated tumor. In vascular-targeted PDT mode, therefore, the light irradiation should be applied during photosensitizer infusion and continued after the completion of infusion so as to center the light delivery period around the peak of the plasma concentration of photosensitizer and hence, putatively, should be maximizing the PDT effect.[40]

5.3. *Immune Responses*

PDT-induced immune responses and particularly antitumor-specific immunity have been studied in various animal models. PDT causes noticeable but short-term and reversible immune suppression under certain experimental conditions. However, unlike chemotherapy and irradiation therapy, PDT does not cause severe negative effects on the host immune system. Substantial evidences have demonstrated that antitumor-specific immune process and enhancement of host immune system might play important roles in secondary cytotoxicity, long-term tumor control and/or complete tumor response to PDT, although these effects are not necessarily lethal to all tumor cells or relevant to the initial tumor ablation.[41-44]

Animal studies show that pro-inflammatory damages formed in cellular membranes and the blood vessel walls of treated sites start to recruit neutrophils, mast cells and monocytes/macrophages after PDT. These cells can also release more inflammatory mediators to enable massive recruitment of immune cells to tumor site. These immune cells and nonspecific immune effector cells have a profound impact on the destruction of tumor.[45-47] PDT can also activate the expression and production of several cytokines, such as IL-1β, IL-2, IL-6, IL-10, TNF-α, and G-CSF. They can play important roles in regulating host immune response involving both lymphoid and non-lymphoid cells.[48] Recent human trial suggests that PDT could enhance the uptake and presentation of tumor antigens by tumor-associated antigen presenting cells (APC) and could ensure lymphocyte involvement.[49] Complement-activating agents may further enhance the antitumor effect of PDT.[50] Recent studies also demonstrate that PDT-treated tumor cell lysates can be effective tumor vaccines, although the mechanism for enhancement of host anti-tumor immune responses to PDT vaccines is still unclear.[51] These advances in understanding PDT-induced immune responses might lead to an attempt to optimize PDT-mediated antitumor modality through the modulation of important inflammatory/immune mediators.

6. PDT Dosimetry

Acceptable clinical PDT, either as an alternative or an adjuvant, needs accurate prediction of the biological effect.[52] In contrast to the well-established treatment-planning system in radiation therapy, where absorbed dose not only has an unambiguous definition but also correlates with the resultant biological effect, there is a lack of well-accepted dose definition and calculation in PDT.[53] As mentioned earlier, PDT relies on three factors: light, drug, and oxygen. Although over-treatment may cause relatively minor adverse consequences compared with radiation therapy, the threshold property of the PDT treatment response means that under-treatment that does not reach the threshold dose will result in treatment failure.[54] The aim of the dosimetry is that the combination of light fluence distribution, drug concentration and distribution, and tissue oxygenation results in maximum tumor ablation while adjacent normal tissue can be spared to an acceptable degree. To reach this goal entails a full understanding of the distribution of light, drug, and oxygen in the target tissue.

PDT dosimetry can be generally categorized in four classes.[55] Each class has its own intrinsic merits and limitations. At present, there is no single dose metric that can address all dosimetry problems. Choice of dosimetry depends on the type and location of tumor, type and characteristic of drug and light, desired accuracy and instrumentation complexity, etc.

(i) Direct dosimetry uses singlet oxygen ($^1O_2$), the putative toxic agent, as PDT dose metric. It is found that cell kill correlated strongly with the cumulative $^1O_2$ luminescence (at wavelength of 1270 nm) and allowed direct estimation of the $^1O_2$ per cell required to achieve a specific cell damage.[56] However, unless there is a substantial reduction in cost and/or complexity of the instrumentation required, direct $^1O_2$ monitoring is unlikely to become a routine tool in clinical or even preclinical PDT.[57] Hence, its most important roles may be as a gold standard for other simpler dosimetry techniques and to answer critical photobiological questions. Indirect means of $^1O_2$ monitoring might be potentially useful.

(ii) Biological dosimetry is a true photobiological dosimetry in that it uses measurable change(s) in tissue that is correlated to the direct result of PDT treatment. Methods such as CT or MRI, laser Doppler, PET, electrical impedance spectroscopy, and bioluminescence imaging have been used to investigate their potential roles in assessing PDT-induced biological effects.[58-62] It is yet unclear whether any of these specific techniques could be used to predict the tissue responses and the outcome.

(iii) Implicit dosimetry uses an implicit surrogate which is indicative of response of PDT treatment. This method is appealing since only one quantity needs to be measured. Fluorescence photobleaching and NADH autofluoresence are convenient quantity to measure.[63] Although the biological effects of Foscan-mediated PDT can well be predicted by photobleaching,[64] correlation of the implicit surrogate with PDT response using other PDT drugs may not be well. For example, Photofrin can bleach through non singlet oxygen mediated mechanism.[65] Photobleaching is not a reliable dose metric for ALA/PpIX-mediated PDT.[66]

(iv) Explicit dosimetry involves the measurements of fluence, drug concentration and tissue oxygenation. It is proposed that the energy absorbed by the drug per unit tissue volume would be a predictor of biological response.[67] Cumulative PDT dose, activated at a particular wavelength for the drug, can be expressed as the time integral of light fluence rate φ(q,t), and absorption coefficient $\mu_{ap}(q,t)$ of the photosensitizer:

$$Dose_{PDT}(q,T) = \int_0^T \varphi(q,t)\mu_{ap}(q,t)dt \qquad (1)$$

where T is the total light irradiation time, q is the generalized spatial coordinate. Individual measurement of light distribution, drug concentration and pO_2 level is nontrivial since each varies dynamically and inter-dependently.

Interstitial PDT is feasible for treating deeply-seated solid tumor, in which light delivery and detection are most efficiently accomplished through implanted optical fibers. The commonly encountered sites are prostate,[8,12,17,68,69] head and neck,[70] liver,[71] lung[72] parenchyma.

Generally, in treating bulky solid tumor, light with longer wavelength is preferred due to its capability of deeper penetration depth, resulting in more access to the tumor.[73] Even with the longer wavelength light, the size of the tissue such as prostate gland cannot be covered by a sufficient light intensity using a single transurethral source. The commonly used light sources are cylindrical diffuse fibers, resembling the radioactive seeds used in brachytherapy. It is also possible that multiple locations of the cylindrical fibers are needed in order to cover the whole prostate. The knowledge of light distribution in the target tissue is important because the generation of cytotoxic species depends on how much available light that can interact with drug. Quantification of the light distribution requires the accurate characterization of optical properties. It is found that both the distributions of absorption and scattering coefficients in the human prostate at 732 nm are not uniform.[68] There is no reason that this is not true for other wavelengths. Weersink et al. found large inter-prostatic variations but little intra-prostatic variations in optical properties of 760 nm.[17] This distribution also changes during the treatment. Clearly, this is the evidence that pre-set values of fluence before the treatment is not sufficient for predicting the subsequent fluence distribution. The knowledge of optical properties is needed and they might change during the treatment.

Another factor is the drug concentration. Current clinical prescription only specifies this quantity in terms of the amount of drug given to a patient per body weight.[74] The simplest method to measure the drug concentration is to detect the fluorescence emitted by the drug, using a single optical fiber as both a source and a detector.[75] The drug concentration can also be measured with spatially-resolved detection technique exploiting the absorption of the drug with multiple implanted interstitial optical fibres.[68] In those methods, light distribution was calculated using an appropriate light propagation model with a priori known optical properties. Assumption was made that the optical properties of tissue and drug concentration distribution were homogenous (or partially homogenous), which is not necessarily true.

The role of oxygen in PDT has been well recognized since as early as late 1980's.[37] It was found that hypoxic tumor cells were protected from PDT damage. Therefore, complete explicit dosimetry clearly needs a term describing the oxygen consumption during the PDT session. This is particularly important in the situation where high light fluence and high drug concentration are used. Measurement of oxygen level can be accomplished either using Eppendorf or spectroscopic measurement of absorption coefficient, from which oxy- and deoxy-hemoglobin concentration and drug concentration can be extracted using spectral deconvolution method such as singular value decomposition method. Zhu et al showed that oxygen level depends on location.[74] Weersink et al. observed significant blood oxygenation changes during vascular PDT.[17] They also indicated that assumption was made that blood was uniformly distributed. Wang et al recently developed a mathematical model describing spatial and temporal dynamics of oxygen consumption and transport in PDT *in vivo*.[76] Their calculations demonstrate that intercapillary heterogeneity of drug contributes significantly to the distribution of photodynamic dose.

The above discussion indicates that there exists a gap between the current clinical protocols and the actual response. Hence, it is not surprising that it has drawn criticism that clinical PDT is still largely based upon empirical dose escalation trials without much consideration of the individual variations amongst patients. The challenge in explicit dosimetry is to find a proper method and procedure that allows us to accurately measure the light, drug and oxygen level in real-time (or nearly real-time) such that local PDT dose can be calculated as per Eq. (1). The adjustment of light distribution and/or drug concentration can be made accordingly depending on whether the PDT dose exceeds the threshold dose or not.

7. Future Prospects

There is a strong and increasing interest and research effort internationally focused on developing new photosensitizers, exploring PDT mechanisms at molecular level, enhancing PDT efficacy with combined modality and evaluating potential clinical indications. Although regulatory approvals for the clinical use of PDT photosensitizers and light applicators now exist in many countries around the world, the total number of approved clinical indications is still limited. It is expected that the involvement of pharmaceutical industry and research institute will continue to launch numerous clinical trials to evaluate applications of PDT in conjunction with or as a replacement for traditional methods for treating solid tumors. Looking into the future, it is expected that combined modality and individualized treatment plan will become an essential component of PDT practice in the treatment of solid tumors over the next decade.

Acknowledgments

This work is supported by a NIH Grant (CA43892).

References

1. Z. Huang, *Technol. Cancer Res. Treat.* **4**, 283 (2005).
2. S.H. Selman, Photodiag. Photodyn. Ther. **4**, 26 (2007).
3. P.J. Muller and B.C. Wilson, *Lasers Surg. Med.* **38**, 384 (2006).
4. Z. Huang, *Photodiag. Photodyn. Ther.* **3**, 71 (2006).
5. M. Biel, *Lasers Surg. Med.* **38**, 349 (2006).
6. J.G. Moser, In *Photodynamic Tumor Therapy - 2nd & 3rd Generation Photosensitizers*, pp.3-8. Harwood Academic Publishers, London (1997).
7. G.M. Loewen, R. Pandey, D. Bellnier, B. Henderson and T. Dougherty, *Lasers Surg. Med.* **38**, 364 (2006).
8. J. Jankun, L. Lilge, A. Douplik, et al. *J. Urol.* **172**, 739 (2004).
9. K.L. Du, R. Mick, T.M. Busch, et al. *Lasers Surg. Med.* **38**, 427 (2006).
10. Treatment of Age-related Macular Degeneration with Photodynamic Therapy (TAP) Study Group. *Arch. Ophthalmol.* **119**, 198 (2001).
11. Z. Qin, K. Li, L. Ren and X. Liu, *Photodiagn. Photodyn. Ther.* **4**, 53 (2007).
12. Z. Huang Z, Q. Chen, D. Luck, et al. *Lasers Surg. Med.* **36**, 390 (2005).
13. J. Trachtenberg, A. Bogaards, R.A. Weersink, et al. *J. Urol.* (2007) [Epub ahead of print]
14. D. Preise, O. Mazor, N. Koudinova, et al. *Neoplasia.* **5**, 475 (2003).
15. J.V.E. Roche, C. Whitehurst, P. Watt, J.V. Moore and N. Krasner, *Lasers Med. Sci.* **13**, 137 (1998).
16. H.J. Nyst, R.L. van Veen, I.B. Tan, et al. *Lasers Surg. Med.* **39**, 647 (2007).
17. R.A. Weersink, A. Bogaards, M. Gertner, et al. *J. Photochem. Photobiol. B.* **79**, 211 (2005).
18. A. Juzeniene, P. Juzenas, L.W. Ma, V. Iani and J. Moan, *Lasers Med. Sci.* **19**,139 (2004).
19. T.S. Mang, *Photodiag. Photodyn. Ther.* **1**, 43 (2004).
20. M.H. Schmidt, D.M. Bajic, K.W. Reichert II, et al. *Neurosurgery.* **38**, 552 (1996).
21. R.A. Lustig, T.J. Vogl, D. Fromm, et al. *Cancer.* **98**, 1767 (2003).
22. J. Chen, L. Keltner, J. Christophersen, et al. *Cancer J.* **8,** 154 (2002).
23. N.E. Martin and S.M. Hahn, *Photodiag. Photodyn. Ther.* **1**, 123 (2003).
24. Z. Huang, Q. Chen, A. Shakil, et al. *Photochem. Photobiol.* **78**, 496 (2003).
25. Q. Chen, Z. Huang, H. Chen, J. Beckers and F.W. Hetzel. *Photochem. Photobiol.* **76**, 197 (2002).
26. Y. Vakrat-Haglili, L. Weiner, V. Brumfeld, et al. *J. Am. Chem. Soc.* **127**, 6487 (2005).
27. T.M. Sitnik, J.A. Hampton and B.W. Henderson, *Br. J. Cancer.* **77**, 1386 (1998).
28. B. W. Henderson, T.W. Busch, L. A. Vaughan, et al. *Cancer Res.* **60**, 525 (2000).
29. A. Curnow, J.C. Haller and S.G. Bown. *Photochem. Photbiol. B.* **58**, 149 (2000).
30. A. Maier, U. Anegg, B. Fell, et al. *Lasers Surg. Med.* **26**, 461 (2000).
31. A. Maier, F. Tomaselli, U. Anegg, et al. Eur. *J. Cardiothorac. Surg.* **18**, 649 (2000).
32. F. Tomaselli, A. Maier, O. Sankin, et al. *Lasers Surg. Med.* **28**, 399 (2001).
33. F. Tomaselli, A. Maier, H. Pinter, H. Stranzl and F.M. Smolle-Juttner. Eur. *J. Cardiothorac. Surg.* **5**, 549 (2001)
34. S.K. Bisland, L. Lilge, A. Lin, R. Rusnov and B.C Wilson, *Photochem. Photobiol.* **80**, 22 (2004).
35. C.J. Gomer, M. Luna, A. Ferrario, S. Wong, A.M. Fisher and N. Rucker, J. Clin. *Laser Med. Surg.* **14**, 315 (1996).
36. W.M. Star, J.P. Marijnissen, A.E. van den Berg-Blok, and H.S. Reinhold. *Prog. Clin. Biol. Res.* **170**, 637 (1984).
37. B.W. Henderson and V.H. Fingar, *Cancer Res.* **47**, 3110 (1987).
38. V.H. Fingar, *J. Clin. Laser Med. Surg.* **14**, 323 (1996).
39. C. Abels. *Photochem. Photobiol. Sci.* **3**, 765 (2004).
40. Z. Huang, Q. Chen, D. Luck, et al. *Lasers Surg. Med.* **36**, 390 (2005).
41. M. Korbelik, *J. Clin. Laser Med. Surg.* **14**, 329 (1996).
42. G. Canti, A. De Simone and M. Korbelik, *Photochem. Photobiol. Sci.* **1**, 79 (2002).
43. F.H. van Duijnhoven, R.I.J.M. Aalbers, J.P. Rovers, O.T. Terpstra and P.J.K Kuppen, *Immunobiol.* **207**, 105 (2003).
44. D. Nowis, T. Stokłosa, M. Legat, T. Issat, M. Jakóbisiak and J Gołąb, *Photodiag. Photodyn. Ther.* **2**, 283 (2005).
45. N. Yamamoto, S. Homma, T.W. Sery, T.W. Donoso and J.K. Hoober, *Eur. J. Cancer.* **27**, 467 (1991).
46. G. Krosl, M. Korbelik and G.J. Dougherty, *Br. J. Cancer.* **71**, 549 (1995).

47. M. Korbelik, G. Krosl, J. Krosl and G.J. Dougherty, *Cancer Res.* **15**, 5647 (1996).
48. S.O. Gollnick, S.S. Evans, H. Baumann, et al. *Br. J. Cancer.* **42**, 1772 (2003)
49. E.S. Abdel-Hady, P. Martin-Hirsch, M. Duggan-Keen, et al. *Cancer Res.* **61**, 192 (2001).
50. M. Korbelik, J. Sun, I. Cecic and K. Serrano, *Photochem Photobiol Sci.* **3**, 812 (2004).
51. S.O. Gollnick, L. Vaughan, B.W. Henderson, *Cancer Res.* **62**, 1604 (2002).
52. W.M. Star, *Phys. Med. Biol.* **42**, 763 (1997).
53. M.S. Patterson and B.C. Wilson BC, In *The modern technology of radiation oncology,* Chp 23, Medical Physics Publishing, Madison, (1999).
54. AAPM Report No. 88, *Photodynamic therapy dosimetry*, AAPM task group report of the general medical physics committee of the science council. Medical Physics Publishing, (2005).
55. B.C. Wilson, M.S. Patterson and L. Lilge, *Lasers Med. Sci.* **12**, 182 (1997).
56. M.J. Niedre, A.J. Secord, M.S. Patterson and B.C. Wilson, *Cancer Res.* **63**, 7986 (2003).
57. S.J. Davis, L. Zhu, A. Minhaj, D.I. Rosen, P.B. Keeting and T. Hasan, *Proc. SPIE.* **4952**, 140 (2003).
58. W.T. Yeung, T.Y. Lee, R.F. Del Maestro, R. Kozak and T. Brown, *J Neurooncol.* **14**, 177 (1992).
59. M.A. Herman, D. Fromm and D. Kessel, *J. Photochem Photobiol.* **52**, 99 (1999).
60. V. Berard, R. Lecomte and J.E. van Lier, *J. Environ. Pathol. Toxicol. Oncol.* **25**, 239 (2006).
61. A. Molckovsky and B.C. Wilson, *Phys. Med. Biol.* **46**, 983 (2001).
62. E.H. Moriyama, SK. Bisland, L. Lilge and B.C. Wilson BC. *Photochem Photobiol.* **80**, 242 (2004).
63. B.W. Pogue, J.D. Pitts, M.A. Mycek, et al. *Photochem Photobiol.* **74**, 817 (2001).
64. J.S. Dysart, G. Singh and M.S. Patterson, *Photochem Photobiol.* **81**, 196 (2005).
65. J.C. Finlay, S. Mitra, M.S. Patterson and T.H. Foster, *Phys. Med. Biol.* **49**, 4837 (2004).
66. J.S. Dysart and M.S. Patterson, *Photochem. Photobiol. Sci.* **5**, 73 (2006).
67. L.O. Svaasand and W.R. Potter, In *Photodynamic therapy: Basic principles and clinical applications.* pp. 369-385, Marcel Dekker: New York, (1992).
68. T.C. Zhu, S.M. Hahn, A.S. Kapatkin, et al. *Photochem. Photobiol.* **77**, 81 (2003).
69. T.C. Zhu, J.C. Finlay and S.M. Hahn, *J. Photochem. Photobiol. B.* **79**, 231 (2005).
70. I.B. Tan, H. Oppelaar, M.C. Ruevekamp, et al. *Head Neck.* **21**, 434 (1999).
71. T.I. Vogl, K. Eichler, M.G. Mack, et al. *Eur. Radiol.* **14**, 1063 (2004).
72. D.I. Fielding, G.A. Buonaccorsi, A.J. MacRobert, et al. *Chest.* **115**, 502 (1999).
73. M.R. Arnfield, J.D. Chapman, J. Tulip, M.C. Fenning and M.S. McPhee, *Photochem. Photobiol.* **57**, 306 (1993).
74. T.C. Zhu and J.C. Finlay. *Photodiag. Photodyn. Ther.* **3**, 234 (2006).
75. J.C. Finlay, T.C. Zhu, A. Dimofte, et al. *Photochem. Photobiol.* **82**, 1270 (2006).
76. K.K. Wang, S. Mitra and T.H. Foster, *Med Phys.* **34**, 282 (2007).

OPTICAL IMAGING AND SPECTROSCOPY OF SUPERFICIAL TISSUE

SP MORGAN†, IM STOCKFORD, JA CROWE, BR HAYES-GILL

School of Electrical and Electronic Engineering, University of Nottingham, University Park, Nottingham, NG7 2RD, UK.

Three techniques used for imaging superficial tissue are described. Firstly a novel polarized light capillaroscope has been developed for imaging the microcirculation. The capillaroscope has been used to make *in vivo* measurements of sickle cell disorder sufferers. Secondly hyperspectral imaging for measuring oxygen saturation is described. The accuracy of such measurements is affected by the non-linear relationship between scattering and absorption and it is demonstrated that polarization techniques can be used to make the relationship more linear, thus improving accuracy. Finally the use of smart CMOS optical sensors for laser Doppler blood flowmetry is described. Through the design of a 4x4 pixel imaging array with on-chip processing the potential for full field laser Doppler blood flow imaging is demonstrated.

1. Introduction

Imaging superficial tissue is important for many applications. Examples include inflammatory responses, sickle cell anaemia, tissue engineering, cancer diagnosis, wound healing, skin diseases and plastic surgery. Three techniques for imaging and spectroscopy of superficial tissue are described; polarization sensitive capillaroscopy, hyperspectral imaging for oxygen saturation measurements and full field laser Doppler flowmetry using smart CMOS sensors.

2. Polarization Sensitive Capillaroscopy

2.1. *Method*

When imaging biological tissue, reflections from the surface of the sample results in a degradation in image quality. Orthogonal polarization imaging [1] and side stream dark field illumination [2] are commonly used to remove this specular component. In orthogonal polarization imaging co-polar illumination and cross-polar detection is used to remove unwanted surface effects. In side stream dark field illumination, illumination from an annulus around the detector is performed, light then ‘back illuminates’ the tissue of interest, minimizing the surface effects. Monte Carlo simulations have been used to compare the imaging resolution and signal to noise ratio. Figure 1 shows the imaging response of both systems to an edge embedded within a scattering medium.

Due to the superior performance of side stream dark field illumination, a polarization sensitive version of this device has been developed (figure 2).

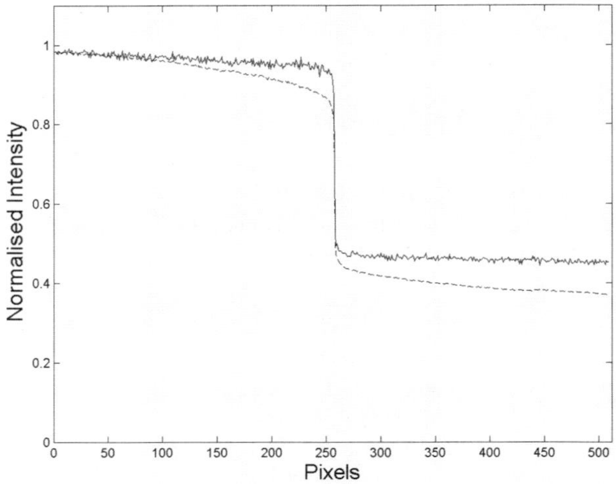

Figure 1 Imaging response of orthogonal polarization imaging (dashed line) and side stream dark field illumination (solid line) to an edge embedded at a depth of 0.4 mean free paths within a scattering medium

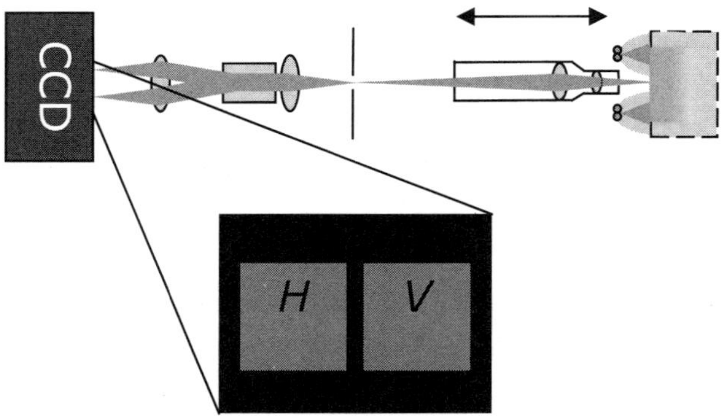

Figure 2 Polarization sensitive capillaroscope

† contact author, steve.morgan@nottingham.ac.uk

Light illuminates the tissue from a ring of LEDs (note the illumination is not polarized) which then back-illuminates the blood cells of interest. The light is then imaged by a x10 microscope objective onto a CCD camera via a Wollaston prism. The Wollaston prism projects two images (horizontal and vertically polarized) onto the CCD camera. The two images are aligned and distortions are corrected by correlating sub-images. From this a single polarization difference image can be obtained.

Our main application is in the detection of linear dichroic cells in sickle cell anaemia sufferers. Sickle cell disorder is a genetic disorder that affects the red blood cells. When the cells become deoxygenated they polymerise and become stiff, which results in painful crises and organ damage. Currently there is no *in vivo* assessment of the number of sickled cells. Previous work [3] has shown polarization difference signals of ~3% in *in vitro* studies.

A device was constructed (figure 3) which was used to take clinical measurements on the lower lip of sickle cell disorder sufferers.

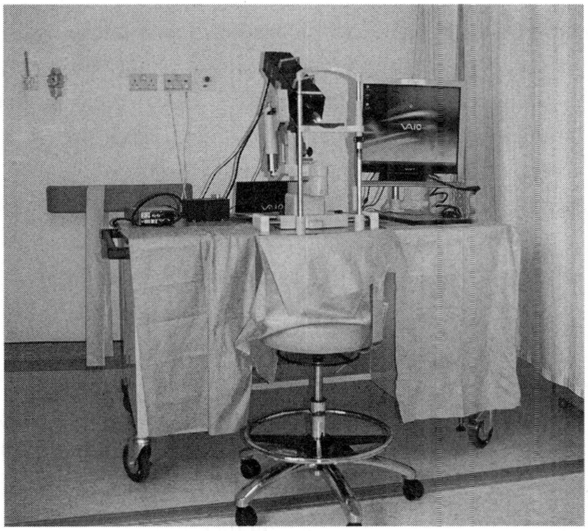

Figure 3 Clinical instrument based upon polarization sensitive capillaroscope

2.2. *Results*

The sensitivity of the system to differences in polarization from controlled components (e.g. polarizers) was demonstrated to be ~0.5% which is more sensitive than the 3% difference signal observed *in vitro*. A typical image is shown in figure 4. To date, however, the evidence on whether sickled cells can be observed through their dichroism signal is inconclusive. It is possible that the instrument is not sensitive enough, in terms of dynamic range, resolution or image alignment, to observe such effects. It is also possible that there are clinical reasons for dichroic signals not to be detectable *in vivo*. For example the cells may be fully oxygenated on the lower lip and may not be polymerized. This remains an open question and will be the study of future research.

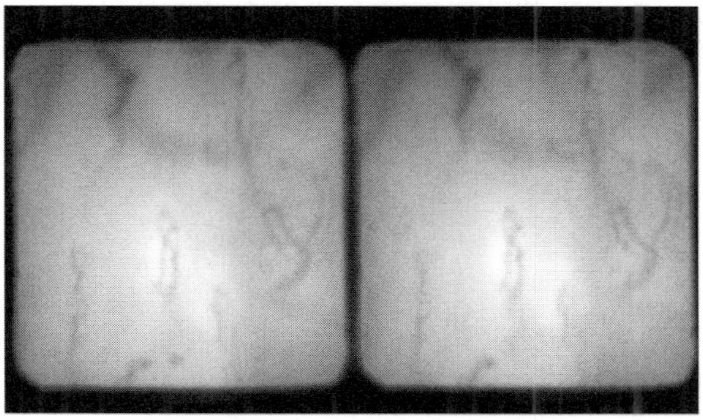

Figure 4 Horizontally and vertically polarized images from the lower lip of a sickle disorder sufferer

3. Hyperspectral Imaging

3.1. *Method*

In the absence of scattering the attenuation, A, of a sample is linearly related to the total absorption coefficient, μ_a. The absorption coefficient itself is a linear sum of the products of the specific absorption coefficients of the constituent absorbers and their concentrations. That is;

$$A = \mu_a d \tag{1}$$

$$\mu_a = \sum_{i=0}^{M} \alpha_i c_i \tag{2}$$

Where A is the attenuation, d is the thickness of the medium (in mm in this paper), μ_a is the absorption coefficient in mm^{-1}, α_i is the molar absorptivity with units of L mol^{-1} mm^{-1} and c_i is the concentration of the compound in solution, expressed in mol L^{-1}.

However in the presence of scattering the relationship between attenuation and absorption becomes non-linear and errors are introduced into the interpretation of the measurements (figure 5). This non-linearity produces errors in spectrophotometric measurements when the Lambert-Beer is assumed, even if an offset to allow for A_s is included as a model of the form $A = A_s + \mu_a d$ [4]. Therefore techniques that could overcome the effects of scattering so that accurate quantitative spectroscopy can be performed would be of widespread use.

An approach that can be easily applied to reduce the effects of scattering is polarization subtraction [5]. Weakly scattered light maintains its original polarization state and a simple subtraction of co- and cross-polarization channels results in extraction of the weakly scattered component. In spectroscopy this results in the relationship between attenuation and absorption becoming more linear.

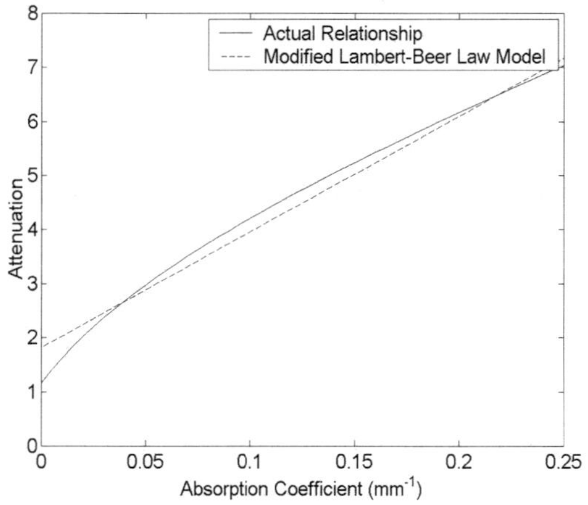

Figure 5 Non-linear relationship between attenuation and absorption in the presence of scattering

Two absorbing species with a concentration ratio of 2 within a scattering medium were simulated using Monte Carlo simulations. Fits to the modified Lambert-Beer Law were carried out to estimate the ratio.

3.2. *Results*

For the case shown in figure 6 two wavelengths were fixed while a third wavelength was varied in the fitting routine. Figure 6 shows estimates of the concentration ratio obtained using polarized light data (both circular and linear) and total intensity measurements. The results demonstrate that linear polarization is most effective at making the relationship between attenuation and absorption more linear and results in more accurate estimates of the ratio of two absorbers R'.

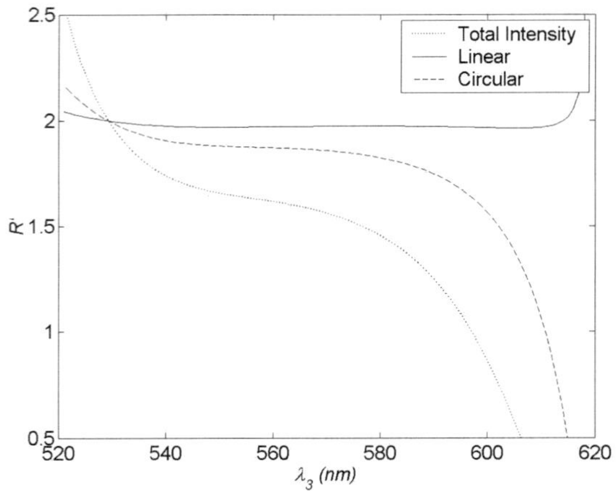

Figure 6 Estimate of the concentration ratio of two absorbers via fits to the modified Lambert Beer Law (actual ratio = 2)

4. Full Field Laser Doppler Flowmetry

4.1. *Method*

Recently, full field laser Doppler perfusion imaging has been implemented using a commercial CMOS image sensor coupled with a digital signal processor (DSP) [6]. The advantage over scanning laser Doppler imaging is that movement artifacts are reduced and the scanning speed of the system is increased due to the absence of moving scanning components. However, a data bottleneck exists between the camera and the signal processing unit as all data needs to be transferred to a separate processing unit to be processed and compromises have to be made to achieve acceptable performance. For example in their most recent work, Serov *et al* [6] have demonstrated full field laser Doppler imaging but employ a sampling frequency of 8KHz, usually a sampling frequency of ~40kHz is considered appropriate for laser Doppler blood flowmetry. Furthermore, as a general purpose CMOS image sensor is used, the capabilities and specifications of the camera are set by the manufacturer.

A custom made camera design offers several advantages over commercial cameras as the specifications can be tailored to the signals of interest. An important advantage of custom made sensors is that on-chip processing allows the data bottleneck that exists between the photodetector array and processing electronics to be overcome, as the *processed* data can be read out from the image sensor to a PC or display at a low data rate. This provides the potential for a large number of parallel processing channels to be implemented each being sensitive to high frequency fluctuations but with a low data readout rate from the sensor.

One sensor design is shown in figure 7. This consists of an array of 4x4 photodiodes with analogue on-chip processing. Each pixel contains a photdiode, transimpedance amplifier, a band pass filter and a frequency weighted filter. Outputs from the sensor are dc, blood concentration and flow values. The experimental set up used to obtain blood flow imaging is shown in figure 8.

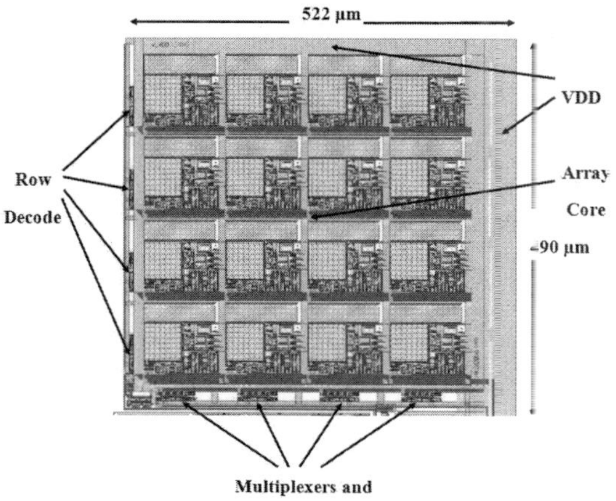

Figure 7 A 4x4 CMOS array with on-chip laser Doppler processing

4.2. *Results*

The sensor has been fully characterized using a modulated light source and this will form the basis of a future publication. Here we simply provide a single result demonstrating that between frames a blood flow change can be observed. Figure 9 shows ten measurements taken on an unoccluded finger, an occluded finger and a static tissue phantom at a single pixel in the array. These signals can be easily differentiated.

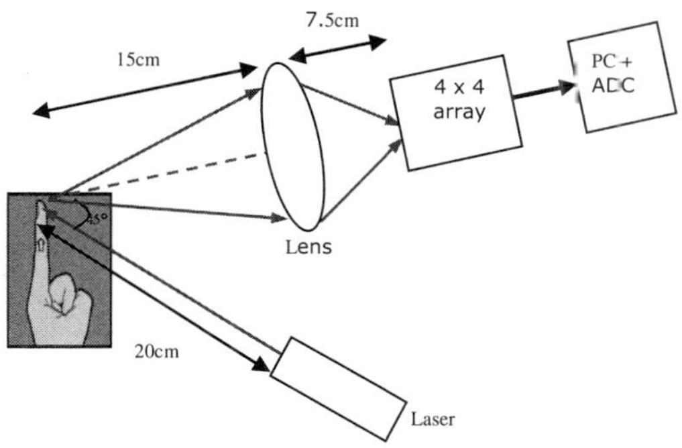

Figure 8 Experiment configuration

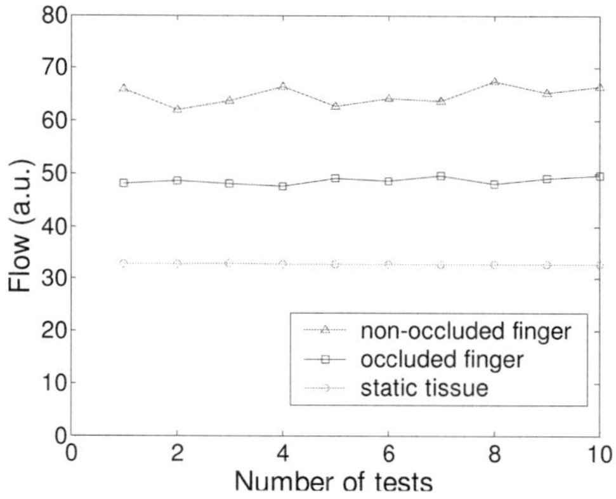

Figure 9 Comparison of measurements taken on a non-occluded finger (upper line) occluded finger (middle line) and a static tissue phantom (lower line)

5. Conclusions

Three novel methods for imaging superficial tissue have been demonstrated. The first, a polarized light sensitive capillaroscope has been shown to be capable of detecting 0.5% differences in polarization images. To date the evidence is inconclusive as to whether it is an appropriate device for monitoring sickle cell disorder *in vivo*. The second methods combines hyperspectral imaging with polarized light imaging to make the relationship between absorption and attenuation more linear, thus making measurements of the ratio of to absorbers e.g. in oxygen saturation measurements, more accurate. Finally the use of custom made smart CMOS optical sensors for laser Doppler blood flow imaging has been demonstrated.

Acknowledgments

This work has been funded by Department of Health (Neat Programme), Big Lottery Fund (via Sickle Cell Society), Royal Society, and the Engineering and Physical Sciences Research Council. Thanks also to M Pickstone, JG Walker, GF Clough, MK Church, D He, PI Rodmell, Q Gu, C Kongsavatsak, B Lu, G Stabler, NBE Sawyer, N Hoang and L Bo.

References

1. W. Groner, , J. W. Winkelman, A. G. Harris, C. Ince, et al., *Nat. Med.* **5**(10), pp1209-1213 (1999).
2. C. Ince, *Crit. Care* **9**(suppl 4), S13-S19 (2005).
3. DA Beach, C Bustamante, KS Wells, and KM Foucar, *Biophys. J* **53**, pp449-456 (1988).

4. R. N. Pittman and B. R. Duling, J. Appl. Phys. 38, 315 (1975).
5. B Lu, SP Morgan, JA Crowe and IM Stockford, Applied Spectroscopy 60: 1157-1166 (2006).
6. A. Serov, B. Steinacher and T. Lasser, Opt. Express. **13**, 3681 (2005).

PHOTOACOUSTIC AND THERMOACOUSTIC IMAGING FOR BIOMEDICAL APPLICATIONS

DA XING

MOE Key Laboratory of Laser Life Science and Institute of Laser Life Science, South China Normal University, Guangzhou 510631, China

Email: xingda@scnu.edu.cn

Laser-based photoacoustic imaging and microwave-based thermoacoustic imaging, combining the advantages of both the high image contrast that results from electromagnetic absorption and the high resolution of ultrasound imaging, could be the next successful generation imaging techniques in biomedical application. It can provide an effective approach of tissue structure and functional images to study the architectures, physiological and pathological properties and metabolisms of biological tissues. This paper is focused on photoacoustic and thermoacoustic imaging application in biomedical research. We have designed and assembled the integrative prototype B-scan photoacoustic tomography system, which can fast obtain the photoacoustic images sequence by conveniently moving the hybridized scanning head. It has the potential to provide a novel and effective approach for high contrast vascular imaging. We also have developed a fast MITT system with the MLTAS, and gained two-dimensional thermoacoustic imaging of biomedical tissues. Comparing to other existing technologies and algorithm, our system was characterized by rapidness and convenience. The experimental results demonstrates that both photoacoustic and thermoacoustic images with multiple contrasts can reflect the absorption of electromagnetic energy. It can provide an effective approach of tissue structure and functional images to study the architectures, physiological and pathological properties and metabolisms of biological tissues which are stimulated using laser and microwave pulses, respectively.

1. Introduction

Laser-based photoacoustic imaging and microwave-based thermoacoustic imaging, combining the advantages of both the high image contrast that results from electromagnetic absorption and the high resolution of ultrasound imaging, could be the next successful generation imaging techniques in biomedical application [1-3]. It can provide an effective approach of tissue structure and functional images to study the architectures, physiological and pathological properties and metabolisms of biological tissues [4-7].

The absorption of electromagnetic energy causes thermal expansion and induces acoustic waves. This effect was first discovered in 1880, referred to as photoacoustic (PA) effect which was induced by light stimulation. In biomedical engineering, PA signal can be utilized to image biological tissues. And thermoacoustic imaging was also developed by replacing laser pulse with microwave pulse in our research. Both microwave and visible or near infrared (NIR) laser light are non-ioning radiation and safe for biomedical applications. This paper is focused on photoacoustic and thermoacoustic imaging application in biomedical research. We have designed and assembled the integrative prototype B-scan photoacoustic tomography system, which can fast obtain the photoacoustic images sequence by conveniently moving the hybridized scanning head. We also have developed a fast MITT system with the MLTAS, and gained two-dimensional thermoacoustic imaging of biomedical tissues. Comparing to other existing technologies and algorithm, our system was characterized by rapidness and convenience.

2. Materials and Methods

2.1 Photoacoustic Imaging System

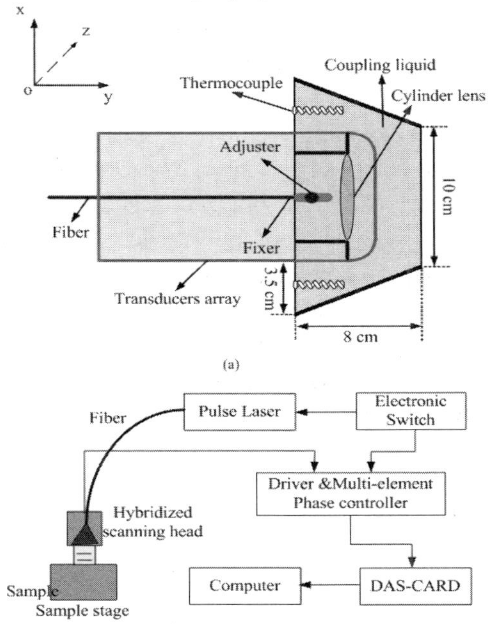

Figure 1. (a). Planform of the novel hybridized scanning head. (b). Experimental setup of the integrative prototype B-scan photoacoustic tomography system.

The integrative prototype B-scan photoacoustic tomography system includes a fiber (core diameter 600, NA=0.22), a transducer array (PL-21; SIUI, China) with 320 vertical transducers and an ultrasonic coupling medium (glycerin and water). The dimensions of the hybridized scanning head are 3.5 cm, 8 cm and 10 cm, as shown in Figure. 1(a); and the height of the hybridized scanning head along the z-axis is 4 cm. The schematic of the experimental setup is shown in Fig.1 (b). A ND: YAG laser (Brilliant B, Bigsky) with wavelength of 1064 nm, output of 8 ns pulse width at 40 mJ/pulse, and a repetition rate of 20 Hz is used to irradiate the sample. The resonance frequency of the transducer array is 7.5 MHz and the scanning width of the array is 49mm. A built-in cylinder acoustic lens, made of silicon rubber with a focal length of 3.5 cm, was used to select the two-dimensional image plane and suppress out-of-plane signals. The fiber and the transducer array were fixed in a box and the angle between them was adjustable. The laser symmetrically irradiated the target tissues through a cylinder optical lens that focused the laser beam along the z-axis to effectively reduce potential damage to tissues. The ultrasound velocity in the ultrasonic coupling medium was adjusted to be as the same as that in the fatty tissue (1547 m/s). Glycerin (Glycerine, glycerol) is a colorless, odorless, noncorrosiveness and nontoxic liquid and can dissolve in water in any proportion, so the glycerin was chosen as the acoustic coupling medium. The glycerin concentration used in our experiments was 11.2%. The coupling medium is transparent relative to the wavelength of the laser. The coupling medium at the front of the hybridized scanning head was sealed with a waterproof polyethylene film; the gap between the film and the sample surface was filled with ultrasonic gel (DOVE, China) to ensure a good acoustic coupling; the distance between the film and the transducer array was 3 cm. The laser and ultrasonic waves can pass through the film with little attenuation. There were two thermocouples placed at the side of the transducer array to maintain a steady temperature during the experiment.

2.2 Thermoacoustic Imaging System

The schematic of the thermoacoustic imaging system is shown in Fig. 2. A microwave generator (BW-1200HPT, China) provides microwave pulses for thermoacoustic auditory generation. The pulses have the following properties: pulsewidth of 0.5 us, pulse-frequency of 1.2 GHz and incident energy density of 0.45 mJ/cm^2. The microwave pulses are coupled into a rectangular waveguide with a cross section of 12.7 mm x 6.3 mm and irradiated to a sample uniformly. The sample is placed on a plastic stage and immersed in a plexiglass tank, which is filled with transformer oil for better coupling of acoustic waves.

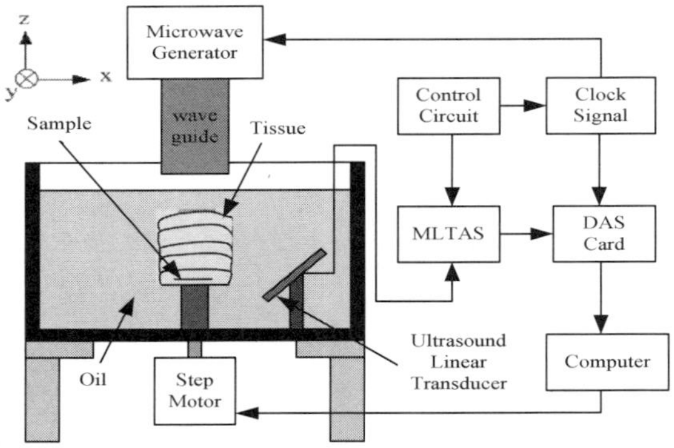

Figure 2. Diagram of the thermoacoustic imaging system.

3. Results

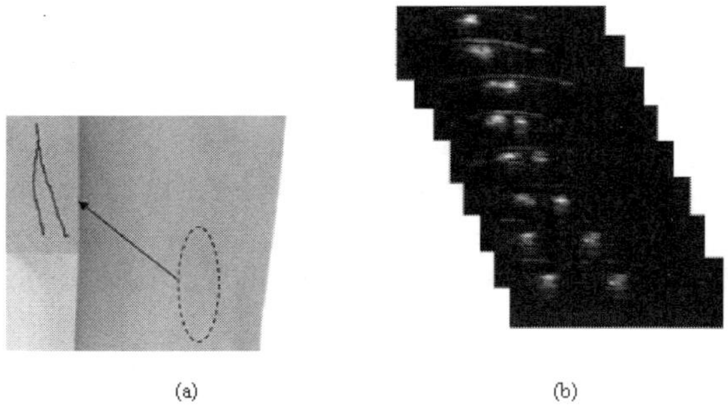

Figure 3. The photoacoustic image of two joining blood vessels on the forearm of a human volunteer[8]

Fig.3 (a) is the photograph of the scanned area, and Fig.3 (b) shows the reconstructed PA image of the blood vessels, which consists of eight slices. The image sequence was acquired by moving the hybridized scanning head 2.1 cm with a step of 3 mm along the forearm at a position 8.7 cm away from the wrist. It can be seen that the cross section of the blood vessels can be imaged with high contrast and spatial resolution, even the joined location of the vessels. Besides the blood vessels, the skin has also been imaged, due to the difference in optical absorption between the skin and the coupling medium.

A PA image of a mouse brain acquired noninvasively with the skin and the skull intact is shown in Fig. 4(a). The PA reconstructed image corresponds to the cortical surface of the mouse brain, where the vascular distribution is evidently visualized. Owing to the different optical absorptions between blood vessels and the background parenchyma, the major characteristic tissue structures in mouse brain are clearly distinguished. The vascular network of the PA reconstructed image is in excellent agreement with the open-skull anatomical photograph Fig. 4(b). This experiment testified that PAI has the ability for noninvasively imaging of brain cortical vasculature.

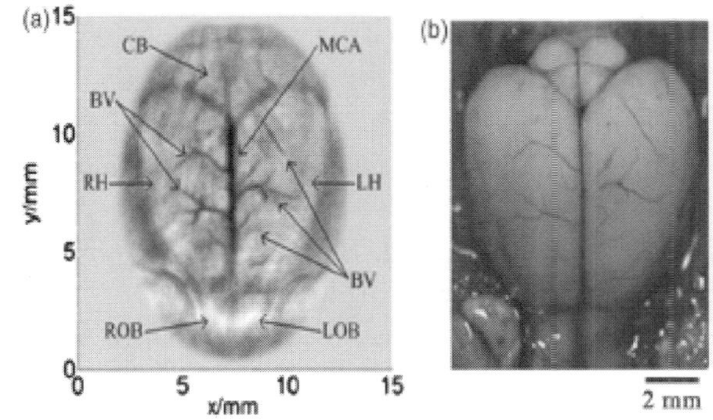

Figure 4. PA images of mouse brain *in vivo*. (a) Noninvasive image of a mouse brain acquired using the PAI system with the skin and skull intact. Detailed brain tissue structures: CB, cerebellum; BV, blood vessel; LH, left hemisphere; LOB, left olfactory lobe; MCA, middle cerebral artery; RH, right hemisphere; and ROB, right olfactory lobe. (b)Open-skull photograph of the mouse brain surface acquired postmortem [9].

A mouse brain was imaged noninvasively with the PAI system before and after its left carotid artery was occluded. On a gray scale, where darker areas represent higher absorbed optical energy, the images clearly show a diminished blood volume pattern after carotid artery occlusion. A rectangular region in the left cerebral cortex was selected, as marked with a dotted frame in the images of Fig. 5(a) and 5(b), corresponding to the anticipated ischemia during the carotid occlusion. Compared with the preocclusion image in Fig. 5(a), the image in Fig. 5 (b) represents a clear map of local ischemic region, where the blood vessels provided a weak PA signal. The decrease in signal intensity of the blood vessels within the dotted frame primarily resulted from the decline of blood flow and blood volume. Hence, amplitude differences as displayed by the images indicate real

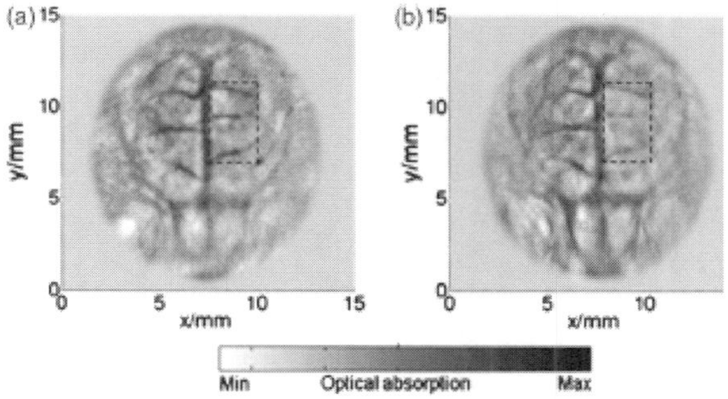

Figure 5. PA monitoring of cerebrovascular occlusion. (a) PA image of brain cerebral cortex surface before carotid ligation. (b) PA image of brain cerebral cortex surface during carotid artery occlusion. Dotted frame corresponds to the anticipated ischemic region [9].

changes in signal strength. The observed fractional change in absorbed optical energy density, A' /A [A and A'] are the optical absorptions per unit volume before and after artery occlusion, respectively. The images and the regional changes of the PA signal intensity strongly suggest that brain blood vessel occlusion can be monitored with the PAI system. After the imaging experiment, a phenomenon of hemiplegy of the right limbs of the mouse was found after its recovery from anesthesia. This is consistent with brain function: The symptoms are usually on the side of the body opposite to the side of the brain where the damage has occurred.

We made a piece of quadrate homogeneous pork fat tissue with a thickness of 4.5 cm. Next, we placed parallel two simulating blood vessels on the fat tissue as shown in Fig. 6(a). The simulating blood vessels were made of silicon rubber tubes with 0.8 mm diameter, filled with chicken blood. The length of the blood vessels and the distance between them was 3.1 cm and 2.3 cm, respectively. Then we covered the simulating blood vessels with a same thickness fat tissue. In order to finalize the design, the sample was cooled at room temperature for several minutes. Fig. 6(b) is the thermoacoustic tomography image of the simulating blood vessels with phase-controlled reconstructed algorithm. From the reconstructed image, the image of simulating blood vessels was corresponded well with the picture of sample. But the background noise of the reconstructed image is not as well as that of the images above, because the thick fat tissue results in the attenuation of thermoacoustic signal intensity and microwave energy.

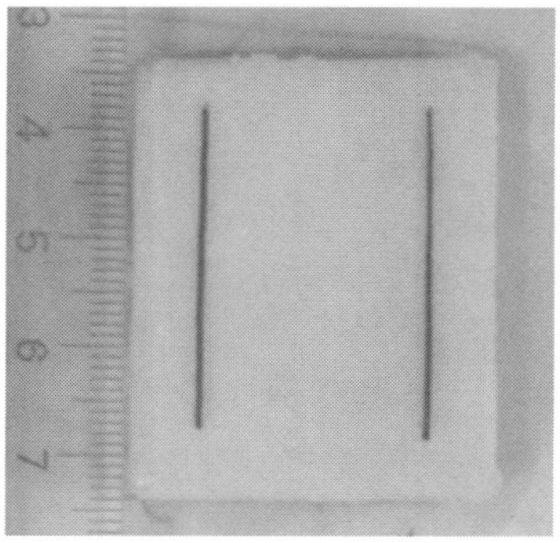

(a)

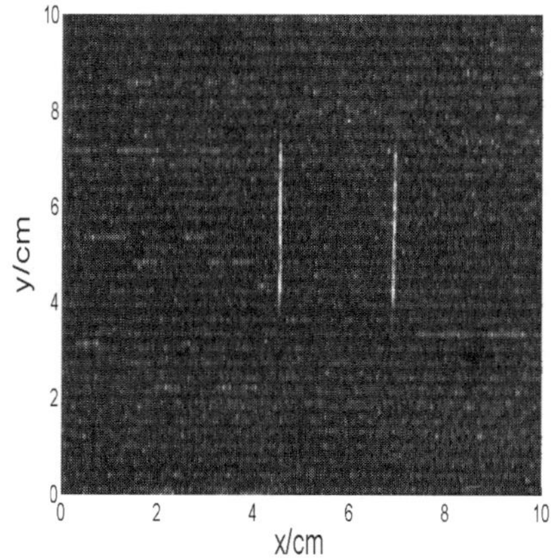

(b)

Figure 6. Cross section of the tissue sample; and (b) Thermoacoustic tomography image of the simulating blood vessels [10].

4. Conclusions

This paper is focused on photoacoustic and thermoacoustic imaging application in biomedical research. Comparing to other existing technologies and algorithm, our system was characterized by rapidness and convenience. The experimental results demonstrates that both photoacoustic and thermoacoustic images with multiple contrasts can reflect the absorption of electromagnetic energy. It can provide an effective approach of tissue structure and functional images to study the architectures, physiological and pathological properties and metabolisms of biological tissues which are stimulated using laser and microwave pulses, respectively.

Acknowledgments

This research is supported by the National Natural Science Foundation of China (30470494; 30627003) and the Natural Science Foundation of Guangdong Province (7117864).

References

1. Yaguang Zeng, Da Xing and Yi Wang, *Opt. Lett.*, **29(15)**, 1760-1762(2004).
2. Xiang Liangzhong, Xing Da, Gu Huaimin, et al, *Chin. Phys. Lett.*, **24 (3)**, 751 (2007).
3. X. Wang, Y. Pang, G. Ku, et al, *Nat. Biotechnol.*, **21**, 803-806 (2003).
4. L.ZH. Xiang, D. Xing, H.M Gu, et al, *Journal of Biomedical Optics,* **12**, 014001-8(2007).
5. Sihua Yang, Da Xing, Yeqi Lao, Diwu Yang, Lvming Zeng, Liangzhong Xiang, and Wei Chen *Appl. Phys. Lett.*, **90(24)**, 24392(2007).
6. W. D. Yang, D. Xing, H. M. Gu, Y. Tan and L. M. Zeng, *Appl. Phys. Lett.*, **87**, 194101-3(2005).
7. Xiang Liangzhong, Xing Da, Gu Huaimin, Zhou Feifan, Yang Diwu, Zeng Lvming, Yang Sihua, *Chin. Phys. Lett.,* **24 (3)**, 751 (2007).
8. W. D. Yang, D. Xing, Y. Tan, H. M. Gu and S. H. Yang, *Appl. Phys. Lett.*, **88**, 174101-3(2006).
9. Sihua Yang, Da Xing, Quan Zhou, Liangzhong Xiang and Yeqi Lao, *Med. Phys.,* **34(8)**, 3294-3301(2007).
10. Lüming Zeng, Da Xing, Huaimin Gu, Diwu Yang, Sihua Yang and Liangzhong Xiang, *Chin. Phys. Lett.,* **23(5)**, 1215-1218(2006).

MEASURING TISSUE OPTICAL PROPERTIES IN VIVO USING REFLECTANCE-MODE CONFOCAL MICROSCOPY AND OCT

S.L. JACQUES, R. SAMATHAM, D.S. GAREAU, D. LEVITZ, N. CHOUDHURY

Oregon Health & Science University, 3303 SW Bond Avenue, Portland, OR 97229, USA
E-mail: jacquess@ohsu.edu

The confocal collection of reflected photons in either a confocal microscope or an optical coherence tomography system can be used to specify the optical scattering properties of a tissue, the scattering coefficient ($\mu_s[cm^{-1}]$) and the anisotropy of . scattering (g, $[dimensionless]$). The reflectance (R) as a function of the depth of foucus (z) behaves as $R(z) = \rho exp(-\mu z)$, where ρ is the reflectivity and μ is an attenuation coefficient. The (ρ, μ) maps uniquely into (μ_s, g).

Keywords: confocal reflectance; optical properties

1. Introduction

The ability to separately measure the scattering coefficient ($\mu_s[cm^{-1}]$) and the anisotropy of scattering (g) is difficult, especially when measuring an in vivo site that can not be excised for bench-top measurements. The scattering properties ($\mu_s andg$) can characterize the ultrastructure of a biological tissue (nuclear size, mitochondra, cytoskeletion, collagen fibers, density of membranes) without needing an added contrast agent. This report describes the use of reflectance-mode confocal scanning laser microscopy (rCSLM) to measure optical properties. rCSLM is the same as optical coherence tomography (OCT) when the OCT is conducted in focus-tracking mode. The experimental measurement involves translating the depth of focus, z, of an objective lens, down into a tissue. As depth z increases, the reflected signal R decreases due to attenuation by the tissue scattering. The experimental data behaves as a simple exponential,

$$R(z) = \rho exp(-\mu z) \quad (1)$$

where ρ is the local reflectivity [dimensionless] and $\mu[cm^{-1}]$ is an attenuation coefficient. The relationship between (ρ, μ) and (μ_s, g) is:

$$\mu = (\mu_s a(g) + \mu_a)2G(NA)aba : appeq2 \quad (2)$$

and

$$\rho = \mu_s L_f b(g, NA) \quad (3)$$

where $a(g)$ is a factor that drops from 1 to 0 as g increases from 0 to 1 (determined by Monte Carlo simulations) allowing photons to reach the focus despite scattering, $G(NA)$ is a geometry factor describing the average photon pathlength that depends on the numerical aperture (NA) of the lens, L_f is the axial extent of the focus, and $b(g, NA)$ is the fraction of scattered light that backscatters into the lens for detection which depends on both g and NA. The absorption μ_a is usually too low to be a significant factor, but if significant the Eq. 2 shows how it affects μ.

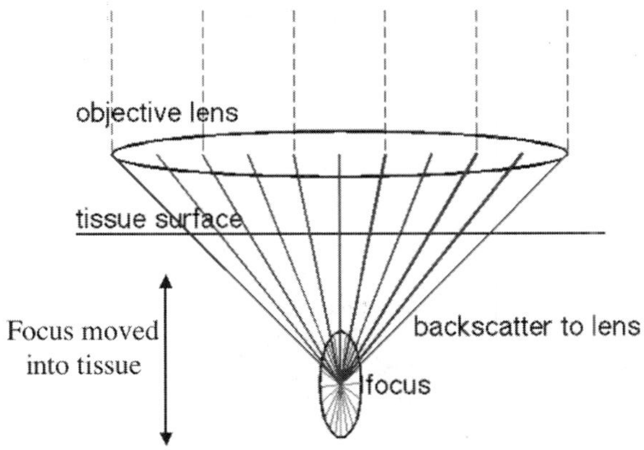

Fig. 1. The focus is scanned versus depth z. Light backscattered from focal volume is recollimated for later focus through pinhole to reach detector.

2. Experimental

The focus is scanned axially down into the tissue. The reflected light from the focus is recollimated by the objective lens, for later refocus through a pinhole, or into an optical fiber, for detection. Hence, the measurement is confocal detection. Some experiments used a reflectance mode confocal microscope using a 0.90 NA objective lens, operating at 488 nm wavelength. Other experiments used an optical coherence tomography system operating at 1310 nm and in focus-tracking mode, in which the focus of the objective lens was co-located with the coherence gate of the interferometric system.

The resulting data behaves as an exponential decay.

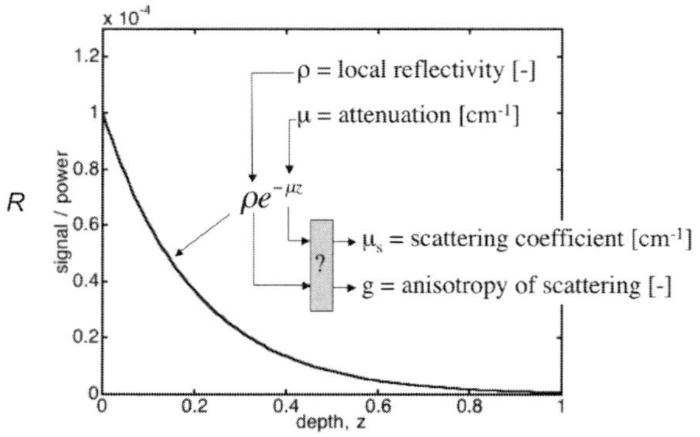

Fig. 2. The reflectance falls exponentially as the focus is scanned down into the tissue, to yield $R(z)$.

Phantom tissues were prepared using 2% agarose gels with polystyrene microspheres at concentrations of 1% to 4% volume fraction. These phantoms were measured to provide a calibration of the systems tested. Figure 3 shows the grid with measurements

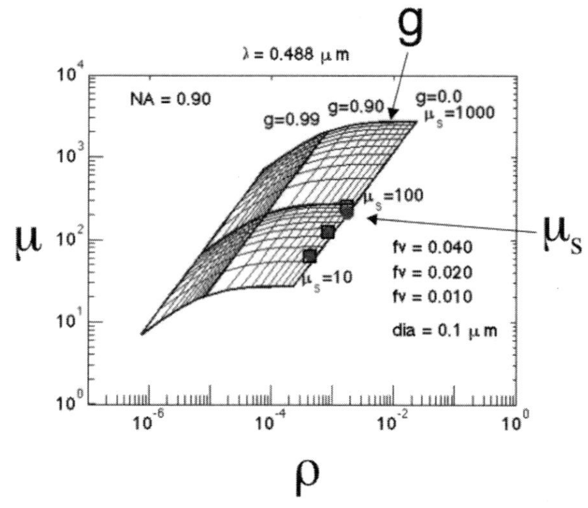

Fig. 3. The μ and ρ grid maps into μ_s and g. Phantoms made with polystyrene microspheres is shown, using $488-nm$ wavelength.

of the agaros gels with polystyrene microspheres at various concnetrations.

The experimental 488 nm measurements on fresh mouse tissues are summarized in Fig. 4. The results show the range of optical properties encountered in brain (white and gray matter), liver, muscle and skin. Also a measurement on cultured cells is shown.

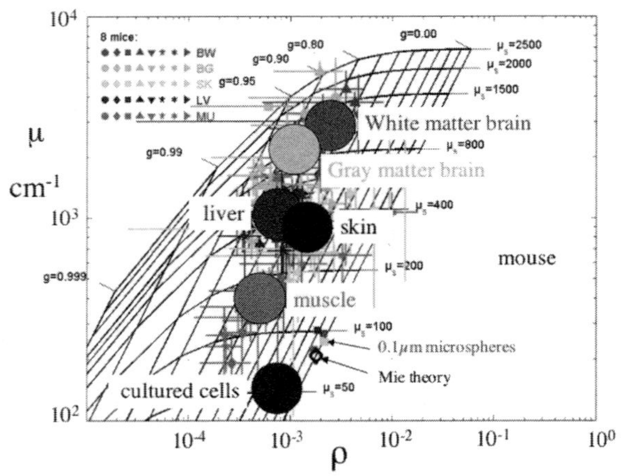

Fig. 4. The results of measurements on freshly excised mouse tissues.

3. Conclusion

The method outlined here is able to measure the scattering properties of a tissue non-invasively in vivo. It is not necessary to excise a tissue sample to measure the optical properties. This advantage offers the ability to characterize tissues in their native state.

The method also allows the $R(z)$ scan to be conducted for each x, y position in a field of view. Hence, a 2-D image of a tissue with x, y resolution of variation in optical scattering properties is possible.

The optical properties of μ_s and g can map into the apparent particle size and density distribution in a tissue, which is able to characterize the ultrastructure of a tissue. Hence, image contrast can depend on tissue structure.

THE ENHANCEMENT OF CONFOCAL PROBING WITH OPTICAL CLEARING

I.V. MEGLINSKI[a]
Cranfield Health, Cranfield University,
Cranfield, MK45 4DT, UK

V.V. TUCHIN
Department of Physics, Saratov State University,
Saratov, 410026, Russia

Current report represents an assessment of the effective depth of skin tissues optical clearing that can be reached with the chemicals administration of superficial skin tissues. A solution of glycerol is taken in the capacity of innocuous osmotic agent. Diffusion of this bio-compatible chemical agent into the skin temporary pushes water out of the tissues, and the matching of the refractive indices of skin structural elements is caused. This temporary decreases scattering and increases transparency of topical skin layers that allows for unrestricted light to permeate deeper into the skin. The results of experiments and modeling show that signal spatial localization offered by a confocal probe in the skin tissues during their clearing is useable for the monitoring of deep reticular dermis and improving the image contrast and spatial resolution. The discussion of skin tissues optical properties and their changes due to diffusion of glycerol into the skin is given. Optical properties of tissues and their changes due to chemical administration are estimated based on the results of experimental *in vitro* study with the rat and human skin.

1. Introduction

The unique characteristics of optical/laser methods are of great interest to researchers working in various areas of biology and medicine [1,2]. Nevertheless, the problem of implementing these techniques in clinical practice in order to solve a wide range of actual diagnostic tasks remains unresolved. The difficulties in clinical application are due to highly and anisotropy scattering of the probing laser radiation in most of biological tissues. Furthermore, random inhomogeneous variations of the optical properties of the topical skin layers act like a screen, which keeps stray the optical radiation out of the deep penetration into the human body [3]. The mathematical expression for the propagation of an optical radiation describing these conditions is complex. Practically it is extremely difficult to distinguish regular waves corresponding to internal structure of the medium or individual characteristics of the scattering particles.

Diffusive injection of some chemical agents in biological tissues produces the effect of the light scattering reducing [4-6]. The preliminary experiments on the rat skin [5,7-9] where glucose, glycerol, trazograph, polyethylene glycol, cosmetic lotions and gels were taken in the capacity of innocuous bio-compatible chemical agents showed that diffusion of the chemicals into skin temporary pushes water out of the upper skin tissues. The matching of the refractive indices of the structural elements of skin cells is caused by both processes – opposing diffusion of water and chemical agent with higher than water refractive index. This temporary increases the transparency of upper skin layers that, respectively, allows the unrestricted light to permeate deeper into tissues. The augment of the skin layer transparency typically increases for 10-40 minutes, depending on the chemical agent. Then due to the physiological response of the organism water content of skin tends back to initial level and the transparent skin returns to normal stage. Nevertheless, this temporal gap of scattering reduction can significantly improve the efficiency of skin image reconstruction techniques, such as confocal microscopy or optical coherence tomography (OCT), for instance, and provide information, which is not available for the dermatologists and other researchers at the present time. Thus, assessment of depth that we can be reached by optical probing during skin tissues optical clearing is required.

2. Materials and Methods

A stochastic numerical modeling was used to study the spatial localization of output signal offered by a

Correspondence: i.meglinski@cranfield.ac.uk and/or tuchin@sgu.ru confocal probe within the human skin tissues. The basis for this technique is the Monte Carlo (MC) approach [10,11] that excludes energy conservation problem inherent to the previous MC algorithms [12-14]. The details of the technique are justified shortly. The confocal probing is simulated following the ray-tracing approach coupled with MC [15] for a multi-layered medium. The layers correspond to the randomly inhomogeneous distribution of blood, chromophores and various pigments within skin tissues [16-18], that produce spatial variation their optical properties, i.e. scattering coefficient μ_s, absorption coefficient μ_a, anisotropy factor g and refractive index n [1]. Boundaries of the layers are simulated as the wavy surfaces (Fig.1) [10,11] with respect to the anatomical peculiarities of human skin cells structure, as well as chromophores, pigments and blood vessels spatial distribution [16-18].

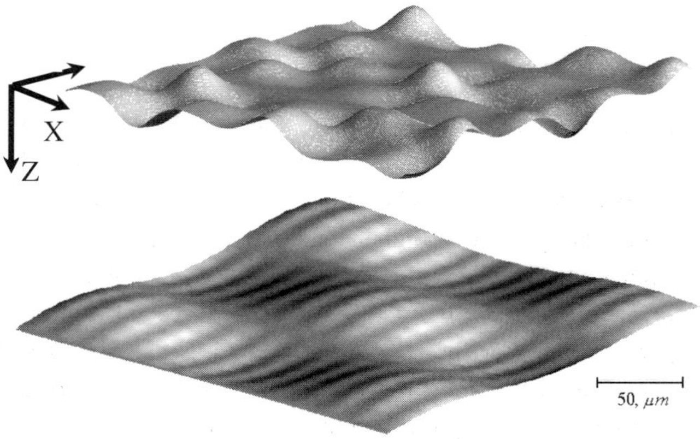

Figure 1. This surface plots shows the mathematical form used to describe boundaries between the skin layers in MC model, corresponded to a cross-section of epidermal-dermal junction (upper) and boundary between reticular dermis and subcutaneous fat (lower).

We subdivide epidermis by two sub-layers called non-living and living epidermis. Non-living epidermis or stratum corneum (about 20 μm thick) consists of only dead squamous cells, which are highly keratinized with a high lipid and protein content, and has relatively low water content [17]. This content defines the density of average absorption and scattering cross-sections for the layer. Living epidermis (~ 100 μm thick) contains most of the skin pigmentation, mainly melanin, which is produced in the melanocytes. Large melanin particles such as melanosomes (> 300 *nm* in diameter) exhibit mainly forward scattering. Whereas, melanin dust, which particles are small (< 30 *nm* in diameter) has the anisotropy in the scattering profile, and optical properties of melanin particles (30-300 *nm* in diameter) may be predicted by the Mie theory. Random spatial distribution of melanin particles in living epidermis affects both the scattering and absorption properties of human skin [19].

Reticular dermis is vascularised layer and the main absorbers are the blood-bone pigments haemoglobin, carotene and bilirubin. The optical properties of derma, mainly absorption, depend on the distribution of blood containing haemoglobin.

Following the distribution of blood vessels in skin [16,18] we subdivide dermis by four layers: papillary dermis (~ 150 μm thick), upper blood net plexus (~ 100 μm thick), reticular dermis (~ 1150 μm thick) and deep blood net plexus (~ 100 μm thick). Scattering properties of dermal layers are mainly defined by the fibrous structure of the tissue, where collagen fibres are packed in collagen bundles and have lamellae structure [16]. The light scatters on both single fibres and scattering centres, which are formed by the inter-placement of collagen fibres.

The subcutaneous fat (~ 6000 μm thick) is formed by aggregation of white fat cells adipocytes (~ 120 μm in diameter) containing stored lipid oil in the form of single droplets of triglyceride.

Table 1. Optical coefficients of the skin layers used in the simulation.

	Skin layer	μ_s, mm^{-1}	μ_a, mm^{-1}	g	n
1	*Stratum corneum*	80	0.1	0.8	1.53
2	*Living epidermis*	35	0.15	0.8	1.34
3	*Papillary dermis*	30	0.07	0.9	1.4
4	*Upper blood net dermis*	25	0.01	0.95	1.39
5	*Reticular dermis*	20	0.07	0.76	1.4
6	*Deep blood net dermis*	30	0.12	0.95	1.39
7	*Subcutaneous fat*	15	0.07	0.8	1.44

The refractive index of outside ambient (air) is n_0=1.

Table 1 outlines the summarized optical properties of skin layers. These properties are corresponded to the intrinsic optical parameters of normal human skin (*Caucasian* type) at 633 *nm* wavelength. The refractive index of external medium is 1.0 (air), whereas for the deepest area it is assumed the same to subcutaneous fat.

3. Results and Discussion

Current model was used to calculate axial focusing profiles of the detecting signal distribution [11,15]. The diffusion of exogenous chromophores or bio-compatible chemical agents into the skin produces the matching of refractive indices of skin tissues scattering elements and interstitial fluid of its tissue [1]. Thereupon, the path-length in a forward direction for the incident radiation increases as the un-restriction of incident light on the refractive index mismatching boundaries decreases. This plays significant role for the shallow optical probes [10,20]. It should be pointed out, that direct measurements of the optical properties of the skin layers described above are impossible. We estimate the optical properties changes in the skin layers based on their chromophores content and micro structure [16-18] and based on the experimental images obtained with standard 4-laser Carl Zeiss confocal scanning microscope for normal skin (Fig.2).

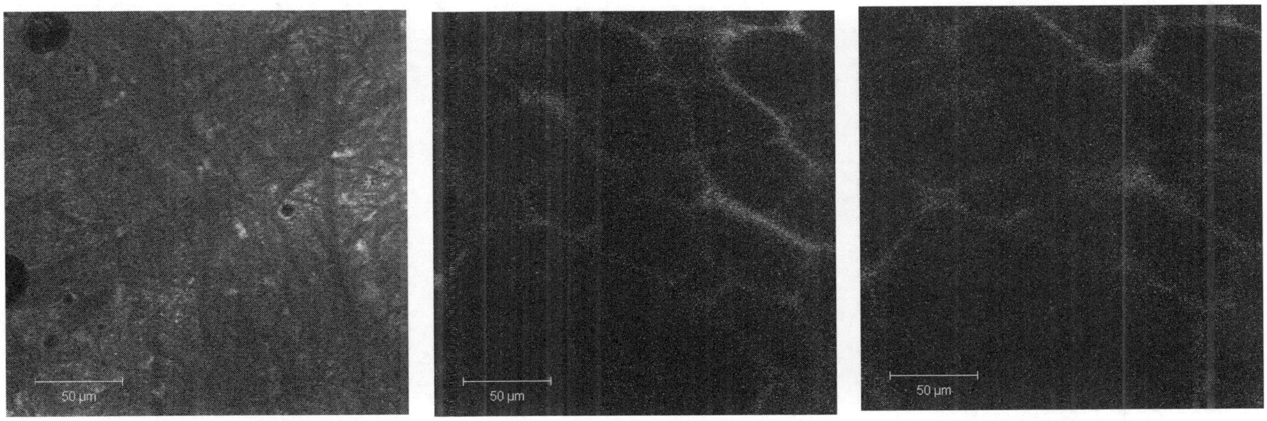

Figure 2. The confocal images of skin at normal conditions (no optical clearing is applied), obtained with standard 4-laser Carl Zeiss confocal scanning microscope.

Stratum corneum greatly clears in the first minute of the process (Fig.3). It is connected mainly with an immersion of dead cells just after administering of agent. Then diffusion is dramatically reduced due to barrier produced by dermal-epidermal junction. Intra-dermal diffusion of chemical agents is characterised by retarding in the upper blood net plexus that seems to be explained by blood and lymph vascular net distribution [16,18,22]. In deep skin layers character of glycerol diffusion are more homogeneous and significantly slow. We estimate the average value of glycerol diffusion coefficient in the reticular dermis as 1.05×10^{-4} mm^2/sec at the $36.9°C$, and its variations in the layers as $0.5\text{-}2\times10^{-4}$ mm^2/sec. These data are evaluated based on the model of free diffusion, Gladstone and Dale law [1] and experimental results [8,21].

Difference in blood and lymph vessels distribution, cells structure and packing of the collagen fibre bundles within the skin produces the complex oscillations in the diffusion of glycerol (n=1.45) and other agents within the skin tissues. We assumed that scattering properties of blood contained layers are changed non-liner due to blood flow (and less due to lymph flow), which periodically wash away a part of an osmotic agent. In the model we also emphasize, that dermal-epidermal junction and variations of size and packing of topical skin cells and collagen fiber bundles produce the spatial variations in a glycerol concentration. Schematically it is represented in Fig.4.

An obtained change in the confocal images is proportionally related to the effective photons pathlength distributions and/or to the glycerol concentration changes, respectively. Profiles of the detecting signal are simulated separately for each stage of glycerol diffusion as it depends on the optical properties of tissues and probing geometry.

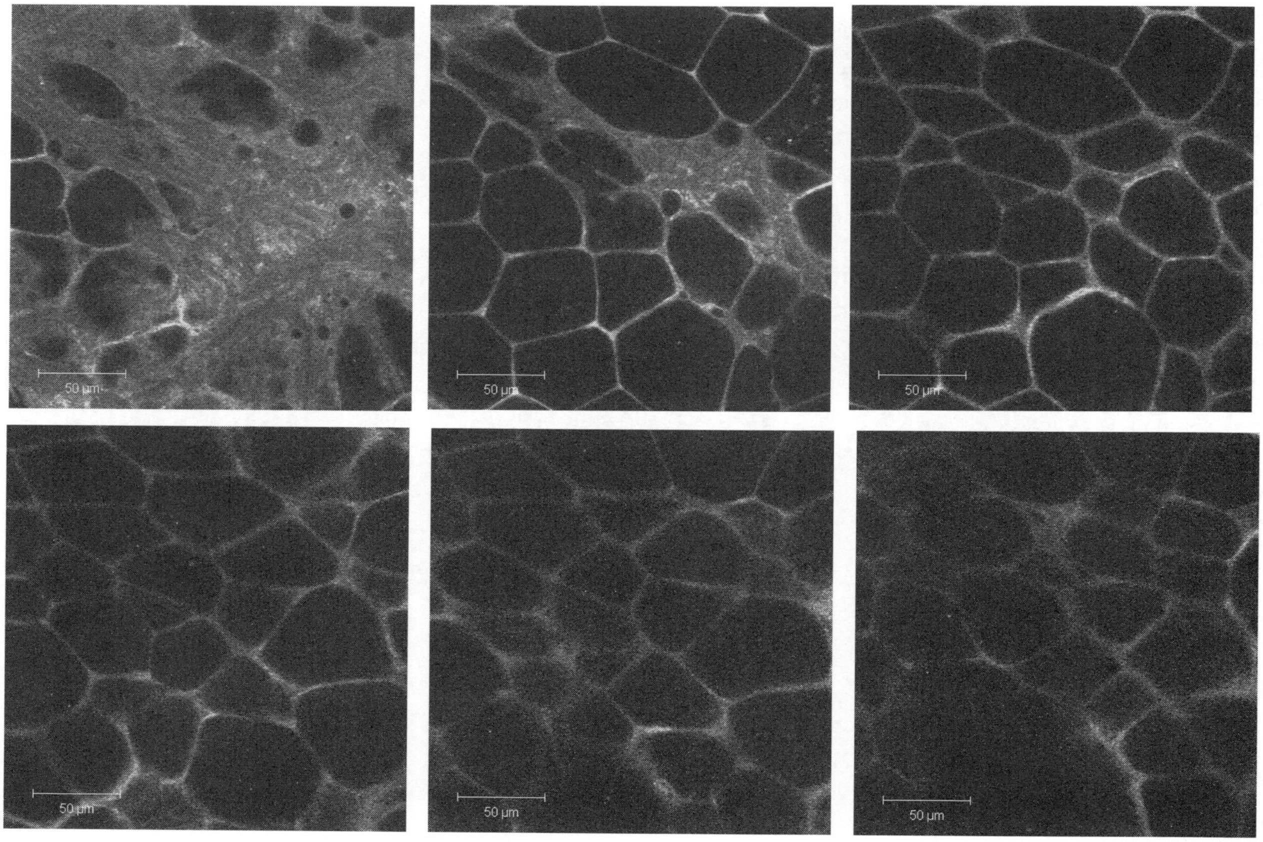

Figure 3. The confocal images of skin with optical clearing.

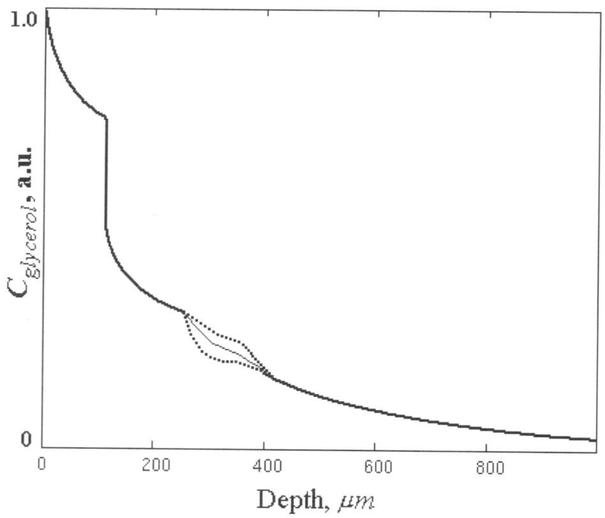

Figure 4. Schematic profile of glycerol concentration changes in the outer layers of skin. Dots are represented variations of glycerol concentration in upper blood net dermis due to blood flow.

Figure 5 presents simulated profiles of the detected signal distributions for a confocal probe focusing at 300 μm into the skin. Parameters of the probe used in the simulation are: lens diameter - 5 *mm*, focal length - 10 *mm*, the pinhole diameter - 10 μm. Optical properties corresponding to the skin layers properties are presented in Table 1. The results show (see Fig.5-a) that at normal stage of skin the main signal is collected from its topical area (20 μm thick).

This illustrates the screening affect of skin layers, and how quick the topical skin layers keep stray the optical radiation out of the deep penetration into the deep tissues. Nevertheless, for shallow focusing of confocal probe significant part of the detecting area with a total thickness about 150 μm is localized around focal plane.

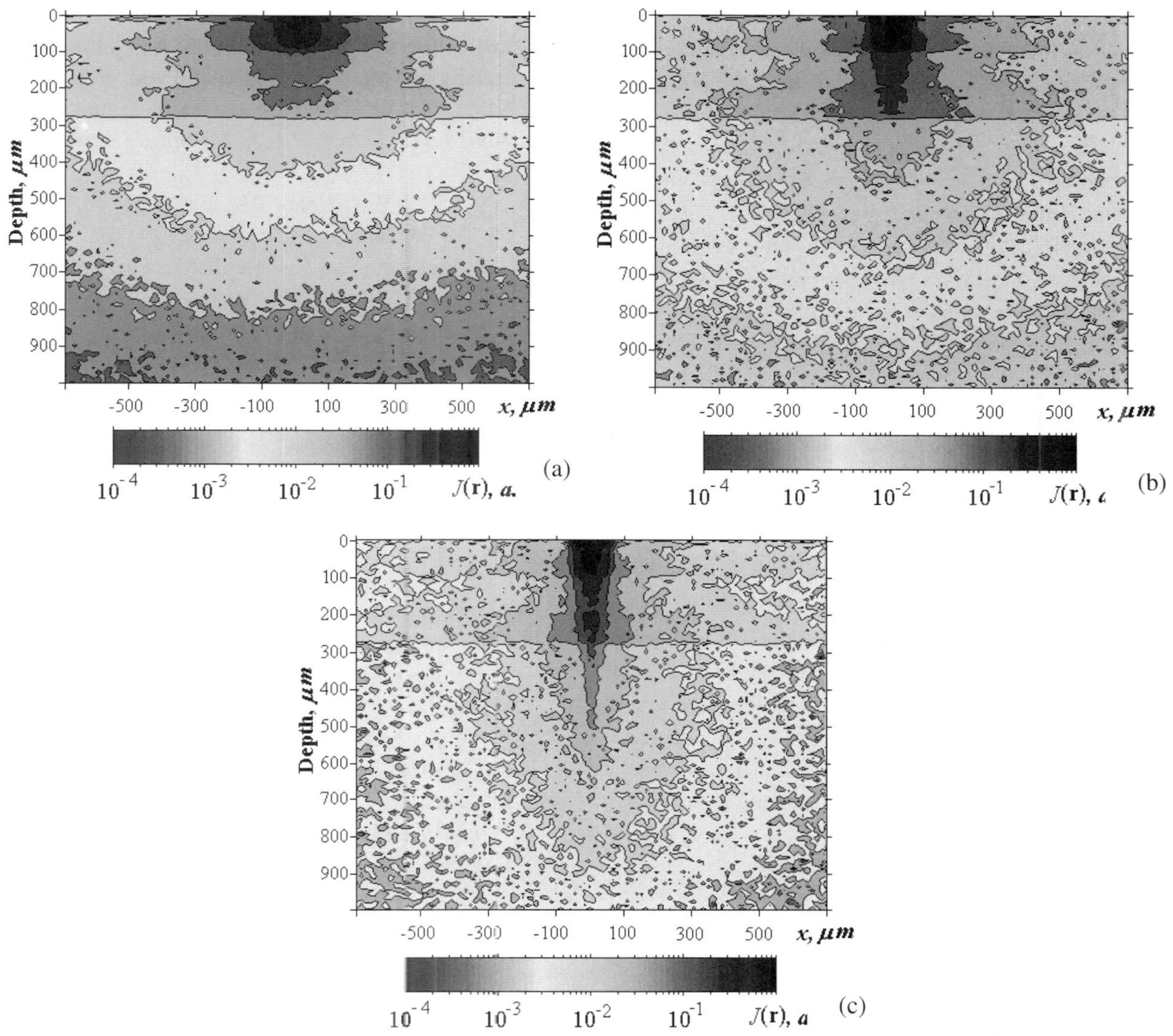

Figure 5. The axial profile of confocal probing within the skin. Optical properties of skin layers are presented in Table 1.

Optical changes of topical skin layers by the glycerol diffusion distort the screening effect. Decreasing of scattering of topical skin layers ($\mu_s^{(1)}$=20 mm^{-1}, $\mu_s^{(2)}$=15 mm^{-1}, $\mu_s^{(3)}$=25 mm^{-1}, n_1=1.45, n_2=1.4, other optical properties are unchanged, see Table 1) significantly increases detecting signal focusing depth (see Fig.5-b). However, main detected signal is collected from superficial layers of skin, but the influence of epidermis and topical layers of dermis are still significant. We ascribe these optical changes to the 10 minutes of topical glycerol diffusion into the skin tissues. Posterior matching of refractive indices (20 minutes later) essentially increases the localization of signal at 900 μm and even deeper (see Fig.5-c). These results are presented for $n_{1\text{-}5}$=1.4, $\mu_s^{(1)}$=15 mm^{-1}, $\mu_s^{(2)}$=10 mm^{-1}, $\mu_s^{(3)}$=10 mm^{-1}, $\mu_s^{(4)}$=12 mm^{-1}, $\mu_s^{(5)}$=15 mm^{-1}, and other optical properties are unchanged (see Table 1). Influence of topical layers of dermis is decreased (see Fig.5-b and Fig.5-c to compare). Spatial localization of detector signal offered by a confocal probe is localized mainly around the focal plane area for all depths of focusing (see Fig.5-c).

4. Summary and Conclusions

Using a simple numerical method we have estimated the effective depth for a confocal probing during the skin tissues clearing by glycerol. The results of the simulation show that the spatial localization offered by a confocal probe in skin due to its clearing are potentially useable for monitoring upper dermal layers, i.e. three times deep then an original probing depth. Present computational model could not consider the dynamics of the skin tissues optical properties changes, whereas it is an important factor. Nevertheless, in frame of simulation we estimated skin layers optical properties based on the experimental measurements of skin tissues glycerol diffusion, and taken into account the observed structure of skin tissues. We emphasize that hypodermal injection of bio-compatible chemical agents increases the speed of the tissue clearing as this agent diffuses straight into the dermal layers with the exception of the barriers. Whereas, in a topical superficial administering of skin tissues the diffusion processes are slow and time of clearing significantly (~ 10 times) long. The obtained results may be useful to evaluate the capabilities of skin imaging systems based on the confocal probing. Current approach has the potential to dramatically improve the specificity of visual diagnostic techniques, such as colposcopy, fluorescence microscopy, optical coherence tomography (OCT) and other medical diagnostic techniques and could be useful in laser surgery as well.

Acknowledgements

Authors are very grateful to Dr A.N. Bashkatov, Dr. E. Genina, Dr. A. Pravdin and Mr. L Dolotov (Saratov State University, Saratov, Russia) and Mr. B. Veksler (Cranfield University, Cranfield, UK) for useful discussions and experimental assistance during this study. Authors also acknowledge the financial support of grant "Leading Scientific Schools" No. 00-15-96667 of the Russian Basic Research Foundation and by Award No. REC-006 of the U.S. Civilian Research & Development Foundation for the Independent States of the Former Soviet Union (CRDF).

References

1. V.V. Tuchin, "Tissue Optics: Light Scattering Methods and Instruments for Medical Diagnosis" (Bellingham, WA: SPIE Press **TT38**, 2000).
2. V.V. Tuchin, "Handbook of Optical Biomedical Diagnostics" (Bellingham, WA: SPIE Press, 2002).
3. D. Sliney and M. Wolbarsht, "Safety with Lasers and Others Optical Sources. A comprehensive Handbook" (New York: Plenum Press, 1980).
4. V.V. Tuchin, I.L. Maximova, D.A. Zimniakov, I.L. Kon, A.H. Mavlutov, A.A. Mishin, *J. Biomed. Opt.*, **2**, 401 (1997).
5. G. Vargas, E.K. Chan, J.K. Barton, H.G. Rylander III, A.J. Welch, *Laser Surg. Med.*, **24**, 133 (1999).
6. S.G. Proskurin and I.V. Meglinski, *Laser Phys. Lett*, **4**, 824 (2007).
7. V.V. Tuchin, A.N. Bashkatov, E.A. Genina, *et al.*, *Proc. SPIE*, **3863**, 10 (1999).
8. A.N. Bashkatov, E.A. Genina, I.V. Korovina, *et al.*, *Proc. SPIE*, **4224**, 300 (2000).
9. I.V. Larina, E.F. Carbajal, V.V. Tuchin, M.E. Dickinson, K.V. Larin,, *Laser Phys. Lett*, **6**, 476 (2008).
10. I.V. Meglinsky, S.J. Matcher, *Med. Biol. Eng. Comput.*, **39**, 44 (2001).
11. I.V. Meglinski, S.J. Matcher, *Opt. Spectrosc.*, **91**, 654 (2001).
12. M. Keijzer, S.L. Jacques, S.A. Prahl and A.J. Welch, *Laser Surg. Med.*, **9**, 148 (1989).
13. I.V. Yaroslavsky, V.V. Tuchin, *Opt. Spectrosc.*, **72**, 505 (1992).
14. S.L. Jacques, L. Wang, L.-Q. Zheng, *Comput. Meth. Prog. Bio.*, **47**, 131 (1995).
15. J.M. Schmitt, A. Knüttel and M. Yadlowsky, *JOSA A*, **11**, 2226 (1994).
16. W. Montagna, A.M. Kligman, K.S. Carlisle, "Atlas of Normal Human Skin" (New York: Springer Verlag, 1992).
17. G.F. Odland, "Physiology, Biochemestry and Molecular Biology of the Skin", Edited by L A Goldsmith (Oxford: Oxford University Press, 1991, Vol.I, p.3).
18. T.J. Ryan, "Physiology, Biochemestry and Molecular Biology of the Skin", Edited by L A Goldsmith (Oxford: Oxford University Press, 1991, Vol.II, p.1019).
19. M.R. Chedekel, "Melanin: Its Role in Human Photoprotection", Eds. L. Zeise, M.R. Chedekel and T.B. Fitzpatrick (Overland Park: Valdenmar, 1995, p.11).
20. J.Q. Lu, X.-H. Hu and K. Dong, *Appl. Opt.,* **39**, 5890 (2000).
21. A.N. Bashkatov, E.A. Genina, Y.P. Sinichkin and V.V. Tuchin, *Proc. SPIE*, **4623**, 144 (2002).
22. K.A. Holbrook and P.H. Byers, *Am. J. Med. Genet.*, **34**, 105 (1989).

Topic I

Tissue Optics and Diffuse Optical Imaging

REFRACTIVE INDEX OF HEMOGLOBIN AND ALBUMIN SOLUTIONS INCUBATED WITH GLUCOSE

OLGA S. ZHERNOVAYA, VALERY V. TUCHIN

Institute of Optics and Biophotonics, Saratov State University,
83 Astrakhanskaya str., 410012 Saratov, Russia

Measurement of glycated proteins is widely used for routine monitoring of long-term glycemic status in patients with diabetes mellitus. Glycated proteins level are used both as an index of mean glycemia and as a measure of risk for development of diabetes complications. The level of glycated hemoglobin reflects the mean glycemia during the preceding 2-3 months, while glycated plasma proteins (mainly glycated albumin) reflect mean free glucose level during much more short time, about 2-3 weeks. Since glucose and glycation of proteins affect optical properties of blood, optical methods may be suggested for estimating glycated hemoglobin and glycated albumin amount in blood.

Keywords: glycated hemoglobin, glycated albumin, refractive index, Abbe refractometer

1. Introduction

Over the past decades diabetes mellitus had an explosive increase in the number of people worldwide and took its place as one of the dominant threats to human health in the 21st century[1]. The prevention of diabetes and control of its complications is main goal because of the huge morbidity and mortality associated with the disease[2,3]. The measure of diabetes complications, including retinopathy, neuropathy and cardiovascular disease, represents by the level of glycated (glycosylated) hemoglobin and albumin.

Hemoglobin and albumin are the main blood proteins that react with glucose forming glycated hemoglobin and glycated albumin. The level of glycated hemoglobin reflects the mean glycemia during 2-3 months, the average lifespan of erythrocyte whereas the level of glycated albumin introduces the mean glycemia for about the period of 2-3 weeks. These parameters indicate average amount of free glucose in blood over quite a long period of time, whereas the concentration of free glucose in blood fluctuates widely during a day[4]. The most reactive amino group of hemoglobin is the NH_2-terminus of the β-chain[5]. The principal glycation site of albumin is Lys-525[6,7]. HbA1c is the major component of HbA1, which is formed by a non-enzymatic irreversible process of combination of aldehyde group of glucose with the amino-terminal valine of the β-chain of hemoglobin. This process comprises a sequence of non-enzymatic reactions (Fig. 1), known as the Maillard reactions[8]. The first is the rapid but reversible formation of an aldimine (or Schiff base), whereas the second one is the considerably slower formation of a stable ketoamine via a process known as the Amadori rearrangement. The ketoamine accumulates over the life of the erythrocyte and forms the main part of the glycated hemoglobin. The next stage of the reaction is formation of hemoglobin-advanced glycation end products (Hb-AGE)[9].

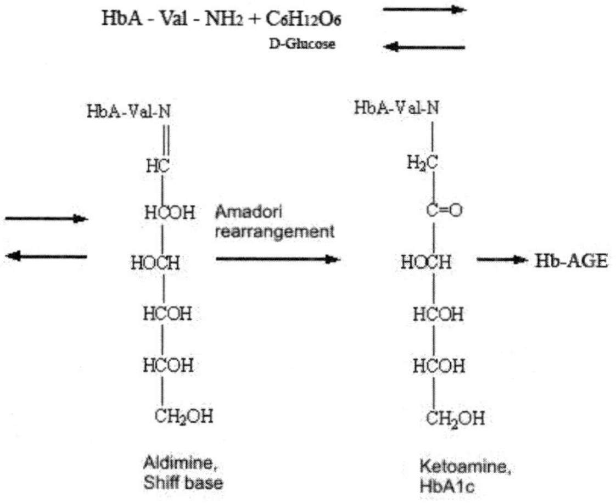

Fig. 1. Nonenzymatic formation of HbA1c and Hb-AGE from hemoglobin and glucose.

Inherently less than 7% of hemoglobin[10] and 6-15% of total serum albumin[3] are glycated in normal conditions, and the percentage of glycated hemoglobin and glycated albumin rises with the increasing of free glucose concentration. The rate of glycated hemoglobin formation is proportional to free blood glucose concentration. The increase of HbA1c for 1% corresponds to increase of mean plasma glucose for a 35-mg/dl (1.95-mmol/l)[4].

There are many different glycated hemoglobin assay methods in current use[11]. Most methods quantify hemoglobin A1c, defined as hemoglobin A with glucose attached to the NH2-terminus valine of one or both β-chains. Nevertheless, obtained results can variate depending on the particular hemoglobinopathy and assay methodology. In this case new methods for assessing long-term glycemic control are required.

Recently several optical diagnostic techniques have been tested for monitoring free glucose and glycated hemoglobin levels in blood[12,13]. Fluorescence measurements have shown that in vitro glycated hemoglobin gave the small rise of fluorescence intensity after 7 weeks of incubation, and at 11 weeks of incubation the increase was substantial, which finally reached threshold at 17 weeks for all samples[13]. The threshold intensity varied linearly with the concentration of glucose. In addition, colour change was noticed from red to dark brown after 2 months of incubation, which perhaps indicated the advanced stage of glycation. Optical coherent tomography (OCT) has been applied for refractive index measurements of hemoglobin-glucose solutions[12]. The disadvantage of OCT and OCT-based techniques is low accuracy of refractive index measurements (about 10^{-2}-10^{-3} for 1mm cuvette). The more precise measurements of refractive index of highly absorbing and scattering fluids have been achieved by Intracavity Laser Refractometry in Reflection (ILRR)[14]. This method allows to measure refractive index with up to 10^{-8} sensitivity and time of measurement of several microseconds.

In this paper we present the study of refractive index changes of water solutions of hempoglobin and albumin during incubation with glucose. Measurements were taken using Abbe refractometer at 589 nm. Our experiments show that the changes of refractive index caused by glycation of hemoglobin may be observed using refractive index measurements.

2. Materials and Methods

Samples of water solutions of hemoglobin with concentration 25 g/l and with water solution of albumin with concentration 40 g/l were studied. D-glucose dissolved in the distilled water was added to water solutions of hemoglobin (Agat-Med, Russia) and albumin (Biomed, Russia). The glucose concentrations were chosen in a range from 0 to 1000 mg/dl with a step 100 mg/dl.

Refractive index measurements of water solutions of hemoglobin and albumin with glucose were performed by Abbe refractometer (IRF-454B2M, Russia) at 589 nm. The light emitting diode (LED) at the wavelength 660 nm was used in Abbe refractometer as a light source. It provides more contrast visibility of the dark/lightened field border of Abbe refractometer and, therefore, more precise and stable measurements of refractive index of highly absorbing hemoglobin samples due to a high brightness of quasi-monochromatic LED radiation. Measured values of refractive index correspond to wavelength 589 nm because of Amici prism action. The accuracy of the measurements is about 5×10^{-4}.

3. Results and Discussion

Initial concentration of water solutions of hemoglobin is 120 g/l. This solution was dissolved in distilled water to obtain lower concentration of hemoglobin. The refractive index of the initial hemoglobin solution can be calculated from an equation:[15]

$$n_{Hb+H_2O} = n_{H_2O}V_{H_2O} + n_{Hb}V_{Hb} \quad (1)$$

whence

$$n_{Hb} = \frac{n_{Hb+H_2O} - n_{H_2O}V_{H_2O}}{V_{Hb}} \quad (2)$$

where n_{Hb} is the refractive index of the initial hemoglobin solution, V_{H_2O}, V_{Hb} is the volume fractions of distilled water and initial hemoglobin solution, respectively, n_{Hb+H_2O} is measured refractive index of hemoglobin solution in the samples (hemoglobin concentration is 130 g/l), n_{H_2O} is measured refractive index of distilled water.

The refractive index of glucose solution n_{gl} can be found from the expression: [16]

$$n_{gl} = n_{H_2O} + 0.1515 \cdot C_{gl} \quad (3)$$

where C_{gl} is glucose concentration, g/ml.

The refractive index of solution supposing non-interacting hemoglobin and glucose molecules can be estimated from the equation:[15]

$$n_{calc} = n_{Hb}V_{Hb} + n_{gl}V_{gl} \quad (4)$$

where V_{Hb}, V_{gl}, is the volume fractions of hemoglobin and glucose solutions, respectively.

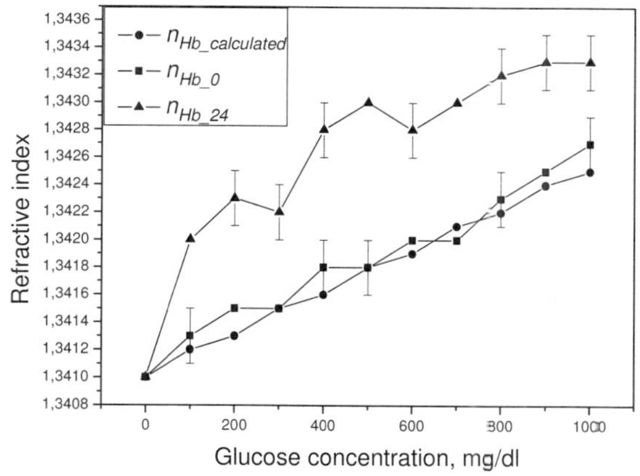

Fig. 2. Experimental (n_{Hb_0}, n_{Hb_24}) and calculated ($n_{Hb\_calculated}$) values of refractive indexes of water solutions of hemoglobin with glucose. Incubation time for n_{Hb_24} is 24 days, n_{Hb_0} correspond to the refractive index measurements of freshly prepared samples.

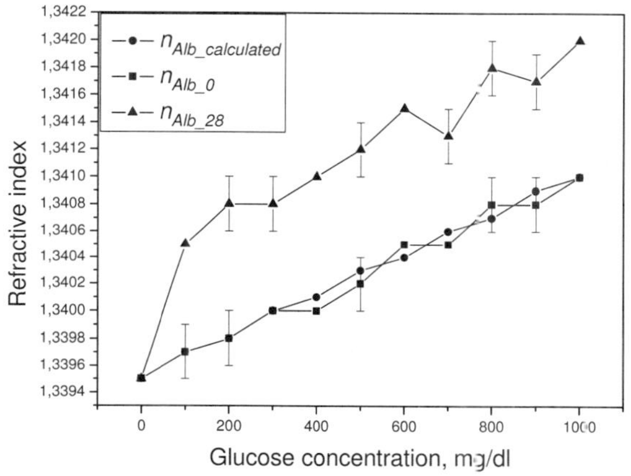

Fig. 3. Experimental (n_{Alb_0}, n_{Alb_28}) and calculated values ($n_{Alb\_calculated}$) of refractive index of albumin solutions with glucose. Incubation time for n_{Alb_28} is 28 days, n_{Alb_0} correspond to the refractive index measurements of freshly prepared samples.

Figures 2 and 3 represent the experimental results of the refractive index measurements of water solutions of hemoglobin and albumin with glucose and the results of theoretical calculations assuming that protein and glucose molecules are non-interact to each other. At the beginning of incubation measured refractive index values of hemoglobin and albumin solutions were in close agreement with theoretical values of refractive index, which were calculated supposing non-interacting protein and glucose molecules. But after rather long time of incubation (24 days for hemoglobin solutions and 28 days for albumin solutions) the excess of measured values of refractive index above calculated data was observed for samples with glucose, whereas the refractive index of protein solutions without glucose did not change (see Fig. 2, 3). This difference is hypothesized to be due to glucose binding to protein molecules. The difference between measured and calculated values (non-interacting protein and glucose molecules) of refractive index may be explained by assumption of chemical interaction between glucose and protein molecules and formation of glycated hemoglobin and glycated albumin.

The experimental data presented at Figures 2 and 3 well agreed with the classical theory of light dispersion in condense matter:[17]

$$n = 1 + \alpha N \frac{q^2}{m} \quad (5)$$

where α is the wavelength dependent coefficient,
N is the number of molecules,
m is the molecule's mass,
q is the molecule's charge.

The equation (5) shows that refractive index of hemoglobin solution at the addition of different glucose concentrations depends also on the molecular weight and charge: with the elevation of glucose concentration the increase of R-group charge of hemoglobin grows quicker than the elevation of molecular weight at glucose binding, $\Delta(q^2)>\Delta m$, therefore, the refractive index of hemoglobin solutions with glucose increased. Such results were obtained for water solution of albumin with glucose (see Fig. 3), where the presence of glucose in samples leads to considerable increase of refractive index after 28 days of incubation compare to the samples without glucose.

There are also some factors which effect refractive index of water solutions of hemoglobin, such as pH value[18] and oxygen saturation[19]. Besides, high concentration of glucose increases hemoglobin affinity to oxygen[20]. Our measurements were taken considering 100% hemoglobin oxygenation, and our experiments show that pH does not change during incubation[21]. Apparently, the excess of measured values of refractive index above calculated data is explained by assumption of chemical interaction between glucose and protein molecules and formation of glycated hemoglobin and glycated albumin.

4. Conclusions

Our experiments have shown that the method of refractive index measurements based on Abbe refractometer allows determining the changes caused by glycation of proteins. However, the method based on Abbe refractometer is inapplicable for refractive index measurements of highly concentrated hemoglobin solutions and blood because of strong scattering and absorption. The results of experiments have shown the considerable excess of measured refractive index values above calculated data after long time of incubation with glucose, which apparently due to glucose binding to proteins.

Acknowledgments

Studies presented in this paper were supported by grant of Federal Agency of Education of RF № 1.4.06, RNP.2.1.1.4473 and by CRDF BRHE grant RUXO-006-SR-06.

References

1. P. Zimmet, K. G. M. M. Alberti, J. Shaw, "Global and societal implications of the diabetes epidemic", *Nature*, **414**, 782-787, 2001.
2. Diabetes Control and Complications Trial Research Group, *N. Engl. J. Med.*, **329**:977–86, 1993.
3. J. F. Day, S. R. Thorpe, J.W. Baynes, "Nonenzymatically Glucosylated Albumin: *in vitro* preparation and isolation from normal human serum", *J. Biol. Chem.*, **254**, No. 3, 595-597, 1979.
4. C. L. Rohlfing, H.-M. Wiedmeyer, R. R. Little, J. D. England, A. Tennill, D. E. Goldstein, "Defining the relationship between plasma glucose and HbA1c", *Diabetes Care* **25,** 2, 2002.
5. R. Shapiro, M. J. McManus, C. Zalut, H. F. Bunn, "Sites of Nonenzymatic Glycosylation of Human Hemoglobin A", *J. Biol. Chem.*, **255**, No. 7, 3120-3127, 1980.
6. N. Iberg, R. Fliickige, "Nonenzymatic Glycosylation of Albumin *in vivo*: identification of multiple glycosylated sites", *J. Biol. Chem.*, **261**, No. 29, 13542-13545, 1986.
7. N. Shaklai, R. L. Garlick, H. F. Bunn, "Nonenzymatic Glycosylation of Human Serum Albumin Alters Its Conformation and Function", *J. Biol. Chem.*, **259**, No. 6, 3812-3817, 1984.
8. L. C. Maillard, "Actions des acides amines sur les sucres: Formation des melanoidines par voie methodique", *C.R. Acad. Sci.*, **154**:66–68, 1912.
9. H. Vlassara, M. R. Palace, "Diabetes and advanced glycation endproducts", *J. Intern. Med.*, **251**, 87-101, 2002.
10. American Diabetes Association: Standards of medical care for patients with diabetes mellitus (Position statement), *Diabetes Care* **24** (Suppl. 1):S33–S43, 2001.
11. A.Stott, I. F. Casson, G. J. Higgins, "Glycated haemoglobin assays. Approaches to standardization of results", *Diabetic Medicine*, **18,** 274-279, 2001.
12. E.I. Galanzha, A.V. Solovieva, V.V. Tuchin, R.K. Wang, S.G. Proskurin, "Application of optical coherence tomography for diagnosis and measurements of glycated hemoglobin", *Proc SPIE*, **5140**, 125-132, 2003.
13. N. Vigneshwaran, G. Bijukumar, N. Karmakar, S. Anand, A. Misra, Autofluorescence characterization of advansed glycation end products of hemoglobin", *Spectrochimica Acta Part A*, **61**, 163-170, 2004.
14. S. A. Gonchukov, Yu. B. Lazarev, "Laser Refractometry in Medicine and Biology", *Laser Phys.*, **13**, No. 5, 749-755, 2003.
15. B.A. Fikhman, *Microbiological refractometry*, Moscow, 1967.
16. J.S. Maier, S.A. Walker, S. Fantini, M.A. Franceschini, E. Gratton, "Possible correlation between blood glucose concentration and the reduced scattering coefficient of tissues in the near infrared", *Opt. Lett.*, **19**, 2062–2064, 1994.
17. M.V. Volkenschtein, *Molecular Optics,* Moscow, 1951.
18. G. Mazarevica, T. Freivalds, A. Jurka, "Properties of erythrocyte light refraction in diabetic patients", *J. Biomed. Opt.*, **7**, 244-247, 2002.
19. D.J. Faber, M.C.G. Aalders, E.G. Mik, B.A. Hooper, M.J.C. van Gemert, T.G. van Leeuwen, "Oxygen saturation-dependent absorption and scattering of blood", *Phys. Rew. Lett.*, **93,** No. 2, 2004.
20. N.V. Maksimova, A.V. Nagovitsyn, and G.V. Maksimov, "The role of ion transport in regulation of affinity of hemoglobin to oxygen at diabetes", *Bulletin of Exp. Biol. and Med.*, **133**, 4, 2002.
21. O. S. Zhernovaya, V.V. Tuchin, "Measurements of refractive index of hemoglobin mixed with glucose at physiological concentrations", *Proc. SPIE*, **6163**, 61631Q, 2006.

DYNAMICS OF VISIBLE ABSORBANCE SPECTRUM OF HEMOGLOBIN SOLUTION INCUBATED WITH GLUCOSE

EKATERINA N. LAZAREVA

Research Educational Institute of Optics & Biophotonics, Saratov State University, 83 Astrakhanskaya Str., 410012 Saratov, Russia

VALERY V. TUCHIN

Institute of Precise Mechanics and Control of RAS, Saratov 410028, Russia

This study is focused on the determination of the absorbance of oxyhemoglobin solutions at different concentrations of glucose (from 0 to 1000 mg/dl with a step 100 mg/dl) and hemoglobin (3.2g/l) incubation with glucose from 3 hours to a few weeks. The absorbance was determined within the wavelength range from 500 to 900 nm. Measurements of the absorbance spectra have been performed using a double-beam double-wavelength spectrophotometer.

Key words: glucose sensing, glycated hemoglobin, near-infrared spectroscopy

1. Introduction

It is known that glucose interacts with plasma proteins and hemoglobin in erythrocytes. Glycated hemoglobin is the result of an irreversible non-enzymatic fixation of glucose on hemoglobin (glycation) of the beta chain of hemoglobin A.[1-8] Glucose combines with the α amino group of the valine residue at the N-terminus of 18 globin chains to form an aldimine compound (Schiff base). This reaction is reversible, and dissociation to native hemoglobin and glucose occurs readily. Internal rearrangement of the aldimine intermediate by the Amadori reaction yields a stable ketoamine derivative.[9] 60% of the glucose is bound to the N-terminal valines of the β chains. A small amount of glucose binding occurs at the α chain N-terminal valines.[10] In fact, glycation of hemoglobin occurs over the entire 120 day life span of the red blood cell, but within this 120 days recent glycaemia has the largest influence on the HbA1c value. Theoretical models and clinical studies suggest that a patient in stable control will have 50% of his HbA1c formed in the month before sampling, 25% is in the month before that, and the remaining 25% in months from two to four.[7] More recent clinical experience suggests that in practice a fall from an elevated HbA1c can be demonstrated within a few days of a change in therapy and certainly within 2 weeks or 4–8 weeks.[11] Glycated hemoglobin is a useful marker for long-term glucose control in diabetics.[9,10]

There are more than 20 techniques of determination of glycated hemoglobin, such as cation-exchange chromatography, electrophoresis, affinity chromatography and immunoassays, but each of these techniques measures a different fraction of the glycated hemoglobin. All methods have shortcoming, for example, affinity chromatography to be less sensitive to changes in temperature and pH.[4,5,7] Therefore, the search of quick, exact, and high sensitive methods for measurement of all fractions of glycated hemoglobin in blood is important.

In particular, the optical method would give the promising results.[2] Knowledge of the optical properties of blood is important not only for many diagnostic and therapeutic applications in laser medicine, but also for routine medical diagnostics.[12] They can be described by the optical parameters such as absorption coefficient μ_a, scattering coefficient μ_s, and anisotropy factor g. It is known that the complex refractive index of deoxyhemoglobin and oxyhemoglobin is required to calculate the optical properties of oxygenated and deoxygenated blood.[13] The imaginary part of the complex refractive index, characterized by the absorption constant k, is proportional to the absorption coefficient.[12] It was shown recently by G. Mazarevica and coauthors of Ref. 14 that the refractive index of hemoglobin of dried erythrocytes is differed at the normal and increased level of glucose, which is due to glucose binding to hemoglobin.[14] Early we also showed that value of the real part of the complex refractive index of hemoglobin solutions at different concentrations of glucose increased after one week of incubation.[15]

The increase in affinity of hemoglobin to oxygen at glucose was observed by G.V. Maksimov and coauthors.

They have assumed that observable changes at action of glucose on erythrocytes are caused by conversion from Fe^{+2} to Fe^{+3}, or transition of atom of iron from high-spin to low-spin state.

This study is focused on the determination of the absorbance of oxyhemoglobin solutions of concentration at different concentrations of glucose (from 0 to 1000 mg/dl with a step 100 mg/dl) and hemoglobin (3.2g/l) incubation with glucose from 3 hours to a few weeks. The absorbance was determined within the wavelength range from 500 to 900 nm. Measurements of the absorbance spectra have been performed using a double-beam double-wavelength spectrophotometer.

2. Materials and Methods

The absorbance was determined at the wavelength range from 500 to 900 nm. Measurements of the collimated absorbance spectra have been performed using double-beam double-wavelength spectrophotometer (BioPribor, Russia). An error of installation of lengths of waves, within the limits, was ±0.5 nm. A spectral interval was 2nm.

An error of measurements of optical density, within the limits, was: in a range 0.05-1B — ±0.02, in a range 1-2B — ±0.2. A constant of time of the channel of registration was 1.28 sec. Noise level in a maximum of sensitivity of a photodetector at a constant of time of a registering path 1sec and an allocated spectral interval of 2 nm was 0.0002.

Sources of radiation are lamps DDS-30 and KGM12-40. The scheme of the experimental setup is shown in Fig. 1.

Experiments were performed using standard control hemoglobin water solution (120 g/l) applied for calibration of blood/hemoglobin analyzers. On the basis of control hemoglobin solution less concentrated samples of hemoglobin solution of 1.6 g/l were prepared using distillate water. The choice of concentration proves opportunities of the spectrometer used for measurement of transmission coefficient in this work. The different amounts of D-Glucose (from 0 to 1000 mg/dl with a step 100 mg/dl) were added to hemoglobin solution samples. These concentrations of glucose comply with the physiological standard.[4] Samples were stored in vacuum cuvettes at temperature 6±1°C. All experiments have been performed at room temperature, 22.1±0.7°C.

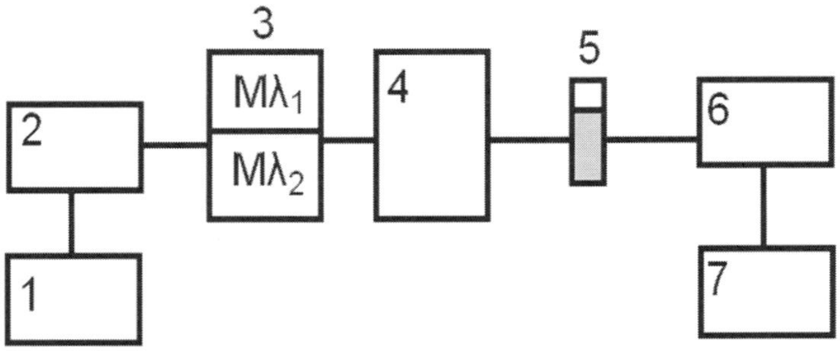

Fig. 1. Experimental setup for measurements of the absorbance spectra: (1) power source; (2) source of radiation; (3) monochromator; (4); (5) cuvette; (6) photoroltaic cell power source; (7) PC.

3. Experimental Results and Discussion

The absorbance spectra of hemoglobin solution of concentration of 1.6 g/l in the wavelength range of 500-900 nm are shown in Fig. 2 (for oxygenated and fully deoxygenated hemoglobin after: a) 3 hours; b) 7 days; and c) 14 days of hemoglobin storage in cuvette). The typical absorbance maxima are at 542 nm, 577 nm and 633 nm for oxygenated hemoglobin and at 553 nm for deoxygenated hemoglobin after three hours of storage. Absorbance of oxyhemoglobin solutions at 633 nm indicates the formation of methemoglobin.[17] There are four isosbestic points, where absorbance is independent on the oxygenation level. It is clearly seen at the wavelength 805 nm, 653 nm, 617 nm and 538 nm. The isosbestic point at 805 nm was used in our experiments for referencing. Such referencing makes possible to account for changes of total hemoglobin concentration, which may have take place during prolonged measurements and sample storage. The positions of the absorbance maxima of deoxyhemoglobin and of oxyhemoglobin were not changed after 7 and 14 days of storage in cuvette.

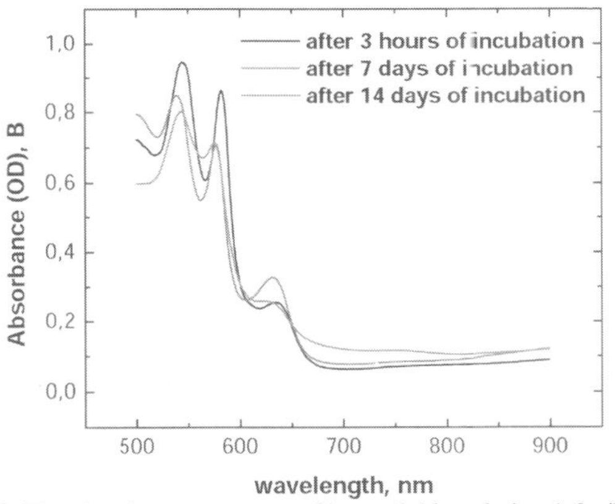

Fig. 2. The absorbance spectrum of hemoglobin solution 1.6 g/l.

The absorbance spectra of oxyhemoglobin solutions at different concentration of glucose are shown in Fig. 3. There are two isosbestic points (621 and 805 nm) on the absorbance spectra of oxyhemoglobin solutions measured at different concentrations of glucose.

The deformation of hemoglobin spectra caused by production of glycated hemoglobin in the course of hemoglobin incubation with glucose can be evaluated using parameter "θ" introduced as the slope of spectral curve for absorbance of hemoglobin solution in the vicinity at glucose concentration.

The changes in spectra of absorbance of oxyhemoglobin at different concentrations of glucose may be caused by two processes: glycation and oxygenation efficiency at glucose action. The slope of spectral curve for absorbance of hemoglobin solution in the vicinity at glucose concentration for isosbestic points at 805 and 621 nm are shown in Figs. 4a and 4b. The tendency increase of "θ" with glucose concentration for isosbestic point 805 nm and for isosbestic point of 621 nm can indicate the increase of absorbance at the wavelength 633 nm. As absorbance on 633 nm corresponds to some concentration of methemoglobin in solution it is possible to assume, that glucose increases affinity of hemoglobin to oxygen and promotes process of oxidation of atom of iron in hem and to conversion of oxy forms of hemoglobin to met form.[17,18] Value of "θ" for oxyhemoglobin without glucose has remained constant during its storage up to 14 days, thus spectral deformation in the vicinity of isosbestic point may be associated with hemoglobin glycation process which takes at least several hours of incubation to be seen.[9]

a)

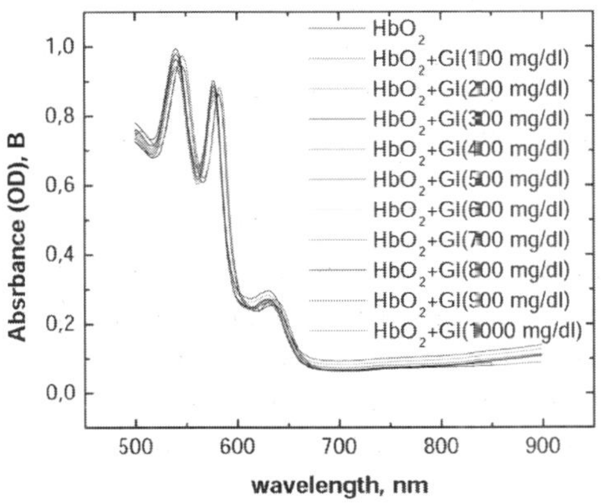

b)

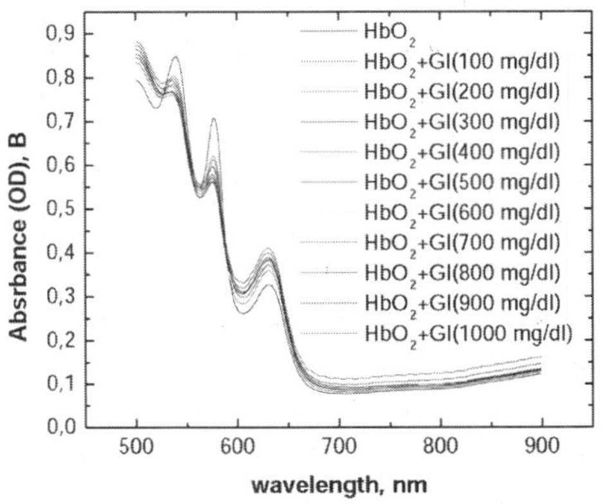

c)

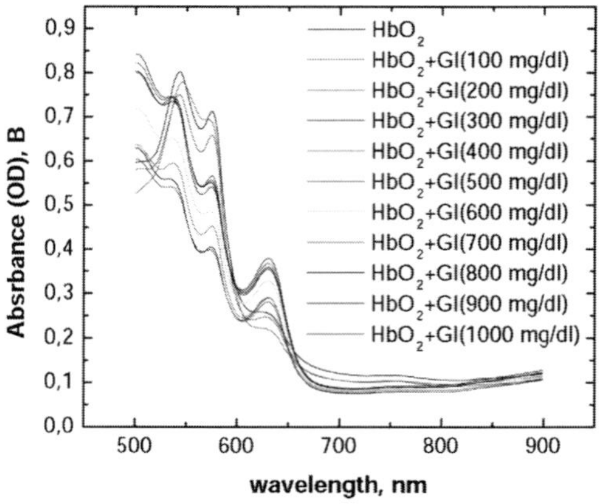

Fig. 3. The absorbance spectra of oxyhemoglobin solutions at different concentrations of glucose after: a) 3 hours; b) 7 days; and c) 14 days of incubation in cuvette.

a)

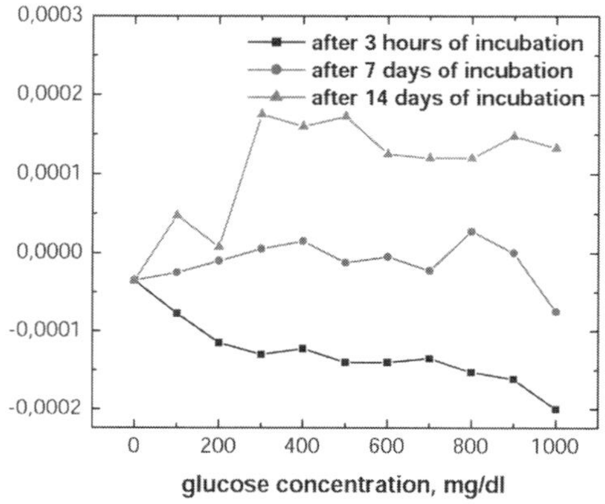

b)

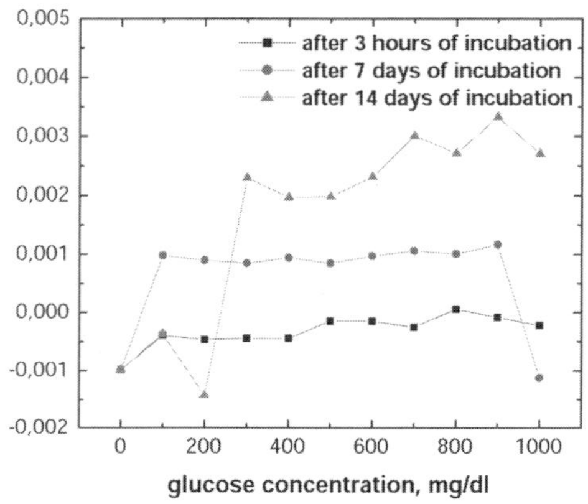

Fig. 4. The slope of spectral curve for absorbance of hemoglobin solution in the vicinity at glucose concentration: a) for isosbestic point at 805 nm; b) for isosbestic point at 621 nm. The average estimated error is ±0.004.

The increase in affinity of hemoglobin to oxygen at glucose was observed by the authors of Ref.19. They have assumed that observable changes at action of glucose on erythrocytes are caused or conversion from Fe^{+2} to Fe^{+3}, or transition of atom of iron from high-spin to low-spin state. As in low-spin states Fe (II) or Fe (III) high-energy orbitals are free, the atom of low-spin iron has the smaller size, than atom of high-spin iron. This dependence of the size of atom of iron on a spin condition plays the important role in definition of geometry of system “ion-macrocycle”. As diameter of "cavity" in the center of a macrocycle makes 2.02 Å, and in low-spin Fe (II) - and Fe (III) states near by 1.91 Å in this case atoms of iron are built in “a cavity of a kernel” and are localized in a plane of a macrocycle. On the contrary, high-spin Fe (III) leaves a plane on distance nearby 0.3 Å, and high-spin Fe (II) - on 0.7 Å as are not built in ""a cavity of a kernel.” In connection with this, in patients with diabetes high extracellular concentration of glucose gave rise to redistribution of electronic density between C=C and C-N binds of a molecule of porphyrin. In this case the displacement of position of atom of nitrogen from the center of a macrocycle to atom of carbon occurs. Such redistribution of electronic density through a macrocycle leads to expansion of “a kernel of a cavity” and changes localization of atom of iron in a plane of a macrocycle that leads to increase of binding of O_2 in a molecule of hemoglobin and promotes development hypoxia.[19]

In this study we also have noticed, that glucose may increase affinity of hemoglobin to oxygen. In Fig. 5 of absorbance spectra of oxyhemoglobin measured for different concentrations of glucose after 3 hours (Fig. 5a), 7 days (Fig. 5b), and 14 days (Fig. 5c) of incubation are shown. It is not observed visible changes of absorbance of oxyhemoglobin at different concentrations of glucose after 3 hours and 7 days of incubation, changes of absorbance of solutions after 14 their days of incubation however are appreciable. The increase of absorbance of oxyhemoglobin on the wavelength 633 nm and its reduction on the 542 nm and 577 nm as a function of glucose concentration is seen. The increase of absorbance on 633 nm is due to increase of methemoglobin concentration in solutions and indicates the change of affinity of hemoglobin to oxygen. The reduction of absorbance of oxyhemoglobin solutions at 542 nm and 577 nm indicates the oxidation of iron in them.[17,18]

a)

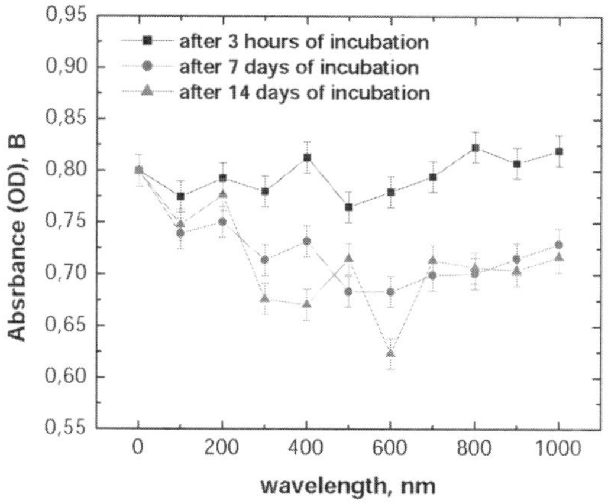

b)

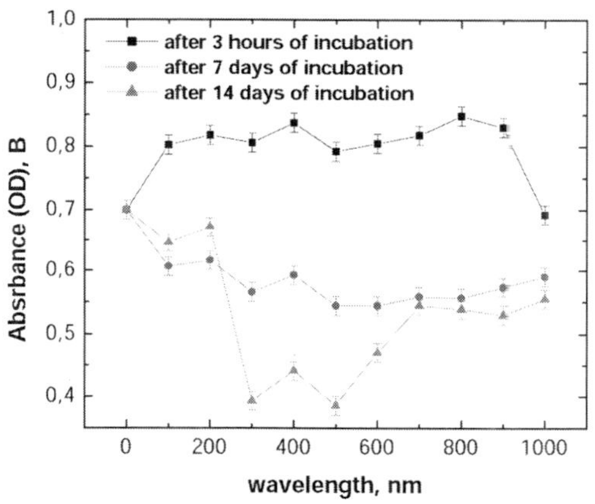

c)

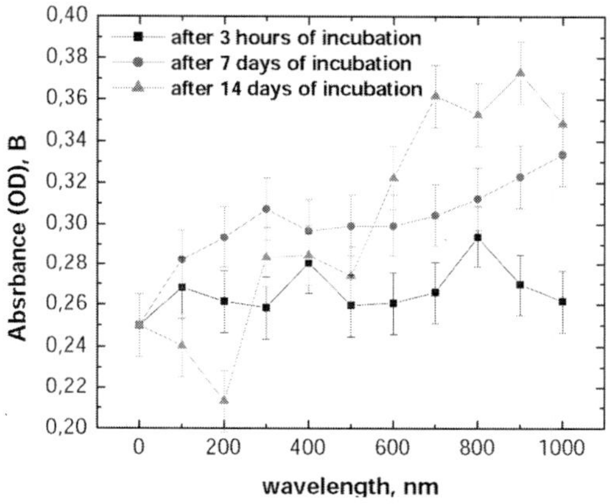

Fig. 5. The absorbance of oxyhemoglobin solution at different glucose concentrations: a) on the wavelength 542 nm; b) on the wavelength 577 nm; c) on the wavelength 633 nm.
The average estimated error is ±0.015

4. Conclusions

Measurements of the complex refractive index of deoxyhemoglobin and oxyhemoglobin is required to calculate the optical properties of oxygenated and deoxygenated blood.[13] The imaginary part of the complex refractive index, characterized by the absorption constant k, is proportional to the absorption coefficient.[12] It was shown that changes in the absorbance spectra of oxyhemoglobin in the region from 500 to 900 nm can be caused by the processes of hemoglobin glycation. In this study early discovered by authors of Ref. 19 the increase of affinity of hemoglobin to oxygen at glucose concentration increase has been found. The changes in the absorbance spectra of oxyhemoglobin at different concentrations of glucose observed in this study also can be explained by the conversion of oxyhemoglobin to methemoglobin. These investigations provide a basis for further research of changes of refractive index and other optical parameters of hemolyzed blood, erythrocytes, hemoglobin and plasma protein (albumin, globulin) solutions at different concentrations of glucose and investigation of effect of hemoglobin oxygenation on refractive index of blood at different level of glycemia.

Acknowledgment

The authors are grateful to V.I. Kochubei for the help in the experiments. The research was supported by grants of Federal Agency of Education of RF № 1.4.06, RNP.2.1.1.4473 of the RF Program "The Development of Scientific Potential of the High School (2006-2008)".

References

1. V.V. Tuchin, R.K. Wang, E.I. Galanzha, J. B. Elder, and D. M. Zhestkov, "Monitoring of glycated hemoglobin by OCT measurement of refractive index," Conference Program CLEO/QELS, Baltimore, p.120, 2003.
2. J.P. Goulle, C. Lairoix, D. Bouige, "Glycated hemoglobin a useful post-mortem reference marker in determining diabetes," Forensic. Sci. Int. 128, 44-49, 2002.
3. T. Lahousen, R.E. Roller, R.W. Lipp, W.J. Schnedl, "Determination of glycated hemoglobins (HbA1c)," Wien. Klin. Wochenschr 114, 301-315, 2002.
4. Kenneth Zierler, "Whole body glucose metabolism metabolism," Am. J. Physiology 276, E409–E426., 1999.
5. Erin K. Cloherty, Karen S. Heard, and Anthony Carruthers, "Human Erythrocyte Sugar Transport is Incompatible with Available Carrier Models," Biochemistry 35, 10411-10421, 1996.
6. Erin K. Cloherty, Kara B. Levine, and Anthony Carruthers, "The Red Blood Cell Glucose Transporter Presents Multiple, Nucleotide-Sensitive Sugar Exit Sites," Biochemistry 40, 15549-15561, 2001.
7. Eric S. Kilpatrick, "Glycated haemoglobin in the year 2000," J. Clin. Pathol. 53, 335-339, 2000.
8. T. Lahousen, R.E. Roller, R.W. Lipp, W.J. Schnedl, "Determination of glycated hemoglobins (HbA1c)," Wien. Klin. Wochenschr 114, 301-315, 2002.
9. Jan Peacock, "Glycosylated haemoglobin measurement and clinical use," J Clin Pathol, 37(8), 841-851, 1984.
10. Jan Goodall, "HbA 1c Standardisation Destination. Global IFCC Standardisation. How, Why, Where and When," Clin Biochem Rev. 26, 5-20, 2005.
11. S.L. Jeffcoate, "Diabetes control and complications:the role of glycated haemoglobin, 25 years on," Diabetes UK. Diabetic Medicine 21, 657–665, 2003.
12. Moritz Friebel, "Determination of complex refractive index of highly concentrated hemoglobin solutions using transmittance and refractance measurements," J. Biomed. Opt. 10(6), 064019 (5), 2005.
13. Dirk J. Faber, Maurice C.G. Aalders, Egbert G. Mik, Brett A. Hooper, Martin J. C. van Gemert, and Ton G. van Leeuwen, "Oxygen Saturation-Dependent Absorption and Scattering of Blood," Phys Rev Lets 93, 028102 (4), 2004.
14. G. Mazarevica, T. Freivalds, A. Jurka, "Properties of erythrocyte light refraction in diabetic patients," J. Biomed. Opt. 7, 244-247, 2002.
15. Valery V. Tuchin and Ekaterina N. Lazareva , «Monitoring of hemoglobin glycation using spectral and refraction measurements», Qingming Luo, Lihong Wang,, and Valery V. Tuchin (Eds.), Photonics and Imaging in Biology and Medicine, Qingming Luo, Lihong V. Wang , Valery V. Tuchin, Min Gu (Eds.), Proc. SPIE 0000, Bellingham, WA, USA, 2007.
16. Moritz Friebel and Martina Meinke, "Model function to calculate the refractive index of native hemoglobin in the wavelength range of 250–1100 nm dependent on concentration," Appl. Opt. 45(12), 2838-2842, 2006.
17. Jinming Huang, Zhou Zou, Daniel B. Kim-Shapiro, Samir K. Ballas and S. Bruce King, "Hydroxyurea Analogues As Kinetic and Mechanistic Probes of the Nitric Oxide Producing Reactions of Hydroxyurea and Oxyhemoglobin," J. Med. Chem. 46, 3748-3753, 2003.
18. Marianne Boylston, Deborah Beer, "Methemoglobinemia: A Case Study", Critical Care Nuse, Vol 22, No. 4, august, 50-55, 2002;
19. N.V. Maksimova, A.V. Nagovitsyn, G.V. Maksimov, "Role of ionic transport in regulation of hemoglobin affinity for oxygen in diabetes mellitus," Bull Exp Biol Med., 133(4):396-8, Apr 2002.

RESEARCH ON THE MOIRE CHROMATOGRAPHY TECHNIQUE AND ITS APPLICATION

TIANZE LI, HONGMENG YAN, BOXUE TAN, ZHENMEI LI, SHUYUN WANG

School of Electric and Electronic Engineering, Shandong University of Technology, Zibo 25504, China

Expound the Moire chromatography technique and the optical characteristics when the light crosses the 3D air flow field. Based on this, taking grating, CCD camera, optical filter and diaphragm etc as the main device, a multi-direction Moire chromatography optical system is designed. The system is consist of five detective light paths, in each detective light paths there is parallel and same periodic grating and filtering minipore placed on the lens focal length plane, the grating is used to adjust the space length and included angle between the two gratings; and the filtering minipore is used to remove the unorderly high frequency intermixed in the Moire fringe. Utilize the computer escalator method to convert the tested value, the vivid interference fringe of tested parameter can be obtained. In the end analyze and discuss the application of this system in precision measurement and medicine project.

1. Introduction

There have been much literature and news on utilizing the optics interference method to study the compressible air density field and plasma diagnostic in 3D air flow, wind tunnel and shock tube. But the researches on these field are only limited in the symmetrical 2D or 3D distributing so far, thereinto only one directional interference pattern, there will be quantitative analysis about the temperature and density field of 2D and axial symmetrical distributing, however, it is difficult to realize the symmetric in practice. For dissymmetrical distributing temperature and density field, multi-directional interference patterns are need to realize the 3D temperature and density field. So the key problems of technology is that how to build the moiré deflection system and computer tomography arithmetic in order to realize the back analysis on temperature and density field.

In this paper, the optics interference and Moire chromatography technique theory are expounded firstly, and based on that a multi-directional Moire chromatography system is designed, and then the parameters (temperature and pressure etc) of 3D air flow field are tested using the computer tomography arithmetic and Moire chromatography technique. The system will have wide application foreground in medicine project.

2. Basic Principle of Moire Chromatography Technique

2.1. *Beam Deflect Effect*

Fermat principle points out that when light wave spreads in dissymmetrical medium, its wave surface (equiphase surface) will turn into aberrance. In symmetrical medium, the light is beeline; in dissymmetrical medium, the light is curve, namely, while the light passes the dissymmetrical refractive index field, the deflect effect of light is produced. It is shown in fig.1.

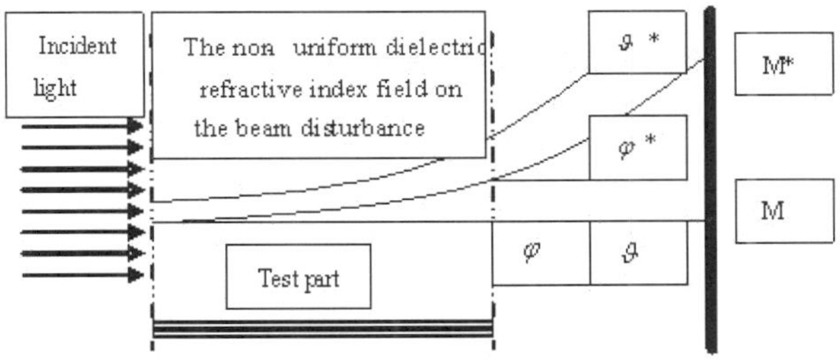

Fig1 Deflection of Light beam in non-uniform medium refractive index field

2.2. *Formation and Characteristic of Moire Pattern*

Make the two gratings surface congruent, in order to make the bar-line to form the wee angle, then when the light shines the crack, the bright and dark Moire pattern will be produced, it is shown in Fig.2. The relation and main characteristic among space, bar pitch and angle of Moire pattern are that: (1)average effect, because Moire pattern is produced by plenty of reticles of gratings, it has some average function to the scale error of grating; (2) magnification function, the two gratings move a bar-pitch comparatively, Moire pattern move a space also; (3)corresponding relation, there is the relation between the mobile volume, mobile direction of Moire pattern and displacement volume, displacement direction of grating scale. Utilizing the functions, the angle of light through the temperature field and density field can be obtained accurately, then the temperature and density etc of air can be get by means of refraction angle. If the tested parameters high-temperature air flow field, density etc are dissymmetric or asymmetric, the data of Moire pattern can be measured form different direction and different angle, simultaneity we can back-analyse distributing of the temperature field and density field by using the computer tomography technique.

2.3. *Deflection Analysis of Moire Pattern on the Transformation Coordinate*

In the fig.3, it shows that there is a angle ξ between the two gratings on the two coordinates. On the *x-y* coordinate, the density distributing of temperature field is $\rho\ (x,y)$, and on y'-ξ coordinate, the deflection distributing of Moire pattern is $\omega\ (y', \xi)$. The data of Moire pattern is tested by means of parallel beam around the experiment area shining. The deflection of pattern is direct proportion to the whole density grads of on the normal observation light path direction. The relation between density $\rho(x,y)$ and Moire deflection $\omega\ (y', \xi)$ is :

$$\omega(y',\xi) = \int_{-R}^{R} \frac{\partial f(x,y)}{\partial y'} dx' \qquad (1)$$

Where $f(x,y)$ is the refractive index

$$f(x,y) = \frac{\beta\Delta[\rho(x,y) - \rho_s]}{\theta\rho_n} \qquad (2)$$

Where β is degree which the air refractive index depends on the intensity; Δ is the distance of the two grid line; θ is the angle of the two grid line; ρ_s is density of tested temperature field; ρ_n is the density of temperature field under standard circumstance; R is the curvature radius of tested area.

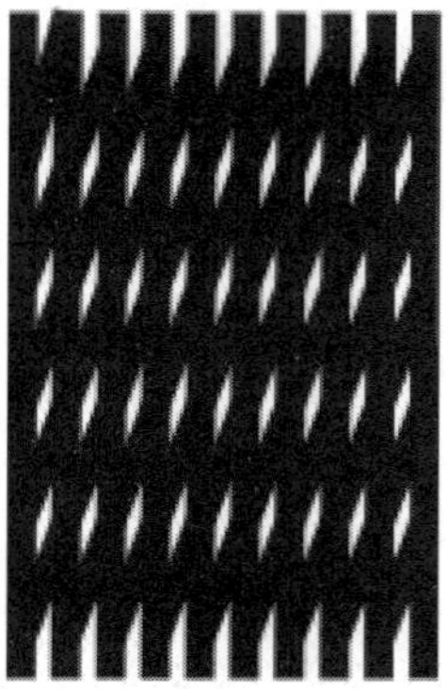

Fig.2 Moire stripes

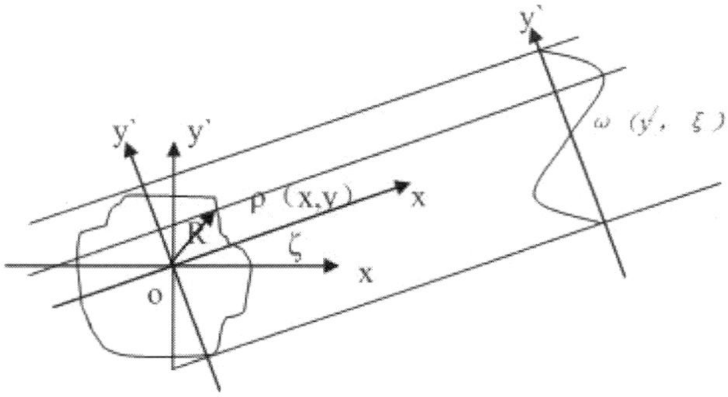

Fig.3 Fixed and turning coordinate systems for the test field

3. Computer Simulation Algorithm

If the higher precision is wanted, the proper algorithm is needed. We will rebuild the refractive index in formula (1). Divide the rebuilding temperature field into $L=M*N$ pixels, the space is d_x,d_y, expanse the $f(x,y)$ into series :

$$f(x,y) = \sum_i \sum_j f_{ij} b_{ij}(x,y) \qquad (3)$$

The deflection angle of line beam is

$$\omega(y',\xi) = \sum_i \sum_j f_{ij} \int \frac{\partial b_{ij}}{\partial y'} dx' \qquad (4)$$

Suppose $G_{ij}(y',\xi) = \int \frac{\partial b_{ij}}{\partial y'} dx'$ then

$$\omega(y',\xi) = \sum_i \sum_j f_{ij} G_{ij}(y',\xi) \qquad (5)$$

3.1. *Series* $\sin c$ *Expansion Method*

Supposed that b_{ij} expanse according to series

$\sin c$:

$$b_{ij}(x,y)=\text{sinc}[\frac{1}{d_x}(x-id_x)]\text{sinc}[\frac{1}{d_y}(y-jd_y)] \quad (6)$$

From the sample theorem

$$\overline{f}(x,y)=\sum_{i=0}^{M-1}\sum_{j=0}^{N-1}\overline{f}(id_x,jd_y)\sin c[\frac{1}{d_x}(x-id_x)]$$

$$.\sin c[\frac{1}{d_x}(y-jd_y)] \quad (7)$$

Where $\overline{f}(x,y)$ is the approximation of $\overline{f}(x,y)$.Put the formula (7) into line beam difference base equation of single line beam, we can obtain:

$$\sum_{i=0}^{M-1}\sum_{j=0}^{N-1}\overline{f}(id_x,jd_y)\int\sin c[\frac{1}{d_x}(x-id_x)].$$

$$\sin c[\frac{1}{d_y}(y-jd_y)]dx'=\psi(p_m,\xi_n) \quad (8)$$

$$\sum_{i=0}^{M-1}\sum_{j=0}^{N-1}\overline{f}(id_x,jd_y)\int_{-\infty}^{\infty}\sin c(\frac{x-id_x}{d_x}).$$

$$\sin c(\frac{y-jd_y}{d_y})(1+a_n^2)^{1/2}dx=\psi(p_m,\xi_n) \quad (9)$$

From $x'=x-id_x$, $y'=y-jd_y$, then the line beam equation of parameter (p_m,ξ_n) is

$$y'=a_nx'+b'_{mn} \quad (10)$$

Then the formula (9) will turn into

$$\sum_{i=0}^{M-1}\sum_{j=0}^{N-1}\overline{f}(id_x,jd_y)\int_{-\infty}^{+\infty}\sin c(\frac{x'}{d_x})$$

$$.\sin c(\frac{a_nx'+b'_{mn}}{d_y})(1+a_n^2)^{1/2}dx'=\psi(p_m,\xi_n) \quad (11)$$

If $a_n\neq 0$, then

$$(1+a_n^2)^{\frac{1}{2}}\sum_{i=0}^{M-1}\sum_{j=0}^{N-1}\overline{f}(id_x,jd_y)\int_{-\infty}^{+\infty}\sin c(\frac{x'}{d_x})$$

$$.\sin c[\frac{a_m}{d_y}(q_{mn}-x')]dx'=\psi(p_m,\xi_n) \quad (12)$$

Where $q_{mn}=-b'_{mn}/a_n$, use the symmetry of $\sin c$, by means of 2D fourier transform formula (12) can be turned into:

$$\int_{-\infty}^{\infty}\sin c(\frac{x'}{d_x})\sin c[\frac{a_n}{d_x}(q_{mn}-x')]dx'$$

$$=F^{-1}[\frac{|d_x||d_y|}{|a_n|}rect(\frac{d_yf_y}{a_n})] \quad (13)$$

Where $rect(x)\equiv\{^{1,|x|\leq 1/2}_{0,|x|>1/2}$ is the fourier transform of $\sin c(x)$.Summing up the above, we can export:

$$b'_{mn}=p_m\sec(\xi_n)+id_x\tan(\xi_n)-jd_y \quad (14)$$

$\overline{f}(id_x,jd_y)$'s algebraic equations is

$$\sum_{i=0}^{M-1}\sum_{j=0}^{N-1}W_{ij}(p_m,\xi_n)\ \overline{f}(id_x,jd_y)=\psi(p_m,\xi_n) \quad (15)$$

Thereinto

$$W_{ij}(p_m,\xi_n)=\begin{cases}(1+\tan^2\xi_n)^{1/2}d_x\sin c[(p_m\sec\xi_n \\ +id_x\tan\xi_n-jd_y)/d_y] \\ (0\le|\tan\xi_n|\le d_y/d_x) \\ (1+\tan^2\xi_n)^{1/2}(d_y/|\tan\xi_n|)\sin c \\ .[p_m\sec\xi_n+id_x\tan\xi_n-jd_y)/d_x\tan\xi_n] \\ (d_y/d_x<|\tan\xi_n|<\infty \\ d_y\sin c[(p_m+id_x)/d_x] \\ (\tan\xi_n\to\pm\infty)\end{cases} \qquad (16)$$

3.2. *Computer Network Algorithm*

In fig.4, the refractive index *f(x,y)* is a group of rectangular function, the sample value $f(id_x, jd_y)$'s weight is a rectangular function

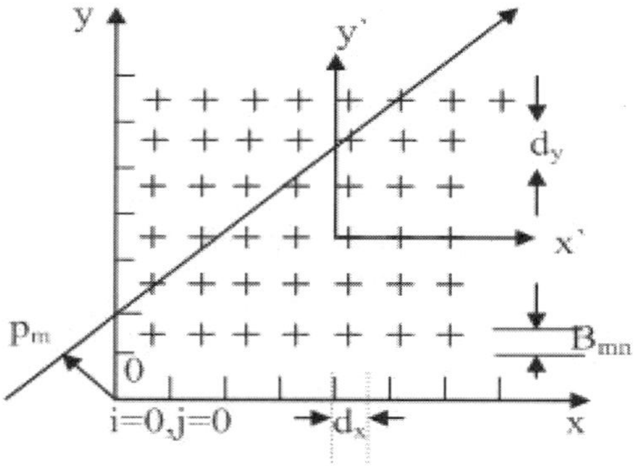

Fig.4 Rectangular function image of Refractive index $f(x, y)$

$rect[(x-id_x)/d_x]\ rect[(y-jd_y)/d_y]$, then 3D high temperature air flow field 's refractive index *f(x,y)* is

$$\overline{f}(x,y)=\sum_{i=0}^{M-1}\sum_{j=0}^{N-1}\overline{f}(id_x,jd_y)rect[\frac{1}{d_x}(x-id_x)]$$

$$.rect[\frac{1}{d_y}(y-jd_y)] \qquad (17)$$

Where $\overline{f}(id_x, jd_y)$ can be get form the formula (15). And the coefficient W_{ij} can be obtained by eq.(18):

$$W_{ij}(p_m,\xi_n)=\begin{cases}(1+\tan^2\xi_n)^{1/2}d_x \\ ......|B'_{mn}|\le(d_y-d_x|\tan\xi_n|)/2 \\ ......and\ |\tan\xi_n|>d_y/d_x \\ [(1+\tan^2\xi_n)^{1/2}d_y/|\tan\xi_n| \\ ......|B'_{mn}|\le(d_x|\tan\xi_n|-d_y)/2 \\ ......and\ |\tan\xi_n|>d_y/d_x \\ [1+\tan^2\xi_n)^{1/2}/|\tan\xi_n|][(d_x|\tan\xi_n|+d_y)/2-|B'_{mn}|] \\ ......|\frac{d_y-d_x|\tan\xi_n|}{2}|<|B'_{mn}|\le\frac{d_y+d_x|\tan\xi_n|}{2} \\ d_y..........|C'_n|<\frac{d_x}{2}...and...|\tan\xi_n|=\infty \\ 0...........|B'_{mn}|>(d_x|\tan\xi_n|+d_y)/2\end{cases} \qquad (18)$$

Where $C'_n = p_n + id_y$ B'_{mn} is given form eq.(14).

According to different observation angle θ and difference radial value y', the deflection angle value of light beam can be obtained

$$A = K_1 \times K_2 \qquad (19)$$

Where K_1 is the observation data of different direction; K_2 is the observation direction data according to difference radial value. These data will make up of a linear equational group of unknown f_{ij}

$$B = W_{ij} f \qquad (20)$$

Where B is A deflection angle data vectors; f is $L = M \times N$ unknown refractive index data vectors, W_{ij} is matrix consisting of $A \times L$ coefficient. Adopting iterative rebuild arithmetic, we can obtain:

$$f^{(0)} = W^Z B \qquad (21)$$

$$\begin{cases} f^{(p)} \\ (\sum_n^{MN} W_{mn}=0) \\ f^{(p)}+\lambda^{(p)}\dfrac{\omega_{mp}-\sum_n W_{mn}f_n^{(p)}}{(\sum_n^{MN}W_{mn})^2}W_{m_p} \\ (\sum_n^{MN} W_{mn}\ne 0)\end{cases} \qquad (22)$$

Where p is iterative time; W^Z is transposed matrix; W_{mp} is the vector making up of the m_t line elements of matrix W_{ij} ; ω_{mp} is the deflection angle of corresponding light line; $\lambda^{(p)}$ is relaxation parameter.

For higher temperature air flow field, the calculation method of refractive index difference is :

$$f(x, y) = [1 - T_0 / T_D(x, y)](n_0 - 1) \qquad (23)$$

Where T_0 is air temperature at room; $T_D(x, y)$ is temperature of simulation temperature field. After obtaining the refractive index difference distribute of simulated temperature field, the deflective angle of light line $\omega(y', \xi)$ can be get.

4. The Design of Multi-Direction Moire Deflection System

The dissymmetrical temperature air flow field need to analyze multi-direction coinstantaneous note interference pattern in order to obtain the 3D space distribution of temperature field refractive index. D.W. Sweeney , C.M. West and P.T.R. duloric have utilized the multi-direction holographic interferometry to measure dissymmetrical natural convection pinnate flow 3D temperature field distribution, Maruyama, H.Krauss and H.M.Hertz etc use Mach-Zehnder interferometer to obtain the multi-direction deflection pattern of tested 3D temperature field by means of rotating the tested 3D field in turn, but the method above could not reflect the natural situation of tested 3D temperature field truly. In this paper CCD, grating are adopted as the main components to design an optical path simple, easy to cause large-caliber 3D temperature field multi-directions MOIRE chromatography deflection system, it is shown in fig.5

In the design of light path, there are five detective light. Each detective light path has parallel-placed same period grating G. Adjust the space and angle of the two gratings, the Moire horizontal reference stripes would be clearer and the stripes space would be suitable. The filtering hole on the lens focal plane can avoid unorderly high frequency part interfusing Moire pattern.

5. Experiment and Discussion

5.1. *Utilize the Multi-Direction Moire Deflection System to Do a Simulation Test of Dual-Core Flame*

In the experiments, the simulative dissymmetrical temperature field Moire deflection pattern is get from the five different directions. The obtained projection data are suitable for the completeness data rebuilding. According to the multi-direction Moire deflection system, the dual-core flame temperature distribution isomap by means of using the computer simulation arithmetic.

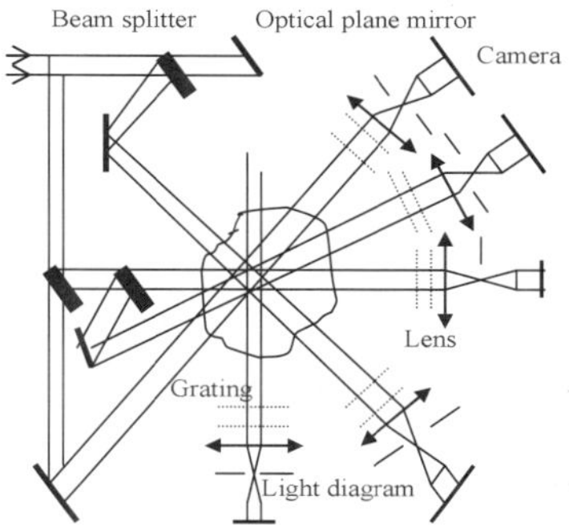

Fig.5 Multi-direction Moire deflection system

5.2. *The Stability Requirement of Experimental Device*

Moire chromatography technique take the deflection angle of light beam as the projection data, but the measurement of deflection angle do not like the measurement of optical path difference in interferometry which need the reference beam, and the stability requirement of experimental device do not so harsh, so it is convenient for test in locale.

5.3. *The Observation Angle of Multi-Direction Moire Deflection System*

The observation angle of Multi-direction Moire deflection system is wide, it can record the multi-direction interference pattern in the range of 180°. In the interference pattern, the interference stripes is sharp, resolution is high, and because each light path is separate, the performance of earthquake is good.

5.4. *The Characteristic of Multi-Direction Interference System*

When most multi-direction interference system reappearance the temperature field distribution quantificationally on the base of obtained the multi-direction interference pattern, they all suppose that the light is beeline through the tested temperature field. In fact, the medium density can cause the light to be deflection in the light path. The deflection can not be ignored, when we analyze the interference pattern, and the imaging lens action of observation stripes pattern or screen stripes pattern must take into account and adjust by means of use the imaging lens. Moreover, Multi-direction Moire deflection system measure the refractive index of tested field by means of deflection effect of tested field when the light is through the tested field, so this method avoids the error caused by light deflection.

6. Application Prospect of Multi-Direction Moire Chromatography Optical System

6.1. *The Application in Aerodynamics and Fluid Visual Observation Measurement*

The multi-direction Moire chromatography technique can be used to study the aerodynamics, fluid visual observation, plasma diagnostics, heat conduction and measurement of material transfer etc. If the object is in the form of force, the light path will change, then according to that, the change of transparent object can be analyzed by means of this system.

6.2. *Application of Moire Technique in Monitoring of Dam Deformation*

CT technique (short form of Computerized tomography) is a kind of technique which can rebuild the 2D image of object's specific levels and form the 3D image according to the 2D image above. In the recent years, the technique having been applied in the field of industry, earth physics, dam monitoring etc. Italy and Japan apply it in the state diagnosis of dam, the security checks and engineering treatment of dam have good results. Because the CT technique can elicit the 3D image which can reflect the material's distribution and location of bug truly, we can apply it to control the dam geological structure, guess the distribution of faultage broken area, and the range and extent of rock relaxation before digging the tunnel and after. CT technique exerts important action in the location, construction and operation period of dam, it not only reduces the complexity of device, but also improves the security of dam, this technique can afford important gist for the internal behavior detection, defect search and aging evaluation.

6.3. *Application in Medical Engineering*

The application prospect of Moire chromatography optical system which combines computerized tomography technique will wide in medical area. It can make the doctors observe person's body, discuss the computer photography technique application in improving the chest tumor diagnosis and recognize the etiology, pathogenesis, diagnosis and prevention measures.

References

1. K.Tabea, H.Shirai, et al., Temperature and/or Density Measurements of Asymmetrical Flow Fields by Means of the Moire-Schlieren Method[J]. JSME Int. J., Ser. II, 1990, 33(2): **249**.
2. Yan Dapeng, Zhu zhaoqing, He Anzhi. Real-time tomography of three-dimensional temperature field. SPIE. **1993**.
3. D.W.Sweeney and C.M.Vest. Measurement of tree-dimensional temperature fields above heated surfaces by holographic interferometry. Int. J. Heat Mass Transfer. 1974, 17: **1443-1454**.
4. Qi Kang, Yiqing Gao, Hanquan Ding et al., Study on 3-D arc plasma fields with holographic interferometry. Chin. J. Lasers (E.E), 1992, 1 (3): **269-276**.
5. G.W.Faris, R.L. Byer. Three dimensional beam deflection optical tomography of a supersionic jet. Appl. Opt., 1988, 27 (24): **5200-5212**.

MONTE CARLO SIMULATION OF POLARIZED LIGHT SCATTERING IN TISSUES

TIANLIANG YUN *, WEI LI, XIAOYU JIANG, HUI MA

Laboratory of Optical Imaging and Sensing, Graduate School at Shenzhen, Tsinghua University, Shenzhen 518055
Key Laboratory for Molecular and Nanosciences of Education Ministry, Department of Physics, Tsinghua University, Beijing 100084
** ytl05@mails.tsinghua.edu.cn*

We investigate the propagation of polarized light in fibrous tissues such as muscle and skin. The myofibrils and collagen fibers are approximated as long cylinders and the tissue phantom is composed of spherical and cylindrical structures. We apply Monte Carlo method based on this phantom to simulate and analyze polarization imaging process of muscle. The good agreement between the simulation results and the experimental results validate the assumption of the phantom composition. This paper also presents how to describe the fiber orientation distribution and tissue anisotropy according to three parameters derived from the polarization imaging.

1. Introduction

The biological tissues are highly-scattering medium, so the background noise arising from diffusive photons has a great impact on most of biophotonics imaging technologies. Polarization imaging plays an important role in eliminating the background noise by selecting photons backscattered from the superficial layers But there has not been an effective theory to describe polarized light propagating in the tissues.

In 1995, the source code of Monte Carlo program of photon transport in multilayered tissues was published [1], in which the scatterers were regarded as spheres. Since then, Monte Carlo simulation has increasingly become an important and useful method in investigating the propagation of photons in turbid medium, and many research papers in this field have been reported. Some recent research in Monte Carlo simulation focus on the polarization characteristic of photons [2][3]. However, the sphere assumption is not valid for all scatters when considering the real structure of tissues related with the polarization effect, for example, myofibrils in muscle are approximate cylinders. The scattering properties of cylindrical particles has been reported by several researchers [4][5]. In this paper, we present a tissue phantom composed of spherical and cylindrical particles and the simulation results and analysis show its feasibility in studying the propagation of the polarized light in turbid medium.

2. Monte Carlo Simulation Method

We complete the Monte Carlo program based on our previous version that simulated the polarized light propagation in the isotropic medium composed of spherical scatterers. The program starts with launching millions of photons, then tracks and records the photon one by one, until all the photons are absorbed or escape from the medium. Before tracking the photon, we calculate a series of scattering matrixes of spherical and cylindrical scatters respectively, which depend on the size parameter, the complex index of refraction of the particle, the incident and scattered angle of the photon. According to an analytical solution of the Maxwell equations [6], we obtain the phase function of an infinitely long cylindrical scatter at oblique incidence. The calculations show that the light scattered by an infinitely long cylinder at an angle ξ is a conical wave, with the cylinder as the axis of the cone having a half angle ξ [seeFig.1]. The scattering intensity around the cone for a certain ξ depends on the scattering direction defined by θ. Thus, the phase function of infinitely long cylinder, which is related to both ξ and θ, can be calculated from the corresponding scattering matrixes $M(\xi,\theta)$[see Fig.2]. Then the direction of propagation o n the cone is determined according to the phase function. In the Fig.2, the calculations are performed for the cylinder having a radius of 1μm, a refractive index inside and outside the cylinder of 1.4 and 1.35 respectively. It shows that with the angle ξ increasing, light tend to propagate in the incident direction (θ=0).

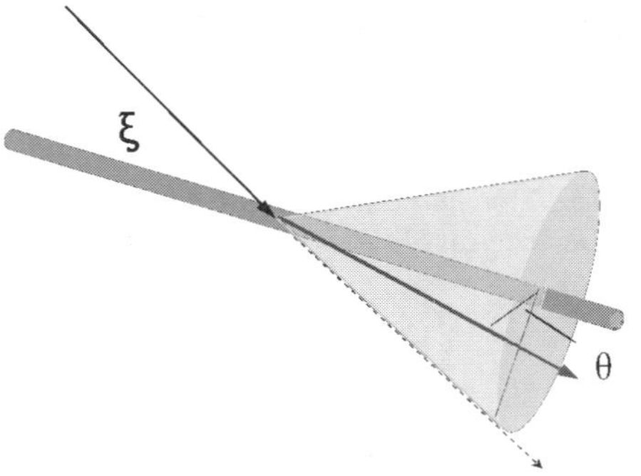

Fig. 1. Scheme of light scattering characteristics by an infinitely long cylinder.

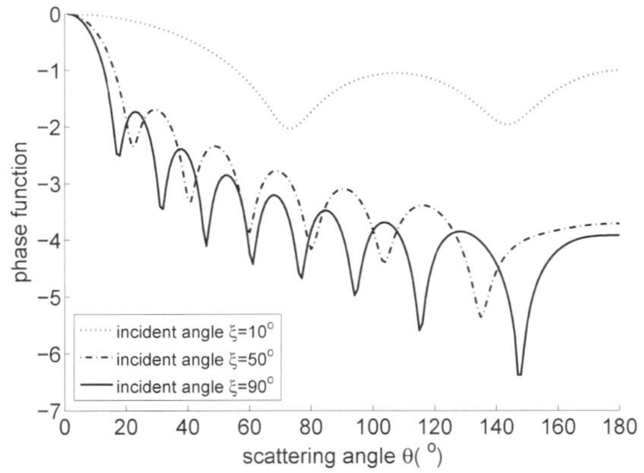

Fig. 2. Scattering phase functions of an infinitely long cylinder for different incident angles ξ.

The initial feature of photons, including location, direction of propagation and polarization state defined by the Stokes vector S = [I, Q, U, V], are determined by given conditions. The photon is moved a propagation distance Δs that is calculated according to the total scattering coefficient, the absorption coefficient of the media, and the random number. Then according to the direction of propagation, we can determine the next scattering location of the photon. The type of scatterer hit by photon depends on the proportion of scattering coefficients corresponding to different scatterers. The scattering coefficient of cylinders is related to the incident angle ξ [7].

$$\mu_s(\xi) = Q_{sca}(\xi) \cdot d \cdot C_A \tag{1}$$

Where Q_{sca} is the scattering efficiency, d is the diameter of cylinder, and C_A is the density of the cylinders. The $\mu_{s,cyl}$, we mentioned below is a given value when ξ=90° ($\mu_{s,cyl} = \mu_{s,cyl\ (\xi=90^\circ)}$) and other values of $\mu_s(\xi)$ can be calculated for the given function of Q_{sca}. The total scattering coefficient of the media can be approximately the sum of scattering coefficient of two types of scatters ($\mu_{s,total} = \mu_{s,sph} + \mu_{s,cyl}$). Additionally, the arrangement of the cylindrical scatters in real tissues isn't exactly parallel but a Gaussian distribution along a direction, which is described with FWHM Δη in simulations.

When the type of scatterer is determined, in addition to the incident and the scattering direction obtained from the phase function, the corresponding scattering matrix is selected and then $S' = M \times S$ is calculated. Repeat the steps above for a photon, until its life ends when it passes through a boundary or when its weight W value falls below a threshold (0.01).

3. Validation of the Monte Carlo Simulation Program

In this section, we concentrate on the part of program when the tissue phantom contains cylindrical scatterers only ($\mu_{s,sph} = 0$). We did some simulations and compared the results to what had been published.

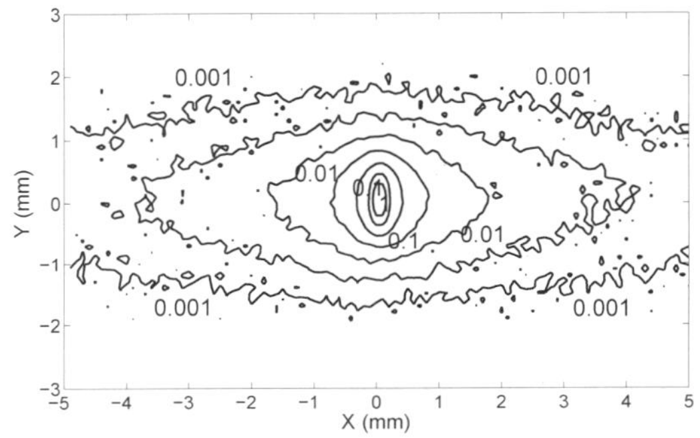

Fig. 3. R(x, y) pattern obtained from Monte Carlo simulations of a semi-infinite turbid medium containing cylinders along the axis with Δη=15°.

Fig.3 shows the spatially resolved reflectance R(x, y) obtained from the Monte Carlo simulation of a medium that contains cylinders only with r =1μm, Δη = 15° along x axes, $\mu_{s,cyl}$ = 300mm^{-1}. The main axes of the intensity contour ellipses are perpendicular to the cylinders of the tissue at small distances (<1mm) but parallel to the cylinders at large distances (>1mm). The similar result was published in 2004 [4], so this simulation result confirms the confidence of our Monte Carlo program.

4. The Monte Carlo simulation of polarization imaging

We apply the program to simulate the experiment of polarization imaging. The experiment setup is illustrated in Fig.4 [8].

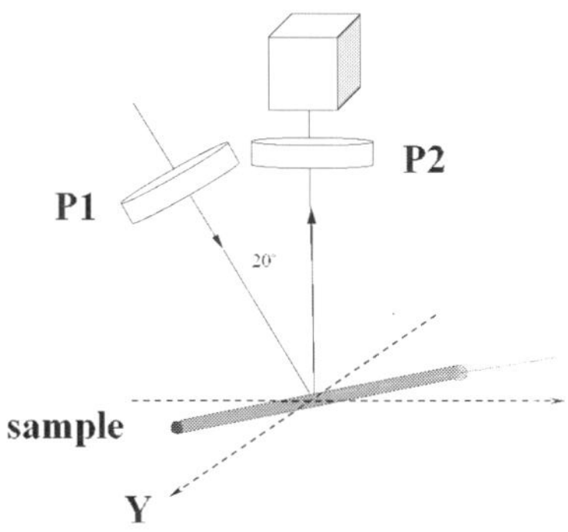

Fig. 4. Scheme of experimental setup.

In the experiment, the muscle tissues are laid on X-Y plane and the myofibrils of that are aligned parallel to X axis. The collimated 650*nm* light from a 1*W* LED passes through a linear polarizer P1 and then illuminates the sample at an angle of 20° to the normal direction of the sample surface. Backscattered photons from the sample are focused after passing though a linear polarizer P2 and recorder by a CCD. We can acquire a series of images $I(\theta_i,\theta_s)$ corresponding to different combinations of θ_i and θ_s (θ_i and θ_s represent the incident and detection polarization angles) by rotating P1 and P2 separately. Linear Differential Polarization (LDP) is calculated as the detected intensity difference with two orthogonal polarization analyzers, respectively [8].

$$LDP(\theta_i,\theta_s) = I(\theta_i,\theta_s) - I(\theta_i,\theta_s+\pi/2) \quad (2)$$

We plot the normalized LDP(θ_i ,0) as functions of θ_i, that is a cosine curve as shown in Fig.5

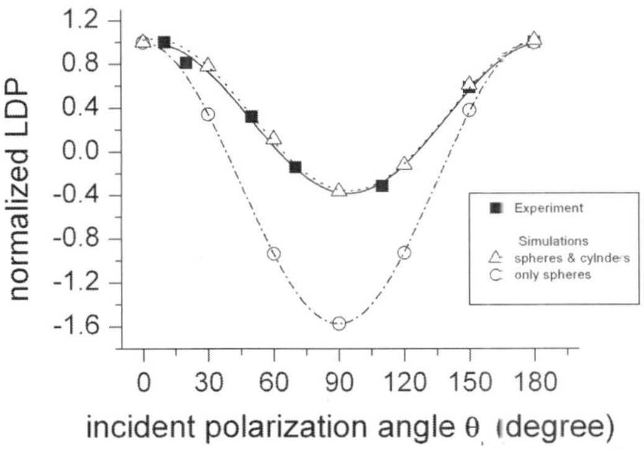

Fig. 5. Dependence of Linear Differential Polarization to incident polarization angle.

In the simulation, we studied two tissue phantoms of muscle respectively: (1) Spherical scatterers only. (2) Spherical and cylindrical scatterers. For the phantom (1), the minimum of LDP equals to -1 in the case of normally incident light and less than -1 in the case of inclined incident light, which don't agree with the experimental results. But for the phantom (2), the simulation results can match experiment well by varying the proportion of the two scatterers (20% spheres, 80% cylinders), as shown in Fig.5.

According to a great mount of data from simulations and experiments, we find that LDP can be expressed as [8]:

$$LDP(\theta_i,\theta_s) = \sqrt{A\cos(4\theta_s - \phi_1) + B}\cos(2\theta_i - \phi_2(\theta_s)) + C\cos(2\theta_s - \phi_3) \quad (3)$$

In the experiment, the fibers are aligned at a preset angle, and then the $LDP(\theta_i,\theta_s)$ data series at every pixel of image can be obtained. We can obtain the experimental value of parameter *A, B, C,* ϕ_1 and ϕ_3 by curve fitting according to Eq.(3), and the similar processing is completed in simulation to get the above parameters. The results imply that $\phi_3/2$ can indicate the orientation angle of cylindrical component in tissue, as shown in Fig.6.

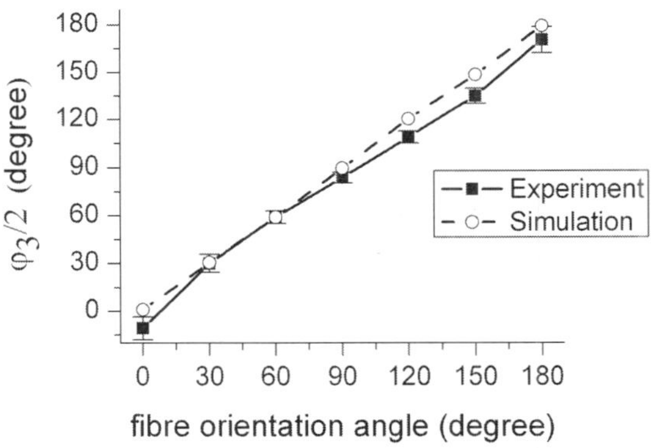

Fig. 6. Dependence of $\phi_3/2$ to orientation of the cylindrical scatterers.

In Fig.7, the distribution of the orientation of fibers ($\phi_3/2$) at all pixels of the sample shows that the fibers in tissue have a Gaussian distribution, which validate the assumption in simulation.

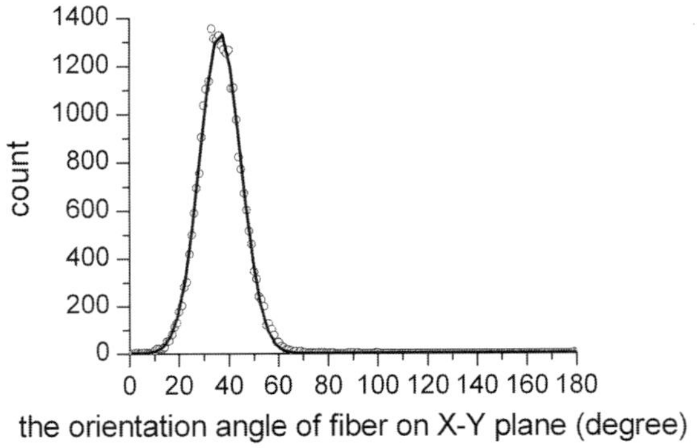

Fig. 7. Distribution of the orientation of fibers in a 1.76mm×1.76mm area of sample.

In addition, we find that A/B derived from Eq.3 is unaffected by sample orientation and also insensitive to tissue clearing. We think that A/B maybe be related to the structural anisotropy of the sample. The comparison of experimental results for different biological tissues is shown in Table 1.

experimental results and the idea that the variation of A/B can quantitatively figure out the anisotropy of fibrous structure.

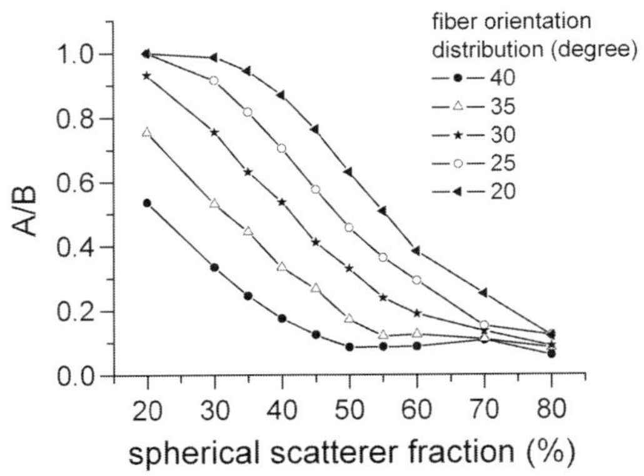

Fig. 8. Dependence of A/B to the fraction of spherical scatterers and the width of the Gaussian distribution.

Table 1. Comparison of different samples.

Group	Sample	A/B (Average±RSD)
1	Porcine skeleton muscle with regular alignment of fibers parallel to sample surface	0.95 ± 7.4%
2	Porcine skeleton muscle with irregular alignment of fibers parallel to sample surface	0.63 ± 33%
3	Gallinaceous skeleton muscle with regular alignment of fibers parallel to sample surface	0.96 ± 4.2%
4	Gallinaceous skeleton muscle with irregular alignment of fibers parallel to sample surface	0.65 ± 38%
5	Porcine fatty tissue	0.17 ± 53%
6	Porcine liver	0.07 ± 57%

In table 1, we can see that for the sample with prominent fibrous structures like porcine skeleton muscle, the value of A/B is larger than that of known isotropic tissue like porcine liver. Compared with samples in group 1 and 3, those in group 2 and 4 have smaller A/B due to the irregular alignment of the fibers, which reduces the structural anisotropy.

To validate the assumption about A/B, we apply the Monte Carlo simulation with tissue phantom composed of spherical and infinitely long cylindrical scatters. Similarly, A/B is unaffected by the cylinders' orientation but sensitive to the ratio of scattering coefficients of two types of scatterers and the width of the Gaussian distribution of fibers. With the increasing of the proportion of spherical scatterers or the width $\Delta\eta$ of the Gaussian distribution, the tissue phantom becomes less anisotropic and A/B decreases, as shown in Fig.8. Therefore, the simulation method validates the

5. Conclusion

In summary, we present and validate the model composed of spherical and infinitely long cylindrical scatters describing the fibrous tissues. We apply Monte Carlo simulation based on this model to simulate and explain quantitatively the experimental results in polarization imaging of fibrous tissues such as muscle and skin. This tissue phantom is suitable to approach the real sample by varying several parameters, such as the orientation distribution of the cylindrical scatters, the proportion and the radius of the two scatters. The correlation between the Monte Carlo simulation and experimental results of muscle show that the model can be used to reveal structure characteristics of tissues qualitatively and even quantitatively from two aspects: (1) The orientation of cylindrical structure in tissues can be calculated by fitting backscattering polarization difference when varying incident polarized angles, and the experimental data processing demonstrates the fibers are arranged along one direction and have a Gaussian distribution; (2) We find that the value of A/B is related to the proportion of the cylindrical scatters and the width of orientation distribution, so that the variable in the model, A/B, can be a indicator of anisotropy of fibrous tissues. We believe that some pathologic information about skin or muscle, correlating with fiber orientation distribution and the amount of fibers [9], can be figured out by measuring the variation of A/B. Theses above findings show exciting potentials of this anisotropic model and polarization imaging parameters for therapeutic and diagnostic applications.

Acknowledgement

This work has been supported by National Natural Science Foundation of China (grant 60578003), and Ministry of Science and Technology (grant 2006CB70570).

References

1. L. V. Wang, S. L. Jacques and L. Zheng, Computer Methods and Programs in Biomedicine **47**, 131 (1995).
2. J. Ramella-Roman, S. Prahl and S. L. Jacques, Optics Express **13**, 4420 (2005).
3. J. Ramella-Roman, S. Prahl and S. L. Jacques, Optics Express **13**, 10392 (2005).
4. A. Kienle, F. Forster and R. Hibst, Optics Express **29**, 2617 (2004).
5. A. Kienle and R. Hibst, Physical Review Letters **97** (2006).
6. C. F. Bohren and D. R. Hu®man, Absorption and Scattering of Light by Small Particles (John Wiley and Sons, New York, 1983).
7. A. Kienle and C. Wetzel, Journal of Biomedical Optics **12** (2007).
8. J. Xiao-Yu, Z. Nan, H. Yong-Hong and M. Hui, Progress in Biochemistry and Biophysics **34**, 659 (2007).
9. J. Ranasinghesagara and G. Yao, Optics Express **15**, 3998 (2007).

DETERMINATION OF THE RELATIONSHIP BETWEEN RELATIVE FRACTIONAL BLOOD VOLUME AND LIGHT PROPAGATION IN PORT WINE STAINS TISSUE

TIAN YONG-BIN, CHEN PING*, LIN LIE, CAI BING-TAO, ZHANG SI-WEN

Institute of Modern Optics, Nankai University, 300071 Tianjin, P.R.China
**E-mail: chping@nankai.edu.cn*

HUANG ZHENG

Radiation Oncology Department, University of Colorado at Denver and Health Sciences Center, Colorado, USA
E-mail: zheng_huang@msn.com

The vascular tissue of port wine stains (PWS) birthmarks in the papillary layers of dermis is intricate and has great intra- and inter-patient variations. The average relative fraction of blood inside the PWS blood vessels at the depth of 100 to 150 microns is often greater than 60 percent. According to previous studies, the blood has a great light absorption effect on some wavelengths used for the photodynamic therapy (PDT). In order to determine the relationship between blood volume fraction (BVF) and light propagation, as well as the light penetration depth, we revised the conventional Monte Carlo method to simulate the light propagation in PWS tissue with an emphasis on the effect of the blood volume. For instance, during the simulation we choose different absorption and scattering coefficient parameters according to the probability roulette produced by the BVF at certain depths, and thereby to simplify the simulation of complicated PWS blood vessel structure. The simulation result of the fast radiant power attenuation might provide useful information for the light dosimetry of PWS PDT of different wavelengths.

Keywords: Port wine stains (PWS); Monte Carlo; Photodynamic therapy (PDT)

1. Introduction

PWSs are vascular malformations in the skin which make the affected areas with a red discoloration. These vascular malformations are often occurred on the face or neck, which are disfiguring and put the patients at a social disadvantage. The cause of PWS is the complicated vascular tissue under epidermis. In previous studies, researchers have done some three-dimensional reconstruction of PWS vascular tissue,[1] giving a clear three-dimensional view of PWS vessels. Results of the histological reconstruction revealed that there are multiple clusters of small diameter (10-50 μm) blood vessels in dermal region, which present ruleless variety.

The preferred treatment of PWS is laser therapy, either thermal or PDT. In order to select more effective laser dosimetry, it is essential to estimate the light distribution in vascular tissue before treatment. When light propagates in tissue, it will be absorbed and scattered. Tissue optical properties, including absorption coefficient(μ_a), scattering coefficient(μ_s) and scattering anisotropy(g) are commonly used to describe the absorption and scattering effects. Epidermis, dermis, and blood have very different optical properties. Previous studies showed that blood has the highest absorption effect. The μ_a of blood is round 300 cm^{-1}, while the μ_a of epidermis and dermis is round 20 and 2.3 cm^{-1}, respectively.

Monte Carlo simulation has been used to understand the laser-tissue interaction, and to simulate how light distributes in tissue. MCML (Monte Carlo simulation for multiple layers), a software developed by Wang and Jacques, is used widely.[2] Previous simulations have generally involved a simplified skin geometry which just consists of slabbed layers of epidermis and dermis, and regular arrays of cylindrical blood vessels.[3] Since PWS tissue is much more complicated, more proper models are needed for better understanding of light propagation in PWS tissue.

We have developed a unique method to simulate light distribution in PWS tissue, by analyzing relationship between light fluence and the BVF at certain depth of dermal tissue. We modified the C code of MCML and developed a program for our simulation.

2. Method

According to the study of Smithes *et al*, we educed a function of BVF versus depth of skin. Then we modified the code of MCML and did our simulations.

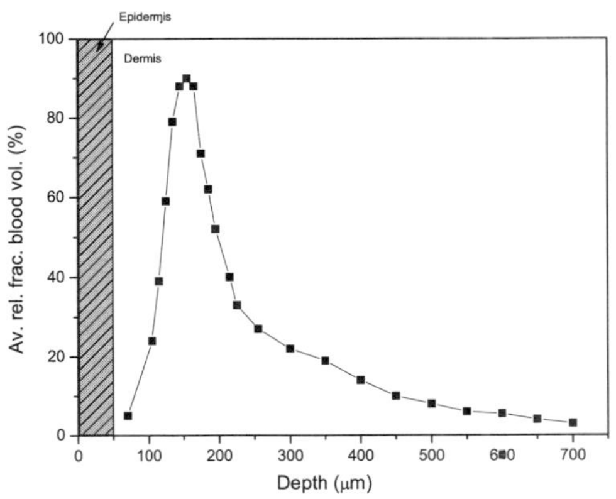

Fig. 1. Average relative fractional blood volume versus depth(d) in PWS tissue. We referred and sampled the experimental data of Smithes *et al.*

2.1. *Average Fractional Blood Volume*

Smithes *et al* calculated the BVF, i.e. average fractional blood volume, by doing three-dimensional reconstruction of PWS tissue.[1] We referred their results as one sample, to get the idea of PWS BVF versus depth, see Fig. 1. At the depth of 100 to 300μm under the surface of skin, blood fraction is above 20%. The peak is 90%, at the depth of 160μm.

We fitted the curve using Lorentz model, and educed the function of BVF versus depth, as Eq. (1). $y0$, xc, w and A are Lorentz fitting parameters, which are listed in Table 1.

$$BVF(d) = y0 + (2A/\pi)\frac{w}{4(d - xc)^2 + w^2} \quad (1)$$

We can calculate BVF at certain depth using this function and then generate a probability roulette for our simulation, since BVF is actually the probability of photon comes across blood when they are at certain depth.

Table 1. Lorentz Parameters

Params	Values	Errors
$y0$	0.07829	±0.0279
xc	0.01597	±0.00024
w	0.00749	±0.00089
A	0.00992	±0.00112

2.2. *Modification of Monte Carlo Method*

According to the Monte Carlo routine, every time the loop runs, we launched one photon into the tissue and set its first step length. When the photon reaches dermis, the fact is that, the blood fraction equals the possibility the photon comes across with blood. So we developed a probability roulette according to BVF to determine whether this photon is in blood or dermis. We can then choose different μ_a and μ_s for the photon's following movements. The pseudo code of the roulette is as follows:

```
d = GetCurrentDepth
bvf = CalculateBVF(d)
r = GenerateRandomNum
if (r > bvf)
  CurrentOpticalProperty(DERMIS)
else
  CurrentOpticalProperty(BLOOD)
```

The flow chart of modified Monte Carlo method is as Fig. 2. The main difference from traditional Monte Carlo method is the probability roulette, which can determine the relationship between the BVF and light distribution, as described above.

3. Simulation

Modified Monte Carlo simulation was performed under the following conditions. Light source was a Gaussian beam with 1 J radiance and 0.1 cm $1/e^2$ radius. The wavelength was 577nm, under which the optical properties are listed in Table 2[4,5].

Thicknesses of epidermis and dermis are 50 and 650 μm in our simulation, respectively. For the normal tissue, we estimated that BVF is approximately 10% in order to get a rough idea.

Table 2. Optical Properties (577nm light source)

Tissue	$\mu_a(cm^{-1})$	$\mu_s(cm^{-1})$	g	n
Epidermis	18.5	480	0.787	1.37
Dermis	2.2	131	0.787	1.37
Blood	354	468	0.995	1.33

Source code was compiled by gcc, and program was performed on Debian GNU/Linux 3.1, Pentium III (Coppermine) 936MHz.

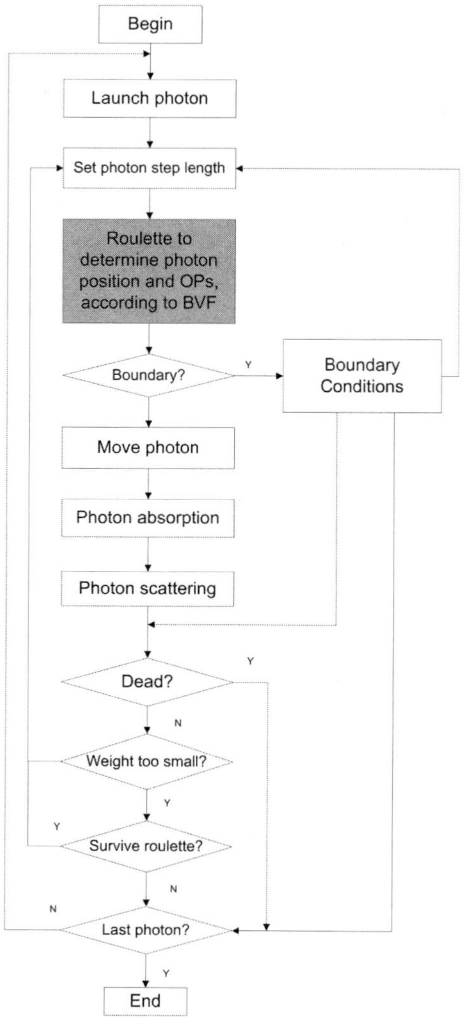

Fig. 2. Flow chart of modified Monte Carlo method to simulate light propagation in PWS tissue.

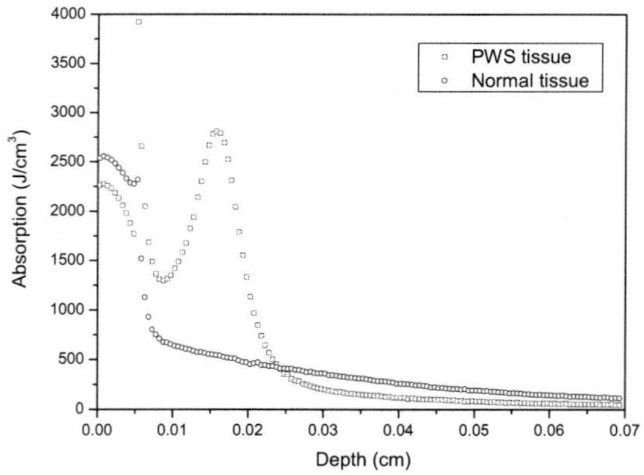

Fig. 3. Light absorption versus depth, in PWS and normal tissue, where r = 0.001 *cm* in cylindrical coordinate. Light source is 577nm Gaussian beam with 1 J radiance and 0.1 *cm* $1/e^2$ radius.

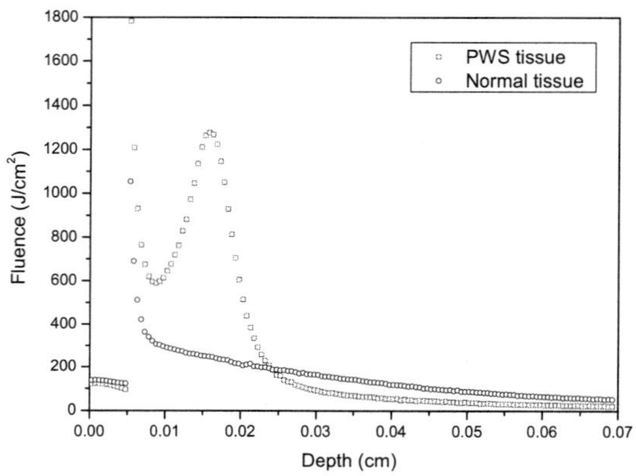

Fig. 4. Light fluence versus depth, in PWS and normal tissue, where r = 0.001 *cm* in cylindrical coordinate. Light source is 577nm Gaussian beam with 1 J radiance and 0.1 *cm* $1/e^2$ radius.

4. Result and Discussion

Results were obtained using CONV,[6] a program for calculating the infinite beam convolution. Light absorption and fluence are calculated and plotted both in PWS and normal tissue. See Fig. 3 and Fig. 4.

The result shows that PWS tissue typically has a strong light absorption effect. The peak of the absorption is at the depth of near 0.015 *cm*. This must be due to the high BVF and big absorption coefficient μ_a of blood at the depth close to 0.015 *cm*. Light fluence increases along with the BVF too. Result reveals that light fluence in PWS tissue is not simply descending along the depth. Meanwhile, as we can expect, in normal tissue light absorption and fluence both decrease regularly versus the depth.

Besides, in the depth less than 0.005 *cm*, i.e. the epidermis region, light absorption in normal tissue is larger than in PWS tissue (Fig. 3), this is probably due to the boundary reflectance diversity.

5. Conclusion

Monte Carlo method is a probability computing method. It has been adopted in many areas. Our modification is based on probability theory too. When we simulate as many as 1,000,000 photons, it can always give a reasonable statistical result.

In PWS PDT research, it is the pre-work to calculate light fluence to do some mathematic modeling. However, due to the complicated structure and variety of PWS vascular tissue, it is impossible to do a precise prediction. Our Monte Carlo model can be used to do a satisfying calculation along the z (depth) direction. Besides, light penetration depth can be obtained too. We expect our model could give guidance for selection of more effective laser dosimetry.

Acknowledgments

This work is supported by the National Natural Science Foundation of China (No. 60508C04), Doctoral Program Foundation of Institutions of Higher Education of China (No. 20060055024).

References

1. D. J. Smithies, M. J. C. van Gemert, M. K. Hansen, T. E. Milner and J. S. Nelson, *Phys. Med. Biol.* **42**, 1843 (1997).
2. L.-H. Wang, S. Jacques and L.-Q. Zheng, *Computer Methods and Programs in Biomedicine* **47**, 131 (1995).
3. D. J. Smithies and P. H. Butler, *Phys. Med. Biol.* **40**, 701 (1995).
4. G. W. Lucassen, W. Verkruysse, M. Keijzer and N. J. C. van Gemert, *Lasers Surg. Med.* **19**, 345 (1996).
5. R. M. Adrian and E. A. Tanghetti, *Dermatol. Surg.* **24**, 71 (1998).
6. L.-H. Wang, S. L. Jacques and L.-Q. Zheng, *Computer Methods and Programs in Biomedicine* **54**, 141 (1997).

TIME-RESOLVED OPTICAL PARAMETERS DETERMINATION FOR THE FINITE THICKNESS HOMOGENEOUS AND LAYERED TISSUE

BIYING YU, HUI LI*

Key Laboratory of Optoelectronic Science and Technology for Medicine, Ministry of Education, Fujian Normal University, Fuzhou, Fujian 350007, China

A multi-distance measurement, two measurements of reflectance or transmittance at two different distances, with a single parameter (reduced scattering coefficient $\mu_s^{'}$) fitting method and a double parameters ($\mu_s^{'}$ and amplitude factor A or $\mu_s^{'}$ and amplitude factor $A^{'}$) fitting method on the basis of the least square Newton fitting algorithm is used to derive the optical parameters for the finite thickness homogenous tissue respectively. It can be concluded that the rest terms in expressions of diffusion reflectance and transmittance for the corresponding tissue geometries that subtract the exponential term contained source-detector distance r depend on the detected position weakly based on the multi-distance measurements. And some discussions about the discrepancy induced by those two fitting methods are included. Additionally, the probable structure information for the finite thickness two-layered slab tissue can be obtained from the diffusion reflectance and transmittance measurements according to the discrepancies of fitting $\mu_s^{'}$ and the sign (positive or negative) of the constants A and $A^{'}$.

1. Introduction

Time-resolved optical measurements have been a subject of interest for the determination of optical properties of tissue. The diffusion light detected from tissue surface contains tissue optical parameters information[1-3]. So the tissue optical parameters can be obtained indirectly by fitting the experimental data or the Monte Carlo (MC) simulation[4,5] data to the diffusion reflectance formula and/or transmittance formula. Then the optical parameters resolved tissue structure, the physiologic and pathologic state information can be obtained finally. Therefore, the study of tissue optical properties and how to extract tissue optical parameters with accuracy play a significant role in optical examine and diagnosis.

Currently, three boundary conditions, the zero boundary condition, the partial current boundary condition and extrapolated boundary condition, are employed to obtain the diffusion equations generally[6-8]. So the optical parameters obtained are influenced by the boundary condition which is employed in the diffusion equations. A method proposed by *R.K.Wang et al.*[9] was demonstrated for determining the optical parameters from time-resolved measurement at two positions on the surface of the semi-infinite homogeneous tissue. The derived reduced scattering coefficient ($\mu_s^{'}$) was not affected by whatever boundary condition was applied.

In many practical situations, biological tissues have layered structures. The approximate solutions of diffusion theory for the layered tissues are difficult to obtain and the optical parameters are difficult to abstract from the diffusion equations accurately for the increase of the parameter numbers waited for fitting and the complication of equations employed to fit. A multi-distance measurement, two measurements of reflectance at two different distances, proposed by *F.Martelli et al.*[10] on the basis of the method proposed by *R.K.Wang et al.*[9] has been employed to obtain the optical parameters of semi-infinite homogeneous and layered tissue under an assumption that the diffusion coefficient (D) is independent of the absorption coefficient (μ_a), which is not agree with the conclusions of *D.J.Durian et al.*[11] and *R.Aronson et al.*[12] that the influence of μ_a on D can not be ignored in diffusion theory. But it seems a reasonable assumption for the conditions of tissue with $\mu_s^{'} \gg \mu_a$. Then $\mu_s^{'}$ can be derived from the logarithm of the ratio between two measurements of reflectance at two different distances based on a Levenberg-Marquardt algorithm. Additionally, the tissue structure information can also be obtained.

The multi-distance measurement is applied to investigate the finite thickness tissue with the diffusion reflectance and transmittance measurements in this paper. Then the optical parameters of tissue two sides can be obtained under an assumption that tissue is non-absorbed for convenience. In Sec.2, the multi-distance measurement method is displayed. In Sec.3, the multi-

* Email: hli@fjnu.edu.cn

distance measurement with a single parameter fitting (μ_s') method and a double parameters fitting (μ_s', A or μ_s', A') method respectively on the basis of the least square Newton fitting algorithm is used to derive the optical parameters for the finite thickness homogenous tissue, and some discussions about the discrepancy induced by those two fitting method are included. The double parameters fitting (μ_s', A or μ_s', A') method is applied to investigate the tissue with two-layers, then tissue structure information can be derived from the fitting results. In Sec.4, some conclusions are made based on the fitting results.

2. Multi-distance Measurement Method

The time-resolved diffusion reflectance and transmittance from a finite thickness homogeneous slab tissue with thickness d are[7,8]

$$R(r,t) = h_{homo\text{-}slab}(d, \mu_a, \mu_s', n, t) \times \exp(-\frac{r^2}{4Dct}) \quad (1)$$

$$T(r,t) = h_{homo\text{-}slab}'(d, \mu_a, \mu_s', n, t) \times \exp(-\frac{r^2}{4Dct}), \quad (2)$$

where c is the speed of light, diffusion coefficient $D = 1/3\mu_s'$ and r is the source-detector distance. $h_{homo\text{-}slab}$ and $h_{homo\text{-}slab}'$ are independent on the measurement positions but dependent on the tissue properties and the boundary condition employed. They can be removed by a mathematics processing method, i.e. the ratio between two measurements of reflectance and transmittance at two different distances. Then equations (1) and (2) can be transformed as follows[10]

$$\ln(R(r_2,t)/R(r_1,t)) = -(r_2^2 - r_1^2)/4Dct, \quad (3)$$

$$\ln(T(r_2,t)/T(r_1,t)) = -(r_2^2 - r_1^2)/4Dct, \quad (4)$$

in which subscripts 1 and 2 refer to the different measurement distances, and $r_2 > r_1$.

Similarly, the time-resolved diffusion reflectance and transmittance from a finite thickness slab tissue with two layers are[13]

$$R(r,t) = h_{layered\text{-}slab}(d_0, d_1, \mu_{a0}, \mu_{a1}, \mu_{s0}', \mu_{s1}', n, t) \times \exp(-\frac{r^2}{4D_0ct}), \quad (5)$$

$$T(r,t) = h_{layered\text{-}slab}'(d_0, d_1, \mu_{a0}, \mu_{a1}, \mu_{s0}, \mu_{s1}, n, t) \times \exp(-\frac{r^2}{4D_1ct}) \quad (6)$$

where $h_{layered-slab}$ and $h_{layered-slab}'$ are the complicated functions of the optical properties of the two layers, the subscripts 0 and 1 refer to the first layer and the second one respectively.

Then equations (5) and (6) can be transformed as follows[10]

$$\ln(R(r_2,t)/R(r_1,t)) = -(r_2^2 - r_1^2)/4D_0ct, \quad (7)$$

$$\ln(T(r_2,t)/T(r_1,t)) = -(r_2^2 - r_1^2)/4D_1ct. \quad (8)$$

F.Martelli et al. has demonstrated that the function h for semi-infinite layered medium is dependent on r weekly. Then a constant factor A which is independent on time is added to the transformed equation. The same as the semi-infinite medium, A and A' may can be added to the reflectance and transmittance transformed equations respectively for the finite thickness two layered tissue as follows[10]

$$\ln(R(r_2,t)/R(r_1,t)) = A - (r_2^2 - r_1^2)/4D_0ct, \quad (9)$$

$$\ln(T(r_2,t)/T(r_1,t)) = A' - (r_2^2 - r_1^2)/4D_1ct. \quad (10)$$

Optical parameters of the finite thickness homogenous tissue and two-layered tissue can be derived from equations (9) and (10) based on a nonlinear least square fitting algorithm. For finite thickness homogenous tissue, a single parameter fitting (μ_s') method and a double parameters fitting (μ_s', A and μ_s', A') method are used to derive the optical parameters from the reflectance and transmittance measurements respectively. For finite thickness two-layered tissue, the double parameters fitting method is used to abstract the optical parameters from the reflectance and transmittance measurements respectively.

3. Optical Parameters Measurements and Discussions

A time-resolved Monte Carlo (MC) simulation algorithm, a source emitting a $\delta(t)$ pulse is engaged, is employed to simulate light distribution in medium. The photon numbers of the MC simulations are 5×10^8, anisotropy factor $g = 0.8$, refractive index $n = 1.4$ and $\mu_a = 0$ for both geometry media. The data of two group source-detector distances $r_1 = 5.25mm$, $r_2 = 10.25mm$ and $r_1 = 10.25mm$, $r_2 = 15.25mm$ are employed on the basis of the least square Newton fitting algorithm, which is faster than the Levenberg-Marquart algorithm at the same PC and the results of them are equivalent nearly. The fitting range is from the position of the peak value P_{peak} (T_{peak}) to the position of $P_{peak}/100$ ($T_{peak}/100$) at r_2.

3.1. *Results of the Finite Thickness Homogeneous Slab*

The single parameter ($\mu'_{s,fit-s}$) and the double parameters ($\mu'_{s,fit}$, A or $\mu'_{s,fit}$, A') fitting results of the reflectance and transmittance measurements for the finite thickness homogeneous slab media are as Tab.1 shown with the variations of μ'_s and d.

Table 1 Single parameter and double parameters fitting results for the reflectance and transmittance measurements for the finite thickness homogeneous slab media

Measurement fashions	μ'_s (mm-1)	d (mm)	$\mu'_{s,fit-s}$ (mm-1)	$\mu'_{s,fit}$ (mm-1)	A_{fit} (mm-1)	A'_{fit}
R_{11}	0.5	3	0.40p	0.43p	0.05	—
R_{12}	0.5	3	0.41p	0.46p	0.09	—
T_{11}	0.5	3	0.40p	0.43p	—	0.05
T_{12}	0.5	3	0.42p	0.46p	—	0.10
R_{21}	0.5	6	0.45	0.48	0.03	—
R_{22}	0.5	6	0.46	0.49	0.05	—
T_{21}	0.5	6	0.45	0.47	—	0.02
T_{22}	0.5	6	0.46	0.48	—	0.03
R_{31}	1	3	0.91	0.94	0.04	—
R_{32}	1	3	0.93	0.98	0.09	—
T_{31}	1	3	0.92	0.98	—	0.08
T_{32}	1	3	0.92	1.00	—	0.11
R_{41}	1	4	0.94	0.99	0.05	—
R_{42}	1	4	0.95	1.02	0.09	—
T_{41}	1	4	0.94	0.98	—	0.04
T_{42}	1	4	0.95	0.98	—	0.03
R_{51}	1.5	3	1.43	1.49	0.07	—
R_{52}	1.5	3	1.44	1.49	0.06	—
T_{51}	1.5	3	1.42	1.45	—	0.01
T_{52}	1.5	3	1.44	1.46	—	0.03
R_{61}	1.5	4	1.46	1.49	0.02	—
R_{62}	1.5	4	1.46	1.53	0.07	—
T_{61}	1.5	4	1.45	1.54	—	0.08
T_{62}	1.5	4	1.47	1.46	—	-0.01
R_{71}	2	2	1.91	1.93	0.02	—
R_{72}	2	2	1.94q	1.79q	-0.20	—
T_{71}	2	2	1.90	1.99	—	0.10
T_{72}	2	2	1.95q	1.51q	—	-0.69
R_{81}	2	3	1.94	1.97	0.03	—
R_{82}	2	3	1.96	1.96	0	—
T_{81}	2	3	1.94	2.00	—	0.06
T_{82}	2	3	1.97q	1.91q	—	-0.06

R_{ij} and T_{ij} are indicate the reflectance and transmittance measurements respectively. The subscripts i and j refer to the different slab and the different measurement positions respectively. Where $j=1$ indicates the measurement positions are $r_1 = 5.25mm$ and $r_2 = 10.25mm$, $j=2$ indicates the measurement positions are $r_1 = 10.25mm$ and $r_2 = 15.25mm$. Subscripts p and q refer to the relative big discrepancy of the fitting curve and the MC data and the relative few data useable respectively.

From Tab.1, for the single parameter fitting, the fitting accuracies are improved for the reflectance and transmittance measurements in homogeneous media with the increase of d (other parameters unchanged) when $\mu_s^{'}$ is equal to 0.5, 1, 1.5 and 2 mm^{-1} respectively. And the results of two group measurements of the reflectance/transmittance are almost equivalent for the same medium. Similarly, the influences of increased $\mu_s^{'}$ on the fitting accuracy are the same as the increased d does. As a whole, there are some deviations more or less between the fitting curve and MC data with the single parameter fitting method. And they are increased when the tissue optical thickness is small, while they are decreased though the useable MC data are reduced when the tissue optical thickness is big. For both conditions, more fitting errors will be increased.

For the double parameters fitting, the influences with the variations of $\mu_s^{'}$ and d on the fitting accuracies are similar to the conditions with the single parameter fitting method except that the discrepancies of fitting results between those two group measurement positions of the reflectance/transmittance are increased by the double parameters fitting method especially when $i = 7,8$. Besides, the fitting accuracies are spoiled for the media with big optical thickness where the useable MC data are decreased. And the constant factors A and $A^{'}$ are negative or zero when $\mu_s^{'} = 2mm^{-1}$ and $j = 2$.

A comparison is made between the single parameter fitting and the double parameters fitting for the finite thickness homogeneous slab. And some conclusions can be obtained. The discrepancies between the fitting curves and the MC data always exist for the single parameter fitting method especially when the optical thickness of tissue is relative small, while they exist only when the optical thickness is relative small for the double parameters fitting method. Besides, the fitting results are more accuracy for the double parameters fitting method than the results of the single parameter fitting method except that $\mu_s^{'} = 2mm^{-1}$ and $j = 2$. So the added constants A and $A^{'}$ are helpful to improve the fitting accuracies. It means that the rest term in expressions of diffusion reflectance and diffusion transmittance for the corresponding tissue geometry that subtract the exponential term contained source-detector distance r depends on the detected position weakly based on the multi-distance measurements. As an example, the fitting curves and the MC data of R_{41} are shown in Fig.1 (a) and Fig.1 (b) to declare the differences between those two fitting methods. The agreement of fitting curve and the MC data for the double parameters fitting method is better than the single parameter fitting method especially when $t < 175 ps$.

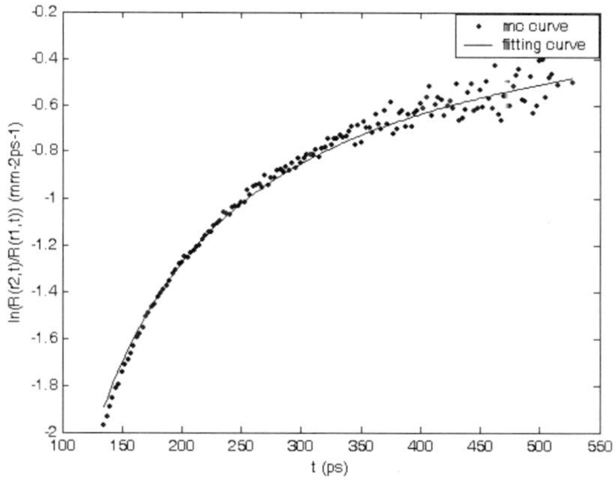

Fig.1(a) Single parameter fitting curve and the MC data for the finite thickness homogeneous slab.

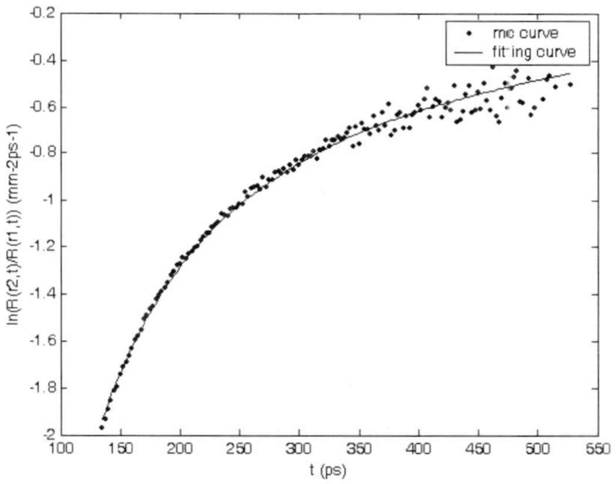

Fig.1(b) Double parameters fitting curve and the MC data for the finite thickness homogeneous slab.

Additionally, the optical thickness of tissue plays a significant role in the fitting results with accuracy. The fitting errors are relative big when the tissue optical thickness is relative small especially when the measurement fashions are R_{11} and relative big especially the measurement fashions are T_{72} and T_{82} . The reasons may be that the detected data are not agree with the diffusion theory so good for many photons can not experience enough scattering for the tissue with relative small optical thickness, and the detectable photons are reduced by the strong scattering for the reduced useable photon numbers and the increased data fluctuation. So the data of the detected positions relative away from source for the optical thickness relative small will be chose to use, while selective criterion of the detected positions for the optical thickness relative big are opposite. And the constant factors A and A' are negative or zero, which indicate that the media are strong scattering and the optical thicknesses are relative big.

3.2. *Results of the Finite Thickness Two Layered Slab*

The double parameters fitting results of the reflectance ($\mu'_{s0,fit}$ and A) and transmittance ($\mu'_{s1,fit}$ and A') measurement for the finite thickness two-layered slab medium are as Tab. 2 shown with the variations of μ'_{s0}, μ'_{s1}, d_0 and d_1. The subscripts 0 and 1 refer to the first layer and the second one respectively. There are two group measurement positions employed, i.e. $r_1 = 5.25mm$, $r_2 = 10.25mm$ and $r_1 = 10.25mm$, $r_2 = 15.25mm$.

Table2 Double parameters fitting results for the reflectance and transmittance measurements of the finite thickness two-layered slab media

Measurement Fashions	μ'_{s0} (mm^{-1})	μ'_{s1} (mm^{-1})	d_0 (mm)	d_1 (mm)	$\mu'_{s0,fit}$ (mm^{-1})	$\mu'_{s1,fit}$ (mm^{-1})	A_{fit}	A'_{fit}
R_{11}	0.5	1.5	1	4	0.99	—	-0.06	—
R_{12}	0.5	1.5	1	4	1.00	—	-0.07	—
T_{11}	0.5	1.5	1	4	—	1.23	—	0.04
T_{12}	0.5	1.5	1	4	—	1.16	—	0.003
R_{21}	0.5	1.5	4	4	0.57	—	-0.07	—
R_{22}	0.5	1.5	4	4	0.60	—	-0.06	—
T_{21}	0.5	1.5	4	4	—	0.93	—	0.06
T_{22}	0.5	1.5	4	4	—	0.85	—	0.04
R_{31}	0.5	1.5	4	1	0.55	—	0.04	—
R_{32}	0.5	1.5	4	1	0.56	—	0.05	—
T_{31}	0.5	1.5	4	1	—	0.59	—	0.06
T_{32}	0.5	1.5	4	1	—	0.57	—	0.04

From Tab.2, the thickness of d_0 / d_1 is changed by keeping the thickness of other layer d_1 / d_0 unchanged with $\mu'_{s0} = 0.5mm^{-1}$ and $\mu'_{s1} = 1.5mm^{-1}$ as $i = 1,2$ / $i = 3,2$ shown. The reduced scattering coefficient fitting results of every measurement fashions are shifted to the μ'_{s0} / μ'_{s1} by increasing d_0 / d_1. The constant factors $A_{fit} < 0$ and $A'_{fit} > 0$ for $i = 1,2$, while they are both positive for $i = 3,2$. The discrepancies between the fitting results and the true values are relative big for the appearance of other layer especially the optical thickness of other layer is relative big.

The thickness of d_0 / d_1 is changed by keeping the thickness of other layer d_1 / d_0 unchanged when μ'_{s0} and μ'_{s1} are equal to two of four data, i.e. 0.5, 1, 1.5 and 2 mm^{-1} (The fitting results are not shown). The changed trends of $\mu'_{s0,fit}$ and $\mu'_{s1,fit}$ are the same as the conditions above. Meanwhile, the constant factors A_{fit} and A'_{fit} are negative in some conditions. The fitting curve is not agrees with MC data so good when the tissue optical thickness is relative small, while the useable MC data are reduced and the data fluctuation is increased when the tissue optical thickness is relative big.

Some conclusions can be made to declare the tissue probable structures from the fitting results obtained. If the discrepancy between $\mu'_{s0,fit}$ and $\mu'_{s1,fit}$ is small, it may mean that the optical thickness of one layer is relative small when it is compared with that of the other one or the optical parameters between two layers are equivalent almost. If the discrepancy between $\mu'_{s0,fit}$ and $\mu'_{s1,fit}$ is relative big, then the optical parameters distribution of the two-layered tissue can be concluded that $\mu'_{s0,fit} > \mu'_{s1,fit}$ / $\mu'_{s0,fit} < \mu'_{s1,fit}$, which indicates that the scattering ability of first/second layer is bigger than the second/first one. If the constant factor A_{fit} is negative at R_{ij}, it may mean that $\mu'_{s0} < \mu'_{s1}$ and the second layer is the strong scattering medium respect to the first one, meanwhile the optical thickness of the second layer is relative bigger than that of the first one. If $A'_{fit} < 0$ at T_{i2}, it may mean that one layer of tissue is strong scattering medium with relative big optical thickness. So the tissue probable structure information can be obtained from the diffusion reflectance and transmittance measurements according to the discrepancy of fitting μ'_s between two layers and the sign (positive or negative) of constants A and A' for the finite thickness two-layered slab tissue. But more attentions must be pay for the conditions when the optical thickness is relative small and relative big which will increase the fitting errors.

4. Conclusions

A multi-distance measurement method is displayed in this paper. The optical parameters of tissue can be derived from the finite thickness homogeneous tissue and finite thickness two-layered tissue on the basis of the least square Newton fitting algorithm. For the homogeneous tissue, the single parameter fitting (μ'_s) method and the double parameters fitting (μ'_s, A or μ'_s, A') method are used to derive the optical parameters from the finite thickness homogenous tissue respectively. It can be concluded that the rest term in expressions of diffusion reflectance and diffusion transmittance for the corresponding tissue geometry that subtract the exponential term contained source-detector distance r depends on the detected position weakly. Additionally, for the two-layered tissue, the tissue probable structure information can be obtained from the diffusion reflectance and the diffusion transmittance measurements according to the discrepancy of fitting μ'_s from the first and the second layers and the sign (positive or negative) of constants A and A'.

Acknowledgments

This work was supported in part by the Natural Science Foundation of Fujian under Grant No.A0520001 and the Program for New Century Excellent Talents in University under Grant No.NCET-04-0615 by the Ministry of Education of China, and in part by the National Natural Science Foundation under Grant No.60578056.

References

1. F.F.Jobsis. Science. **198**, 1264 (1977).
2. M.S.Patterson, B.Chance, and B.C.Wilson, Appl. Opt. **28**, 2331 (1989).
3. J.C.Hebden, A.Gibson, T.Austin, R.M.Yusof, N.Everdell, D.T.Delpy, S.R.Arridge, J.H.Meek and J.S.Wyatt, Phys. Med. Biol. **49**, 1117 (2004).
4. L.-H. Wang, S. L. Jacques, and L.-Q. Zheng. Computer Methods and Programs in Biomedicine. **47**, 131 (1995).
5. L.H.Wang, J. Opt. Soc. Am. **A15**, 936 (1998).
6. R.C.Haskell, L.O.Svaasand, T.T.Tsay et al., J. Opt. Soc. Am. **A11**, 2727 (1994).
7. M.S.Patterson, B.Chance, and B.C.Wilson. Appl. Opt. **28**, 2331 (1989).
8. A.Kienle and M.S.Patterson., J. Opt. Soc. Am. **A14**, 246 (1997).
9. R.K.Wang and Y.A.Wickramasinghe, Appl. Opt. **37**, 7342, (1998).
10. F.Matelli, A.Sassaroli, Y.Yamada and G.Zaccanti, Opt. Let. **25**, 1508 (2000).
11. D.J.Durian, Opt. Let. **23**, 1502 (1998).
12. R. Aronson, and N. Corngold, J. Opt. Soc. Am. **A16**, 1999, 1066 (1999).
13. F.Matyrlli, A.Sassaroli, Y.Yamada and G.Zaccanti, J. Opt. Soc. Am. **A19**, 71 (2002).

FURTHER STUDY ON BIREFRINGENCE FEATURE OF CARTILAGE BY CONVENTIONAL OPTICAL COHERENCE TOMOGRAPHY

YINQI FENG* †

Opto – Mechatronic Equipment Technology Beijing Area Major Laboratory
Beijing Institute of Petrochemical technology, Beijing 102617, P.R. China

ZHENG CHANG

Department of Mechanical Engineering
Beijing Institute of Petrochemical technology, Beijing 102617, P.R. China

To diagnose cartilage diseases or monitor the cartilage growth, articular cartilage with an animal model is studied by use of a conventional Optical Coherence Tomography (COCT) system. Changes of cartilage properties with dehydration, mechanical action and chemical action are presented and analyzed. The results show that layer structures of cartilage tissues on the COCT images provide birefringence information resulting primarily from the tissues' linear or fibrous structures, and a COCT system may be capable of monitoring cartilage birefringence determined by the collagen fiber alignment and its packing density.

1. Introduction

Arthritis is a most wide spread disease in the world, which describes the different type of inammatory or degenerating diseases that damage joints. The structural and morphological changes in articular cartilage such as swelling, thinning or disorganization of the cartilage indicate cartilage damages. Collagen organization will be affected by the degradation of cartilage. In tissue engineering collagen organization will alter during the production of tissue constructs. It will be very important to develop a convenient and inexpensive imaging technique in order to diagnose cartilage diseases or monitor the cartilage growth.

Optical coherence tomography (OCT) is an important and recently developed non-invasive medical image modality that can reveal the subsurface' microstructures of biological tissues. High spatial resolution (1 ~ 10 μm) and high scanning speed (approaching the video rate) were achieved over the past few years [1]. It has been shown to be suitable for non-invasive two- dimensional tomographic imaging of microstructures beneath the tissue surface[2] and has the potential to improve the diagnostic limits of the currently available imaging techniques, allowing a wide range of clinical disorders to be addressed at an early stage.

Recently polarization-sensitive OCT (PSOCT)[3, 4] emerged as a very attractive branch of OCT, in which the polarizing nature of light was considered and more information about optical properties of biological tissues is provided. Results of these studies reveal the importance of polarization as contrast mechanism and a characterization of the optical polarization properties of the tissue. However, along the progress of PSOCT development there exist some controversial discussions in recent literature about the possibility of obtaining birefringence information of biological tissues with conventional OCT (COCT)[5, 6, 7]. Yang et al. have shown that fiber-based COCT may be able to extract birefringence information of well-organized tissues[8].

To investigate experimentally articular cartilage using a COCT system with an animal model, the effects of various cartilage properties on COCT images due to the changes of dehydration, mechanical action and chemical action were studied. All these effects will make the birefringence of cartilage tissue change, which could be monitored by OCT images.

* This work is supported by a 2007 grant from General Program of Beijing Natural Science Fund and a 2006 grant (KM200610017005) from Science Research Development Program of Beijing Education Council.
† Work partially supported by the grants from the returned scholars and students funding programs of Ministry of Education of China and Beijing Ministry of Personnel.

2. Method and Materials

A COCT system used in this study, which is made by a light source, a fiber-optical Michelson interferometer and some electronic parts, is presented in Figure 1. The light source is superluminescent diode delivering an output power of 1 mW; it is a broadband source. Its central wavelength is 820nm and its bandwidth is 25nm. The principle of the interferometer can be analyzed in terms of theory of two-beam interference for low coherent light. The sample is positioned in one arm (called sample arm) of the interferometer. An interferometric signal is detected when the optical path length of light backscattered from the sample matches to that from the reference within the source coherent length. By lateral (x direction) and longitudinal (z direction) scanning, two- dimensional OCT images are constructed by directing low coherence light at the sample and detecting the reactions from various internal structures by the use of an optic-fiber integrated system that map the amplitude of light backscattered from the sample.

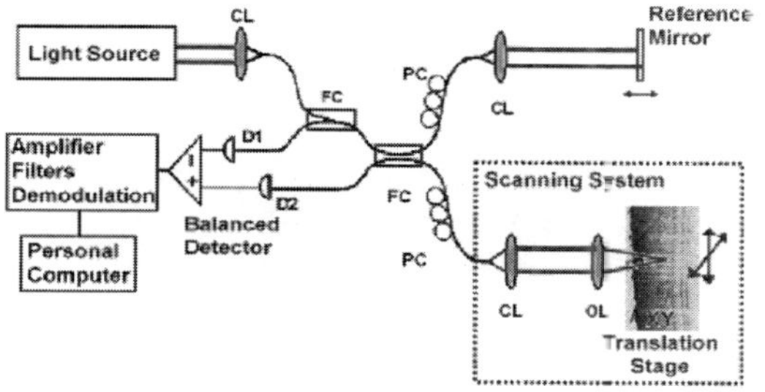

Figure 1. A conventional optical coherence tomography system.

For the system we used, the imaging axial (z direction) resolution in air was measured at 12μm, the transverse resolution (x direction) was measured at 16μm. Polarization fibers are used to provide the maximum obtainable resolution for the given spectral width of the source. The images are presented in a cross-sectional format with the horizontal axis representing the dimension along the surface of tissue and the vertical axis representing depth below the surface.

Knee joint, the largest and most complex joint in the body, has four bones: femur, patella, tibia and fibula. Its cartilage is a semi-rigid form of the supporting tissue that is made of collagen fiber. As cartilages on the different parts of the joint are very different, we concentrate our study on the cartilage of femur with chicken. The COCT images in this study were taken around the flat area on the chicken femur, not the condyle area. The fresh samples of chicken cartilage used in this study were obtained from Superstores. In some measurements, 1% of a collagenase solution (DMSO enzyme) was applied to cartilage specimens at 37 °C for 30 minutes in order to digest partially the organization of collagen fibers.

3. Results and Discussion

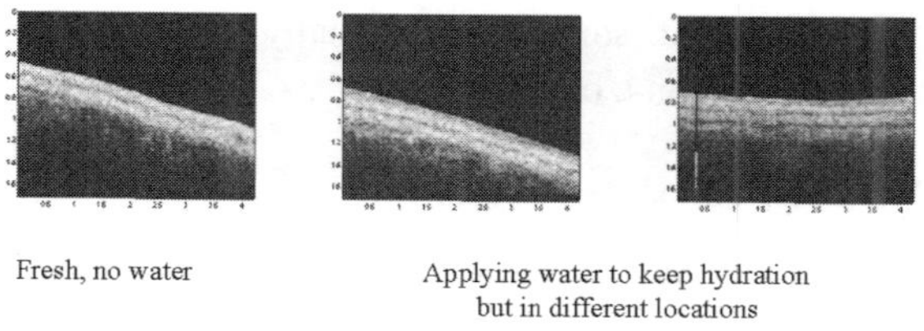

Figure 2. The COCT images of chicken cartilage.

The images of chicken cartilage obtained with the COCT system show layer structures in Figure 2. But the resolution of the COCT system is not higher enough to distinguish collagen fiber matrix itself. The only explanation is that the layer structures on COCT images are induced by the cartilage birefringence that is given by the difference in refractive indices along the length of the fibers and those along the cross section. Organized tissue such as cartilage could be assumed as a linearly birefringent slab with birefringence Δn. According to statistics optics, the light traveling in the tissue can be decomposed into two directions on mutually perpendicular polarized lights, called as fast- and slow-axis components. Traveling a distance d through the tissue, the component along the slow-axis of the tissue undergoes a single-pass phase retardation of $\delta = (360/2\pi)\Delta n k_0 d$ degrees relative to the component polarized along the fast-axis[7]. Here $k_0 = 2\pi/\lambda_0$ is the wave number associated with the central wavelength λ_0 of the light source. Therefore, the backscattered light from the sample arm, consisting of two components with a phase retardation δ relative to each other, interferes with the light from reference arm and forms the interference fringes under the condition that the optical path-length difference between the two arms are matched to within the coherence length of the light source used. Obviously, because the intensity rather than the superposed wave is detected, it would produce minimum intensities if the phase difference between the fast- and slow-axis is $\delta = (2N+1)\pi/2$, and the maximum intensities is provided if $\delta = N\pi$ (N=0, 1, 2, ⋯)[7]. This may explain the images with layer structures in birefringence tissues as shown in Figure 2.

We found in experiments that the dehydration, chemical and mechanical actions on cartilage tissues will well affect the layer structures of COCT images. Actually these effects make the collagen fiber organizations and packing densities within cartilage tissues change and lead the birefringence or the phase information of the cartilage tissue changes. In the following we will study experimentally how the COCT images reflect some of these changes.

3.1. *The Effect of Dehydration*

To investigate the effect of dehydration, an area on a chicken femur was imaged from fresh at a half-hour intervals in Figure 3. These images demonstrate that the inner parts of the cartilage on the femur become bright with the elapse of time but the layer structures of the images become unclear or disappear in these intervals. This means that the penetration depths of lights and the phase information of the cartilage were changed by dehydration.

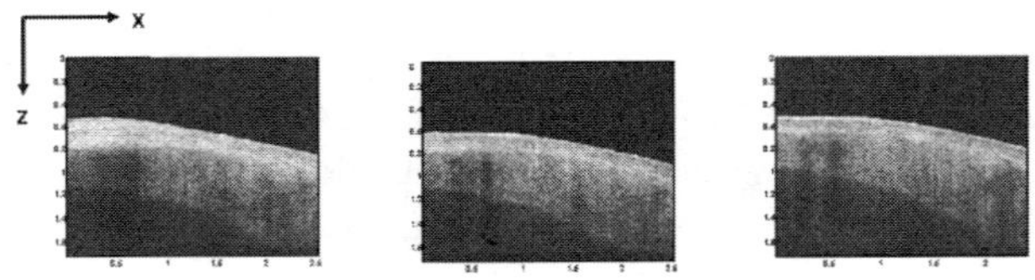

Figure 3. The effect of Dehydration: scanning each OCT image every a half hour.

The depth of penetration for a near-infrared light into a biological tissue mainly depends on the scattering characteristics of the tissue. The scattering of tissue is determined by the ratio of refractive indices of tissue scattering centers and tissue ground substance. If the mismatch between scattering centers and ground substance is decreased, the scattering of tissue will become less. Due to the changes of concentration of ground substance, water loss would increase the refractive index of the ground substance of the tissue. Therefore more refractive index matching is created by dehydration in the cartilage tissue, which result that the decrease in scattering and the increase in penetration depth are fulfilled. That is why the inner part of the cartilage on the chicken femur got more lights in and became bright.

Because of changes of the concentration of ground substance when water loss happens, the density and orientation of the collagen fibers change as well, which leads to the increasing of the refractive indices of the ground substance of the tissue. Then the birefringence of the cartilage as well as the phase difference becomes smaller when water losses happens, which results the layer structures of the images unclear or disappear as shown from Figure 3. In another word, by changes of layer structures of the COCT images, we can tell if packing densities within the cartilage tissues change.

3.2. *The Effect of Mechanical Action*

As shown in figure 3, lights are not easy to go much deeper into cartilage tissues, so the layer structures of the images just can be seen on the surface of cartilage tissues. This may be caused by the difference of properties between their top surfaces and inner parts. It seems that a cartilage tissue has a very strong reflectance on its top surface.

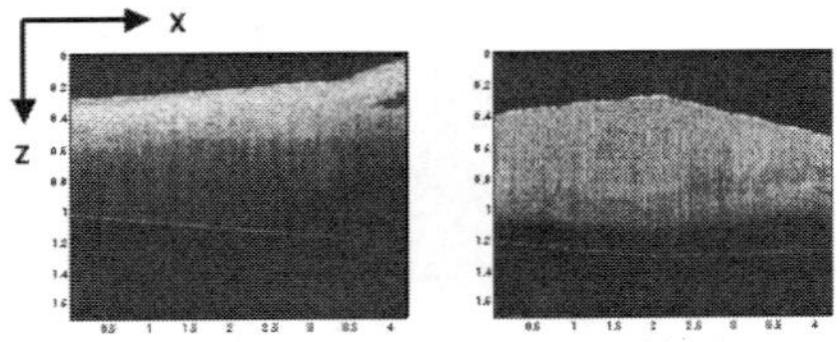

(a) Top (b) upside down

Figure 4. The Effect of Mechanical Action.

In order to study if there are different properties between the top surface and inner part of the cartilage tissue, we scanned a fresh cartilage section cut from a chicken femur. The section of cartilage was scanned from top surface and upside down separately when it was placed on a glass slide. The results are shown in Figure 4. Clearly, the penetration depth of light is deeper when scanning was started from the opposite of the top surface for the cartilage section. That means, at least, the optical property of cartilage tissue at its top surface is different from that of inner part. Another reason may be that the top surface of the section had stronger reflection than the opposite of the section. From figure 4, we notice that there is no layer structure on the image of cartilage. That means the birefringence may be lost after cutting. Mechanical action—cutting may affect the collagen fiber's orientation and density, in another word, the birefringence of the cartilage and the penetration depths of light are changed by the mechanical action.

3.3. *The Effect of Chemical Action*

As there may be the difference of properties between their top surfaces and inner parts, we expect there would be layer structures at the more inner part of a cartilage tissue on COCT images if lights could go in more deeper. For the proof of this expectation, we corroded the top surface using 1% of collagenase solution (DMSO enzyme) and saw what happened.

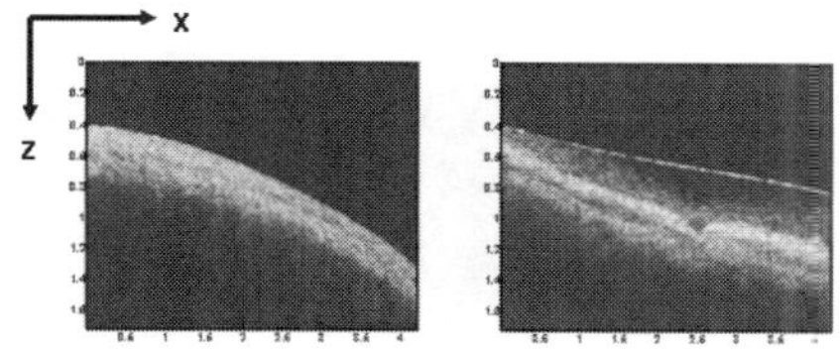

Figure 5. The effect of Enzyme action: (a) before using enzyme; (b) after using 30 minutes.

Figure 5 shows the image results before and after he application of an enzyme (DMSO). The image in Figure 5(b) was obtained after 30 minutes of applying DMSO enzyme to the chicken cartilage tissue. And because of the action of the enzyme, the top surface of the cartilage becomes easily penetrative. Its layer structures on the image are lost at the top but appeared at the inner part of the cartilage. The reason for this phenomenon is that the action of the enzyme made the collagen fiber's orientation disordered or the linear or fibrous structure damaged at the top surface of the cartilage tissue, which led the refractive index along any direction same. Therefore the birefringence at the top surface of the tissue as well as the layer structures on its OCT image lost by this chemical action. At the same time, by the action of the enzyme in the area, the mismatch between scattering centers and ground substance is decreased and the depth of penetration for lights going into the cartilage tissue is also increased. Due to the limited action time, much more inner linear or fibrous structures of the cartilage didn't been changed by this enzyme and so its birefringence. Then the layer structures were shown on the COCT image in Figure 5(b) as more lights can reach this inner part of the cartilage tissue. This phenomenon means that the cartilage is damaged or not can be told by the layer structure of its COCT image. In another word, the COCT images can be used to monitor the collagen fiber's orientation.

4. Conclusion

Cartilage birefringence resulting primarily from the linear or fibrous structures of a background medium has been investigated experimentally by use of a conventional optical coherence tomography. In this study, we focused on images of the cartilage of chicken femur tissue appearing layer structures observed from COCT measurements. The layer structures on the COCT images of cartilage change with dehydration, mechanical and chemical actions are observed, which may change collagen fibre's orientation and density. Accordingly, that the conventional OCT system is capable of monitoring cartilage birefringence that is determined by the collagen fibre alignment and its packing density are concluded.

Although the birefringence is one of the relative phenomena of polarisation, a COCT system is not capable of providing all tissue polarisation properties. For the full characterisation of polarisation properties of bio-tissues, a PSOCT has to be used. The present work is the foundation of the future investigation of cartilage tissues with a polarisation sensitive OCT system.

Acknowledgments

The author is grateful to Dr. Ricky Wang for his helps with the experiments that had been done by the authors when in the United Kingdom. Yinqi Feng's Email address is fengyinqi@bipt.edu.cn.

References

1. Drexler W, Morgner, U, Krtner, FX, Pitris C, Boppart SA, Li XD, Ippen EP and Fujimoto JG. In vivo ultrahigh resolution optical coherence tomography. *Opt. Lett.* **24**, 1221 (1999).
2. Wang RK and Elder JB. High resolution optical tomographic imaging of soft biological tissues. *Laser Physics*, **12**, 611 (2002).
3. De Boer JF, Milner TE, Van Gemert MJC, and Nelson JS. Two-dimensional birefringence imaging in biological tissue by polarization- sensitive optical coherence tomography. *Opt. Lett.*, **22**, 934 (1997).
4. Everett MJ, Schoenenberger K, Colston BW, Jr., and Da Silva LB. Birefringence characterization of biological tissue by use of optical coherence tomography. *Opt. Lett.*, **23**, 228 (1998).

5. Saxer CE, de Boer JF, Park BH, Zhao Y, Chen ZP, and Nelson JS. High-speed fibre-based polarization-sensitive optical coherence tomography of in vivo human skin. *Opt. Lett.*, **25**, 1355 (2000).
6. Hansen KA, Barton JK and Weiss JA. Optical coherence tomography imaging of collagenous tissue microstructure. *Pros. SPIE*, 3914, (2000).
7. Feng Yinqi, "The Study on Birefringence Feature of Cartilage by Conventional Optical Coherence Tomography" *ACTA Laser Biology Sinica*, **Vol. 15**, No. 6, 655 (2006).
8. Yang Y, Wu L, Feng Y, and Wang RK. Observations of Birefringence in Tissues from Optic-fibre Based Optical Coherence Tomography. *Measurement Science and Technology* **14**, 41 (2003).

FLOW-INJECTION ENHANCEMENT CHEMILUMINESCENCE DETERMINATION OF HUMAN SERUM ALBUMIN BASED ON FCLA-$^1O_2$ REACTION

WEI XU, YANCHUN WEI, DA XING, SHINGMING LUO, QUN CHEN†

MOE Key Laboratory of Laser Life Science & Institute of Laser Life Science, South China Normal University, Guangzhou 510631, China

A flow-injection chemiluminescence method using reactive oxygen species (ROS) specific CL probe-FCLA for low level human serum albumin (HSA) detection is proposed in this study. The technique is based on the phenomena that low level HSA can effectively enhance CL emission of FCLA-$^1O_2$ system. In comparison to the CL quenching technique used by others, the proposed method significantly improves the detection sensitivity (ca. 100 folds higher). After optimizing the current apparatus, we have achieved a reliable detection range of HSA between 1×10^{-10}-1×10^{-8} mol L^{-1}. The limit of detection is 4.5×10^{-11} mol L^{-1} (S/N=3) for HSA and the R.S.D. is 3.08% (n=27) for 6×10^{-9} mol L^{-1}HSA. The results are in excellent agreement with that from established clinical technique.

1. Introduction

Human Serum Albumin (HSA), has diverse functions such as playing important roles in maintaining the osmotic pressure of the blood, transportating of various ions, hormones and drugs. HSA concentration is an important marker for various renal disorders and clinical indications. Quantitative analysis of HSA in urine can provide critical information for early diagnosis and early treatment of nephrosis [1, 2].

At present, the method to determine HSA in clinical is radioimmunoassay. Its principle can be explained as that antigen with and without radioactive element can bind competitively specificity antibody. There is some functional relation between antigen and impulse signal emitted by labeled antigen-antibody complexes. Under the same conditions, impulse signal tested can deduce the concentration of antigen in sample according to the functional relation. The apparatus can realizably quantify HSA in urine samples up to10^{-9} mol L^{-1} with limitations of high cost and long time for the analysis. A more economical and less-time consuming method to determine micro-amount human serum albumin is thus desired for both research and potential clinical applications. As for analysis of micro-concentration of HSA, the most commonly used methods are the Lowry [3], CBBC-250 [4, 5], electrochemiluminescence [6], spectrophotometry [7], fluorospectrophotometry [8], Rayleigh light scattering [9,10]methods and chemiluminescence [11]. Among them, chemiluminescence (CL) technique, especially coupled with flow injection analysis (FIA), is considered as the most sensitive and versatile analytical technique. It is characterized by high sensitivity, a large dynamic range of concentrations of the substances determined, minimum background interference, and good reproducibility. CL technique coupled with flow injection analysis using Luminol as chemiluminescence emitters has been widely applied in analytical fields such as protein determination [12, 13], drug analysis [14, 15, 16], environmental monitoring [17], clinical diagnose [18], biomedical detection and biological process. The conventional methods used a quenching effect of proteins for their quantitative measurements [19]. But, our proposed technique was satisfactorily fulfilled to determine low level concentration of HSA based on CL enhancement effect reported by our formerly work [20]. In comparison to the CL quenching technique, the method significantly improves the detection sensitivity (ca. 100 folds higher).

We used FCLA-$^1O_2$ reaction system, not Luminol system. ROS (H_2O_2,$^*OH$, O^{2-}, $^1O_2$) are involved in reaction with Luminol, which indicates Luminol has a poor selectivity. Fluoresceinyl Cypridina Luciferin Analog (FCLA) is highly selective to trap superoxide anion or singlet oxygen [21, 22] and has a more simple mechanism compared to Luminol CL probe.

Using CL enhancement technique coupled with FIA to successfully determine HSA in FCLA-$^1O_2$ system is the purpose of this work. Based on previous work that CL from FCLA-$^1O_2$ system can be strongly enhanced with the presence of HSA [20, 23]. Within a certain range, the increased CL emission is proportional

† Work partially supported by grant 2-4570.5 of the Swiss National Science Foundation.

to the concentration of HSA. A simple and fast flow injection analysis has been developed for the determination of HSA, which has been satisfactorily applied to evaluate clinical urine samples.

It was confirmed by our experiments that CL of FCLA-$^1O_2$ was also enhanced by the addition of IgG and bovine serum albumin, which implied that FCLA-$^1O_2$ system would have a wide application to detect proteins. This is of great importance to broaden analytical fields.

2. Experimental

2.1. Chemicals and Reagents

All reagents were of analytical grade or the best grade available. The water used was deionized and double distilled. Stock solution of 1×10^{-4} molL^{-1} FCLA (Tokyo Kasei Kogyo Co., Tokyo, Japan) was prepared by dissolving 1mg FCLA in 15 ml water deoxygenated by N_2 bubbling and stored at refrigerator (-20□). Working standard solutions were prepared by appropriate dilution of stock solution with PBS7.4.

Human serum albumin (HSA), purchased from Sigma (St. Louis, MO, USA), was used without further purification. Stock solutions of HSA were directly diluted water to a concentration of 1×10^{-5} mol L^{-1} and stored in 4^0C. Working solution was freshly prepared by appropriate dilution with PBS7.4.

Phosphate Buffered Saline (PBS) was prepared by dissolving 8.50 g of sodium chloride, 2.20 g of sodium hydrogen phosphate, 0.20g of sodium dihydrogen phosphate to exactly 1L of water. The pH value of this solution was adjusted to 7.4±0.1. 1×10^{-2} mol L^{-1} hydrogen peroxide solution was prepared daily by diluting 77.6 μL of the product (30%, V/V, Guangzhou chemical reagent co, China) in 100 mL PBS, which was standardized by titration with a standard solution of KMnO4. The stock solution of sodium hypochlorite was 3.5×10^{-2} mol L^{-1}.

2.2. Apparatus and Flow Injection System

The schematic of the flow system used in this work is shown in Fig. 1. There are two peristaltic pumps A and B. pump (A) was used to deliver the flow streams of FCLA and hydrogen peroxide, and pump (B) was used to deliver merged stream of either sample or standard of HSA and sodium hypochlorite. PTFE tube (0.8 mm i.d.) was used as supplying connection of the flow system. FCLA solution (75 μL) was injected into the carrier stream using an eight-way injection valve equipped with a 75 μL sample loop, and then it was merged with the mixture solution of sample solution and sodium hypochlorite, finally reached the flow cell to produce CL emission. The flow cell was a coil of glass tube (1.3 mm i.d.) spiraled to a diameter of 35mm with three turns positioned directly in the front of the detection window of a CR-105 photomultiplier tube (Hammamatsu, Tokyo, Japan). The CL signal produced in the flow cell was detected and recorded with a computerized luminescence analyzer (type MPI-B, manufactured at Remax ElectronicScience and Technology Co. Ltd., Xi'an, China). Data acquisition and treatment were performed with MPI-B software running under Windows XP.

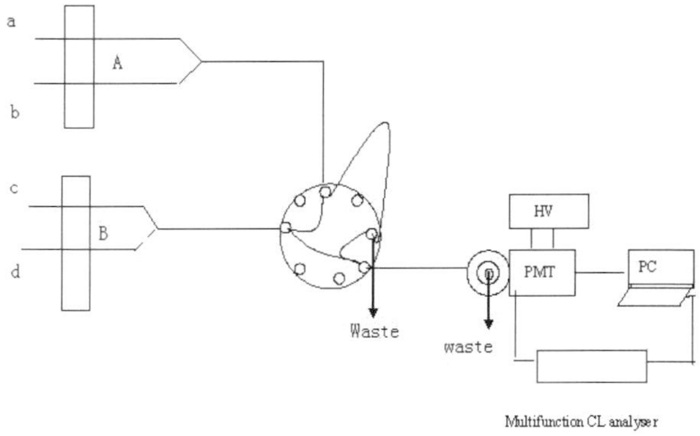

Fig. 1 Schematic diagram of CL flow system

(a) FCLA solution; (b) hydrogen peroxide; (c) sample or standard solution; (d) sodium hypochlorite; PMT photomultiplier tube; HV high voltage power.

2.3 Procedure

To obtain good mechanical and thermal stability of the flow-injecting CL system, the instruments were run for 10min before the first measurement. Flow tubes (a, b, c, d) as shown in Fig. 1 were connected with FCLA solution, hydrogen peroxide, sample or standard solution, and sodium hypochlorite, respectively. Incorporated flow streams of FCLA and hydrogen peroxide were merged with flow streams of the sample solution or standard of HSA and sodium hypochlorite. The CL intensity (peak height) was employed as the quantitative criterion. The determination of HSA was based on the increase in the CL intensity, calculated as

$\Delta I = I - I_0$ Where I was the CL signal of the system in presence of HSA and I_0 was the CL intensity with the absence of HSA.

Inevitably, the physical characteristics of the apparatus would greatly impact the signal generation and collection. The setup was thus optimized after carefully studying relevant parameters. Once the system is optimized, CL intensity corresponding to a range of known concentrations of HSA was measured to establish a titration curve. Various chemicals, known to exist in typical urine samples, were also studied for their effect on the FCLA-CL.

3. Results and Discussion

3.1 Influence of Chemical Variables

The effect of the concentration of hydrogen peroxide on the CL intensity was investigated over the range of 1×10^{-3} to 2.0×10^{-2} mol L^{-1}. It was found that relative CL intensity reached a plateau value when hydrogen peroxide concentration was 1×10^{-2} mol L^{-1}. The signal-noise ratio was over 2 and relative standard deviation (R.S.D) was 2.57% at this concentration. Considering sensitivity and reproducibility, 1×10^{-2} mol L^{-1} was selected as the optimal concentration.

The effect of the concentration of sodium hypochlorite on the CL intensity was examined over the range from 1.5 to 4.5×10^{-2} mol L^{-1}. The CL signal increased as the concentration of sodium hypochlorite was increased up to 3.5×10^{-2} mol L^{-1}. Above which it started to be constant. The RSD (four repeated experiments) was 2.70% at this concentration. Therefore, 3.5×10^{-2} mol L^{-1}NaClO was employed to subsequent experiments. Considered the detection sensitive of FCLA, referred to [20], 1×10^{-8} mol L^{-1}FCLA was selected for the sequence detection.

The length effect of Reaction tube was tested in the 11-70 cm range, as shown in Fig. 2. The signal-noise rate (S/N) was 1.553 when the length of reaction coil was 40 cm; the coil length was 70 cm, S/N decreased to 1.13. This is likely that the reaction occurs very fast. Chemical reagents merged and simultaneity started to emit light. To attain steady CL peak value, mixing reagents uniformly was necessary. For a comprehensive consideration, 40 cm was selected to be the optimum length.

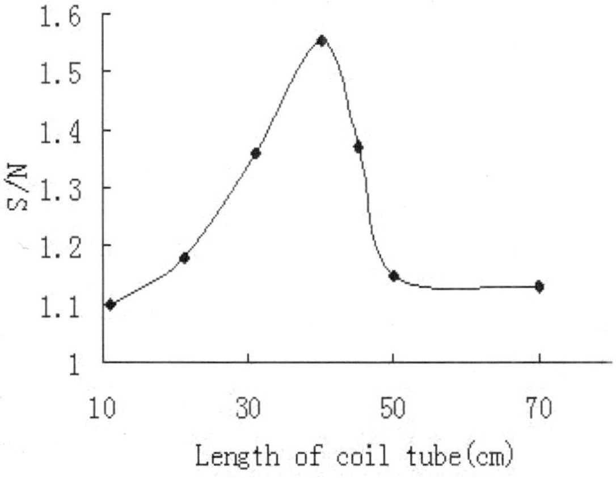

Fig. 2 Effect of the lengh of reaction tube

Conditions: FCLA 1×10^{-8} mol L^{-1}; H_2O_2 1×10^{-2} mol L^{-1}; NaClO 3.5×10^{-2} mol L^{-1}; HSA 1×10^{-9} mol L^{-1}; PMT -800V; sample volume 75 μL; flow rate:2.4 ml min^{-1}.

3.2 *Effect of pH*

The CL intensity was greatly affected by pH, as shown in Fig. 3. With the presence of HSA, CL was examined over a pH range between 3.12-11.0. Given otherwise identical conditions, CL signal was maximum at pH 7.4. Around this peak value, the signal was relatively stable within pH range of 7.3-8.0. Thus, pH 7.4 was selected for the subsequent experiments. Meanwhile several buffer solutions (PBS, H_3BO_3-NaB_4O_7, Tris-HCl) were considered. It was found that CL in PBS solution was the most stable. The RSD was 2.96% while that of H_3BO_3-NaB_4O_7 and Tris-HCl were 3.25% and 4.69%, respectively (5 repeated experiments).

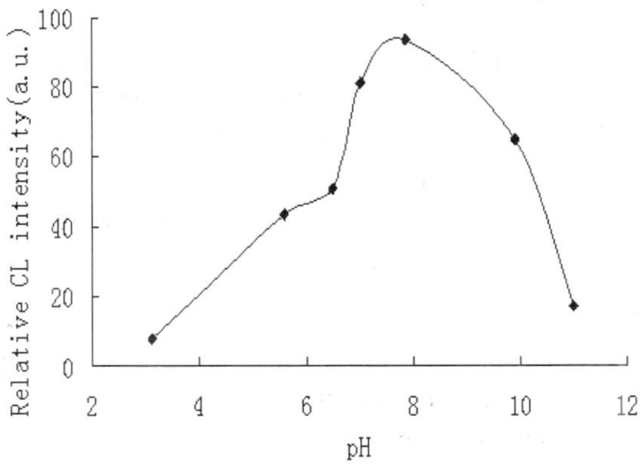

Fig. 3 Effect of pH value on the CL intensity of the system.

Conditions: FCLA 1×10^{-7} mol L^{-1}; NaClO 3.5×10^{-2} mol L^{-1}; HSA 1×10^{-8} mol L^{-1}; PBS7.4; negative high-voltage, -800V; flow rate, 2.4 ml min^{-1}. Every experiment repeated five times.

3.3 *Calibration Curve and Detection Limit*

A representative calibration curve characterizing the relationship between CL and HSA concentration is shown in Fig. 4. The calibration curve obeyed a second order equation: ΔI_{CL}=262.39+38.69lgC_{HSA}+1.84 (lgC_{HSA})^2. The regression coefficient was R^2>0.99. However, the obtained linear dynamic range was 1×10^{-10} to 1×10^{-8} mol L^{-1} and was expressed by the first-order equation: $\Delta I_{CL} = 116.51 + 5.84 lgC_{HSA}$ (R^2=0.99, n=7), with a detection limit of 4.5×10^{-11} mol L^{-1}(S/N=3). R.S.D for the consecutive CL detection of without HSA was 3.28 % (n=11). The RSD for 27 repeated measurements of HSA (6×10^{-9} molL^{-1}) was 3.08% (n=27).

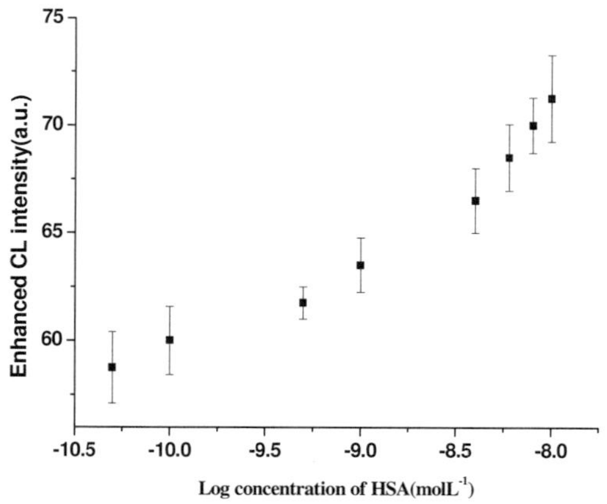

Fig. 4 Calibration curve for lg[HSA]

Condition: FCLA 1×10^{-8} molL^{-1}; NaClO 3.5×10^{-2} mol L^{-1}; H_2O_2 1×10^{-2} mol L^{-1}; PBS7.4; negative high-voltage, -800V; flow rate, 2.4ml min^{-1}; injection volume 75 μL.

3.4 *Interference studies*

The potential interference from basic amino acids, glucose, certain ions commonly found in urine samples to the CL measurement was examined. For all practical reasons, the acceptable deviation of the measurement result should be less than 5% of that with HSA only.

The results indicated that, up to 20 folds of NaCl, $MgSO_4$, NaH_2PO_4,Oxalate, 50-folded Glucose, 10-folds KNO_3 and some amino acids to the clinically observable maximum concentrations of the substance, had no practically significant influence on the CL measurement results.

3.5 *Validation with Urine Samples*

The developed method was applied to the determination of HSA in urine samples. The results were shown in Table 1. The samples were first assayed with conventional clinical technique without any condensation or dilution. The samples were then diluted to its 1:100, so the concentration of HSA in the sample was at a level typically not measurable with the conventional techniques, and then subject to the CL measurement. The HSA concentration was extrapolated from the titration curve obtained with standard samples in the same measurement batch. The resulted HSA concentration was multiplied by 100, to reflect the dilution factor, and compared to that of the conventional measurement. The results show that, our CL technique can recover 97.1 to 105.8% of that by the conventional technique.

Table 1 Determination results for HSA in human urine

Sample	clinical method (10^{-8} mol L^{-1})[a]	proposed method (10^{-8} mol L^{-1})[b]	standard added(10^{-8} mol L^{-1})	Found value (10^{-8} mol L^{-1})[b]	Recovery (%)
1	5.4	5.7±0.24	5	10.6±0.33	102.1
2	13.3	12.7±0.61	10	22.5±0.74	96.7
3	7.7	7.9±0.17	5	13.4±0.66	105.0
4	10.2	9.6±0.55	10	21.4±1.23	105.8

[a] data obtained using clinical technique(radioimmunoassay) in Nanfang hospital (Guangzhou, China)
[b] average of five determinations ±standard deviation(S.D).

Acknowledgments

This research is supported by the National Natural Science Foundation of China (30470494; 30627003) and the Natural Science Foundation of Guangdong Province (7117864) and the US NIH grant PO1-43892

References

1. E. L. Gelam and M. Tabak, *Spectrochimica Acta. A* 2255 (2000).
2. A. Sukowska, J. Równicka, B. Bojko and W. Sukowski, *Journal of Molecular Structure*, **133** (2003).
3. O. H. Lowry, N. J. Rosebrough, A. L. Farr and J. Randall, *Biol. Chem.* **265** (1951).
4. M. M. Bradford, *Anal. Biochem.* 248 (1976).
5. R. Flores, *Anal. Biochem.* 605 (1978).
6. X. L. Wang and W. Sun, *Journal of Instrumental Analysis.* **106** (2005).

7. C. Q. Ma, K. A. Li and S. Y. Tong, *Anal. Lett.* **1021** (1998).
8. N. Li, K. A. Li and S. Y. Tong, *Anal. Lett.* **1763** (1995).
9. Y. Li, L. J. Dong, W. P. Wang, Z. D. Hu and X. G. Chen, *Analytical Biochemistry.* **64** (2006).
10. P. Feng, C. Z. Huang and Y.F. Li, *Analytical Biochemistry.* **83**(2002).
11. C. B. Huang, K.Z hang and Y. X. Ci, *J Biochem Biophys Methods.* **341** (2007).
12. M. R. Alvarezetc, *Journal of luminescence.* **193** (2006).
13. N. T. Deftereos, N. Grekas and A. C. Calokerinos, *Analytica Chimica Acta.* **137** (2000).
14. J. X. Du, L. Hao, Y. H. Li and J. R. Lu, *Analytica Chimica Acta.* **98** (2007).
15. A. Safavi, M.A. Karimi and M.R. H. Nezhad, *Journal of Pharmaceutical and Biomedical Analysis.* **1499** (2003).
16. H. J. Zhou, Z. J. Zhang, D. Y. He and Y. Xiong, *Sensors and Actuators B.* **798** (2005).
17. J. F. Huertas-Pérez, A. M. García-Campaña, L. Gámiz-Gracia, A. González-Casado and M.D.O. Iruela, *Analytical Chimica Acta.* **161** (2004).
18. H. Buschmann, E. Cleve, S. Torkler and E. Schollmeyer, *Talanta.* **145** (2000).
19. Y. X. Li, D. H. Zhao, C. Q. Zhu, L. Wang and J. G. Xu, *Anal Bioanal Chem.* **395** (2002).
20. J. Zhou, D. Xing and Q. Chen, *Photochemistry and Photobiolog.* **1058**(2006).
21. K. Akutsu, H. Nakajima and K. Fujimori, *J Chem Soc perkin Trans.* ***2***:1699.
22. S. Mashiko, N. Suzuki, S. Koga, M. Nakano and T. Goto, *J Biolumin Chemilumin,* **69** (1991).
23. J. Zhou and Y. C. Wei, *Spectroscopy and Spectral Analysis in the press.*

IDENTIFICATION OF GASTRIC MUCOSA LESIONS USING OPTICAL COHERENCE TOMORGRAPHY IN RATS

XIN GONG, CHANGCHUN ZENG, ZHOUYI GUO, HONGLIAN XIONG

Laboratory of Photonic Chinese Medicine, MOE Key Laboratory of Laser Life Science, South China Normal University, Guangzhou 510631, Guangdong Province, China

**Email: gongxin001@126.com*

Abstract Gastric mucosa lesions are typical exosyndrome of gastropathy including edema, erosion, bleeding, necrosis, ulcer etc. To detect the characterization of various gastric mucosa lesions, an optical coherence tomography (OCT) system, which can obtain high resolution cross-sectional images of tissue, was used for the detection of microstructure of gastric mucosa lesions in rats. The gastric mucosa lesions animal models were induced in Sprague-Dawley rats by different methods including indometacin, alcohol and acetic acid. These results showed that there were different forms of gastric mucosa lesions in the induced animal model, pathological changes of gastric mucosa can be clearly observed, and the character of the gastric mucosa lesions can be identified by analysis of OCT imaging in a noninvasive and quick way. Our results indicated that OCT can be used to clearly detect the microstructure of gastric mucosa lesion tissue. Thus, the OCT technology, with an ancillary endoscope, maybe be benefit and helpful as a tool to provide an objective diagnostic method for study on gastric mucosa lesions in the clinical practice.

Key words: Gastric mucosa lesion, gastric ulcer, optical coherence tomography, pathological detection.

1. Introduction

Gastric mucosa lesions is a common disease that to see frequently in the clinic. The several forms of gastritis are extremely common gastric mucosal diseases that represent reactions to a variety of exogenous and endogenous noxious influences. The ingestion of alcohol, indomethacin and the reflux of bile into stomach are important etiologic factors in gastric mucosa lesions[1-3]. There are also many other factors that can lead to Gastric mucosa lesions, such as noise, brain hemorrhage or even mental strain[2-6].

Many kinds of methods were used to detect the injury of stomach such as gastroscope, barium meal, ultrasonic detector etc. In our research, a new technique optical coherence tomography (OCT)[7, 8] was adopted. OCT is a recently developed and non-invasive imaging technique, which provides high-resolution cross-sectional rapid images to the microstructures of biological tissue. It combines light from a low-coherence source with a Michelson interferometer to produce cross-sectional images of tissue structures. Optical coherence tomography systems have been used clinically for ophthalmologic[9], gynecology[10], dermatological[11]and cardiology[12]. However, there is limited information about the application of this modality in the assessment of gastric mucosa lesions. In this pilot study, we evaluated the usefulness of OCT in differentiating various injury of gastric mucosa in rats.

2. Materials and Method

2.1. *Optical Coherence Tomography System*

The OCT system used in this study was set up by our laboratory collaborated with Graduate School at Shenzhen, Tsinghua University, which mainly includes resource module, fiber conduction module, reference and sample arms modules. Simple to introduce, a broadband light source emitted from a super luminescent diode is the central wavelength at 1310 nm and a bandwidth of 50 nm. A visible light source（$\lambda =$ 645nm）is used to guide the probe beam. This OCT system provides an axial resolution of 10~15μm, the transverse resolution of the system is about 25 μm, the signal-to-noise ratio (SNR) of this system is measured at 100dB. The OCT system operation is controlled automatically by computer, scanning an image will takes about 1.0 s, and the scanning depth is about 2-3mm. The data acquisition software is written in Lab VIEW.

2.2. *Animal*

Thirty Sprague-Dawley rats, from the Laboratory Animal Center, Sun Yat-sen University, weighing between 230-260g were used in our study. Animals were divided into four groups including normal group and model groups and maintained in standard laboratory, in which chow and tap water available ad libitum for seven days.

2.3. *Method*

The animals were deprived of food, but were allowed free access to water for 24 hours before the experiments. The rats of the first group were treated with intragastric administration of 75% alcohol for10ml/kg, and we detected the internal surface of their stomach by optical coherence tomography system 4 hours later. The second group was treated almost the same with the first one but the drug we used to induce the gastric mucosa lesions was indometacin and the dosage was 48mg/kg. In the third group, the rat ulcer model was established by 0.5ml 100% glacial acetic injection into the subserosa by operation, some rats were detected by OCT the next day others were detected 7days later to see the healing of the gastric mucosa. The fourth group was the normal group.

3. Result

The first and the second group were both the model of acute gastric mucosal lesion. In the first group, the appearance of gastric mucosa was dark, congested and bleeding. The elasticity was poor with rare mucus, and the mucosal folds were flat or had disappeared. Various injury of gastric mucosa including edema, erosion, bleeding, necrosis, ulcer etc could be seen on the internal surface of the rats' stomach. Girdle-shaped bleedings paralleled with each other from forestomach to antrum almost breakthrough the body of stomach with Many Ecchymosises and erosions among them. Compare with the first group the intragastric administration of indometacin seemed a little mild. Girdle-shaped bleedings also could be seen but not that long and ecchymosises and erosions were not as many as the first group. Using OCT we could able to image and discriminate different injury of the rats' gastric mucosa. A cross-section image of a Girdle-shaped bleeding which caused by alcohol is presented in Fig.1. The injury is more deep and wider than that caused by indometacin in Fig.2. Fig.3 is an image of an area of erosion and Fig.4 is an image of an edema. Both of them were detected from alcohol group. Fig.5 is an image of a bleeding point of indometacin group.

In the third group we intended to set up a model of healing of the injured gastric mucosa. Fig.6 was detected 1 day after operation and Fig.7 was another rat detected 7 days later. Amelioration could be seen form the images of OCT.

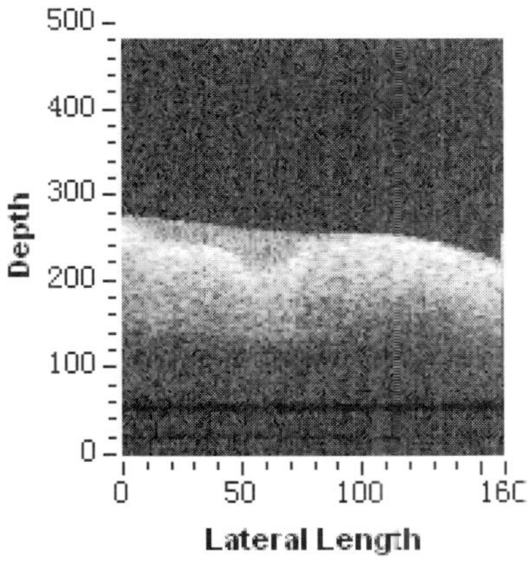

Fig.1

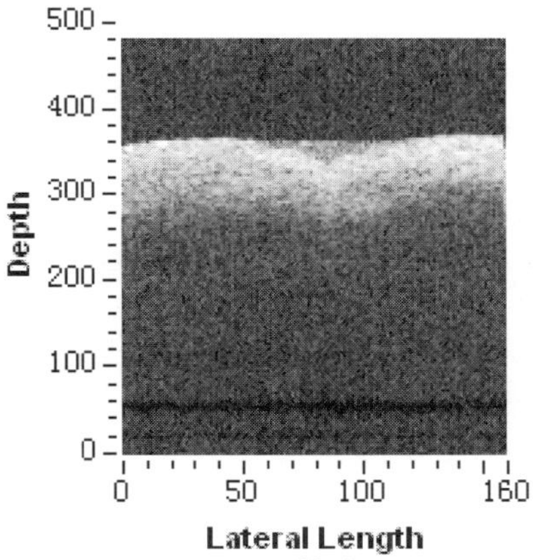

Fig.2

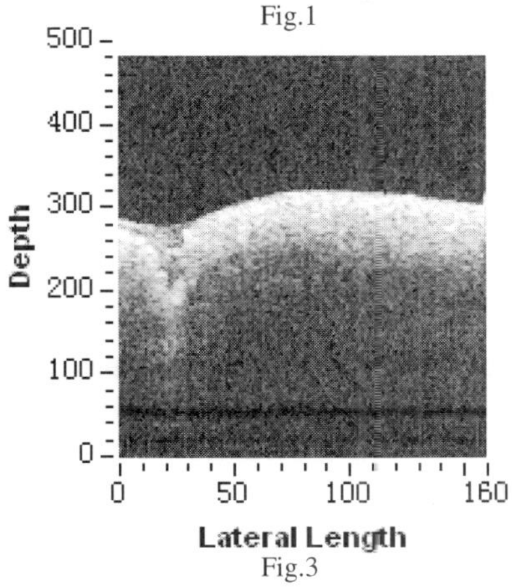

Fig.3

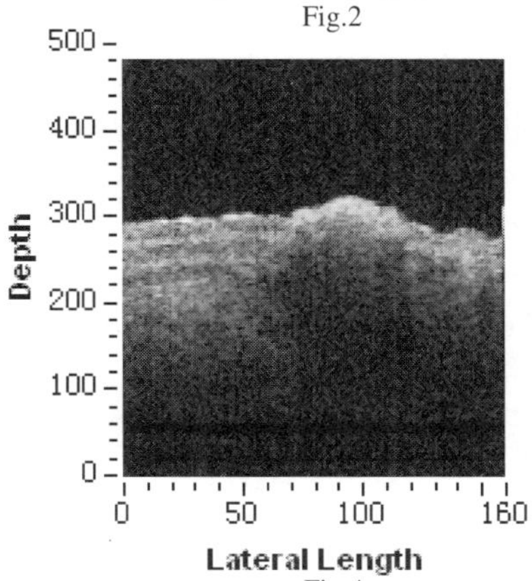

Fig.4

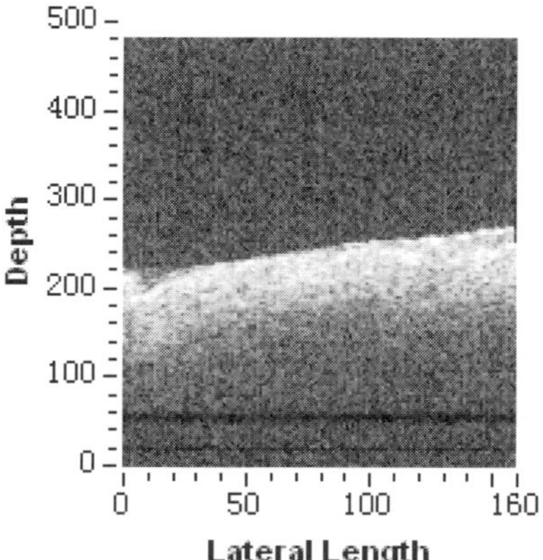

Fig.5

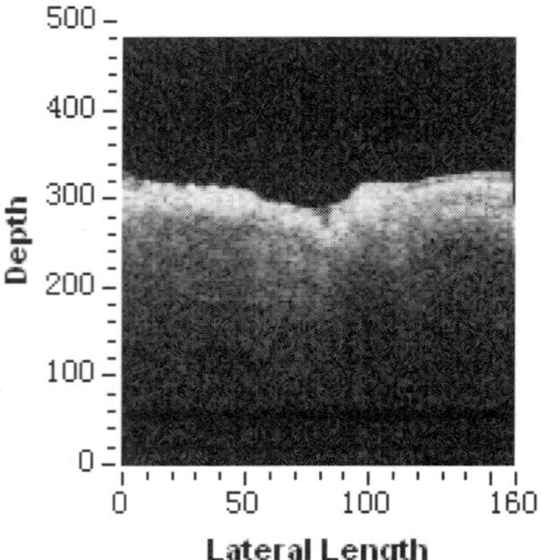

Fig.6

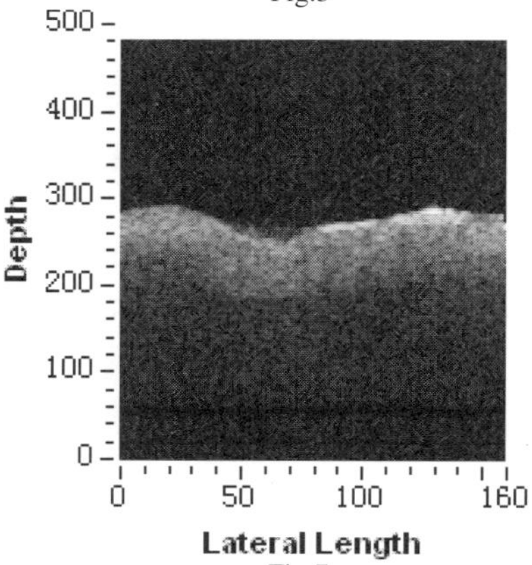

Fig.7

4. Discussion

In the present study we have investigated the practical clinical potential of the OCT system set up by ourselves. It is the initial stage of our study on using OCT system to detect the lesions of alimentary canal mucosa. OCT was used because it is an increasingly popular device in the detection of mucosa lesions and is of general interest for many different medical applications. The spectral properties of biological tissue is multiplicity and nontransparent, which limited the resolution and penetration depths of many image technique. OCT is a non-invasive imaging technique that allows the acquisition of high-resolution, cross-sectional images of biological tissue based on the detection of backscattered light. Thanks to the OCT technique, spectral information of the injuries and the lamellar structures could be clearly detected in a non-harmful and quick way. But the images we got in our study were still detected in vitro, and the next step endoscope which allows us to get images in vivo and real-time will be connected to the OCT system.

5. Conclusion

Optical coherence tomography, a fairly new non-invasive imaging technique, is of great useful in differentiating various injury of gastric mucosa in rats. It maybe be benefit and helpful as a tool to provide an objective diagnostic method for study on gastric mucosa lesions in the clinical practice.

References

1. Qingwen Zhi, Shuhao Wang, Min Chai. Transgenic Mini-tomato and Protection Against Alcohol-induced Gastric Injury. Journal of Genetics and Genomics Volume 34, Issue 8, August 2007, Pages 756-763.H. Müller and B. D. Serot, *Phys. Rev.* **C52**, 2072 (1995).
2. S.-K. Chioua, b, and S. Mandayam. NSAIDs enhance proteasomic degradation of survivin, a mechanism of gastric epithelial cell injury and apoptosis. Biochemical Pharmacology, Volume 74, Issue 10, 15 November 2007, Pages 1485-1495.

3. Omar M. E. Abdel-Salam, a, József Czimmerb, András Debrecenib, János Szolcsányic and Gyula Mózsik Gastric mucosal integrity: gastric mucosal blood flow and microcirculation. An overview. Journal of Physiology-Paris, Volume 95, Issues 1-6, January-December 2001, Pages 105-127.
4. Chaoqun Liu, Tao Sun, Zhongxiao Li, Zhifa Liu, Shanfeng Fu, Jianliang Shen, .Plasma polypeptide hormone levels in rats with gastric ulcer after exposure to intense noise. Chinese Journal of Industrial Hygiene and Occupational Diseases, 2003(01).
5. Yuxin Huang, Guoshi Liu, Shuan wen Li, Qingli Wang. Acute gastric mucosal injury and gut hormone levels in rats stressed by firing noise. World Chinese Journal of Digestology, 2000(01).
6. Yichun Tang, Gencheng Chen, Yiming Tan. Preventive Effects of Radix et Rhizoma Rhei Granule on the Occmrence of Complicated Stress Ulcer of Stomach in Rats After Cerebral Infarction. Journal of Guangzhou University of Traditional Chinese Medicine, 2001(02).
7. D. Huang, E. A. Swanson, C. P. Lin, J. S. Schuman, W. G. Stinson, W. Chang, M. R. Hee, T. Flotte, K. Gregory, C. A. Puliafito, and J. G. Fujimoto. Optical coherence tomography. *Science*, 254, 1178–1181(1991).
8. G. Fujimoto, M. E. Brezinski, G. J. Tearney, S. A. Boppart, B. E. Bouma, M. R. Hee, J. F. Southern, E. A. Swanson, and B. Bouma. Optical biopsy and imaging using optical coherence tomography. Nature Medicine, 9, 970-972 (1995).
9. E. A. Swanson, J. A. Izatt, M. R. Hee. In vivo retinal imaging by optical coherence tomography. *Opt. Lett.* 18 (21), 1864-1867 (1993).
10. N. D. Gladkova, G. A. Petrova, N. K. Nikulin, S. G. Radenska-Lopovok, L. B. Snopova, Y. P. Chumakov, V. Nasonova, G. V. Gelikonov, R.V. Kuranov, A. M. Swrgeev, F. I. Feldchtein. In vivo optical coherence tomography imaging of human skin: norm and pathology. Skin Research and Technology, 6(1), pp. 6-16 (1999).
11. G. J. Tearney, I. K. Jang, B. E. Bouma. Optical coherence tomography for identifying unstable coronary plaque. International Journal of Cardiology, 107, 154-165 (2006).
12. M. Shakhova, F. I. Feldchtein, and A. M. Sergeev. Applications of Optical Coherence Tomography in Gynecology. *Chap. 24 in Handbook of Optical Coherence Tomography*, B. E. Bouma and G. J. Tearney ed. pp. 649-672, Marcel Dekker, NY (2002).

IMAGES OF THE MORPHOLOGY CHANGE OF 30% HYDROGEN PEROXIDE APPLIED TO HUMAN ENAMEL BY MEANS OF OPTICAL COHERENCE TOMOGRAPHY

YIRONG NI; ZHOUYI GUO

Laboratory of Photonic Chinese Medicine, MOE Key Laboratory of Laser Life Science, South China Normal University, Guangzhou Province 510631 China

JINGQUAN CHI; JIANYING CHEN

School- hospital, South China Normal University, Guangzhou province 510631 China

There is still some controversy in the tooth whitening whether hydrogen peroxide bleaching causes the microstructure changes of human enamel. One of the reasons for this controversy is that there is as yet no reliable, non-destructive in vitro method for imaging and assessing the morphology changes in bleached teeth. So we have used an OCT system for investigations of enamel. We have imaged before and after 30% hydrogen peroxide bleaching enamel and demonstrated the detection of the changes in tooth tissue microstructure. We discuss the details of images and its potential and limitations with respect to dental applications.

1. Guidelines

Tooth bleaching was first described in 1848 using oxalic acid; later ranges of peroxides, such as hydrogen peroxide, were used[1]. The first clinical study on at home tooth whitening using an agent of 10% carbamide peroxide was reported in 1989[2]. Now tooth whitening is a popular aesthetic procedures used in modern day dentistry due to its relative ease, efficacy and safety [3]. As a conservative method to whiten natural teeth vital bleaching has gained worldwide popularity. The tooth whitening process involves the direct contact of the whitening product on the surface of the teeth for an extended period of time. However, this direct contact with enamel for prolonged times has increased concerns about the potential adverse effects of active bleaching agents on enamel [4].

Present knowledge of the effects of 30% hydoren peroxide agents on the mineral content of enamel is still controversial [5]and more research examining the effects of bleaching agents on enamel tissues is required [6].So our study was to observe the possible morphology changes in surface of human enamel and prepared for Optical Coherence Tomography (OCT). Although there are many different techniques available to assess unfavourable effects on enamel, some of methods are destructive on enamel samples [7,8]. Optical coherence tomography (OCT) is a new non-invasive imaging technique that can be used to acquisition of high-resolution [9], cross-sectional images of biological tissue based on the detection of backscattered light [10]. The principles behind the OCT method have been described in great detail previously [11-13]. OCT has already been applied in vitro and in vivo to a variety of tissue including the human eye, retina, skin, gastrointestinal tract, and oral cavity ect [14-25]. But it has not been used to evaluate the effects of bleaching agents on enamel previously. Therefore in this study, OCT was used to investigate the possible microstructure changes of the morphology of enamel of 30% hydrogen peroxide cyclic bleaching in vitro.

2. Materials and Methods

2.1. *Preparation of the Specimens*

Twenty human extracted teeth were obtained from school-hospital (south china normal university guangzhou, China). In addition, extracted human incisors and molars selected by visual inspection have been examined. The teeth were cleaned with deionized water to remove any surface debris. Then the teeth were stored in isotonic saline solution and placed in a refrigerator at 4℃ until needed. Typically the storage time was less than 1 week. Imaging was performed through air at room temperature with the orientation of the imaging scan marked on the specimen. When 30% hydrogen peroxide application to the natural enamel for 8h a day followed by immersion into isotomic saline solution for the rest of the day. All specimens were kept in a humid environment at 37 ℃ . To achieve a satisfactory outcome, this treatment was repeated for two consecutive weeks [26].

2.2. *Optical Coherence Tomography System*

The OCT system used in this study was set up by our laboratory collaborated with Graduate School at Shenzhen, Tsinghua University, which mainly includes resource module, fiber conduction module, reference and sample arms modules. Simple to introduce, a broadband light source emitted from a super luminescent diode is the central wavelength at 1310 nm and a bandwidth of 50 nm. A visible light source（λ = 645nm）is used to guide the probe beam. Real-time imaging rates were required for in vitro measurements in order to avoid motion artifacts. No post-processing of these images has been applied. This OCT system provides an axial resolution of 10~15 μ m, the transverse resolution of the system is about 25 μm, the signal-to-noise ratio (SNR) of this system is measured at 100dB. The OCT system operation is controlled automatically by computer, scanning an image will takes about 1.0 s, and the scanning depth is about 2-3mm. The data acquisition software is written in Lab VIEW.

2.3. *OCT Measurements and Evaluations*

Each tooth specimen was scanned several times before and after bleach application by OCT. Sound enamel and bleaching enamel provide same imaging depths. The same scanning parameters were applied in all scans. The data sets from scans were analysed allowing comparisons between the baseline data (first scan), and the datas obtained after bleaching.

3. Results

Fig.1 shows an photograph of a human dental .this incisor was extrated from periodontitis patient who have not taken teeth bleaching before.

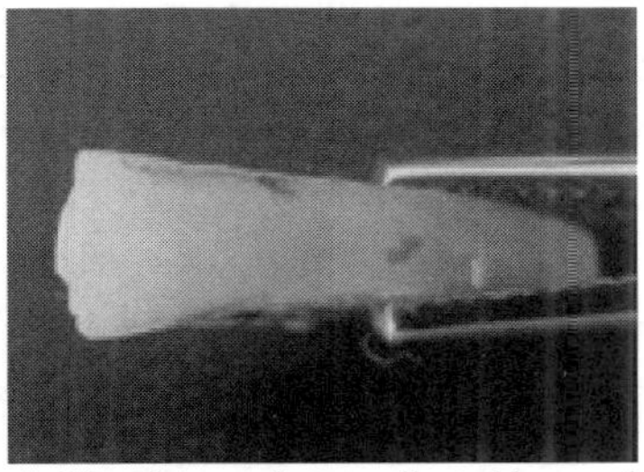

Figure 1. The picture of new collected human incisor

Fig.2 shows an photograph of the same incisor which treated with 30% hydrogen peroxide for 14 days.

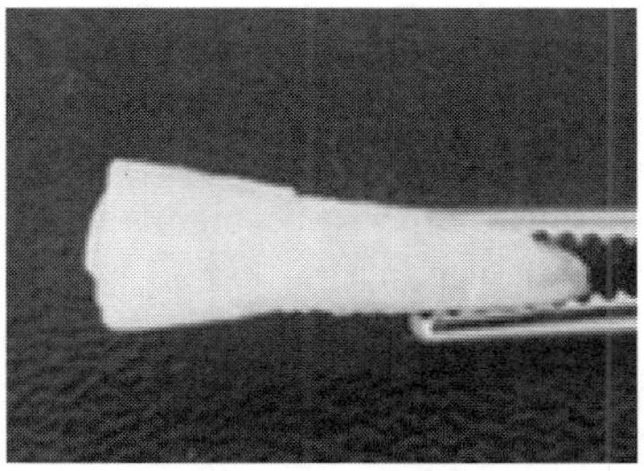

Figure 2. The picture of the incisor treated with 30% hydrogen peroxide for 14 days.

Let us contrasted Fig.1 to Fig.2, we could seen the bleaching incisor was much whiter. But whether it had microstructure changes? OCT system allowed us to detect relevant structural details.

We achieved imagings depth of more than 2 mm in all samples. The light penetrated the tissue in the vertical direction.

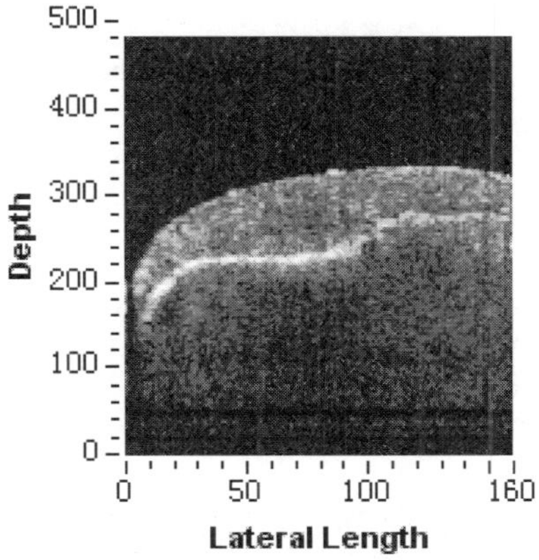

Figure 3. The OCT image of new collected incisor.

Fig. 3 shows a typical in vitro OCT image of an incisor enamel. The penetration depth of the detection of the enamel more than 2 mm deep.

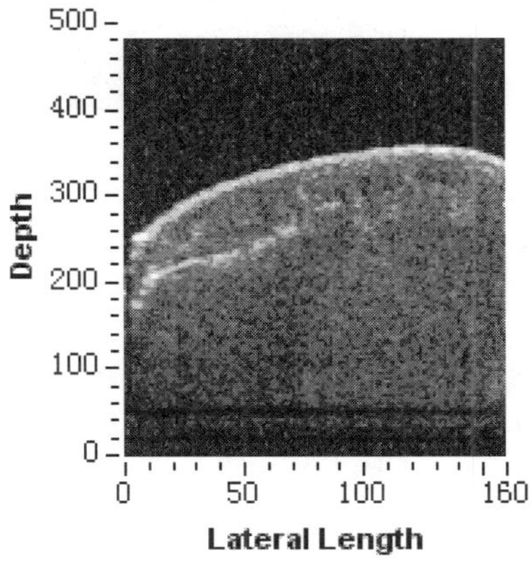

Figure 4. The OCT image of the incisor treated with 30% hydrogen peroxide for 14 days.

Fig.4 shows an image of the same incisor which treated with 30% hydrogen peroxide for 14 days.

As contrasted Fig.3 and Fig.4, we observed that the same position of the same enamel both before and after 30% hydrogen peroxide cyclic bleach application for 14 days. And revealed that there was significant morphology changes in the outer 2 mm of the enamel after bleaching. The demineralised areas are apparent and obscure the enamel underneath.

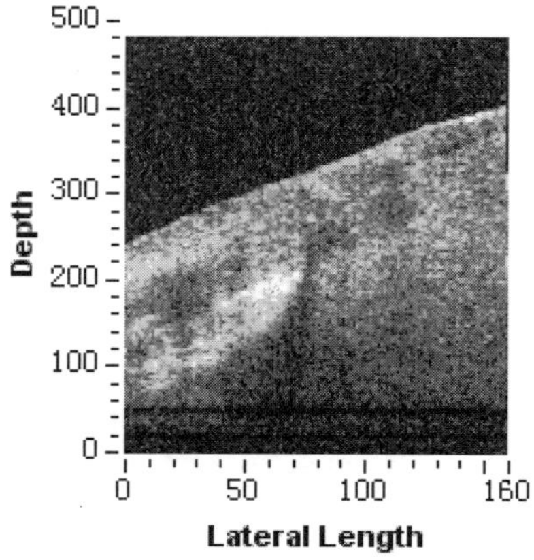

Figure 5. The OCT image of the new collected molar.

Image of an extracted molar which had not treated with bleaching agent was presented in Fig. 5. The penetration depth exceeds 2 mm.

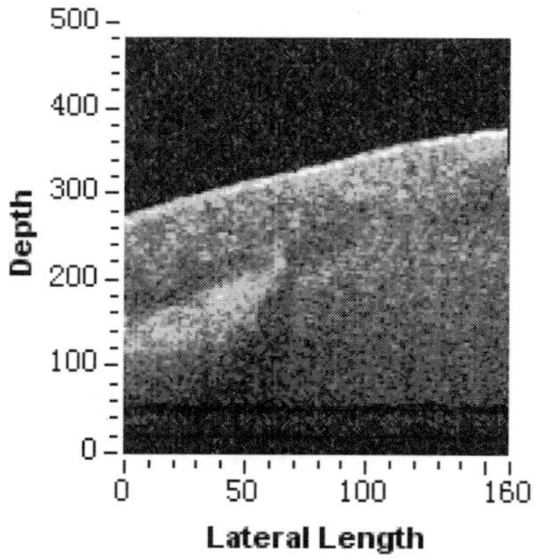

Figure 6. The OCT image of the molar treated with 30% hydrogen peroxide for 14 days.

Fig.6 shows an molar which was treated with 30% hydrogen peroxide for 14 days. The OCT image shows that microstructure had changed in fig.6. We can see that some enamel crystals had obvious demineralization. And the bleaching agent demineralized enamel and some texture of them were destroyed.

4. Discussion

OCT is an increasingly popular device in dentistry. This study suggests that the use of OCT with the non destructive method of selection and detection is indeed a reliable tool to investigate the effects of bleaching agents. The microstructure changes were detectable in early stages but complete visualization was demonstration in finally stages. The images showed that application of 30% hydrogen peroxide on enamel for 14 days for 8 h/day decreased the morphology changes at 2mm below the enamel surface. It can be argued that the application bleaching agent may have made enamel more susceptible to demineralization. And it has already been shown that OCT is possible monitoring the change procedures. Moreover, the capability of imaging the microstructure of enamel allows the early detection of demineralization, which is of increasing importance.

We suggested that further studies with treatment controls should be undertaken to verify this result. It is recommended that application of bleaching agents should be carefully considered in patients susceptible to caries and tooth wear.

Acknowledgments

This project was supported in part by the National Natural Science Foundation of China, (Grant No. 60778047), And Technology Project of Guangdong Province of China (Grant No. 2004B10401031, No. 2005B50101015 and No.2006B35602001)

References

1. Freedman GA, McLaughlin G and Greenwall L, I In Bleaching techniques in restorative, London: Martin Dunitz p. 139. (2001).
2. Haywood VB and Heymann HO, Quintessence Int. **20**, 173-6 (1989).
3. Haywood VB, Contemporary Esthetics and Restorative Practice **3**, 2–9 (1999).
4. Neslihan Efeoglu, David Wood and Candan Efeoglu, J. Dent. **33**, 561–567 (2005).
5. Attin T, Kielbassa M, Schwanenberg M and Hellwig E, J. Oral Rehabil. **24**, 282–6 (1997).
6. Mc Cracken MS and Haywood VB, J. Dent. Res. **24**, 395–8 (1996).
7. Arends J and Bosch JJ. , J. Dent. Res. **71**, 9248 (1992).

8. White DJ, Faller RV and Bowman WD, J. Dent. Res. **71**, 929–33 (1992).
9. D. Huang, E.A. Swanson, C.P. Lin, J.S. Schuman, W.G. Stinson, W.Chang, M.R.Hee, T. Flotte, K. Gregory, C.A. Puliafito, J.G. Fujimoto Science **254**, 1178 (1991).
10. Roland Brandenburg, Bernd Haller and Christoph Hauger. Optics Communications **227**, 203–211 (2003).
11. B.E. Bouma, G.J. Tearney (Eds.), Handbook of Optical Coherence Tomography, Marce l Dekker, NewYork, p. **41** (2002).
12. J.M. Schmitt, IEEE J. Select. Top. Quantum Electron. **5**, 1205 (1999).
13. A.F. Fercher, J. Biomed. Opt. **1**, 157 (1996).
14. B.E. Bouma, G.J. Tearney (Eds.) and Marcel Dekker, Handbook of Optical Coherence Tomography, Marce l Dekker, New York, p. **385** (2002).
15. J.G. Fujimoto, C. Pitris, S.A. Boppart and M.E. Brezinski, Neoplasia **2**, 9 (2000).
16. C.A. Puliafito, M.R. Hee, C.P. Lin, E. Reichel, J.S. Schuman, J.S. Duker, J.A. Izatt, E.A. Swanson, J.G. Fujimoto, Ophtalmology **102**, 217 (1995).
17. A.G. Podoleanu, J.A. Rogers, D.A. Jackson and S. Dunne, Opt. Exp. **7**, 292(2000).
18. J. Welzel, Skin Res. Technol. **7**, 1(2001).
19. A. Pagnoni, A. Knuuttel, P. Welker, M. Rist, T. Stoudemayer, L. Kolbe, I. Sadiq, A. Kligman, Skin Res. Technol. **5**, 83 (1999).
20. N.D. Gladkova, G.A. Petrova, N.K. Nikulin, S.G. Rade nskaLopovok, L.B. Snopova, Y.P.Chumakov, V.A.Nasonova, V.M. Gelikonov, G.V. Gelikonov, R.V. Kuranov,A.M. Sergeev, F.I. Feldchtein, Skin Res. Technol. **6**, 6 (2000).
21. G. Zuccaro, N. Gladkova, J. Vargo, F. Feldchtein, E. Zagaynova, D. Conwell, G. Falk, J. Goldblum, J. Dumot, J. Ponsky, G. Gelikonov, B. Davros, E. Donchenko, J. Richter, Am. J. Gastroenterol. **96**, 2633 (2001).
22. A.V. Shakhov, A.B. Terentjeva, V.A. Kamensky, L.B.Snopova, V.M. Gelikonov, F.I Feldchtein, A.M. Sergeev, J. Surg. Oncol. **77**, 253 (2001).
23. J.G. Fujimoto, S.A. Boppart, G.J. Tearney, B.E. Bouma,C. Pitris, M.E. Brezinski, Heart **82**, 128 (1999).
24. R. Heermann, C. Hauger,P.R.Issing,T. Lenarz, Laryngo Rhino-Otol. **81**, 400 (2002).
25. F.I. Feldchtein, G.V. Gelikonov, V.M. Gelikonov, R.R.Iksanov, R.V. Kuranov, A.M. Sergeev, N.D. Gladkova,M.N. Ourutina, J.A. Warren, D.H. Reitze, Opt. Expr. **32**, 39 (1998).
26. Yoshiki Oshida and Cory B. Sellers, Materials Science and Engineering C **25**, 343-8 (2005).

SPECTRAL STABILITY OF CDSE/ZNS QUANTUM DOTS

CHENG CHENG, QINGHAO ZHANG*, HAIZHEN YAN

College of Science, Zhejiang University of Technology, Hangzhou, 310023, China

The photoluminescence (PL) and photoluminescence excitation (PLE) spectra of CdSe/ZnS nanocrystal are measured using UV-visible spectrophotometer and based on fluorescence spectroscopy we discuss the effects of excitation wavelength, concentration, temperature and deposited period on the spectra. With a long excitation wavelength, the peak wavelength of emission band shows a slight red left (4nm) related to the size distribution of the nanocrystal. The absorption peak wavelength is kept unchanged, and its intensity is enhanced as the concentration of CdSe/ZnS solution increases. A comparison of spectra of CdSe/ZnS nanocrystal carried out at interval of three months shows little change. With the increased temperature of the toluene solution from 333K to 373K, the first absorption peak wavelength redshifts ~1nm, while the emission peak wavelength redshifts ~3nm.

1. Introduction

Recently, the interesting size-dependent optical, electronic, and magnetic properties of semiconductor nanocrystals have been studied extensively [1-4]. They have been widely used for applications, such as biomedical tags, light-emitting diodes, laser, and solar sells. For the II–VI quantum dots (QDs), the CdSe quantum dots prepared by chemical methods are most famous. In fact, the study of quantum dots exactly began from the investigation of CdSe quantum dots enchased in glass in the early 1990s. With the improvement of synthesis method, the size distribution can be controlled accurately, and the emission wavelength from quantum dots can cover a wide range from the blue-green to the red. To improve the optical properties of CdSe quantum dots, the CdSe quantum dots are often covered with ZnS shell [5][6]. The change of external condition will affect the optical properties of CdSe/ZnS quantum dots. A detailed knowledge of the spectral stability of quantum dots is therefore fundamental for future applications of CdSe/ZnS quantum dots in optoelectronic and biological fields. In general, there are many factors that effect the spectral stability of CdSe/ZnS quantum dots, such as pH value of solution as quantum dots used for biomedical tags [7], different gaseous environment [8-9], and temperature of experimental environment [10][11]. However, the effect of change of excitation wavelength on PL spectra of CdSe/ZnS quantum dots is not reported up to now.

In this work, we measure the photoluminescence (PL) and photoluminescence excitation (PLE) spectra of CdSe/ZnS QDs and discuss the effects of excitation wavelength, concentration and solution temperature on the spectra. The effect of defect states in ZnS shell on PL spectra is also investigated.

2. Experiment

2.1. *Material*

Ethanol (in concentration of 99.7%); toluene (in concentration of 99.5%); distilled water; the CdSe/ZnS QDs with diameter of 3.2nm dispersed uniformly in toluene solution, made by Evident Corporation. The concentration of CdSe/ZnS QDs solution is about 1.3mg/ml.

2.2. *Spectral Measurement*

The absorption spectrum is measured by UV-2550 (made by Shimadu, in Japan).By using the UV-2550, the first step is to obtain the absorption spectra of toluene. The second step is to measure CdSe/ZnS QDs solution absorption spectra in a given condition, such as different concentration or different solution temperature. Finally, we deduct background from CdSe/ZnS QDs solution absorption spectra and get absorption spectra of CdSe/ZnS QDs. The emission spectrum is measured by RF-5301(made by Shimadu, in Japan). All of the experiments are carried out at room temperature.

*Corresponding authors. Tel: +86 13588484634.
E-mail addresses: ppdan2005@sina.com.

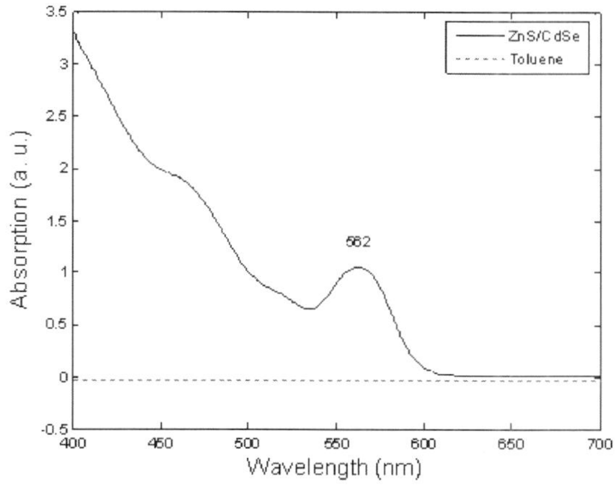

Fig. 1. Absorption spectra of the CdSe/ZnS quantum dot.

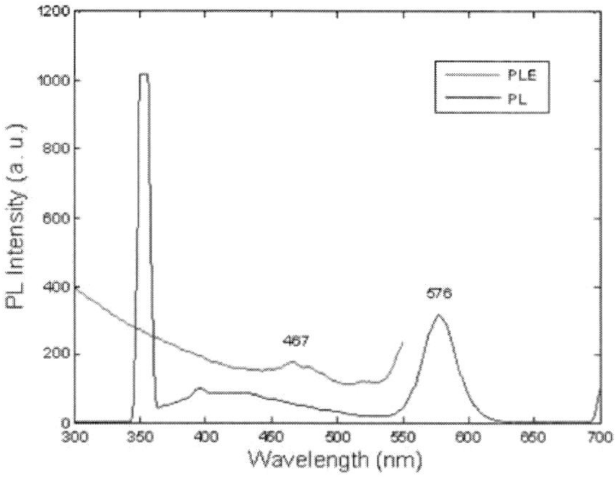

Fig. 2. PL and PLE spectra of the CdSe/ZnS quantum dot.

3. Results and Discussions

The real line in Fig.1 is absorption spectra of CdSe/ZnS QDs and the dashed is absorption spectra of toluene. From Fig.1, we can see the toluene background has little absorption between 400nm and 700nm and CdSe/ZnS QDs has a absorption peak around 550nm. The peak wavelength is about 562nm. The absorption of CdSe/ZnS QDs is enhanced dramatically as the wavelength of incidence light turns to shortwave.

In Fig.2, the excitation wavelength of PL spectra (real line) is about 350nm and first emission peak (around 350nm) in PL spectra results from the excitation light has not been absorbed wholly by QDs and arrives at detector disturbing spectra, therefore, the actual emission peak is located between 500nm-600nm.The peak wavelength is about 576nm with 30nm of FWHM. There is a small peak around 400nm in PL spectra because of the defect states in surface of ZnS shell. The luminescence mechanism of defect states: in the ZnS shell, electron is excited to transit from valance band to conduction band or defect states; when excitation energy is small than the gap between conduction band and valance band, excited electron could not transit to conduction band directly and should be captured by defect states in ZnS shell; finally, electron comes back from defect states through thermalization processes and gives an emission. At present, many researchers have investigated the defect luminescence of ZnS, as shown in reference [12-14]. The excitation wavelength of PLE spectra (dashed) is variable in range of 300nm to 550nm and the measured wavelength is 580nm. From Fig.2, it can be observed in the PLE spectra at 476nm and round 500, which are related to $\mathbf{1}S_{3/2}-\mathbf{1}S_e$ and $\mathbf{1}P_{3/2}-\mathbf{1}P_e$ absorption transitions in the CdSe core.

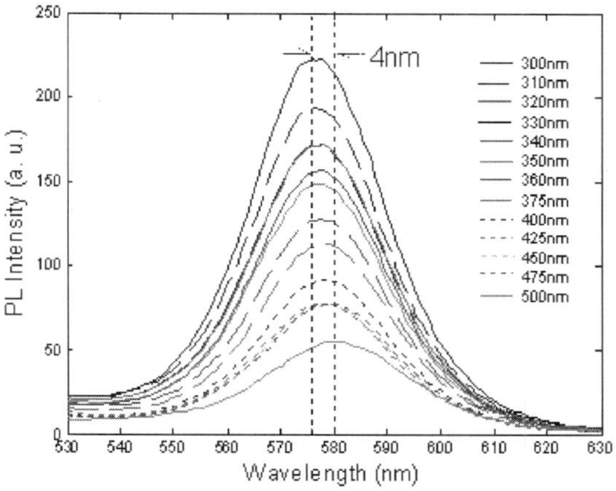

Fig. 3. Measured PL spectra of toluene in different excited wavelengths.

In Fig.3, it is shown that the emission peak wavelength of CdSe/ZnS QDs has a slight move to red, about 4nm, and the intensity of emission peak is decreasing, as the excitation wavelength is increasing from 320nm to 500nm. The main reason results in red shift: the red shift of emission peak wavelength is related to the size distribution of QDs; when excitation energy is equal to energy of the biggest QDs (the smallest optical gap), only these biggest QDs are excited and give contribution to emission line; as excitation energy is increasing and is being to be equal to energy of the smallest QDs (the biggest optical gap), all the QDs should be excited and give contribution to broadening emission line, therefore, superposed emission of QDs are inhomogeneous and the intensity of emission peak is enhancing; in this case, inhomogeneous superposed emission would bring intensity-wavelength relation to be restructured, and form a new emission peak at the biggest intensity location; when excitation energy is increasing (the excitation wavelength is decreasing), more small size QDs are exciting and enhance the intensity of short wave band of superposed emission, thus, emission peak shift to short wave slightly.

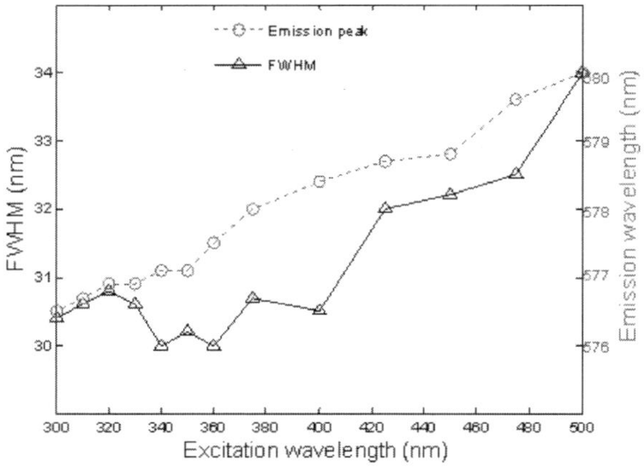

Fig. 4. Emission peak wavelength (empty dots) and FWHM (empty triangles) as a function of the Excitation wavelength.

To describe the effect of excitation wavelength on emission peak of QDs more clearly, we draw out the connection sketch map of excitation wavelength, emission peak wavelength and FWHM, as shown in Fig.4. When excitation wavelength increasing from 320nm to 550nm, the FWHM has a trend of augment, from 30nm to 34nm, did the emission peak wavelength shifts red slightly.

To discuss the effect of excitation wavelength on emission peak of QDs for a further step, we analyse it by energy band theory. A potential energy of zero inside the CdSe core and barrier height of 0.7 eV are assumed[13][16], as shown in Fig.5. When the excitation energy is between 2.2 eV and 3.6 eV, only electron-hole pairs inside CdSe core or carriers would be excited to the energy levels related to the ZnS shell and due to thermalization process in both cases redistributed to the lowest energy level in the CdSe core (which gives the emission at 576nm). However, when the excitation energy is higher than 3.6 eV, created electron-hole pairs in the ZnS shell can recombine nonradiativelly to the defect states and then radiatively giving an emission at 400nm, or both carriers can recombine nonradiatively into CdSe core ground states and then radiatively giving the emission at 576nm.

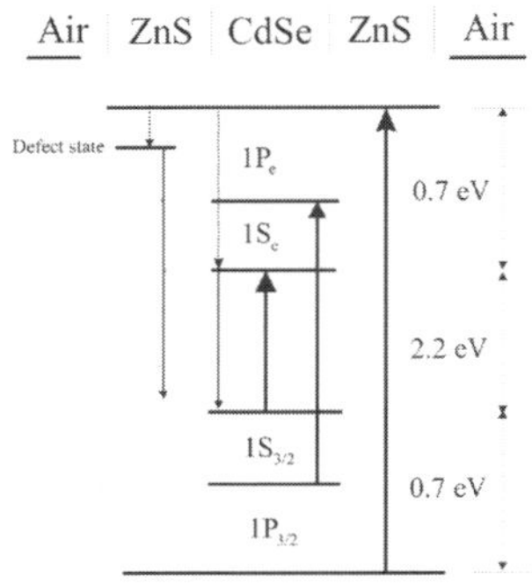

Fig. 5. Energy diagram of the CdSe/ZnS nanocrystal with the absorption (bold arrows), radiative (thin arrows) and nonradiative (dashed arrows) recombination channels.

The concentration of QDs solution has a direct effect on the absorption or emission peak intensity. According to Beer-Lambert law, absorbency of QDs can be written out as $A = \varepsilon C L_d$, where ε is extinction coefficient per mole of the QDs, C is the molar concentration (mol/L) of the QDs, and L is the path length (cm) of the radiation beam used for recording the absorption spectra. In the course of experiment, ε and L is constant. The comparative intensity of absorption and emission peak is rising respectively as C is increasing, as shown in Fig.6 and Fig.7. The wavelength of absorption and emission peak is both keeping unchanged, while the concentration of QDs solution is increasing. It is demonstrated that QDs do not aggregated as the concentration of QDs solution is increasing.

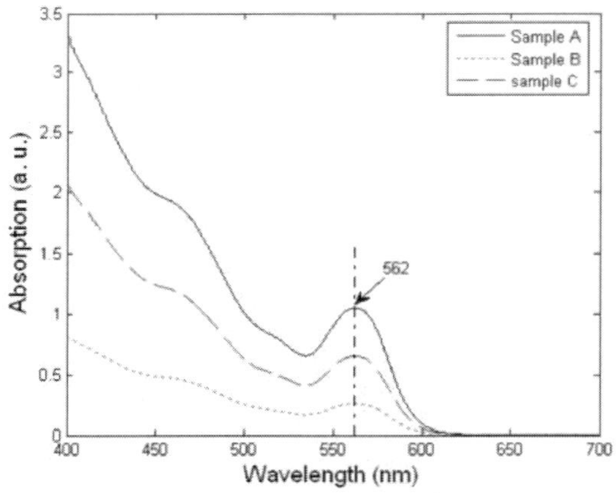

Fig. 6. Measured absorption spectra of the CdSe/ZnS quantum dots with different concentration. Sample A: $1.24\times10^{-1}\,mg/ml$, sample B: $0.31\times10^{-1}\,mg/ml$, sample C: $0.775\times10^{-2}\,mg/ml$.

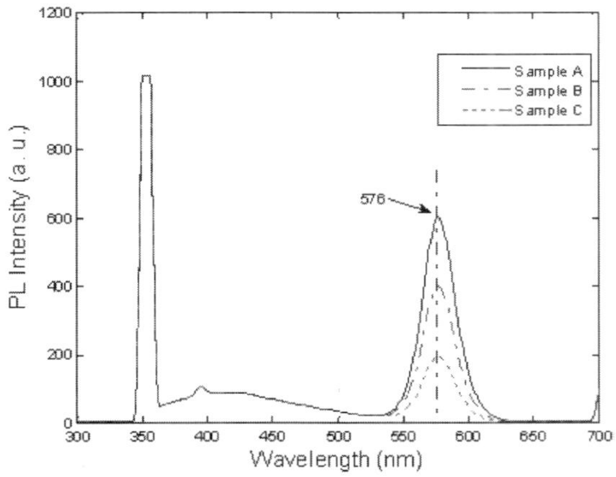

Fig. 7. Measured PL spectra of the CdSe/ZnS quantum dots with different concentration. Sample A： $1.24\times10^{-1} mg/ml$ ，sample B: $0.31\times10^{-1} mg/ml$ ，sample C： $0.775\times10^{-2} mg/ml$.

We heat CdSe/ZnS QDs solution from 333K to 373K, and measure the spectra every 5K, as shown in Fig.8 and Fig.9. The absorption peak of CdSe/ZnS QDs is altered rarely as solution temperature is rising, demonstrating weak effect of solution temperature on optical absorption of QDs. We find the emission peak of QDs is shifting slowly to red and the intensity is declining, when solution temperature is ascending from 333K to 373K. To explain this experimental phenomenon, we utilize Varshni relation, which gives the temperature dependence of the energy gap of bulk semiconductors,

$$E_g(T) = E_g(0) - \frac{\alpha T^2}{T+\beta} \quad (1)$$

where $E_g(0)$ is the energy gap at 0 K, α is the temperature coefficient, and the value of β is close to the Debye temperature θ_D of the material. Experimental data of PL spectra is taken into Eq. (1) and approximately solution is worked out, $\alpha = (2.97 \pm 0.24)10^{-4}$ eV ， $\beta = 278 \pm 30$ K ， which is consistent with the value in reference [11] for bulk CdSe: $\alpha = (2.81 \sim 4.1)10^{-4}$ eV , $\theta_D = (181 \sim 315)$ K . It is illuminated that the red shift of QDs is result from shrink of band gap as solution temperature is ascending. The integrated PL intensity decreases as solution temperature increases from 333K to 373K due to the activation of some nonradiative process, such as radiative relaxation, Auger nonradiative scattering, Forster energy transfer between dots of different dimensions, and thermal escape from the dot.

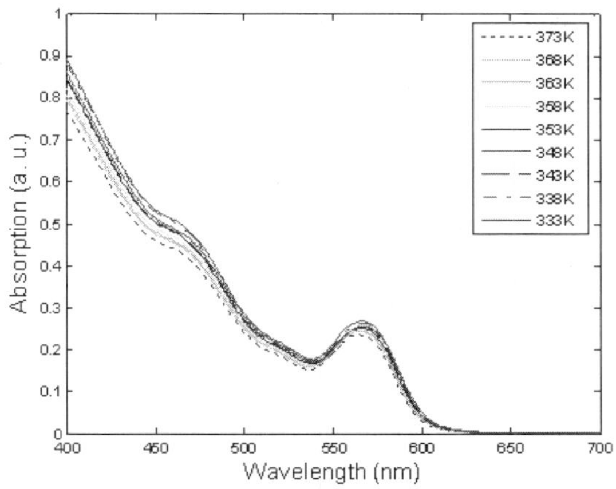

Fig. 8. Measured absorption spectra of the CdSe/ZnS quantum dot at different temperatures.

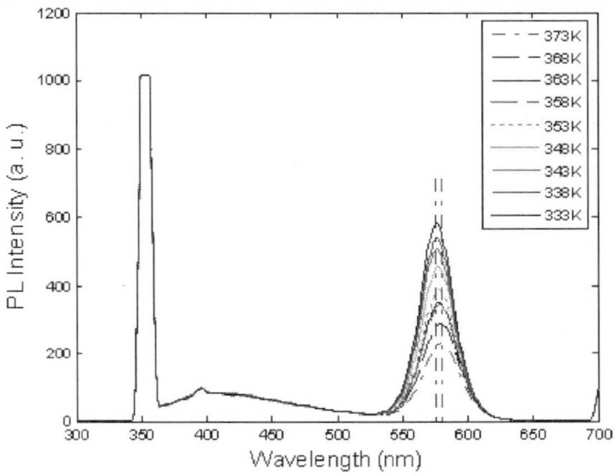

Fig. 9. Measured PL spectra of the CdSe/ZnS quantum dot at different temperatures.

4. Conclusions

In conclusion, photoluminescence (PL) and photoluminescence excitation (PLE) spectroscopy measurements of CdSe/ZnS QDs have been carried out to investigate its optical properties. Effect of excitation energy, concentration and temperature on optical spectra is measured and discussed. The results indicated: When excitation wavelength is longer, the peak position of emission band has a slight red left (4nm) related to the size distribution of QDs. There is a small emission peak in PL spectra is the result of the defect states in the ZnS shell. The first absorption peak shifts obviously and intensity is enhancing as the concentration of CdSe/ZnS solution is increasing. A temperature increase of 333K to 373K in QDs solution, the first absorption peak shifts rarely and the emission peak (around 576nm) moves 3nm to red. These results clarify the properties of light emission from QDs, and are important for future applications of QDs in organic or inorganic photonics and optoelectronic devices.

References

1. P. Bhattacharya, S. Ghosh, A.D. Stiff-Roberts, Annu. Rev. Mater. Res. **34**, 1 (2004).
2. Cheng Cheng, Hang Zhang, Chinese Physics, **55 (8)**, 4139 (2006).
3. FengYuan Zhao, Chunling Zhang, ZhanGuo Wang, Physics. **5**, 327 (2004).
4. ShuMan Liu, Zheng Xu, H.Wageh, Spectroscopy and Spectral Analysis. **6**,908 (2006).
5. Kortan A R, Hull R, Opila R L, J. Am. Chem. Soc. **112(4)**, 1327 (1990).
6. XIE Ying XU, Jing-Juan, YU Jun-Sheng, CHEN Hong-Yuan, Chinese Journal of Inorganic Chemistry. **20 (3)**, 663 (2004).
7. J. Aldana, N. Lavelle, Y. Wang, X. Peng, J. Am. Chem. Soc, **127(8)**, 2496 (2005).
8. Ozasa, K.; Nemoto, S.; Maeda, M.; Hara, M., J. Phys. Chem. B. **108 (18)**, 5507 (2004).
9. Shu, GW et al, Ieee. T. Nanotechnol. **4 (5)**, 632 (2003).
10. Quanqin Dai , Yanli Song et al, Chem. Phys. Lett. **439**, 65 (2007).
11. D. Valerini, A. Cretí et al, Phys. Rev. B. **71**, 235409 (2005).
12. W. G. Becker and A. J. Bard, J. Phys. Chem. **87**, 4888 (1983).
13. J. Bleuse, S. Carayon, and P. Reiss, Physica E. **21**, 331 (2004).
14. S. Yanagida, M. Yoshiya et al, J. Phys. Chem. **94**, 3104 (1990).
15. Al. L. Efros and M. Rosen, Annu. Rev. Mater. Sci. **30**, 475 (2000).
16. B. O. Dabbousi, J. Rodriguez-Viejo et al, J. Phys. Chem. B. **101**, 9463 (1997).

NEAR-INFRARED OPTICAL TOMOGRAPHY IMAGE RECONSTRUCTION APPROACH BASED ON 2-LAYERED BP NEURAL NETWORK

LI TING†

Department of Biomedical Engineering, Nanjing University of Aeronautics and Astronautics
No.29 Yudao Street, Nanjing 210016, P.R.China

LI WEITAO

Department of Biomedical Engineering, Nanjing University of Aeronautics and Astronautics
No.29 Yudao Street, Nanjing 210016, P.R.China

QIAN ZHIYU

Department of Biomedical Engineering, Nanjing University of Aeronautics and Astronautics
No.29 Yudao Street, Nanjing 210016, P.R.China

An image reconstruction approach for optical tomography is presented, in which a 2-layerred BP neural network is used to distinguish the tumor location. The inverse problem is solved as optimization problem by Femlab software and Levenberg-Marquardt algorithm. The concept of the average optical coefficient is proposed in this paper, which is helpful to understand the distribution of the scattering photon from tumor. The reconstructive $\overline{\mu}_s'$ by the trained network is reasonable for showing the changes of photon number transporting inside tumor tissue. It realized the fast reconstruction of tissue optical properties and provided optical OT with a new method.

Keyword: NIR OT, 2-layerred, BP neural network, inverse problem, the average optical coefficient

1. Introduction

Near-infrared optical tomography (NIR OT) is a medical imaging technique in which measurements of near-infrared light transmitted across the body are used to obtain images of optical properties of tissue. It has been emerging as a very promising diagnostic tool in a wide variety of biomedical fields[1,2], since it has the advantages of requiring relatively cost-effective, simple instrument, feasible accuracy and without any physical damage to biological tissues. It employs as a probing source, a near-infrared radiation in the range between 600 and 900 nm.

This technique retrieves the spatial dependent distributions of optical properties inside biological tissues by analyzing light intensities measured at boundary surfaces with a well-established optimization scheme. The difference in absorption and/or scattering between the normal and diseased tissues provides the imaging contrast for tissue diagnostics [3]. In past decades, various algorithms [4-6] have been introduced to enhance the NIR OT technique and great efforts were made to develop more accurate, faster algorithms. However, the development of a reconstruction scheme still remains a challenging problem to be solved before it can be used on extensive basis.

A new image reconstruction algorithm based on 2-layerred BP (back-propagation) neural network (TLBPNN) to NIR OT is proposed in this paper. TLBPNN can be calculated in parallel, so the scale of calculation can be decreased sharply. The more faster and accuracy results can be got by this new method.

2. Theory

2.1. *The Average Optical Coefficient*

The path of light transport in the tissue is based on the optical properties, which are absorption coefficient μ_a (cm^{-1}), scattering coefficient μ_s (cm^{-1}), the anisotropic parameter g . The reduced scattering coefficient can be defined as $\mu_s' = (1-g)\mu_s$ (cm^{-1}). g is always been recognized as 0.9 in experience.

The model of the light transporting in the tissue is shown in the Fig 1(a). Because of the different tissue

† Work supported by grant 30671997 of National Nature Science Foundation of China.
Corresponding author. Zhiyu Qian Email:zhiyu@nuaa.edu.cn

refractive index, the intensity is attenuated when the light goes through the tissue. I_0 is the intensity of line light, I is the intensity of the attenuated light, and x is the scattering distance. In the uniform tissue without considering absorption, the definition of μ_s is [7]:

$$I = I_0 e^{-\mu_s x} \qquad (1)$$

Then we give the definition of average scattering coefficient $\bar{\mu}_s$ (cm^{-1}). When there is a tumor inside the tissue, the intensity of the attenuated light is different from which in the uniform tissue, but we can find a average μ_s, as $\bar{\mu}_s$. The model of the light transporting in the tissue with tumor is shown in Fig 1(b). According to the definition of Eq.(1) , the $\bar{\mu}_s$ is defined as:

$$I = I_0 e^{-\bar{\mu}_s x} \qquad (2)$$

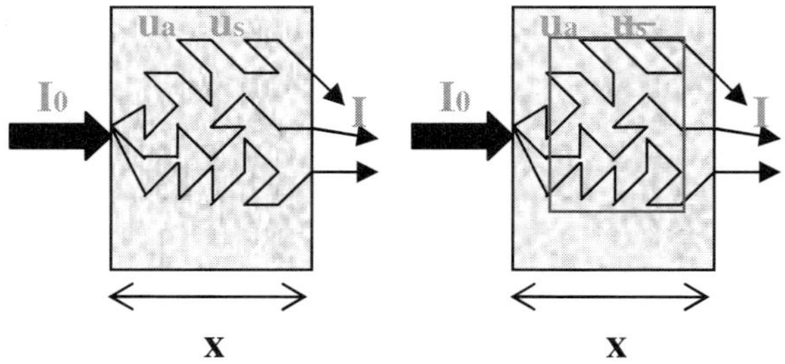

Fig 1 (a) Scattering attenuation of light through the uniform tissue; (b) Scattering attenuation of light through the tissue with tumor

Therefore, according to the same way, we can define the other average optical coefficients, such as $\bar{\mu}_s'$, $\bar{\mu}_a'$.

2.2. *Photon Transport Function*

If in the high scattering tissue media ($\mu_s >> \mu_a$) and the light source is isotropic and continuous intensity, photon transport can be described by steady-state Boltzmanm equation [8] given by

$$-\nabla \cdot \kappa(r)\nabla\Phi(r) + \mu_a(r)\Phi(r) = q \qquad (5)$$

$\Phi(r)$ is the photon density, $\kappa(r) = 1/[3(\mu_a + \mu_s')]$, q is isotropic light source.
The boundary condition should be satisfied the Dirichlet condition, $\Phi(boundary) = 0$.

2.3. *Model for Average Optical Coefficient*

Set a simulated models are shown in the Fig 2. Assumption μ_a and μ_s' of the background are 0.01cm^{-1} and 5cm^{-1} respectively, and μ_a and μ_s' *of* the tumor region are 0.01 cm^{-1} and 10 cm^{-1} respectively. μ_a in the right model is 0.01 cm^{-1}, μ_s' varies from 5 to 15 cm^{-1}.

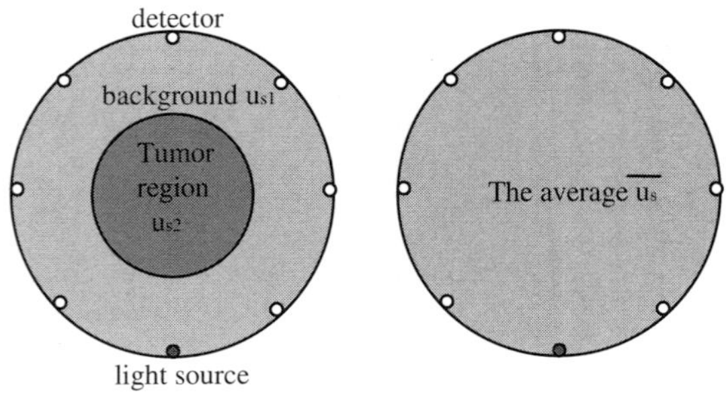

Fig 2 The simulated models

After simulated by Femlab software, as shown in the Fig 3, we can get the attenuated photon intensities I_0 and I_1. I_0 is collected from the left model and I_1 is from the right model of Fig 2. When arrive at minimum error, average μ_s' can be got, the $\bar{\mu}_s'$ is 5.86cm^{-1}. It gives the prove of definition of average optical coefficient.

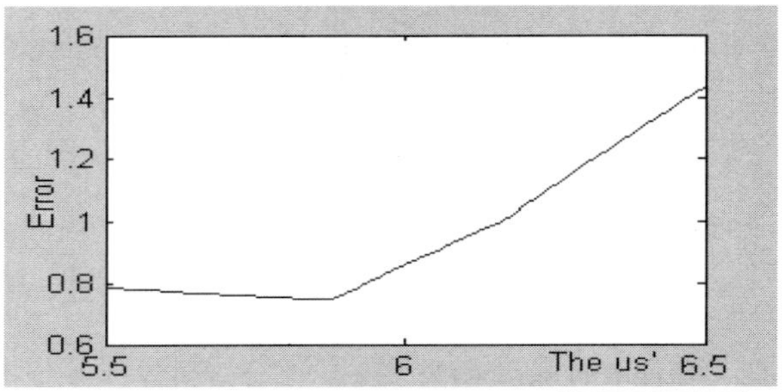

Fig 3 The error curve of I_1 and I_0, Error= $\sqrt{\Sigma(I_1 - I_0)^2}$

3. Reconstruction Method

3.1. *The 2-Layered Model*

The reconstructive object is to recognize the CO4 part which is different from the other seven parts, shown in the Fig 4. Since the reconstructive model has many parts so that the reconstruction is more complex and more slowly computational speed[9]. In this paper, two simpler models are introduced in the Fig 5. Model 1 is the circle tissue divided into four equal sectors as shown in the Fig 5 (a), Model 2 is the sectors rotated 45° as shown in the Fig 5 (b). When the μ_s' in part CO4 is different from the other parts in the Fig 4, then it makes part CO2 in the Fig 5 different from other parts, since its $\bar{\mu}_s'$ are different from others. Then according to the definition of the average scattering coefficient, if the same tissue shows the different optical properties distribution with the two models simulated, with the logic relationship, it's easy to find the tumor part in the reconstructive model.

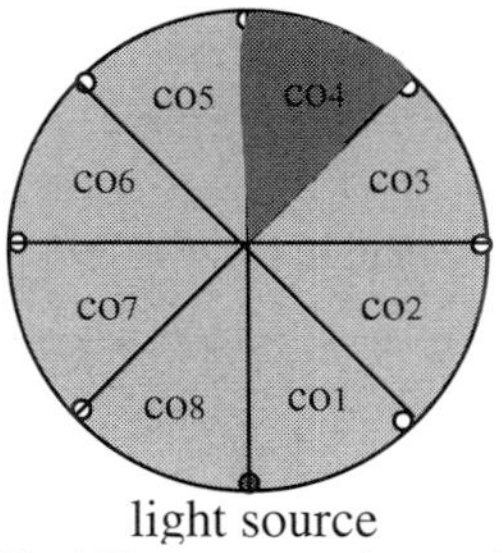

Fig 4 The reconstructive model

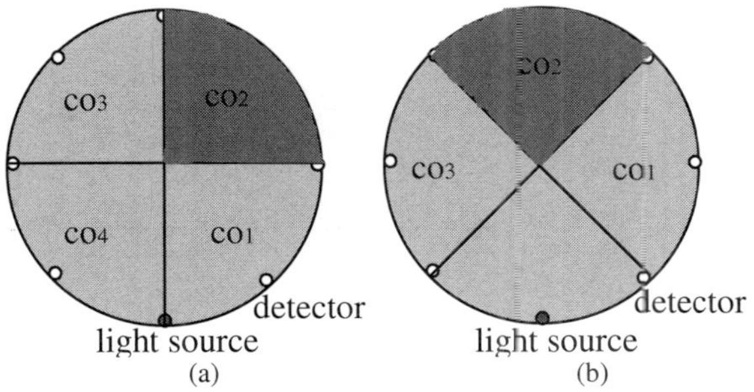

Fig 5 The simulated models (a)Model 1 (b) Model 2

3.2. *2-Layered BP Nerual Network*

The solution of inverse problem is based on the forward problem, so samples set of BP neural network can be obtained from the solver of forward problem. The framework of the two-layerred BP neural network in the training model is as shown in the Fig 6.

In the Fig 5, μ'_{s1}, μ'_{s2}, μ'_{s3} and μ'_{s4} are the reduced scattering coefficient of CO1, CO2, CO3 and CO4, respectively. Their value varies from 5 to 15cm^{-1}. The μ_a is 0.48cm^{-1} in part CO1–CO4. The forward method can be used to get the samples set.

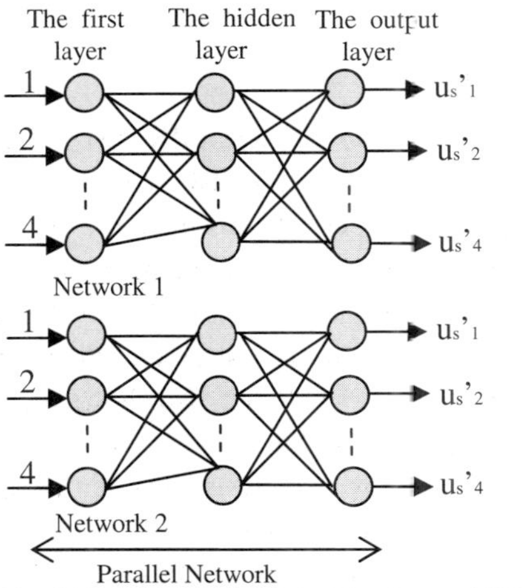

Fig 6 The framework of the Two-layerred BP Neural Network

3.3. *Optimization Algorithm*

The key of inverse problem is how to get reconstruct optimization algorithm, Levenberg-Marquardt (LM) algorithm, is presented in this paper to training BP network[10]. Training error was shown every 25 times and the network stopped training when $m \le 1000$ or $e_k \le 1e-6$.

4. Result and Discussion

4.1. *Reconstruction Results*

There are 1331 training samples which are used to train each of the two networks. After adjusting the network parameters and the training conditions, the trained networks can provide the optimal results. The test samples get from the reconstruction model shown in the Fig 4. The μ'_s in the part CO4 is 6, 7, 10, 15cm^{-1}(The meaning is the same in the tables whose value is compared with the reconstructive values). The others are 5cm^{-1} which is the background. Compared the model with Model 1 and 2 shown in the Fig 4 and 5, each part of the sector models corresponds with the two parts in the model. The reconstructive results from the two networks are shown in the Table 1 and 2. There are four cases in the test. It is obvious that μ'_{s2} of the part CO2 in the Model 1 and 2 is different from the other parts, which means there is a tumor in it.

Table 1 The μ'_s (cm^{-1}) reconstruction map of the sector network

μ'_s	$\overline{\mu'}_{s1}$	$\overline{\mu'}_{s2}$	$\overline{\mu'}_{s3}$	$\overline{\mu'}_{s4}$
6	5.0000	5.5358	5.8316	4.7224
7	5.0000	5.8130	5.9681	4.6444
10	5.0000	6.4236	6.2306	4.5187
15	5.0000	6.8451	6.4418	4.4292

Table 2 The μ'_s (cm^{-1}) reconstruction map of the sector rotated 45° network

μ'_s	$\overline{\mu'}_{s1}$	$\overline{\mu'}_{s2}$	$\overline{\mu'}_{s3}$	$\overline{\mu'}_{s4}$
6	5.0000	6.4380	5.2874	5.1283
7	5.0000	6.7260	5.2507	5.1370
10	5.0000	7.2866	5.2019	5.1434
15	5.0000	7.6990	5.1879	5.1317

Combined with the result of the models, we can conclude the part CO4 shown in Fig. 4 is different from others.

4.2. *Discussion*

Observing the value in the Table 1, $\overline{\mu}'_{s3}$ has remarkable changes because the tumor is near the part CO3 and the scattered photon has made effort. In the Model 1, a portion of the scattered photon is transported into part CO3 which results in the corresponding fiber detector collecting more photon, then the reconstructive $\overline{\mu}'_{s3}$ is larger than 5cm^{-1}. But in the Model 2, the tumor is contained in the part CO2 which the scattered photon is transported in the other segment, then the influence is much small than the Model 1. The other parts can also be discussed in the same reason. Therefore, the distance between the tumor and the around tissue segment decides the reconstructive effort of $\overline{\mu}'_{s3}$ in other parts. Similarly, the effort also depends on the distance from the light source to the tumor. It's obvious that the amount of the photon transported into the tissue near the light source is much more than that far away from the light source.

According to the results reconstructed by the network, it's easy to find the location of the tumor. In order to prove the results , it's reasonable method to compare the test samples with the photon intensity collected from the sector models when μ'_s varying from 5 to 15cm^{-1} in their part CO2 and others are 5cm^{-1}. The result is shown in the Table 3.

Table 3 The μ'_s (cm^{-1}) distribution map of the sector rotated 45° network

μ'_s	$\overline{\mu'}_{s1}$	$\overline{\mu'}_{s2}$	$\overline{\mu'}_{s3}$	$\overline{\mu'}_{s4}$
6	5.0000	5.4474	5.0000	5.0000
7	5.0000	5.8043	5.0000	5.0000
10	5.0000	6.5100	5.0000	5.0000
15	5.0000	7.0439	5.0000	5.0000

When μ'_s increased, the variable curve of $\overline{\mu}'_{s2}$ in the Table 2 and 3 is shown in Fig 7. We can see the trend of two curves is similar, the difference is the size of numerical value which validates the definition of the average optical coefficient. When the tumor with more strongly scattered than the normal tissue, the value of the μ'_s is larger. In this paper, it's only used the data of the sector rotated 45 ° model to study. Hence, the reconstructive results are credible. The error shown in the Fig 8, $\overline{\mu}'_{s2}$ is larger, the error is smaller. Because the μ'_s in the tumor has obvious difference with the background, the scattering function is strong and the photon intensity is distinct.

Fig 7 The trend of the average reduced coefficient

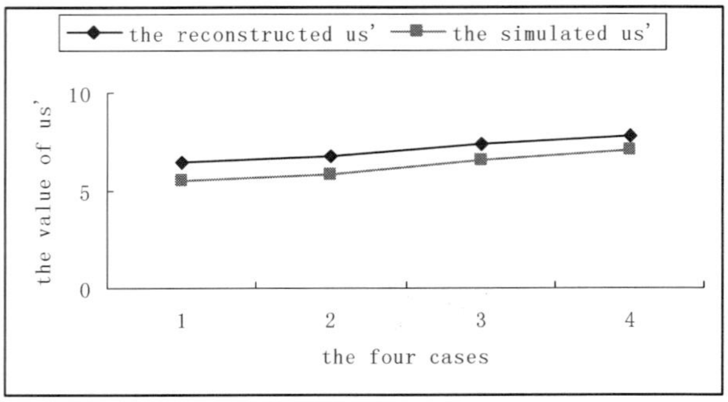

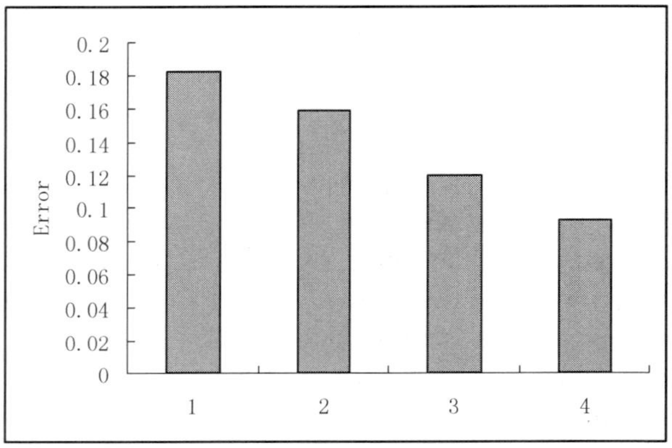

Fig 8 The error between the reconstructive $\overline{\mu}'_{s2}$ and the simulated μ'_s

5. Conclusion

The primary study of the NIR OT reconstructive technique by two-layerred BP neural network was presented in this paper. The inverse problem is solved as optimization problem by the Femlab and Levenberg-Marquardt algorithm. The network can distinguish the location of tumor from the tissue, and the reconstructive $\overline{\mu}'_{s2}$ is not only reasonable but also shows the changes of the transporting photon inside tissue with tumor. According to above analysis, the influence on the reconstructive effort depends on the distance of the location of the tumor and the light source. In this paper, there is only one fiber transmitting the light. The advantage is the measurement instrument which is very simple, and it is useful for the application of the OT real-time. When the μ_s' in the tumor is close to that around the tissue, the reconstructive error is about 18%. It can be improved by using more fibers transporting the light into the tissue.

The scale of calculation trained the network can be decreased sharply because the complex model having 8 parts was divided into two simple Model 1 and 2 with 4 parts. Assumption each part needed the scale of calculation be N, the reconstructive model with 8 parts is N^8, while the model with 4 parts is N^4, then the two models need $2N^4$. The predominance of the fast method is obvious when the reconstructive tissue is more complex. When the network has been trained, the speed of the reconstructive optical coefficient is only several seconds. It realized the fast reconstruction of tissue optical properties and provided optical OT with a new method.

The future work is needed to reconstruct the model with more parts. Because there are relationship between μ_s' and the temperature of biological tissue, so this method will be application in the temperature measurement and the coagulation of the biological tissue[9].

Acknowledgments

This paper was supported by National Nature Science Foundation of China (Grant No.30671997). All the work has been done at the Biophotonics Laboratory of NUAA (www.nuaaphotonics.com).

Reference

1. Bluestone AY, Abdoulaev G, Schmitz CH, Barbour RL, Hielscher AH. Three-dimensional optical tomography of hemodynamics in the human head. Opt Express 2001;9:272 – 86.
2. Nioka S, Yung Y, Shnall M, Zhao S, Orel S, Xie C, et al. Optical imaging of breast tumor by means of continuous waves. Adv Exp Med Biol 1997; 411: 227 – 32.
3. Y. Xu, N. Iftimia, H. Jiang, L. Key, M. Bolster, Opt. Exp.8 (2001) 447.
4. Klose AD, Hielscher AH. Optical tomography using the time-independent equation of radiative transfer—Part 2: inverse model. JQSRT 2002; 72: 715 – 32.
5. Arridge SR, Schweiger M. A gradient-based optimisation scheme for optical tomography. Opt Express 1998;2:213 – 26.
6. Jing Meng, Jiajun Wang, Xianwu Huang, Renjin Liu. Optical tomography reconstruction based on gradient tree: Journal of Quantitative Spectroscopy & Radiative Transfer 102 (2006) 181 – 189.
7. Zhiyu Qian, Xinwei Wang, Hanli Liu. Study of Near Infrared Guidance Technique for 3D Targeted Localization in Neuro-surgery for Parkinson's Disease.
8. Weitao Li, Zhiyu Qian, Huinan Wang. Image Reconstruction Algorithm of Near Infrared Optical Tomography. Chinese Optical Letters Supplement, 2005, 3: 237-238.
9. Weitao Li, Zhiyu Qian, The Research on the Relationship Between the Reduced Scattering Coefficient and Temperature of the Tissue by NIRS Light Scattering. International Symposium on Biophotonics, Nanophotonics and Metamaterials, 2006,10, 16~18.
10. Qiong Wu, Zhiyu Qian, An Optimization Algorithm to Inverse Problem in 2-D Optical Computed Tomography by BP Neural Network. International Symposium on Biophotonics, Nanophotonics and Metamaterials, 2006,10, 16~18.

STUDY ON PORPHYRIN PHOTOBLEACHING PROPERTIES IN PHOTODYNAMIC THERAPY

SHUANG WANG
Department of Physics, Northwest University, Xi' an 710069, P. R. China

XIAOJUN TIAN
Department of Physics, Northwest University, Xi' an 710069, P. R. China

HONGFANG CHANG
Department of Physics, Northwest University, Xi' an 710069, P. R. China

QINGLI HE*
Department of Physics, Northwest University, Xi' an 710069, P. R. China

Based on the dynamic model of photobleaching in Photodynamic therapy , the Porphyrin Photobleaching properties has been studied via using Monte Carlo simulation. It showed that the rate of Porphyrin photobleaching in tumor was increased with the increase of the absorption coefficient and the scattering coefficient. For a given tissue, the photobleaching rate diminished in the thickness of 5mm under tumor light illuminated surface; less than 5mm, the rate reduced gradually; more than 5mm, the rate increased exponentially. That meant from 0 to 5 mm depth of tumor light illuminated surface, the photon was well concentrated, the rate of Porphyrin photobleaching was reduced, the tumor cell could be damaged entirely and effectively by Porphyrin.

1. Introduction

It is well known that Porphyrin-based photosensitizer used in Photodynamic therapy (PDT) can be degraded or modified by light[1-4]. This phenomenon called photobleaching is observed by loss of fluorescence of Porphyrin-type photosensitizer in tissue, solution and cells during light exposure[5, 6]. Many theories and models have been established to explain the photochemistry and photophysical characteristic during Porphyrin-type photosensitizer photobleaching by the method of studying the properties of florescent photoproduct formation.

In this paper, the Porphyrin photobleaching properties has been studied via using Monte Carlo simulation based on the dynamic model of photobleaching in Photodynamic therapy. The Monte Carlo model simulated optical energy deposition inside tissue [7] is used to incorporate into first order bleaching in this study. It offers a more straightforward and simple manner to reveal the photosensitizer concentration changing caused by bleaching. For the further studying, the photobleaching properties in the condition of different absorption coefficient and scattering coefficient are studied in this paper. This may predict the best treatment condition and bleaching rate differences in tissue for PDT using Porphyrin-based sensitizer.

2. The Dynamic Mechanism of Porphyrin Photobleachings

2.1. *The Mechanism of Photobleaching*

In mechanistic terms, the pathway of photosensitizer bleaching is indefinitely clear, as for porphyrin-type photosensitizer, however, two oxidative mechanisms are considered to be principally involved in the photosensitizer modification in tissue, called type I and type II.

Photochemical and Photophysical mechanism of Photobleaching in Photodynamic Therapy Type I mechanism

$$S_0 + h\gamma_e \rightarrow S_1$$
$$S_1 \rightarrow S_0 + h\gamma_f$$

and in Type II mechanism

$$S_0 \rightarrow T_1$$

$$T_1 +^3 O_2 \rightarrow S_0 +^1 O_2$$

$$S_0 +^1 O_2 \rightarrow products$$

$$biom +^1 O_2 \rightarrow products$$

In the type I mechanism the photosensitizer interacts with a bimolecular (or oxygen) with a resulting hydrogen transfer to produce radicals, namely florescent photoproduct formation. In the type II mechanism, an energy transfer between the excited triplet state of sensitizer T_1 and the ground state of oxygen $^3O_2$, generating the first excited state of oxygen, single oxygen $^1O_2$. This zwitterions species is extremely active and can interact with a large number of biological substrate and drug itself.

2.2. *The Photobleaching Kinetics in Spatially Uniform Case Tables*

The photobleaching of photosensitizer under influence of light in general case is described by the following equation [8]:

$$D = D_0 e^{-\beta J} \tag{1}$$

where D is photosensitizer concentration, D_0 is the initial concentration, J is fluence rate in tissue, and β is bleaching rate constant. If the light fluence is not attenuated, the change of fluorescence intensity can reflect the principle of photosensitizer concentration changes. And

$$J = \varphi t = n \cdot h\nu \tag{2}$$

where φ is the laser power density , t is time, $h\nu$ is the power of a photon and n is the photon absorption in tissue. So the change of sensitizer concentration can be described by:

$$\frac{dD}{dt} = k\frac{dD}{dn} = -D_0 \cdot \beta \cdot h\nu \cdot e^{-\beta \cdot n \cdot h\nu} \tag{3}$$

3. Monte Carlo Simulations

In this paper, Monte Carlo method, which was based on macroscopic optical properties that were assumed to extend uniformly over small unites of tissue volume, is used to simulate the optical energy deposition inside tissue. Three kinds of coordinates systems are established to trace photon propagation in tissue[9].

A Cartesian coordinate system, whose origin is the photon incident point on the tissue surface, is used to trace photon movements. The z-axis of it is the normal of the surface pointing toward inside of the tissue; and xy-plane is therefore on the tissue surface. Because the infinitely narrow photon beam is perpendicular to the tissue surface, the problem has cylindrical symmetry. Therefore, a cylindrical coordinate system is established to score internal photon absorption. It share the origin and z-axis with Cartesian coordinate system, and the r coordinate of the cylindrical system is used for the diffuse reflectance and diffuse transmittance as the function of r and α, where α is the angle between the photon exiting direction and the normal to the tissue surfaces. A moving spherical coordinate system, whose z axis is dynamically aligned with the photon propagation direction, is used for sampling of the propagation direction change of a photon. The deflection angle θ and the azimuthally angle, which are due to first sampled scattering, are updated in terms of the directional cosines in the Cartesian coordinates system. Finally, in order to score physical quantities, a two-dimensional grid-system is established in the r and z directions.

3.1. *Photon Launching*

The photon is injected orthogonally onto the tissue at the origin, and the position (x, y, z) is initialized to $(0,0,0)$, the direction cosines (μ_x, μ_y, μ_z) are initialized to $(0,0,1)$, and the weight is set to 1. The ideal Laser light intensity is of Gaussian distribution, but in the same semi-diameter, photon spatial distribution is a circular. Therefore, by the method of random sampling, the round homogeneous distribution and the normal distribution could be calculated where the round homogeneous distribution is

$$R_f = R\sqrt{\zeta_1}$$

and the normal distribution is

$$X_f = \sqrt{-2\ln\zeta_1}\cos(2\pi\zeta_2)$$

Considering the Laser beam radius is much smaller than the radius of tissue, by the method of setting 3 times standard deviation in software, the 99% of launched photo concentrated in the Gaussian distribution and concentrated in the area of 1mm spot, which is the enactment of Laser beam radius,

$$R_f = R\sqrt{\frac{|\zeta|}{3}}$$

where R_f is beam Circular distribution; ζ is the uniformly distributed random number; R is the light spot radius.

3.2. *Photon Random Propagation in Tissue*

When the photon arrived at the surface of tissue, not every photon can incident into tissue media because of specular reflectance If the refractive indices of the outside medium and tissue are n_1 and n_2, respectively, then the specular reflectance, R_{sp} is specified:

$$R_{sp} = \frac{(n_0 - n_1)^2}{(n_0 + n_1)^2}$$

Then the weight is decreased by R_{sp} for the photon entering medium: $w = 1 - R_{sp}$. The photon entered the tissue, move in the random step length, which is determined by:

$$s = \frac{-\ln \zeta}{\mu_t}$$

where ζ is a random number. Once the photon has reached an interaction site, a fraction of the photon weight, Δw, absorbed by the interaction site must be calculated:

$$\Delta w = w * \mu_a / \mu_t$$

After that, the photon with the updated weight is ready to be scattered. The deflection angle, θ ($0 \leq \theta < \pi$), must be sampled statistically. The probability distribution for the cosine of the deflection angle $\cos\theta$ is described by the scattering function, originally proposed by the Henyey and Greenstein:

$$\cos\theta = \left\{1 + g^2 - \left[(1-g^2)/(1-g+2g\zeta)\right]^2\right\}/(2g)$$

After many steps of interaction, the photon random propagation will be terminated when the photon weight decreased to the setting threshold value ($W_{th} = 0.0001$). The technique called Russian roulette is applied to terminate the photon when $w \leq W_{th}$.

4. Results and Discussion

4.1. *Porphyrin Concentration Difference in z-axis Direction Caused by Photobleaching*

Incorporated the results of MC simulation, the Photobleaching kinetics function could suggest the simple properties of Porphyrin concentration changing in 10 cm depth z-direction caused by Photobleaching in tumor model. The tissue optical parameters, used in the simulation, are shown in table 2[10].

Table 2. The tissue optical parameters applied in the simulation

Tissue optical parameters	Value
The absorption coefficient μ_a (mm^{-1})	0.38
The scattering coefficient μ_s (mm^{-1})	28
The anisotropy factor g	0.846
The Refractive index n	1.5

4.2. *Porphyrin Bleaching Rate Difference with Different Tumor Optical Parameters*

This study, not considering the oxygen concentration and local environment, is firstly analysis the Porphyrin concentration changes caused by photobleaching in the condition of different absorption coefficient and scattering coefficient. Figure 1 shows the changes of Porphyrin concentration in tissue during the Photodynamic process, with the increase of absorption coefficient μ_a (μ_a=0.1, 0.2, 0.4 cm^{-1}) and scattering coefficient μ_s (μ_s=20, 40, 60 cm^{-1}) in z direction 0 to 1cm depth respectively.

The rate of Porphyrin bleaching in tumor was increased with the increase of μ_a and μ_s. For the higher absorption coefficient, the photon has higher probability to be absorbed by tissue, and then the drug concentration has a more stationary changing tendency (Fig 1). By a contrast, the tissue with higher scattering coefficient, the photon is more likely scattered by the tissue cells, so in this kind of tissue, the tendency is much intensive. And because more photon is likely being scattering out of the tissue surface, the initial changing rate of photosensitizer is lower (Fig 2). In order to reach the best PDT treatments, the methods, such as solution injection, to modify the tissue optical absorption and scattering properties are needed.

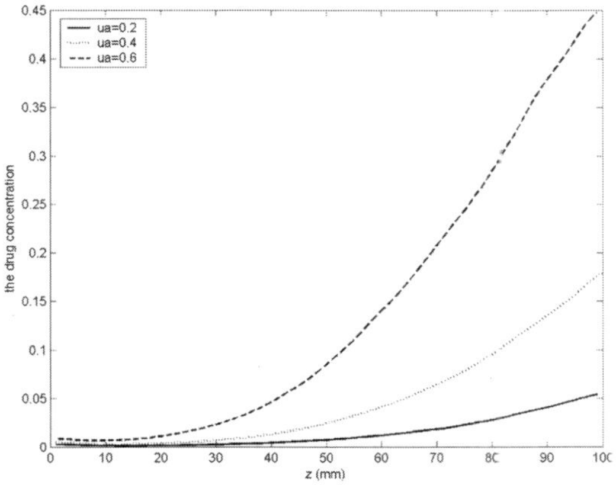

Figure 1. The bleaching rate difference with the increase of absorption coefficient in z direction

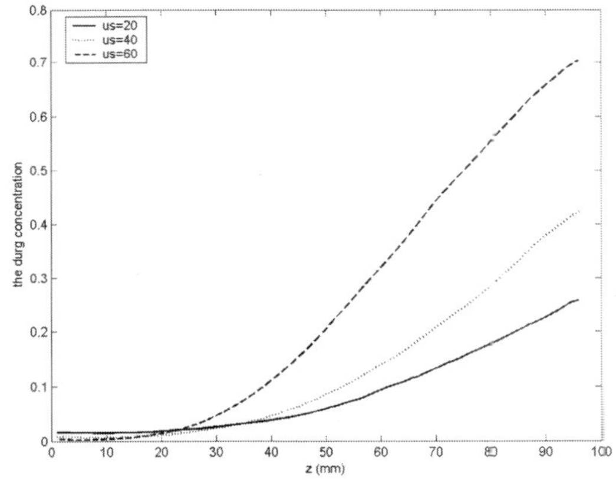

Figure 2. The bleaching rate difference with the increase of scattering coefficient in z direction

5. Conclusion

In this study, the Monte Carlo method is used to simulate the photobleaching properties of Porphyrin based photosensitizer. The results means from 0 to 5 mm depth of tumor light illuminated surface, the photon was well concentrated, the rate of Porphyrin photobleaching was reduced, and the tumor cell could be damaged entirely and effectively by Porphyrin. The chief advantage of this method is that the simple relationship between photosensitizer concentration and photon distribution can be revealed in a straightforward manner [11, 12]. But there are two weaknesses in this simulation. Firstly, according to the references [13, 14, 15], the results meet the experiments data in the early stage of PDT. However, they can not state the properties of photobleaching mediated predominantly by $^1O_2$ in type II mechanism. In this period, triplet state concentration and the aggregation of photon are highest, so the Photobleaching is mainly mediated by the sensitizer triplet reactions. Secondly, the Monte Carlo method offers a dynamic approach to photon distributes in tissue, however, it requires obtaining precise results of many photon histories and it may be expensive in computation time.

References:

1. Moan J, Christensen T and Jacobsen PB, *In: Doiron DR, Gomer CJ (eds) Porphyrin localization and Treatment of Tumor, New York: A Lan R; Liss.* Porhyrin-sensitized photoinactivation of cells in vitro. 1984:419-42.
2. Moan J, *Cancer Lett.* fect of bleaching of porphyrin sensitizers during photodynamic therapy, 1986.33: 45-53.
3. W. G. Roberts, K.M. Smith, J L McCullough and M.W. Berns, *Photochem Photobio.* Skin Photosensitivity and Photo destruction of Several Potential Photodynamic Sensitizers[J], 49: 431- 4 38, 1989
4. Spikes J D, *Photochemistry and photobiology.* Quantum yields and kinetics of the photobleaching of hematoporphyrin, photofin II, tetra-porfrin and urophofrin[J]. 1992, 55: 797-808.
5. Moan J, *Lasers Med Sci.* A change in the quantum yield of photoinactivation of cells observed during photodynamic treatment [J]. 1986, 3: 93–97.
6. Mang T S, Dougherty T J and Potter W R., *Photochemistry and photobiology.* Photobleaching of porphyrin used in photodynamic therapy and implications for therapy. 1987, 45: 501- 506.
7. Lihong Wang, Steven L.Jacques and Ligiong Zheng, *Computer Methods and Programs in Biomedicine.* MCML-Monte Carlo modeling of light transport in multi-layered tissues [J], 1995 (47): 131-146.
8. Alexander A Stratonnikov, Gennady A Meerovich, and Victor B. Loschenov. *SPIE.* Photobleaching of

photosensitizers applied for photodynamic therapy [M]. Vol. 3909 (2000).

9. Lu Yiquan and Li Zhengjia, *J Huazhong Univ of Sci & Tech.* Monte Carlo simulation of Light Transport in Multi-layered Biological Tissues [J]. 2000, 8(28): 325-327.
10. Peters VG, Wyman DR, Patterson MS and Frank GL, *Phys Med Biol.* Optical properties of normal and diseased human breast tissues in the visible and near infra-red. 1990; 35: 1317–34.
11. Svaasand L O and Ellingsen R, *Photochem Photobiol.* Optical penetration in human intracranial tumors [J]. 1985, 41: 73-76.
12. Whitehurst C, Pantelides M L and Moore JV, *J Urol.* In vivo laser light distribution in human prostatic carcinoma [J].1994, 151: 1411-1415.
13. Jarod C Finlay, Soumya Mitra, Michael S Patterson and Thomas H Foster, *Phys Med Biol.* Photobleaching kinectics of Photofrin in vivo and muticell tumor spheroids indicates two simulataneous bleaching mechanisms., 2004, 49: 4837-4860.
14. Alexander A Stratonnikov, Gennady A. Meerovich, and Victor B. Loschenov, *SPIE.* Photobleaching of photosensitizers applied for photodynamic therapy [M]. Vol. 3909 (2000).
15. Jonathan S. Dysart and Michael S. Patterson, *Photochem. Photobiol. Sci.* Photobleaching kinetics, photoproduct formation, and dose estimation during ALA induced PpIX PDT of MLL cells under well oxygenated and hypoxic conditions [J], 2006, 5; 73–81.

SIGNIFICANCE OF QUANTITATIVE DETERMINATION OF PROTOPORPHYRIN IX IN GASTRIC CANCER TISSUES BY HIGH PERFORMANCE LIQUID CHROMATOGRAPHY *

JUN MA*, XN ZHANG, YP ZHU, RE ZHENG †
Physics Department, Ocean University of China, 266100
Qingdao, China

WZ MAO
Department of General Surgery, Qingdao University Medical College Hospital, 266003
Qingdao, China

The utilization of the protoporphyrin IX (PpIX) as an indicator for the assistant cancer distinguishing tool was proved by its quantitative determination in gastric cancer tissues. Tissue homogenates were 40 in number and taken from the normal and abnormal gastric tissues in 10 gastric cancer patients. Emission at 628nm following excitation at 404nm was used to detect PpIX by using high performance liquid chromatography (HPLC). The concentrations of PpIX in the homogenized tissues were calculated according to the PpIX standard solution's linear regression equation and the differences among them demonstrate that the closer the tissue is to the cancer focus, the higher the concentration of PpIX in its homogenate is. The increasing concentration of PpIX in gastric cancer may be considered as an indicator that lays the foundation for the assistant determination of gastric cancer from gastric lining in futurity.

1. Introduction

Spectroscopy and fluorescence photography, including the laser-induced fluorescence (LIF) spectroscopy, have been regarded as the powerful assistant tools for cancer diagnosis for their abilities to tell the malignant part from the normal tissues both in vivo and ex vivo [1-3]. The study of K. Venkatakrishna which was developed a high performance liquid chromatography-laser induced fluorescence (HPLC-LIF) technique to detect and record simultaneously spectra and chromatograms of physiological samples like saliva and serum from normal and oral cancer subjects showed that the possibility of combining these tow techniques to find the characteristic marker of distinguishing the normal and the abnormal tissues [1]. In the autofluorescence spectra of the normal tissue and its abnormal one, one of the obvious differences in a red region, which is from 600nm to 700nm, has been shown in many kinds of cancer disease, oral cancer, gastric cancer and colorectal cancer for instance and it has been assigned to the endogenous protoporphyrin IX (PpIX). In our previous studies [4-5], this kind of red fluorescence was examined much stronger in human gastric cancer tissues than their normal counterparts. In K. Thomas Moesta's research [6], the fluorescence of PpIX was also detected from the colorectal cancers and through the HPLC and the fluorescence spectroscopy analysis, a conclusion that colorectal cancer metastases accumulate diagnostic levels of endogenous PpIX as a result of a tumor-specific metabolic alteration was reached. The HPLC method was also applied in the investigation of Kojiro Onizawaa and the PpIX was intended to the fluorescent substance associated with oral cancer [7]. In the researchment of Mindaugas Tamosiunas [8], the optical differences between necrotic and non-necrotic tumor, hemorrhagic tumor and healthy tissue were exploited by 405nm LED light, and the red fluorescence caused by endogenous porphyrins made these tissues distinguished from each other. Meanwhile, the investigations about PpIX in biological samples using HPLC were carried out widely by domestic researchers as well [9-10]. However, the relationship between the clinical feature of the gastric tissues -- the gastric tumors with different stages and the normal gastric tissue and the quantities of endogenous PpIX in them -- has not established yet, let alone the change of the former with the progression of the tumor stage.

In this study, a method using high-performance liquid chromatography (HPLC) was put into practice to discover the relationship mentioned above and the

*This project is funded by Shandong Province Natural Science Foundation (Y2006G23).
* Corresponding author: Jun Ma, female, professor of Physics Department, Ocean University of China.

quantity of PpIX in each kind of gastric tissues was estimated through the PpIX standard curve as well.

2. Materials and Methods

2.1. *Materials*

The samples were taken from 10 gastric cancer patients, whose cancer were all the advanced stage and infiltrated into the gastric serosa. These samples, each of which was in 2mm thick, were 40 in number and were divided into four groups as follows: the normal tissues 6cm away from the cancer (Group N), the paracancerous tissues 1cm away from the cancer (Group P), the gastric cancer tissues (Group G) and the central cancerous tissues taken from the section plane of the cancerous focus (Group C). The specimens of Group N, P, G were taken from the gastric serosa. The weight range of these tissues was 0.146g-0.472g.

The tissue homogenates were made from the samples mentioned above and extracted by 500μl methyl alcohol. Then the homogenates were put in the centrifuge which was spinning at 10000rps at 4℃ for 6 min during the treatment. A volume of 200μl supernatant was extracted after the centrifugate was processed by the filter paper (ϕ=0.45μm) and analyzed by HPLC in 2 hours.

2.2. *Instruments*

The HPLC system (Water-alliance, USA) consisted of the following: a Venusil XBP-C18 column (5um particle size, 250mm × 4.6mm I.D.) was used for the separation, the elution program was prepared by methyl alcohol, acetonitrile and water with the volume ratio 75:20:5.The flow rate was 1.0ml/min. Emission at 628nm following excitation at 404nm was used to detect porphyrins by the fluorescence detector (Waters TM 474).

2.3. *The Preparation of Standard Solution and its Regression Curve*

The concentration of 0.1μg/ml Protoporphyrin standard solution was made with standard protoporphyrin IX powder (Sigma) and methyl alcohol in volumetric flasks. After oscillation by ultrasonic, the solution was kept at 4°C, avoiding the irradiation of light until use. The volumes of 1, 5, 10, 5, 20μl were acquired precisely from the standard solution and HPLC analysis of these solutions was carried out.

A standard curve was generated by plotting the peak area of PpIX against its weight (μg). The linear regression was applied to obtain the linear regression equation: Y= 1.184 × 106X + 49.919. In the range of 5~100μg/L, a good linear relationship was shown for the correlation coefficient of the equation was 0.99995. The minimal detective concentration of PpIX was 10^{-3} μg/L.

2.4. *Stability and Precision*

20μl of the tissue homogenate was introduced automatically and measured hourly intervals in 8 hours. The peak areas and the RSD were calculated. The 1.95% RSD demonstrated that the specimen was stable in 8 hours.

The precision of the system was tested by measuring PpIX standard solution 10 times successively. 20μl of this solution was handled into the system each time and its peak area was measured from the chromatograph. Then the concentration of PpIX was estimated from the standard curve. The precision was good for the RSD was 0.83%.

2.5. *Recovery of the Specimen*

The normal tissue homogenate was required to prepare three solutions with different PpIX concentrations by the PpIX standard solution: 12.5μg/l, 25μg/l, 50μg/l. Each solution was duplicated. 20μl of this solution was used to examine in the HPLC system and for the following calculation. The recovery of this method was tested by comparing the PpIX concentration measured after the experiment with the one of the induced solution. The recovery ±RSD was 98.1±1.2 %.

3. Results and Discussion

All the 40 tissue homogenates were examined during the experiments. Each time, 20μl of the homogenate was induced to the HPLC system automatically. In the typical chromatographs (Fig. 1, 2, 3 and 4), the retention times were about 4.6min. The weights of PpIX (μg) were deduced according to the peak areas measured from the graphs and the standard curve. Then the concentration of the PpIX (μg/g) in each tissue could be obtained on the basis of the tissue's weight. The PpIX concentration average value of the tissues in each group was listed in Table 1.

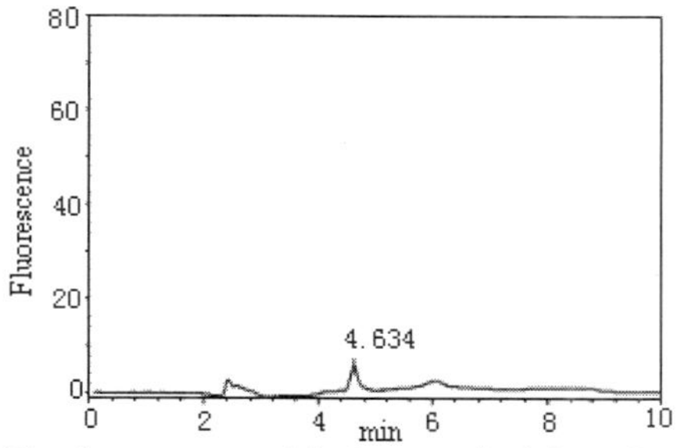

Figure 1. The chromatogram of the homogenized tissue from Group N

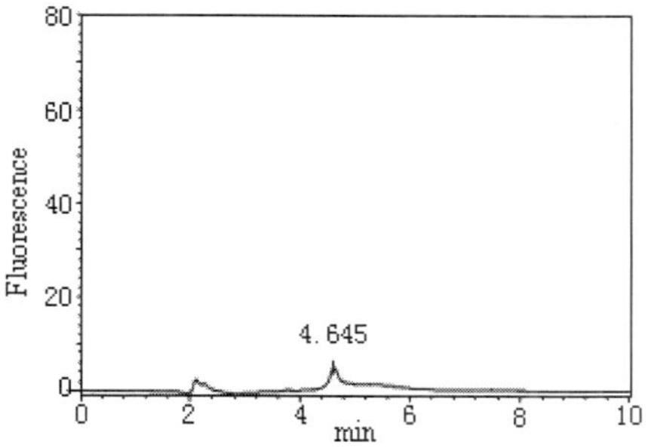

Figure 2. The chromatogram of the homogenized tissue from Group P

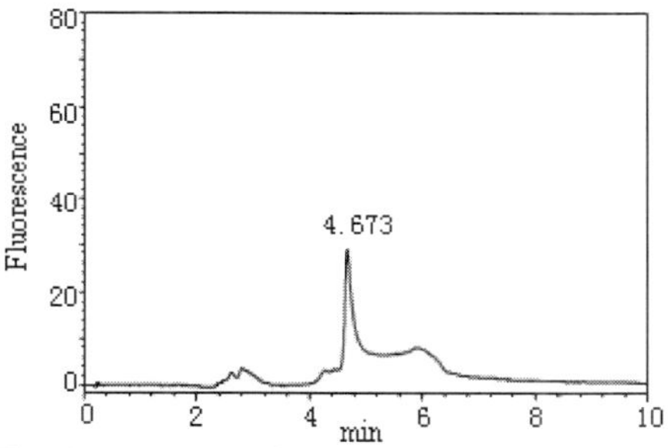

Figure 3. The chromatogram of the homogenized tissue from Group G

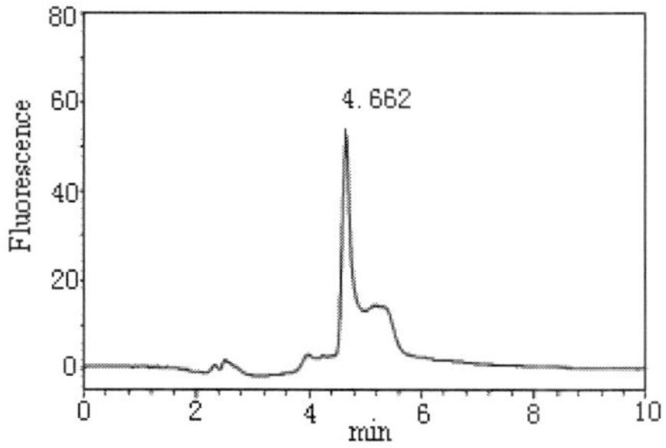

Figure 4. The chromatogram of the homogenized tissue from Group C

Table1 The average concentrations of PpIX in each group

Homogenate tissue	Concentration of PpIX in the tissue (μg/g)
Group N	$3.264 \pm 0.583 \times 10^{-7}$
Group P	$1.137 \pm 0.678 \times 10$-^{5}
Group G	$4.176 \pm 0.971 \times 10^{-4}$
Group C	$2.149 \pm 0.592 \times 10^{-3}$

In all the chromatographs of the four kinds of homogenate tissues, there was an obvious peak around 4.6min in the graphs taken from the three abnormal homogenates, with the exception of the normal ones. This peak was assigned to PpIX for the same retention time was appeared by titrating the PpIX standard solution into the homogenate. Though this peak of the normal homogenate was not as notable as the ones from the homogenates of the gastric cancer tissues and the section plane of the cancerous tissues, we prefer to link this peak to PpIX which could be found in the homogenate tissues of all the groups and the peaks around 4.6min in chromatograms were probably assigned to PpIX.

From the table above, the PpIX concentrations (μg/g) in the four groups were quite different from each other. The ones of Group C and G were much higher than their counterparts of Group N and P. The concrete values of the PpIX concentrations in the four groups were 10^{-3}μg/g in Group C, 10^{-4}μg/g in Group G, 10^{-5} μg/g in Group P and 10^{-7}μg/g in Group N. The differences among them demonstrate that the closer the tissue is to the cancer focus, the higher the concentration of PpIX in it is.

According to the experience of our former laser-induced autofluorescence spectral experiments [5], the optimal excitation and emission light of PpIX are 404nm and 628nm, respectively. The excitation light and emission light were set to 404nm and 628nm respectively in the HPLC system in order to enhance the detective sensitivity.

Conclusion

This study confirmed the fact that the different concentrations of PpIX in normal gastric tissue and its abnormal counterparts were existed in the HPLC analysis of the tissue homogenates. The concentration of PpIX in the cancer homogenate is much higher than the one in the normal homogenate. And it is another proof that PpIX could be regarded as the character of the cancerization for cancer diagnosis was given out through this investigation and this substance could be exploited as a tool for distinguishing the cancer tissue.

Acknowledgments

This research was funded by Shandong Province Natural Science Foundation (Y2006G23). Support was also generously contributed by Qingdao Key Lab of Optics and Photoelectronics, Ocean University of China.

References

1. K. Venkatakrishna, V. B. Kartha and K. M. Pai, et al, HPLC-LIF for early detection of oral cancer, *Curr Sci.* **84**, 551 (2003).
2. M. Pascale and C. Frédéric, Improvement in HPLC separation of porphyrin isomers and application to biochemical diagnosis of porphyrias, *Clin. chem. lab. med.* **44**, 1433 (2006).
3. O. A. Gederaas, K. Berg and I. Romslo, A comparative study of normal and reverse phase high pressure liquid chromatography for analysis of porphyrins accumulated after 5-aminolaevulinic acid treatment of colon adenocarcinoma cells, *Cancer Lett.* **150**, 205 (2000).
4. J. W. Meng, R. E. Zheng, H. L. Miao, R. Wang, X. G. Ren, and C. N. Zhou, Diagnostic Study of Esophagus Cancer using Protoporphyrin IX as the Index Molecule, *Acta Laser Biol Sinica.* **13**, 87 (2004).
5. X. F. Shi, J. Ma and W. Z. Mao, Applying Patial Least-Squares Discriminant Analysis on Autofluorescence Spectra to Identify Gastric Cancer, *Spectrosc Spectral Analy.* **26**, 295 (2006).
6. K. T. Moesta, B. Ebert and T. Handke, et al, Protoporphyrin IX Occurs Naturally in Colorectal Cancers and Their Metastases, *Cancer Res.* **61**, 991 (2001).
7. K. Onizawaa, N. Okamurab and H. Saginoyac, et al, Characterization of autofluorescence in oral squamous cell Carcinoma, *Oral Oncology.* **39**, 150 (2003).
8. M. Tamosiunas, J. Makaryceva and J. Labanauskiene, et al, Autofluorescence of Transplantable Hepatoma A22 (MH-A22): Prospects of Tumor Tissue Optical Biopsy, *Exp Oncol.* **26**, 118 (2004).
9. H. L. Guo, L. Y. Zhou, S. R. Lin, S. G. Ding, L. X. Wang, Z. Jin, Q. M. Geng and Y. M. Zhao, Study on the high performance liquid chromatography with fluorescence detection of gastric juice in the diagnosis of gastric carcinoma, *Chin J Dig.* **21**, 417 (2001).
10. Y. D. Li, HPLC for the determination of the content of zinc porphyrin and protoporphyrin in the serum, *J Guangxi Med University.* **20**, 904 (2003).

DETERMINATION OF TOTAL SUGAR IN INTACT NAVEL ORANGE USING VIS/NIR DIFFUSE REFLECTANCE SPECTROSCOPY *

XINGMIAO CHEN[1], XUDONG SUN[2], YANDE LIU[1†]

1. College of Engineering, Jiangxi Agriculture University, Nanchang 330045, China

AIGUO OUYANG

2. College of Automobile, Jiangxi Blue Sky University, Nanchang 330098, China

The potential of Vis/NIR diffuse reflectance spectroscopy technique for determining total sugar in intact navel orange was discussed. Four types of spectral pre-processing, including average smoothing, multiplicative scatter correction, first and second derivatives, were compared with the original spectra in order to obtain the best prediction results. Calibration and validation models were constructed by partial least squares regression (PLSR) in the wavelength range 450-1750 nm. The best calibration model, with respect to first-derivative spectral pre-processing, gave the prediction results for 1.2133%, 1.7977 and 0.8381 for RMSEP, SDR and r, respectively. The experiment results indicate that the prediction performance should be improved further in order to satisfy the requirement of screening/grading navel orange fruit.

Key words: Vis/NIR diffuse reflectance spectroscopy, pre-processing technique, PLSR, total sugar, navel orange

1. Introduction

Navel orange fruit is grown in a semitropical climate region. In China, Gannan navel orange is one of the most famous fruit. Navel orange fruit typically has high internal quality with respect to sweetness, percent juice and vitamin C. So, it has attracted much attention for both consumers and industries. Total sugar is the property most likely to match the maturity and determines the overall quality of fruits. At present, the classical methods are chemical analyses, which are slow, expensive and need highly qualified staff.

Visible/near-infrared (Vis/NIR) spectroscopy represents a rapid, nondestructive, and relatively easy to implement for measurement of different constituents in various food and agriculture products. Schmilovitch et al. (2000) [1] established relationships between the nondestructive NIR spectral (1200-2400nm) measurements and the major physiological properties and quality indices of mango fruit, such as TSS (r^2=0.9276, SEP=1.223), acidity (r^2=0.6085, SEP=0.161), firmness (r^2=0.8226, SEP=17.14) and storage period (r^2=0.9380, SEP=37.03). Lu (2001) [2] evaluated the potential of NIR reflectance spectra (800-1700nm) for measurement of the firmness (r=0.80, 0.65, SEP=0.55, 0.44) and sugar content (r=0.95, 0.89, SEP=0.71, 0.65) of Hedelfinger cherry and Sam cherry, respectively. Lu et al. (2002) [3] investigated the NIR (900-1500nm) sensing technique in interactance mode for rapid acquisition of spectral information to predict the sugar content (r =0.82, 0.78; SEP=0.56, 0.64) of Empire apple and Red Delicious apple, respectively. Park et al. (2003) [4] investigated the NIR spectroscopic (800-1100nm) techniques with principal component regression (PCR) and Mahalanobis Distance (MD) analysis for determining the soluble solid content (r^2=0.934, 0.966; SEP=0.279, 0.341) and firmness (r^2=0.291, 0.657; SEP=4.920, 7.330) of Delicious apple and Gala apple, respectively. Carlomagno et al. (2004) [5] studied on non-destructive methods for sorting peaches using NIR transmittance spectrometry (730-900nm), correcting discrimination of peaches with a percentage of 82.5%. Ying et al. (2005) [6] studied on the potential of NIR reflectance spectroscopy (800-2500nm) to predict total soluble solids (r^2=0.916, SEP=0.534) and available acid (r^2=0.904, SEP=0.129). Liu et al. (2006) [7] used the FT-NIR spectroscopy (814-1100nm) to measure non-destructively the sugar content (r =0.95, SEP=0.54) and titratable acidity (r=0.74, SEP=0.04) of intact apple fruit. Liu et al. (2007) [8] applied the FT-NIR diffuse reflectance spectroscopy (800-2500nm) to

* This work is supported by New Century Excellent Talents in University (Project No.NCET-2006-557) and National Natural Science Foundation of China (Project No.60468002 and No.30560064)

† Corresponding author: Email: jxliuyd@163.com; Telephone: 0086-791-3828042; Fax: 0086-791-3828042

predict sugar content in apples at different testing distances, and the best model gave the prediction results with r^2=0.8436 and SEP=0.773 at 0mm distance.

The objective of this study was to apply Vis/NIR diffuse reflectance spectroscopy technique for measurement of total sugar in intact navel orange fruit. Different spectral pre-processing methods were also compared with original spectra in order to obtain a better calibration model.

2. Materials and Methods

2.1. *Fruit Samples and Chemical Measurements*

A total of 135 navel oranges with three different varieties (Niu He'er, Nafu Ningna and Pengna), which came form orchards (Xinfeng, China), were purchased at a local market and stored for 3 days at 12^{0}C and 50% relative humidity before Vis/NIR spectrometer measurements were performed. Removed 8 outliers, the sample sets were split into a calibration set with 97 samples and a prediction set with 30 samples. To ensure the adaptability of the model, the samples with high and low concentration value were put in the calibration set, and other samples selection inside each group were performed randomly.

Tissue samples of 20g were cut from each fruit separately, from the marked area in which the Vis/NIR readings had been taken, and were macerated with a commercial juice extractor and filtered. The fruit juice was used for the determination of total sugar by direct titration, defining by the National Standard of China (GB/T 6194-86). All chemical measurements were done in triplicate and the averages of these triplicate measurements, expressed at percentage (%), were used for multivariate analyses.

2.2. *Experiment Equipment and Spectral Acquisition*

The experimental arrangement for testing navel orange fruit included a QualitySpec spectrometer (Analytical Spectral Devices, Inc., USA) with measurement wavelength region from 350 to 1800nm, an external fiber-optic cable with 1.7mm in diameter, a Si detector and an InGaAs detector for 350-1000nm, 1000-1800nm, respectively, tungsten halogen lamp (12V/45W) to provide light source, and a instrument controller to communicate with spectrometer using an Ethernet interface (Fig. 1). Each measured diffuse reflection spectrum was the average of ten individual optical scans divided by a reference spectrum taken from a white Teflon plate.

2.3. *Spectral Pre-processing*

Due to the potential system imperfection, noise, offset and /or bias were produced inevitably in the spectrum, where offset was defined as a constant drift and bias as constant slope along the baseline [9]. For reducing noise, offset and bias, an average-smoother with windows of different sizes was applied, followed by multiplicative scatter correction (MSC), first and second derivatives according to the Savitsky-Golay algorithm, with three points smoothing filter and two-degree polynomial.

2.4. *Data Analyses*

All reflectance measurements were transformed to absorbance (log (1/R)) values to obtain linear correlations of the Vis/NIR values with the concentration of chemical measurements. NIR spectra were collected and transformed by Indico software (version 4.0, Analytical Spectral Devices, INC., USA), from three positions which was marked with a circle beforehand on each navel orange around equatorial position. The averaged log (1/R) spectrum per navel orange was analyzed with partial least squares regression (PLSR) for multivariate calibration with 'Unscrambler v9.5' software (CAMO AS, Trondheim, Norway). Cross-validation following the leave-one-out procedure was performed in order to define the optimum number of factors to be kept in the models, based on the lowest root mean square error of cross-validation (RMSECV).

Model performance was judged by comparing model predictions and measured valued for the data in

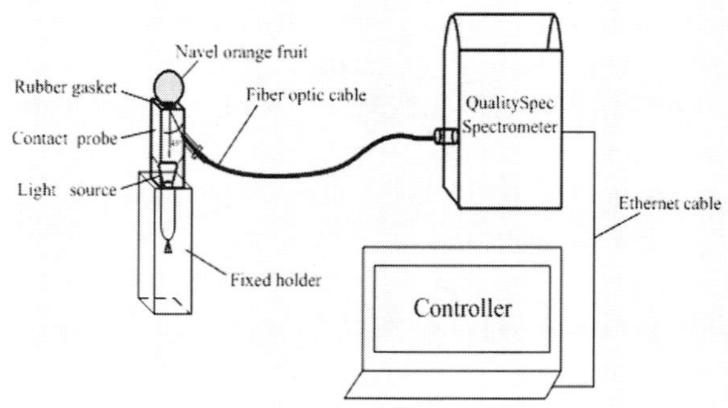

Fig. 1. Schematic diagram of the setup for Vis-NIR measurement of navel orange fruit

the prediction set. The main performance statistics were the correlation coefficient (r), the root mean square error of prediction (RMSEP) and the ratio of data set standard deviation to RMSEP (SDR). An SDR value of 3, which in theory corresponds to r^2=0.89, is generally considered a minimum for any useful sorting/grading purposes [10].

$$r = \sqrt{\frac{\sum_{i=1}^{n_p}(\hat{y}_i - y_i)^2}{\sum_{i=1}^{n_p}(\hat{y}_i - y_m)^2}} \quad (1)$$

$$RMSEP = \sqrt{\frac{1}{n_p}\sum_{i=1}^{n_p}(\hat{y}_i - y_i)^2} \quad (2)$$

$$SDR = \frac{SD}{RMSEP} \quad (3)$$

Where, $\hat{y}_i$ = predicted value of the ith observation; y_i =measured value of the ith observation; y_m = mean value of measured value in prediction set; n_p = number of observations in prediction set, SD=standard deviation in prediction set.

3. Results and Discussion

3.1 *Vis/NIR Spectra and Chemical Analyses*

Fig. 2 shows the representative original spectra of navel orange samples, with low, middle and high concentration, respectively. Considerable noise can be seen in the regions between 350-450 nm and 1750-1800 nm. Thus, the wavelength range of 450-1750 nm was applied to develop the calibration models. Consistent baseline offsets and bias were present in the spectra as well. These are quite common features in Vis/NIR spectra acquired by diffuse reflectance techniques. Nevertheless, theses offset and bias usually produce calibration models which require a larger number of factors or with lower predictive abilities. Thus, it is usually good practice in using spectral pre-processing to remove or reduce these features.

After 8 outliers removal (using studentized residuals, Fig. 3), the summary distributional statistics for calibration and prediction sets are shown in Table 1. The total sugar of navel orange fruit varied from 9.23% to 18.83%, with 14.353% for the mean value and 2.267 for standard deviation. The concentrations of chemical measurements were fairly normally distributed around the mean value, and this was helpful for developing the robust models.

Table 1 Summary distributional statistics for calibration and prediction sets

Sample sets	No.of samples	Minimum (%)	Maximum (%)	Mean (%)	Standard deviation
Total samples	127	9.23	18.83	14.353	2.267
Calibration set	97	9.23	18.83	14.471	2.225
Prediction set	30	9.73	17.70	13.824	2.181

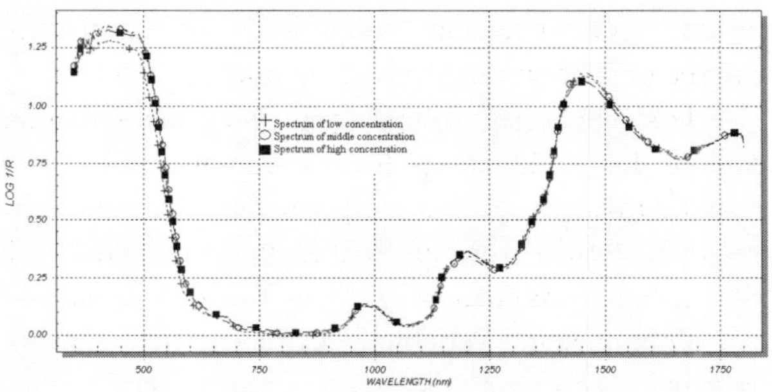

Fig. 2. Original spectra for low, middle and high concentrations of navel orange fruit, respectively

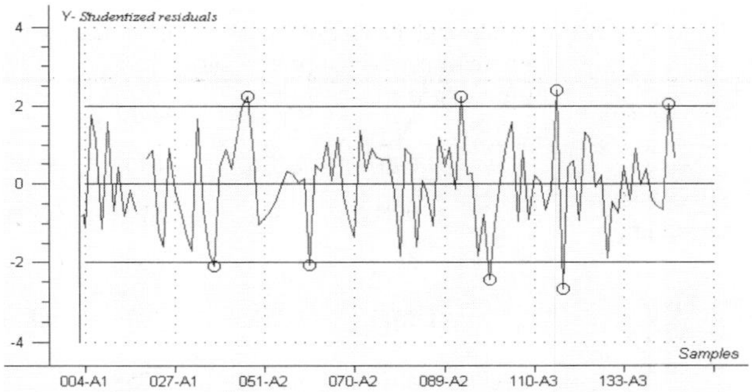

Fig. 3. Studentized residuals chart (the horizontal lines represent studentized residuals with 95% significance)

3.2 *Spectral Pre-processing Analyses*

Table 2 shows the calibration and prediction results with different spectral pre-processing. The model which presented the lowest RMSEP=1.2133 and highest SDR=1.7977 and r=0.8381 values was obtained by pre-processing the spectra using fist-derivative. The prediction model was significant in regression term (r=0.8381), but poor in prediction term (RMSEP=1.2133, SDR=1.7977).

Despite the average smoothing procedure can commonly increase the signal-to-noise ratio, Table 2 shows that the average-smoother with window of three segment sizes do not improve, but weaken the prediction results. This was due to the fact that, high frequency signals which mainly distributed in the fist and last wavelength data had been deleted, and the wavelength range of 450-1750 nm is good enough to progress the calibration models with little noise.

Nevertheless, despite the model obtained with MSC spectral pre-processing could produce the best calibration performance (RMSECV= 1.0582) by using the leave-one-out cross-validation technique, the prediction results were poor. The large error of prediction was probably due to the fact that MSC spectral pre-processing amplified the difference between prediction samples and calibration samples.

A less number of factors (only three factors) were required for building up the calibration model with second-derivative spectral pre-processing. However, the prediction results were lower than those obtained with other spectral pre-processing. This was due to the fact that three latent variables, extracted by second-derivative, were not enough to explain the original spectra, resulting in the poor results for calibration and prediction.

Table 2 Prediction results and number of factors of PLS models with different spectral pre-processing

Pre-processing	PLS factors	RMSECV	RMSEP	SDR	*r*
None	16	1.0907	1.2867	1.6952	0.8097
3-Average smoothing	16	1.0960	1.3042	1.6724	0.8044
MSC	13	1.0582	1.3395	1.6284	0.7969
Fist derivative	**7**	**1.2511**	**1.2133**	**1.7977**	**0.831**
Second derivative	3	1.4011	1.5854	1.3758	0.7224

3.3 *Validation with Unknown Samples*

Based on the above, the best calibration model, with respect to first-derivative spectral pre-processing, was tested with the prediction set consisting of samples not used for calibration. Fig. 4 shows the relationship between the chemical measurements and Vis/NIR diffuse reflectance spectroscopy predictions. Comparison between RMSEP and RMSECV values, small discrepancy indicates that the calibration model had been robustly constructed. Meanwhile, the values of slope and bias parameters were useful for distinguishing systematic errors and studying the correlation between the reference and Vis/NIR methods. For prediction results, slope=0.6845 and bias=0.3178 indicate that prediction accuracy should be improve further in order to satisfy sorting/grating purposes.

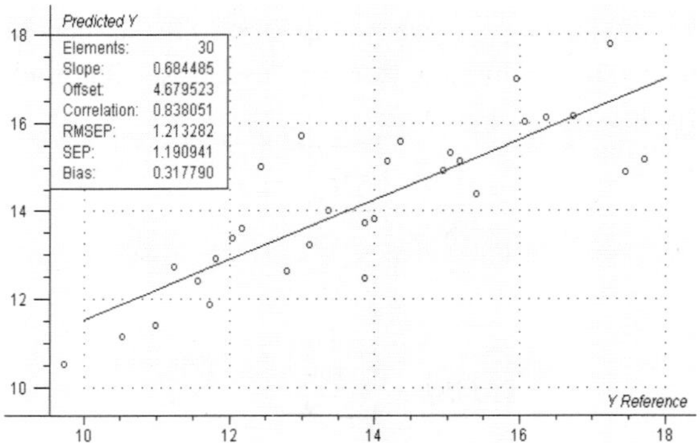

Fig. 4. Scatter plot of testing results for the prediction set with first-derivative spectral pre-processing

4. Conclussions

This study was supported by the Program for New Century Excellent Talents in University (Project No.NCET-2006-557) and Natural Science Foundation of China (Project No.60468002 and No.30560064).

References

1. Z. Schmilovitch, A. Mizrach, A. Hoffman, H. Egozi and Y. Fuchs, "Determination of mango physiological indices by near-infrared spectrometry", *Postharvest Biology and Technology*, **19(3)**, pp. 245-252, 2000.
2. R. Lu, "Predicting firmness and sugar content of sweet cherries using Near-infrared diffuse reflectance spectroscopy", *Transactions of the ASAE,* **44(5)**, pp. 1265-1271, 2001.
3. R. Lu and D. Ariana, "A near-infrared sensing technique for measuring internal quality of apple fruit", *Applied Engineering in Agriculture*, **18(5)**, pp. 585-590, 2002.
4. B. Park, J. A. Abbott, K. J. Lee, C. H. Choi, and K. H. Choi, "Near-infrared diffuse reflectance for quantitative and quantitative measurement of soluble solids and firmness of delicious and gala apples", *Transactions of the ASAE*, **46(6)**, pp. 1721-1731, 2003.
5. G. Carlomagno, L. Capozzo, G. Attolico, and A. Distante, "Non-destructive grading of peaches by near-infrared spectrometry", *Infrared Physics & Technology*, **46(1-2)**, pp. 23-29, 2004.

6. Y. B. Ying, Y. D. Liu, J. P. Wang, X. P. Fu, and Y. B. Li, "Fourier transform near infrared determination of total soluble solids and available acid in intact peaches", *Transactions of the ASAE*, **48(1)**, pp. 229-234, 2005.
7. Y. D. Liu, Y. B. Ying, and H. Y. Jiang, "Rapid determination of maturity in apple using outlet detection and calibration model optimization", *Transactions of the ASAE*, **49(1)**, pp. 91-95, 2006.
8. Y. D. Liu, Y. B. Ying, X. P. Fu, and H. S. Lu, "Experiments on predicting sugar content in apples by FT-NIR Technique", *Journal of Food Engineering*, **80(3)**, pp. 986-989, 2007.
9. A. M. K. Pedro, and M. M. C. Ferreira, "Simultaneously calibrating solids, sugars and acidity of tomato products using PLS2 and NIR spectroscopy", *Analytica Chimica Acta*, **595(1-2)**, pp. 221-227, 2007.
10. V. A. McGlone, R. B. Jordan, and P. J. Martinsen, "Vis/NIR estimation at harvest of pre- and post-storage quality indices for 'Royal Gala' apple", *Postharvest Biology and Technology*, **25(2)**, pp. 135-144, 2002.

VIS-NIR MEASUREMENT OF VITAMIN C IN MANDARIN BY PLS REGRESSION AND WAVELENGTH SELECTION *

XUDONG SUN[1], XINGMIAO CHEN[1], YANDE LIU[1†]

1. College of Engineering, Jiangxi Agriculture University, Nanchang 330045, China

BEIPING GAO[2]

2. Jiangxi Binhu Institute of Agricultural Science, Nanchang 330126, China

A Vis-NIR measurement method for determining vitamin C in mandarin was investigated in the wavelength range of 450-1750nm. Wavelength selection methods were studied and compared by partial least squares regression (PLSR), including different wavelength ranges from visible region, short-wavelength NIR and long-wavelength NIR, and sensitive wavelengths based on X-loading weights and regression coefficients. The calibration model, based on sensitive wavelengths from X-loading weights, showed the prediction results with the regression coefficient (r) of 0.8276, the root mean square error of prediction (RMSEP) of 2.3417 mg/100g. Meanwhile, some sensitive wavelengths which might be related to vitamin C in mandarin fruit were proposed. The experiment results show that some sensitive wavelengths should be selected for building a fast, simple, robust model for the future application.

Key words: Vis-NIR spectroscopy, PLSR, wavelength selection, vitamin C, mandarin fruit.

1. Introduction

Fruit quality is a vital factor affecting its market value, its transportation and storage requirements. Basically, fruit quality indices consist of internal quality, such as sugar content, firmness, vitamin C, and external quality, such as color, shape and appearance. To measure the external quality of fruit, much research has been conducted leading to promising advancements. As a result, there are many types of fruit sorters based on the size or weight of the fruit. However, determining the internal quality of fruit is not as straightforward as external quality measurement. The composition and the texture condition of fruit are usually measured by destructive or invasive approaches, such as chemical or texture analysis. The conventional analytical methods for determination internal quality indices of fruit are highly time-consuming and laboratory-occupying. In addition, on-line applications of fruit quality assessment are limited.

Visible-near-infrared (Vis-NIR) spectrometric technique, combination with sensitive wavelength selection, is a powerful analytical tool, because of its fast detection and simple operation in sampling. Thus, sensitive wavelength selection has been widely applied to Vis/NIR spectrometric technique in order to evaluate various constituents in food and agriculture products.

Carlini et al. (2000) [1] used the wavelengths involving the sub-sampling of the spectral interval from 600 to 1100nm with a fixed step for evaluating of soluble solids content in cherry (r^2=0.97, SEP=0.49) and apricot (r^2=0.89, SEP=0.75) fruits. Ruan et al. (2002) [2] developed a neural network based method for determination of Deoxynivalenol levels in barley using Vis/NIR spectroscopy (400-2500nm). The effects of wavelength interval and ranges on the prediction accuracy of models were examined, and the best prediction accuracy of r^2=0.933 and SEP=3.097 for production sets at a wavelength interval of 2 nm. Chen et al. (2005) [3] presented a new algorithm for simultaneous wavelength selection and outlier detection in multivariate regression of NIR for aqueous (5450-9000cm^{-1}) and tobacco (4000-8000cm^{-1}). The best calibration results, combination of continuous wavelet transform and modified iterative predictors and objects weighting PLS (CWT-MIPOW-PLS), were RMSEP=0.0031 for fructose in aqueous and RMSEP=0.183 for alkaloid in tobacco. Delwiche et al. (2005) [4] developed single- and all combinations of two-wavelength linear discriminant analysis models for

* This work is supported by New Century Excellent Talents in University (Project No.NCET-2006-557) and National Natural Science Foundation of China (Project No.60468002 and No.30560064)

† Corresponding author: Email: jxliuyd@163.com; Telephone: 0086-791-3828042; Fax: 0086-791-3828042.

monochromatic and bichromatic sorting of fusarium-damaged wheat in the wavelength range 410-865nm and 1031-1674nm. Short visible and moderate near-infrared wavelengths produced the highest single-term classification accuracies (at approximately 77% and 83%, respectively). The best two-term models occurred near the wavelengths of 500 and 550nm for the visible region alone (94% accuracy), 1152 and 1248nm for the near-infrared region alone (97%), and 750 and 1476nm for the hybrid region (86%). Yu, et al. (2006) [5] researched the feasibility of NIR (800-2500nm) for discrimination between Chinese rice wine of different geographical origins. The prediction performance of calibration models in different wavelength range was investigated, and the best models gave a 100% classification of wines of the two geographical origins in the range of 1300-1650 nm. Li et al. (2007) [6] used the Vis/NIR spectroscopy (325-1075nm) to measure non-destructively soluble solids content of tea soft drink. The significance of entire wavelengths was analyzed by ANOVA, and five wavelengths were selected with statistical significance at 99% confidence level. The prediction results from the MLR model was r=0.975 and SEP=0.693.

The objective of this study was developed a measurement method for determination vitamin C in mandarin fruit using Vis/NIR spectroscopy. Moreover, this study was particularly aimed at an assessment of the usefulness of some forms of wavelength selection in order to build a more fast, simple and robust model.

2. Materials and Methods

Major headings should be typeset in boldface with the first letter of important words capitalized.

2.1. *Fruit Samples*

A total of 75 mandarins which came from orchards (Xinfeng, China) were purchased at a local market and stored for 2 days at 10^0C and 60% relative humidity before Vis/NIR spectrometer measurements were performed. For experimental and chemical measurement error, 6 samples were deleted as outliers. In fact, 69 samples were used to analyze and develop Vis/NIR models in laboratory. 54 mandarins were selected and used for calibrating the mathematics models, and the remainders were used for evaluating the prediction error. The selection of samples for calibration set was random except for the constraint that the ranges of vitamin C in the calibration samples were wider than the ranges in the validation set.

2.2. *Vis-NIR Diffuse Reflectance Measurements*

The experimental arrangement for testing navel orange fruit included a QualitySpec spectrometer (Analytical Spectral Devices, Inc., USA) with measurement wavelength region from 350 to 1800nm, an external fiber-optic cable with 1.7mm in diameter, a Si detector and an InGaAs detector for 350-1000nm, 1000-1800nm, respectively, tungsten halogen lamp (12V/45W) to provide light source, and a instrument controller to communicate with spectrometer using an Ethernet interface (Fig. 1). Each measured diffuse reflection spectrum was the average of ten individual optical scans divided by a reference spectrum taken from a white Teflon plate.

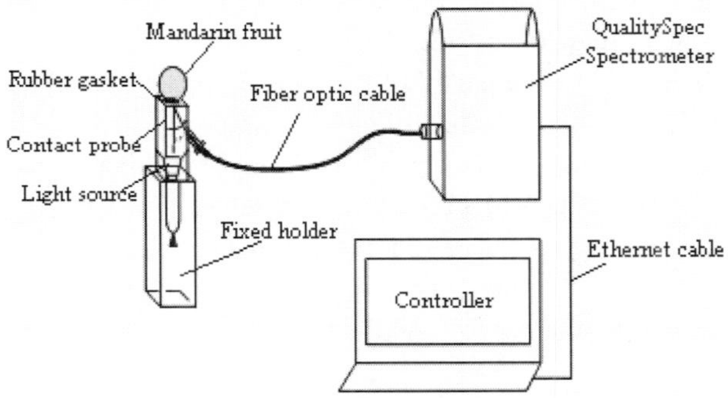

Fig. 1. Schematic diagram of the setup for Vis-NIR measurement of mandarin fruit

Vis/NIR spectra were collected and transformed by Indico software (version 4.0, Analytical Spectral Devices, INC., USA), from three positions which was marked with a circle beforehand on each mandarin around equatorial position. The averaged diffuse reflectance spectrum per mandarin was analyzed with the statistical program for multivariate calibration of 'Unscrambler v9.5' software (CAMO AS, Trondheim, Norway).

2.3. *Concentration Measurements*

Vitamin C of each mandarin sample was titrated manually by 2, 6-Dichloroindophenol titration method, defining by the National Standard of China (GB/T 6195-86). All chemical measurements were done in triplicate and the averages of these triplicate values, expressed as mg/100g, were used for multivariate analysis.

2.4. *Wavelength Selection*

Due to the potential system imperfection, obvious scattering noises could be observed at the beginning and the end of the spectral data. Thus, the first 100 and last 50 wavelength data were eliminated to improve the measurement accuracy. Then, to determine optimal wavelengths, the full range (450-1750nm) was divided into three bands: visible region (450-700 nm), short-wavelength NIR (700-1100 nm) and long-wavelength NIR (1100-1750 nm). At last, to considerate the contributions of some individual wavelengths to the prediction results, the x-loading weights and regression coefficients were used to analyze the sensitive wavelengths.

2.5. *Data Analyses*

Calibration models were developed using partial least squares regression (PLSR) with leave-one-out cross-validation technique. PLSR is a bilinear modeling method where information in original X-data is projected onto a small number of latent variables called PLS components to simplify the relationship between X-data and Y-data for predicting with the smallest number of latent variables. The first step in PLSR is to decompose the matrix, and the model is

$$X = TP + E \tag{1}$$

$$Y = UQ + F \tag{2}$$

Where, T and U are the score matrices of X matrix and Y matrix, P and Q are the loading matrices of X matrix and Y matrix, and E and F are the errors which come from the process of PLSR.

The second step is that T and U are processed by linear regression, and the model is

$$U = TB \tag{3}$$

$$B = (T'T)^{-1}T'U \tag{4}$$

Where, B is the coefficient matrix which come form the process of linear regression.

For prediction,

$$Y_{unknown} = T_{unknown}BQ \tag{5}$$

The performance of the prediction models was quantified by correlation coefficient of prediction (r) and the root mean square error of prediction (RMSEP). r and RMSEP are calculated as follows

$$r = \sqrt{\frac{\sum_{i=1}^{n_p}(\hat{y}_i - y_i)^2}{\sum_{i=1}^{n_p}(\hat{y}_i - y_m)^2}} \tag{6}$$

$$RMSEP = \sqrt{\frac{1}{n_p}\sum_{i=1}^{n_p}(\hat{y}_i - y_i)^2} \tag{7}$$

Where, $\hat{y}_i$ = predicted value of the ith observation; y_i =measured value of the ith observation; y_m = mean value of measured value in prediction set; n_p = number of observations in prediction set.

3. Results and Discussion

3.1 *Features of Vis/NIR Spectra and Chemical Structure*

Fig. 2 shows the raw spectra of all 69 mandarin samples. The trends of spectra were quite similar, however, large variations can be observed among different samples, especially at at the beginning and the end of the spectral data, which could result form external and internal quality indices, such as size, color and chemical components of mandarin samples. Thus, the wavelength range of 450-1750 nm was applied to develop the calibration models. Vis/NIR was sensitive to the concentrations of organics materials, which involved the response of molecular bonds C-H, O-H and N-H. Vitamin C is the organic molecule, and contains bonds C-H, O-H, C-O, C-C, thus, it is possible to use Vis/NIR methods for determination of vitamin C in mandarin. From 450 to 700 nm (visible region), there is a sharp drop in absorbance, and the strongest absorbance exists at 480 nm. From 700 to 1100 nm (short-wavelength NIR), the spectra curves are relatively flat and low except for 960 nm. There is the absorption peak, associated with second overtone of bond O-H. From 1100 to 1750 nm (long-wavelength NIR), the spectra curves begin to ascend until 1450 nm, and then the absorbance curves slide until 1670 nm. At last, the spectra curves re-begin to ascent. There are two strong absorption peaks at 1210 nm and 1450 nm, associated with a combination of second overtone of bond C-H and overtones of bond O-H in H_2O, and fist overtone of bond O-H, respectively.

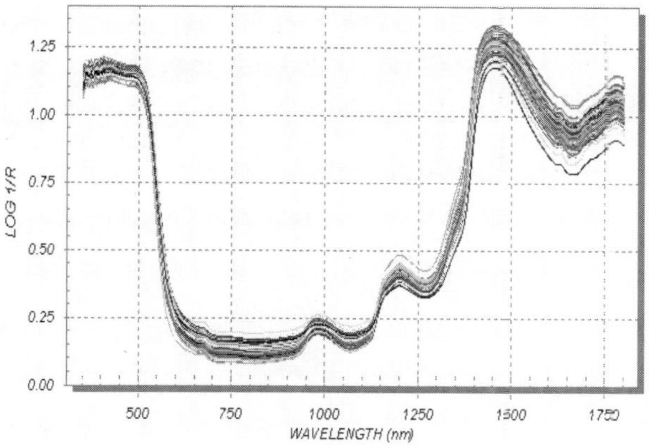

Fig. 2. Vis/NIR spectra of mandarin fruit samples

3.2 *Effect of Wavelength Ranges*

The wavelength is an important Vis/NIR system operation variable. If a narrower range than 450-1750nm is used, data acquisition time can be reduced, more importantly, the model obtained will be more reliable and robust. Fig. 3 shows the prediction results with different wavelength ranges. The best calibration model, based on different wavelength ranges, is the model with wavelength ranges form 1100 to 1750 nm, resulting in *r* of 0.7904 and RMSEP of 2.5368 mg/100g.

3.2 *Effect of Sensitive Wavelengths*

To further examine the effect of wavelengths for prediction results, the contribution of the individual wavelength was analyzed. It would be helpful to examine how vitamin C is simply related to some individual wavelengths so that a better understanding the relationship between chemical concentrations and Vis/NIR spectroscopy of mandarin samples.

For wavelength range from 450 to 1750 nm, the optimum PLS model was obtained with seven latent variables that were recommended by leave-one-out cross-validation technique. The cumulative reliabilities of the first seven latent variables explained 98.20% of the spectral variation of samples. Thus, the relationship between the wavelength variables and the latent variables might imply the sensitive individual wavelengths with vitamin C in mandarin fruit. In Fig. 4 a, the strongest peak of each latent variable in the loading weight curve had been collected as sensitive wavelength, such as 1392 nm (PC1), 1708 nm (PC2), 1388 nm (PC3), 1411 nm (PC4), 552 nm (PC5), 604 nm (PC6), 1475 nm (PC7). The prediction results for PLS model with these seven sensitive wavelengths are

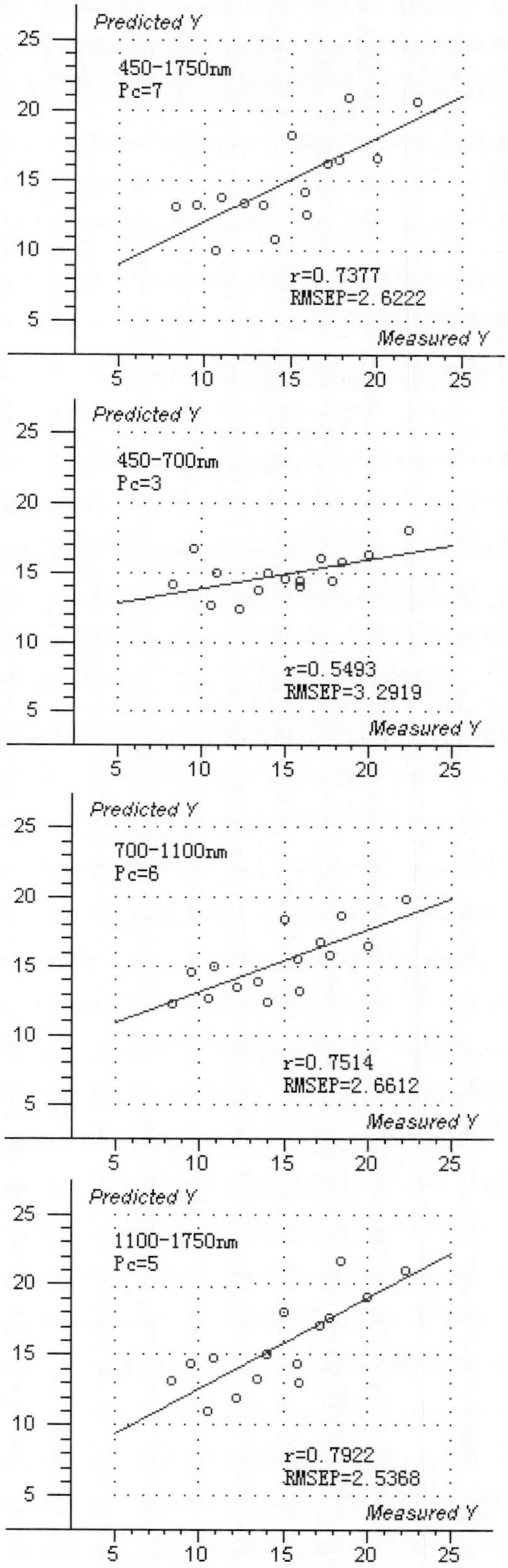

Fig. 3. The correlation between the measured and predicted vitamin C with different wavelength ranges

0.8276 and 2.3417 mg/100g for r and RMSEP, respectively (Fig. 5 a). In Fig. 4 b, the strong peaks and valleys in the regression coefficients curve had also been collected to calibrate the PLS model. The sensitive wavelengths were chosen, such as 519, 548, 676, 1202, 1397, 1475, and 1708 nm. The prediction results with sensitive wavelengths from regression coefficients are lightly worse than those based on X-loading weights, and the r and RMSEP are 0.7968 and 2.3620 mg/100g, respectively (Fig. 5 b).

3.4 *Comparison with Calibration Models*

The results from the above analyses show that the best calibration model could be obtained with some sensitive wavelengths based on X-loading weights, resulting in r and RMSEP of 0.8276 and 2.3417 mg/100g, respectively. The experiment results also show that the prediction performance with sensitive wavelengths from X-loading weights and regression coefficients are better than those with wavelength ranges from visible region, short-wavelength NIR and long-wavelength NIR. Thus, the attempt to find sensitive wavelengths for vitamin C in mandarin fruit could be valuable for enhancing the accuracy and robustness of the calibration models. It is also helpful to develop simple devices for the rapid and accurate measurement of vitamin C in mandarin fruit. Based on the prediction results for sensitive wavelengths from X-loading weights and regression coefficients by PLS, the 519, 540, 552, 604, 676, 1202, 1388, 1392, 1397, 1411, 1475, 1708 nm might be the sensitive wavelengths for vitamin C in mandarin fruit.

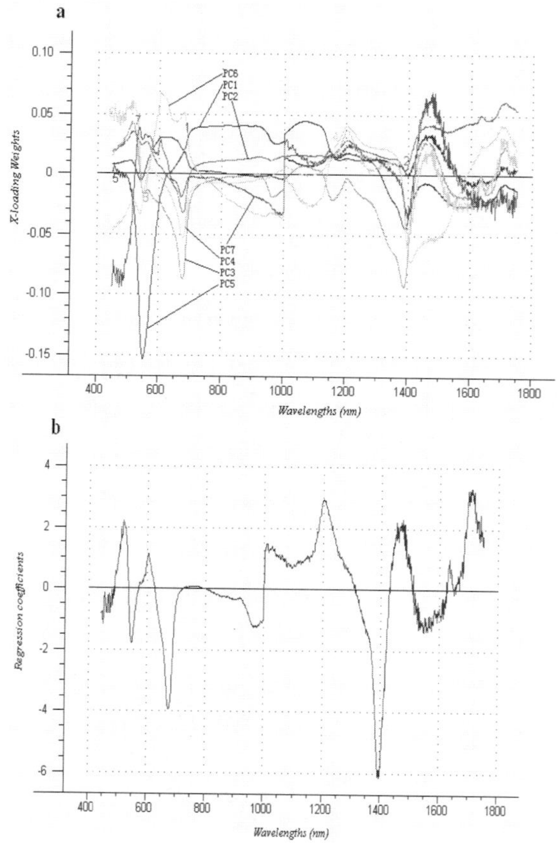

Fig. 4. X-loading weights (a) and regression coefficients (b) for vitamin C in mandarin fruit by PLS

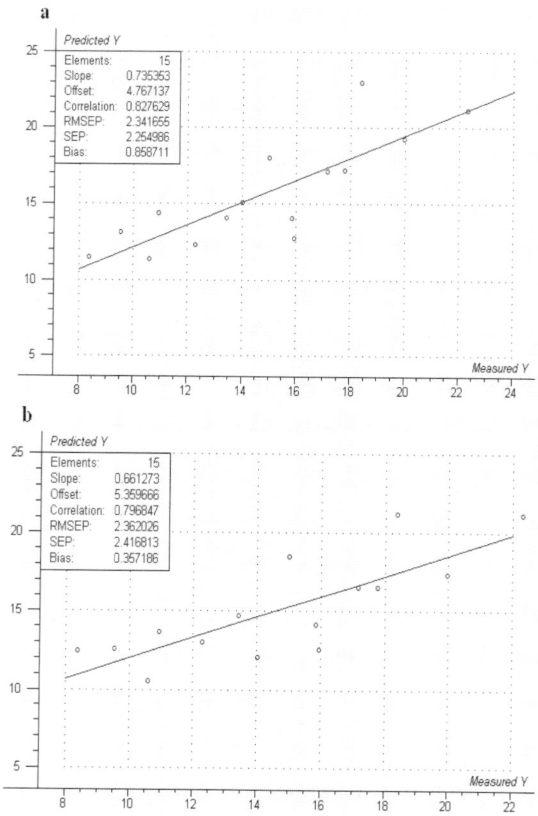

Fig. 5. Prediction results for sensitive wavelengths from X-loading weights (a) and regression coefficients (b) by PLS

4. Conclusions

In this study, a Vis/NIR measurement method was developed to measure vitamin C in mandarin fruit in the wavelength range of 450-1750 nm. Wavelength selection methods were studied and developed the calibration models by PLS, including different wavelength ranges from visible region (450-700 nm), short-wavelength NIR (700-1100 nm) and long-wavelength NIR (1100-1750 nm), and sensitive wavelengths based on X-loading weights and regression coefficients. At the same time, some sensitive wavelengths which might be related to vitamin C in mandarin fruit were proposed on the basis of X-loading weights and regression coefficients by PLS. The best calibration model, based on prediction results, was developed with the sensitive wavelengths from X-loading weights, resulting in *r* and RMSEP of 0.8276 and 2.3417 mg/100g, respectively. The experiment results show that the full wavelengths were used to develop the PLS models, but some sensitive wavelengths should be selected for building a more fast, simple, robust model for the future application.

Acknowledgments

This study was supported by the Program for New Century Excellent Talents in University (Project No.NCET-2006-557) and Natural Science Foundation of China (Project No.60468002 and No.30560064).

References

1. P. Carlini, R. Massantini, and F. Mencarelli, "Vis-NIR measurement of soluble solids in cherry and apricot by PLS regression and wavelength selection", *Journal of Agriculture and Food Chemistry*, **48(11)**, pp. 5236-5242, 2000.
2. R. Ruan, Y. Li, X. Lin, and P. Chen, "Non-destructive determination of deoxynivalenol levels in barley using near-infrared spectroscopy", *Applied Engineering in Agriculture*, **18(5)**, pp. 549-553, 2002.
3. D. Chen, X. G. Shao, B. Hu, and Q. D. Su, "Simultaneous wavelength selection and outlier detection in multivariate regression of near-infrared spectra", *Analytical Sciences*, **21(2)**, pp. 161-166, 2005.
4. S. R. Delwiche, C. S. Gaines, "Wavelength selection for monochromatic and bichromatic sorting of fusarium-damaged wheat", *Applied Engineering in Agriculture*, **21(4)**, pp. 681-688, 2005.
5. H. Y. Yu, Y. Zhou, X. P. Fu, L. J. Xie, and Y. B. Ying, "Discrimination between Chinese rice wines of different geographical origins by NIRS and AAS", *European Food Research and Technology*, **225(3)**, pp. 313-320, 2006.
6. X. L. Li, Y. He, C. Q. Wu, and D. W Sun, "Nondestructive measurement and fingerprint analysis of soluble solid content of tea soft drink based on Vis/NIR spectroscopy", *Journal of Food Engineering*, **82(3)**, pp. 316-323, 2007.

3D IMAGING OF HUMAN BRAIN USING THE LAYER-BASED SIGMOID ADJUSTMENT METHOD

LEI SHANG [1,2†]

1. School of Computer Science and Technology, Shandong University, Jinan, 250013, P.R.China
2. Department of Information Science and Technology, Shandong Institute of Political Science and Law, Jinan, 250013, P.R.China

Diffuse optical tomography (DOT) is an important tool for non-invasive exploration of human brain functions. However, the image quality of DOT is limited by its poor depth resolution. Recently, there has been proposed a layer-based sigmoid adjustment (LSA) method which can effectively improve the depth resolution of DOT. In this paper, we implemented the LSA method in human brain imaging. We did a finger-tapping experiment using our CW5 instrument (TechEn, USA) with one subject. The activated regions were accurately located on the motor cortex in the 3D reconstructed image by using the LSA method. However, the activated regions were clearly biased to the scalp and cerebral spin fluid (CSF) without the LSA method. Further analysis indicates that the activation areas in the reconstructed image using the LSA method were locate near the knob shaped structure of pre-central gyrus. This is coincident to previous studies.

1. Introduction

Brain activations usually accompanied with changes in concentrations of hemoglobins, which are major absorber of near-infrared light in brain tissues, so near-infrared light can be used to detect brain activations.[1] Diffuse optical tomography (DOT) is such a method to detect brain activations by measuring the change in optical density in the activated station comparing to the baseline station.[2-5] Especially, DOT is appropriate for detecting "focal" activations in human brains, such as the activations on the motor cortex.[6-9]

However, most of the works studying the motor cortex using DOT are making 2D images. Parts of the difficulties for making 3D images of motor cortex lie on the time-consuming computation of forward matrix. But the major balk for making 3D images is the poor depth resolution of DOT. It has been shown by some researchers that the activations in the reconstructed 3D images tend to bias to the superficial tissues, so the 3D images can not be used to assert the activated regions in human brain.[10-11]

Zhao et al. have proposed a layer-based sigmoid adjustment (LSA) method to improve the depth resolution of DOT.[12] They pointed out that the major reason for the poor depth resolution of DOT is that it has much lower sensitivity to deep layers than to superficial layers. The brief idea of the LSA method is to reduce the sensitivity contrast by multiplying larger parameters on deep layers and smaller parameters on superficial layers. However, the LSA method has only been validated in simulations. In this paper, we implement the LSA method on human subject in the first time.

The 3D head model of the subject was obtained from the anatomical MR imaging. Then the forward matrix on the head model was calculated using a Monte Carlo method. We reconstructed 3D images of the motor cortex both using the LSA method and without the LSA method. By using the LSA method, the activated regions were accurately located on the motor cortex. However, the activated regions were clearly biased to the scalp and cerebral spin fluid (CSF) without the LSA method. The improvement is mainly because of the better depth resolution by using the LSA method. Further analysis indicates that the activation areas in the reconstructed image using the LSA method were locates near the knob of pre-central gyrus. This is coincident to previous studies.

2. Forward Matrix

We used the multi-centered geometry (as shown in Fig. 1) as the probing arrangement.[13] The multi-centered geometry is a hexagon shaped plastic chip, with 7 sources in the central region and 24 detectors around the sources. The side of the hexagon is 4 cm. The multi-centered geometry can significantly improve the spatial resolution of DOT. To obtain the forward matrix, we laid one vitamin E pill on the position of each optical fiber, and then the multi-centered geometry with vitamin E pills on it was put on the head of human subject. The center of the multi-centered geometry was put on the C3 point in EEG 10-20 system. The anatomical MR imaging of the subject was achieved using a Siemens 3T scanner with voxel size of 1 mm × 1 mm × 1 mm.

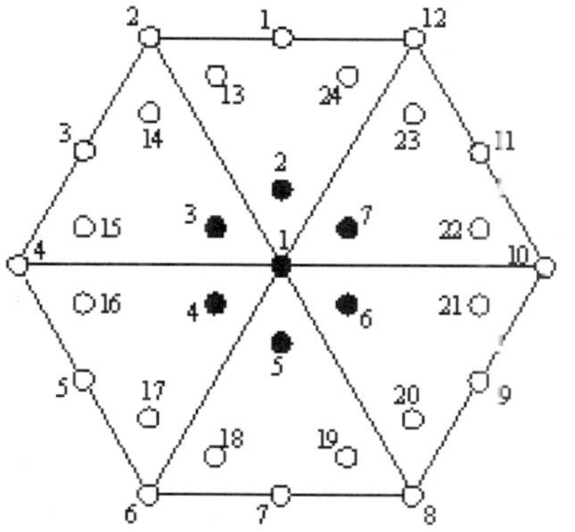

Figure 1. The multi-centered geometry used in this paper. The side of the hexagon is 4 cm. Sources are indicated by dots, and detectors were indicated by open circles.

We then segmented the 3D MR imaging into five tissue types: scalp, skull, cerebral spinal fluid, gray matter and white matter. The air-scalp and the scalp-skull boundaries were confirmed manually. CSF-GM and GM-WM boundaries were extracted by the tissue classification step which was performed with BrainSuite.[14]

The sensitivity matrix was obtained using a Monte Carlo method.[15] The optical properties used for each tissue in the head model were shown in Table 1. Experimentally, near infrared photons can reach a maximum depth of about 3.5 cm beneath the surface of the head, so DOT would be perfectly sensitive to activations in human brain cortex. We used 10^8 photons in each run to get the sensitivity matrix of the 31 optodes (7 sources and 24 detectors). Then we get the sensitivity matrix A of 168 source-detector pairs by multiplying the sensitivity matrix of each source by that of all detectors.

Table 1. Optical properties used in the Monte Carlo simulation. The absorption coefficients and scattering coefficients are in cm^{-1}.

Tissue type	Absorption Coefficient	Scattering Coefficient
Scalp	0.191	6.6
Skull	0.136	8.6
Cerebral spinal fluid	0.026	0.[illegible]
Grey matter	0.186	11.1
White matter	0.186	11.1

The sensitivity matrix A represents the 3D distribution of optical density of the multi-centered geometry. There has some redundant information in matrix A. Firstly, the voxels no deeper than 1 cm from the surface of the head are redundant because the cerebral cortex, which is of our interest, will locate deeper than 1 cm from the surface of head in case of adults. Secondly, the voxels deeper than 3.5 cm from the surface of the head are redundant because too few photons can penetrate more than 3.5 cm in human tissues. The selected partial sensitivity matrix A after removing the redundant voxels are used for the LSA adjustment. The LSA parameter used in this paper is 400. The sensitivity matrix before and after the LSA adjustment is shown in Fig. 2. The sensitivity matrix $\tilde{A}$ after the LSA adjustment and the sensitivity matrix A before the LSA adjustment are both used for the image reconstruction in this paper.

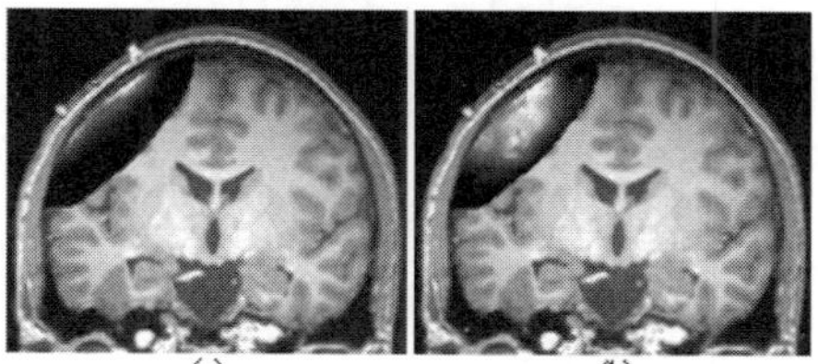

(a) (b)

Figure 2. The sensitivity of one channel (a) before and (b) after the LSA adjustment. The right-above white point on the surface of the scalp is the source point. The left-bottom white point on the surface of the scalp is the detector point.

3. Experiments and Data Processing

We did the finger-tapping experiment using our CW5 instrument (Techen, USA). During the experiment, a single wavelength of 830 nm was used to collect data. Following a complete description of the experiments, written informed consent was obtained from a single human subject before the initiation of the experiments. The experiment was ethically approved by the Institutional Review Board at Department of Information , Shandong Institute of Political science and Law. In the experiment, source 1 in the multi-centered geometry was put on C3 point in the EEG 10-20 system. By doing this, the multi-centered geometry can cover the hand area motor cortex well. In the experiment, the subject was asked to tap his thumb against the other four fingers of right hand for 30 seconds, followed by

30 seconds of rest. Data collection lasted 420 seconds. The finger-tapping frequency is approximately 3 Hz.

The experimental data was bandpass filtered from 0.01 Hz to 0.1 Hz. Then the data collected throughout the 420 seconds were averaged per minute to achieve averaged data for 60 seconds. The mean values of the first 30 seconds were viewed as the optical flux of activated state (Φ) and those of the last 30 seconds were viewed as the optical flux of baseline state (Φ_0). And then the changes in optical density (ΔOD) were calculated with the following equation:

$$\Delta OD = -\ln(\phi/\phi_0) \quad (1)$$

The calculated ΔOD, was used for the image reconstruction.

4. Image Reconstruction

The images of $\Delta\mu_a$ were reconstructed by the Tikhonov regularization method:[16-17]

$$x = A^T(AA^T + \alpha s_{\max} I)^{-1} y . \quad (2)$$

$$\tilde{x} = \tilde{A}^T(\tilde{A}\tilde{A}^T + \alpha \tilde{s}_{\max} I)^{-1} y . \quad (3)$$

Here x and $\hat{x}$ are the reconstructed $\Delta\mu_a$ before and after the LSA adjustment. y is the measured ΔOD. Matrices A and $\tilde{A}$ are the sensitivity matrices before and after the LSA adjustment. And $s_{\max}$ and $\tilde{s}_{\max}$ are the maximal singular values of matrices A and $\tilde{A}$, respectively. The parameter α in the equation was used to restrain the sensitivity of $\Delta\mu_a$ to measurement noise. The noise in the measured data was random and hence, an L-curve method was performed to ascertain the best α for the reconstruction.[18-20]

The reconstructed images are shown in Figs. 3(a) and 3(b), respectively. To better illustrate the image quality of the LSA method, the reconstructed images comparing to the anatomical MR imaging of the subject are shown in Fig. 4.

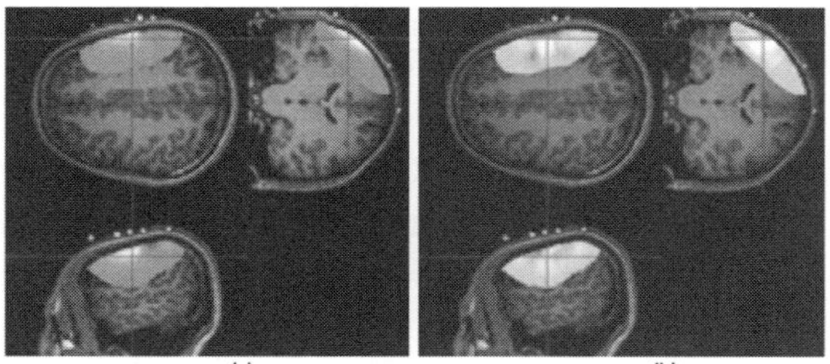

(a) (b)

Figure 3. The 3D activation maps of (a) the DOT method and (b) the LSA method.

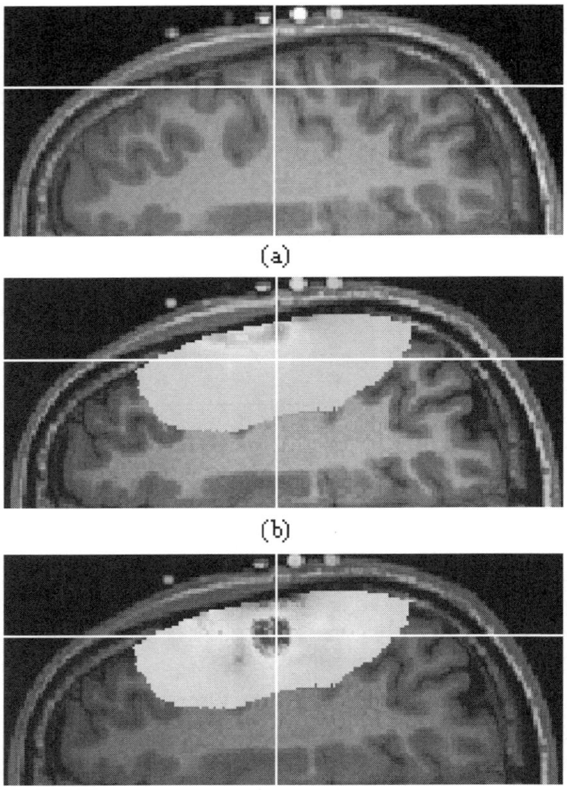

(a)

(b)

(c)

Figure 4. The (a) anatomical MRI of the subject, and reconstructed images by (b) the DOT and (c) the LSA methods. The white crosses in the images indicate the position of central sulcus. The left is the direction of nose. The up is left hand.

5. Results

5.1. *Balance of Photon Density by the LSA Method*

In DOT, photon density descends rapidly as depth increasing. As can be seen in Fig. 2(a), the dark part in the imaging region has small photon density. Contrarily, the bright part near the surface of head has much larger photon density, approximately hundreds times (the number would be much larger if the voxels no deeper than 1 cm from the surface of head was not removed) of that of the dark part. This is mainly because that the photons must reflect many times before reaching deep tissues since human tissues are highly scattering medium for near-infrared photons.

As can be seen in Fig. 2(b), the photon density has been effectively balanced by the LSA method. The contrast of photon density between the superficial layers and the deep layers is not essential. And we can see that a large region has become brighter comparing to Fig. 2(a).

Considering that the photon density acts as the weighting coefficient in the image reconstruction, the voxels in superficial layers and deep layers will have similar importance in image reconstruction by using the LSA method, while the voxels in superficial layers will have much more importance than that in deep layers without the LSA method.

5.2. *Improvement in Image Quality by the LSA Method*

Activated regions are clearly located in the superficial layers, as can be seen in Fig. 3(a). This can be observed from all of the three directions. In Fig. 3(a), voxels no deeper than 1.0 cm have been removed from the imaging region. In case of remaining all voxels from the scalp to the 1.0 cm depth in the imaging region, the activated regions would be located in scalp and skull, which can not be activated during the finger-tapping experiment. This is caused by the poor depth resolution of DOT.

The poor depth resolution of DOT caused several problems. Firstly, the reconstructed images are false because no scalp or skull regions would be activated during the experiment. Secondly, with activated regions located in superficial layers, we could not know which regions are actually activated by the experiment. Thirdly, since activated regions would located in superficial layers whatever the experiment is, we could not recognize the differences in activated regions by different experiments. These problems would make it very difficult to analysis and comprehend 3D images of human brain produced by DOT.

Contrarily, activated regions are clearly located in deeper layers by using the LSA method in Fig. 3(b) comparing to Fig. 3(a). This indicates that the LSA method is effective in improving depth resolution of DOT in human brain imaging.

5.3. *Comparing Reconstructed Image with Anatomical MRI*

To further illustrate the improvement in image quality by the LSA method, we compared the reconstructed images with the anatomical MRI of the subject, as can be seen in Fig. 4. The activated regions are located in the superficial layers in Fig. 4(b). The corresponding region in the anatomical MRI contains part of skull, CSF and part of gray matter. Since skull and CSF can not be activated during the neural activity, it can be clearly told that the activated regions in Fig. 4(b) bias to superficial layers. Furthermore, the hand area motor cortex locates near a knob shaped structure in the pre-central gyrus, as has been revealed by previous work.[21] Since the knob shaped structure is deep in the gray matter, we can assert that the activated region in Fig. 4(b) is falsely located, though it contained part of gray matter.

The activated region in Fig. 4(c) is located deep in the gray matter. The corresponding region in the anatomical MRI is clearly located near the knob shaped structure in the pre-central gyrus, as can be seen in Figs. 4(a) and 4(c). This indicates that the activated region in Fig. 4(c) is properly located on the hand area motor cortex. This is mainly because of the significant improvement in depth resolution by the LSA method. The 3D reconstructed image using the LSA method will be useful in studying human brain functions since it can properly locate the activated regions.

Furthermore, the activated region in Fig. 4(c) spans both the pre-central and post-central gyri. This is because that the motor cortex and sensory cortex are both activated during the finger tapping experiment since fingers touch each other during the experiment, as has been revealed in previous work.

6. Discussions

In this paper, we implemented the LSA method in human brain imaging. Results from the finger tapping experiment showed that by using the LSA method, the activated regions can be properly located on hand area motor cortex. This work indicated that the LSA method could significantly improve the depth resolution in case of human brain imaging, thus made the 3D imaging of human brains possible using the optical method.

In this paper, single subject was used for the experiments because the Monte Carlo simulations were very time-consuming. In the future, we should validate the results in this paper with more subjects. In this paper, we only implemented optical experiments, and there were no comparisons between optical results and fMRI. To further validate the advantages of LSA method, implement the optical and fMRI experiments simultaneously will be necessary.

7. Conclusions

This paper studied the human brain functions using DOT. The forward matrix used in the image reconstruction in DOT was adjusted by the LSA method to balance the sensitivity between superficial layers and deep layers. We implemented the finger tapping experiment using a CW instrument with one subject. In the reconstructed image without the LSA adjustment, the activated region biased to the superficial layers which contained skull and CSF. Contrarily, the activated region could be observed located near a knob shaped structure on the pre-central gyrus by using the LSA method. This indicated that the activated region was properly located on the hand area motor cortex. It could also be observed that the activated region with the LSA method spanned both pre- and post-central gyri. This was mainly because of the finger touches during the experiment.

The LSA method can significantly improve the depth resolution of DOT. This makes the 3D analysis of human brain functions using the diffuse optical method possible.

References

1. F. F. Jobsis, "Noninvasive infrared monitoring of cerebral and myocardial oxygen sufficentcy and circulatory parameters," Science **198**, 1264-1267 (1977).
2. J. P. Culver, A. M. Siegel, J. J. Stott, and D. A. Boas, "Volumetric diffuse optical tomography of brain activity," Opt. Lett. **28(21)**, 2061-2063 (2003).
3. J. C. Hebden, A. Gibson, R. M. Yusof, N. Everdell, E. M. Hillman, D.T. Delpy, S. R. Arridge, T. Austin, J. H. Meek, and J. S. Wyatt, "Three-dimensional optical tomography of the premature infant brain," Phys. Med. Biol. **47(23)**, 4155–4166 (2002).
4. A. Villringer and B. Chance, "Non-invasive optical spectroscopy and imaging of human brain function," Trends Neurosci. **20**, 435-442 (1997).
5. H. Obrig, and A. Villringer, "Beyond the visible—Imaging the human brain with light," J. Cereb. Blood Flow Metab. **23**, 1-18 (2003).
6. A. Maki, Y. Yamashita, Y. Ito, E. Watanable, Y. Mayanagi, and H. Koizumi, "Spatial and temporal analysis of human motor activity using noninvasive NIR topography," Med. Phys. **22**, 1997-2005 (1995).
7. D. A. Boas, K. Chen, D. Grebert, and M. A. Franceschini, "Improving the diffuse optical imaging spatial resolution of the cerebral hemodynamic response to brain activation in humans," Opt. Lett. **29**, 1506-1509 (2004).
8. D. A. Boas, A. M. Dale, and M. A. Franceschini, "Diffuse optical imaging of brain activation: approaches to optimizing image sensitivity, resolution, and accuracy," NeuroImage **23**, 275-288 (2004).
9. D. A. Boas, T. Gaudette, G. Strangman, X. Cheng, J. J. A. Marota and J. B. Mandeville, "The Accuracy of Near Infrared Spectroscopy and Imaging during Focal Changes in Cerebral Hemodynamic," NeuroImage **13**, 76-90 (2001).
10. D. A. Boas and A. M. Dale, "Simulation study of magnetic resonance imaging-guided cortically constrained diffuse optical tomography of human brain function," Appl. Opt. **44(10)**, 1957-1968 (2005).
11. B. W. Pogue, and K. D. Paulsen, "High-resolution near-infrared tomographic imaging simulations of the rat cranium by use of a priori magnetic resonance imaging structural information," Opt. Lett. **23**, 1716-1718 (1998).
12. Qing Zhao, Lijun Ji, and Tianzi Jiang, "Improving the depth resolution of diffuse optical tomography with a layer-based sigmoid adjustment method," Opt. Express **15(7)**, 4018-4029 (2007).
13. Q. Zhao, L. Ji and T. Z. Jiang, "Improving performance of reflectance diffuse optical imaging using a multi-centered mode," J. Biomed. Opt. (Accepted).
14. D. W. Shattuck and R. M. Leahy, "BrainSuite: an automated cortical surface identification tool," Med. Image Anal. **6(2)**, 129–142 (2002).
15. D. A. Boas, J. P. Culver, J. J. Scott and A. K. Dunn, "Three dimensional Monte Carlo code for photon migration through complex heterogeneous media including the adult human head," Opt. Exp. **10(3)**, 159-170 (2002).
16. S. R. Arridge, "Optical tomography in medical imaging," Inverse Probl. **15**, R41-R93 (1999).
17. H. W. Engl, M. Hanke and A. Neubauer, "Regularization of Inverse Problems," Kluwer, Dordrecht, 1996.
18. L. Wu, "A parameter choice method for Tikhonov regularization," Electron. T. Numer. Ana. **16**, 107-128 (2003).

19. P. C. Hansen and D. O'Leary, "The Use of the L-Curve in the Regularization of Discrete Ill-Posed Problems," SIAM J. Sci. Comput. **14**, 1487–1503 (1993).
20. P. C. Hansen, Regularization Tools, a Matlab Package for Analysis and Solution of Discrete Ill-posed Problems, Version 3.0 for Matlab 5.2, Technical Report UNIC-92-03, Danish Computing Center for Research and Education, Technical University of Denmark, June 1992 (Revised June 1998). Available online from http://www.netlib.org/numeralgo/na4. Version 2.1 for Matlab 4.2c is also available on line from http://www.netlib.org/numeralgo/na4-matlab4.
21. W. Penfield and E. Boldrey, "Somatic motor and sensory representation in the cerebral cortex of man as studied by electrical stimulation," Brain **60**, 389-443 (1937).

EFFECT OF SURFACE ROUGHNESS ON DETERMINATION OF OPTICAL PROPERTIES OF SIMULATION BIOLOGICAL SAMPLE

MEIXIU SUN, CHUNPING ZHANG, JIANGUO TIAN

Photonics Center, Institute of Physics, The Key Laboratory of Weak Light Nonlinear Photonics, Ministry of Education and Tianjin Key Lab of Photonics Materials & Technology for Information Science, Nankai University, Tianjin 300457, China

SHENGWEN QI

Dept. of Physics, Dezhou University, Dezhou 253023, China

YEQING SU

Freescale Semiconductor, Inc, Tianjin 300385, China

In order to estimate the effect of surface roughness on determination of the optical properties of biological tissues, the nylon bar, nylon film and Intralipid suspension are used as tissue simulation sample. The optical properties of the nylon with different surface roughness for the rough and smooth incident region of probe beam were determined by a spatially resolved steady-state diffuse reflection technique and goniophotometric measurement technique. And the light distribution in Intralipid suspension with different surface roughness was measurement. Results show that determined tissue optical parameters and measured light distribution are obviously different for different surface roughness of the same tissue and the effect of the surface roughness of the incident region of the probe beam on the determination of tissue optical parameters is predominant. But the tissue optical parameters for the same tissue don't depend on surface roughness. The surfaces of real biological tissues are not smooth and possess a certain grade of roughness, so the determination of tissue optical properties and measurement of light distribution should take surface roughness into account.

1. Introduction

The measurement theory and technique of tissue optical parameters are important for biomedical optics. Knowledge of the optical properties of biological tissue is a prerequisite for many diagnostic and therapeutic applications.[1] The optical properties of biological tissue are modeled by absorption coefficient (μ_a), scattering coefficient (μ_s), reduced scattering coefficient (μ_s') , and anisotropy of scattering (g).[2] Several techniques, for example spatially resolved steady-state diffuse reflectance technique and goniophotometric measurement technique, have developed and applied to measure the optical properties and no one considered the effects of surface roughness when they determined the optical properties used the techniques. The investigation of the interaction of light with tissues can understand the characteristics of light propagation and distribution in biological tissues, which are determined mostly by the tissues' optical properties. [3, 4] By measurement of the light distribution in biological tissues, the scattering characteristics of the tissues can be understood and the optical properties of biological tissues can be obtained.

The rough surface effects become important when the light propagates through the rough interfaces between two media with different refractive indices. [5] In order to estimate the effect of surface roughness on determination of the optical properties of biological tissues and quantitative understanding of its effect on light distribution, the nylon bar, nylon film and Intralipid suspension are used as tissue simulation sample in this paper. Nylon bar and nylon film have strong scattering property like Intralipid suspension. The optical properties of the nylon with different surface roughness for the rough and smooth incident region of probe beam were determined by a spatially resolved steady-state diffuse reflection technique and goniophotometric measurement technique. And the light distribution in Intralipid suspension with different surface roughness was measured.

Our results show that determined tissue optical parameters and measured light distribution are obviously different for different surface roughness of the same tissue and the effect of the surface roughness of the incident region of the probe beam on the determination of tissue optical parameters is predominant. We know that the tissue optical parameters for the same tissue do not depend on surface roughness. The surfaces of real biological tissues are not smooth and possess a certain grade of roughness, so the determination of tissue optical properties and measurement of light distribution should take surface roughness into account.

2. Measurement of Optical Properties

Using the extrapolated boundary and the method of image sources to solve the Boltzmann transport equation in the diffusion approximation for a semi-infinite scattering and absorbing medium and for perpendicular incidence of continuous light, the radial dependence of diffuse reflectance can be expressed as: [6]

$$R(\rho)=\frac{1}{4\pi}\left[\begin{array}{l} z_0(\mu_{eff}+\frac{1}{r_1})\frac{\exp(-\mu_{eff}r_1)}{r_1^2} \\ +(z_0+2z_b)(\mu_{eff}+\frac{1}{r_2})\frac{\exp(-\mu_{eff}r_2)}{r_2^2}\end{array}\right] \quad (1)$$

where

$$z_0=(\mu_a+\mu_s')^{-1},\ z_b=2D\frac{1+R_{eff}}{1-R_{eff}},$$

$$r_1^2=z_0^2+\rho^2,\ r_2^2=(z_0+2z_b)^2+\rho^2,$$

where $D=1/3[(\mu_a+\mu_s')]$ is the diffusion coefficient, $\mu_{eff}=[3\mu_a(\mu_a+\mu_s')]^{1/2}$ is the effective attenuation coefficient, μ_a and μ_s' are the absorption coefficient and the reduced scattering coefficient respectively, ρ is the radial distance from the observation point to incident point of light, z_0 is the depth of one transport mean free path in the medium, z_b is the distance from the extrapolated boundary to the interface between air and the tissue. R_{eff} =0.604 [7] for the nylon is the effective reflection coefficient, representing the fraction of photons that is internally diffusely reflected at the boundary. It can be known from Eq.(1) that the only unknown parameters are μ_a and μ_s'. Therefore we can obtain an estimate for them by fitting Eq.(1) to the measured diffuse reflectance profile.

To estimate μ_s, measurement of the value of anisotropy parameter (g) is required. Goniophotometric measurements of scattering phase function performed on nylon film can be used to determine anisotropy of scattering (g) of nylon bar. One can obtain a better estimation for g from the scattering phase function measured by fitting the scattering phase function to the following phase function composed of two H–G functions [8]:

$$P_{HG-HG}(\theta)=(1/4\pi\omega_0)[\beta(1-g_1^2)/(1+g_1^2-2g_1\cos\theta)^{3/2}$$
$$+(1-\beta)(1-g_2^2)/(1+g_2^2-2g_2\cos\theta)^{3/2}] \quad (2)$$

In this case anisotropy parameter g can be obtained as

$$g=\beta g_1+(1-\beta)g_2 \quad (3)$$

3. Materials and Methods

The experimental setup of diffuse reflectance is shown in Fig.1. A He-Ne laser beam (632.8nm) was chopped (1.04 khz) after through a diaphragm and then was reflected by a mirror. The beam was collimated with a lens and then focus on the surface of test sample for ensuring the narrow collimated beam to incident normally upon the sample surface. Diffuse reflected

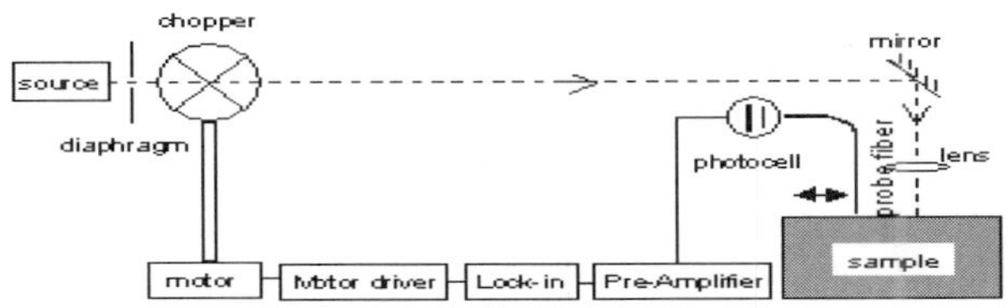

Fig. 1. Schematic of the setup for diffuse reflectance measurements.

light signals from the sample surface received by the optical probe fiber, were transmitted to a photocell. The photoelectrical signals of the photocell were fed into a preceding amplifier and a lock-in amplifier. Moved the probe fiber horizontally and the diffuse reflectance profile at varying distances from the center of the incident beam on the sample surface was recorded.

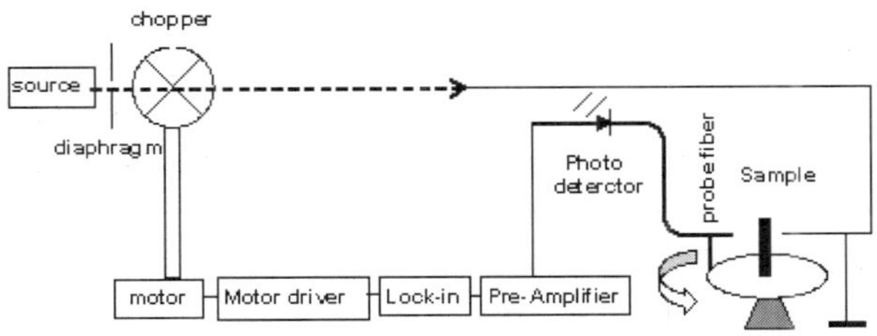

Fig. 2. Schematic of the goniophotometric setup for the measurement of anisotropy parameter *g*.

The distances between the probe fiber end and the sample surface were kept same in the measurement, and then the estimate for μ_s' and μ_a were obtained by using the diffuse equation for reflectance and fitting experimental data with nonlinear regression.

A schematic diagram of the goniophotometric setup for measurement of anisotropy parameter g is shown in Fig. 2. A He–Ne laser beam (632.8nm) was chopped and then injected the nylon film. An optical fiber kept on a rotational stage was used to collect scattered light from different angular positions and then the scattered light signals were transmitted to a photodetector. The photoelectrical signals of the photodetector were fed into a preceding amplifier and a lock-in amplifier. The signals collected from several angular positions (10°–140°) were recorded with lock-in amplifier. To minimize the multiple scattering effects and to make the nylon film suitable to be polished by sand papers and to form different surface roughness, we used samples with thickness of ≤40 μm.

The detection system used in the above mentioned experiment was carefully checked before detection. The sensitivity and linearity of this photoelectric system were satisfying with the measurement.

To obtain the sample with different surface roughness for rough and smooth surface of incident region of probe beam, the sample surfaces of nylon bar and nylon film were polished by the sand papers with the type (mesh number) 120 (90), 50 (100), 180 (120), 240 (160), 320(220), 400 (280), 500 (320), 600 (400), 700 (500), 800 (600) respectively. We measured the surface probfile of the nylon bar and nylon film polished by the sand papers with all the types mentioned above. The corresponding measured values of the surface roughness Ra of nylon bar were 3.194, 2.912, 2.636, 2.409, 1.661, 1.348, 1.253, 0.934, 0.781 and 0.620(μ m) respectively and it of the nylon film were 4.542, 3.961, 3.194, 2.898, 2.619, 2.468, 1.788, 1.277, 0.934 and 0.627(μ m) respectively. Here Ra is rated as the arithmetic average deviation of the surface valleys and peaks expressed in micrometers. Then we made the incident region on the nylon surface smooth with Ra=0.2 μ m, and the incident region was larger than the spot size of the incident beam shown in Fig.3 (smooth surface of the incident region).

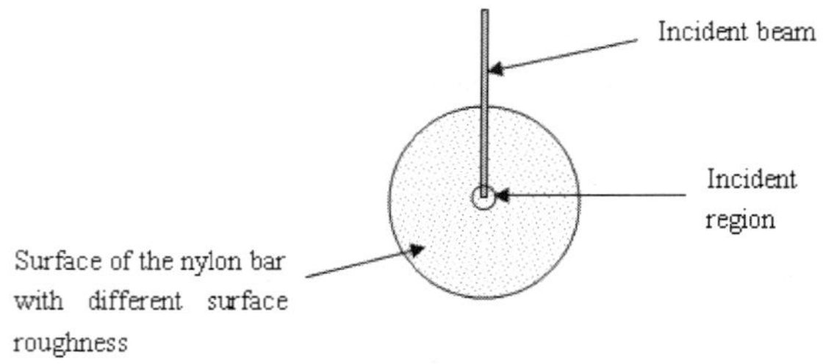

Fig. 3. schematic of incident region.

4. Light Distribution Measurement

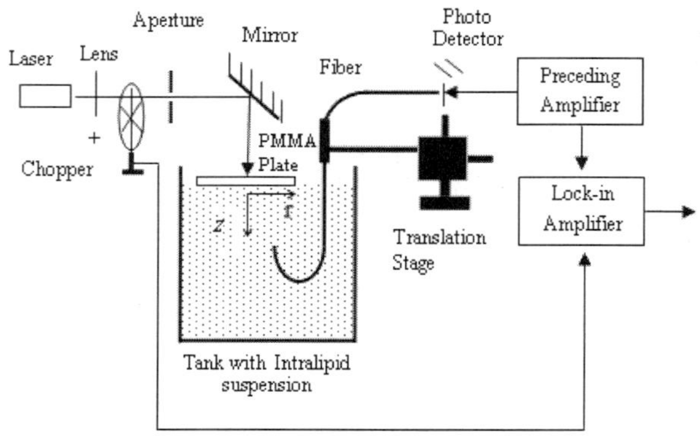

Fig. 4. Experimental setup for determining the light distribution in an Intralipid-20% suspension.

The experimental setup shown in Fig.1 is similar with reference [9]. We chose Intralipid-20% suspension as the simulated material of biological tissue in the experiment. The Intralipid-20% was diluted with concentration 5.18% solution by distilled water. A PMMA plate was put on the upper surface of Intralipid suspension. The upper surface of the PMMA plate is smooth and the undersurface of the PMMA plate is rough, so the interface between the PMMA plate and the intralipid suspension is rough. We polished the undersurface of the PMMA plate to change the roughness of the interface by the sand papers with the type (mesh number) 120 (90), 240 (160), 320 (220) and 800 (600) respectively. Then the surface profiles of the undersurface of the PMMA plate were measured and the corresponding measured values of surface roughness Ra were 1.258, 0.856, 0.673, 0.227μm respectively. The diffusion light signals in Intralipid suspension with different interface roughness received by the optical fiber probe, whose bent mode shown in fig.4 to measure the forward scattering light, were transmitted to a photodiode and amplifier. A fiber translation stage was used to move the optical fiber probe along z and r directions with an error of less than 0.01mm.

5. Experimental Results

The determined optical parameters (μ_s', μ_a, g, μ_s) of the nylon with different surface roughness for the rough and smooth surface of the incident region respectively as a function of the surface roughness Ra are shown in Fig.5 and Fig.6.

The determined reduced scattering coefficient of the nylon bar with different surface roughness for the rough and smooth surface of the incident region respectively decreases with the decrease of the surface roughness and goes to a constant. The determined absorption, scattering and anisotropy coefficients of the nylon with different surface roughness for the rough and smooth surface of the incident region respectively decrease with the increase of the surface roughness.

From Figs.5 and 6, we know that the change degree in the optical parameters of the nylon relative to the surface roughness for the smooth surface of the incident region is obviously less than that for the rough surface of the incident region, so the effect of the surface roughness of the incident region on the determination of tissue optical parameters is predominant

The experimental results for the dependence of the forward-scattering intensity on the distance z and r in the Intralipid-20% suspension with the different interface roughness are shown in Fig.7. When the interface roughness increases, the changes in the measured forward-scattering intensity in Intralipid suspension with z slows, the peaks of the forward-scattering intensity decrease respectively, and positions of the peaks shift slightly to smaller z values respectively; the changes in the measured forward-scattering intensity relative to r accelerate and appear at they maxima, and positions of the peaks shift slightly to larger r values respectively. The curves of the forward-scattering relative to z, r inside the Intralipid-20% suspension for the different interface roughness tend to overlap when z, r increase respectively.

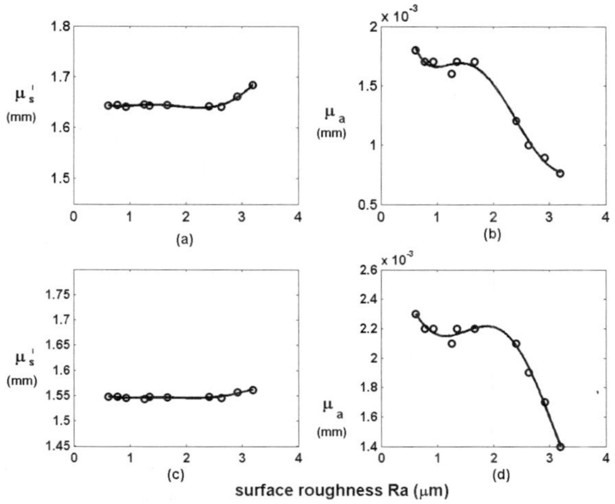

Fig. 5. The reduced scattering and absorption coefficients as a function respectively of surface roughness, for (c) and (d), surface roughness of incident regions is smooth.

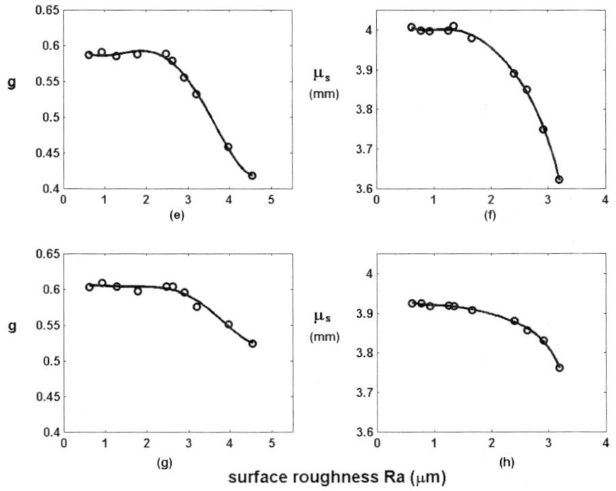

Fig. 6. The anisotropy and scattering coefficients as a function respectively of surface roughness, for (g) and (h), surface roughness of incident regions is smooth.

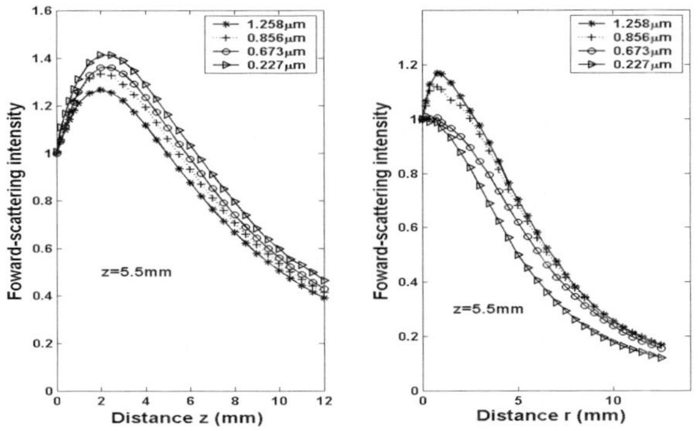

Fig. 7. The dependence of forward-scattering intensity on the distance z, r in intralipid suspension with different interface roughness.

6. Conclusions

The optical properties of the nylon with different surface roughness for the smooth and rough surface of the incident region respectively were determined by the spatially resolved steady-state diffuse reflectance technique and the goniophotometric measurement technique. The diffusion model fitting of the measured reflectance-distance curves and the double H–G function fitting of the measured scattering-angle curves are all based on the assumption of smooth surface even though the surface is rough to estimate the effect of surface roughness on data inversion if surface roughness is not considered. Results show that determined tissue optical parameters and measured light distribution are obviously different for different surface roughness of the same tissue and the effect of the surface roughness of the incident region of the probe beam on the determination of tissue optical parameters is predominant. We know the optical parameters of a diffusive medium are intrinsic to the medium and cannot depend on surface roughness. The surfaces of real biological tissues are not smooth and possess a certain grade of roughness, so the determination of tissue optical properties and measurement of light distribution should take surface roughness into account.

Acknowledgments

Supported by National Nature Science Foundation of China grants (60578020), Nature Science Foundation of Shandong (Y2006A01) and Program for Changjiang Scholars and Innovative Research Team in University.

References

1. A. Kienle, K .Forster, and R. Hibst, Opt Lett. **26**, 1571(2001).
2. N.Ghosh, S.K.Mohanty, S.K.Majumder et al, Appl. Opt. **40**, 176(2001).
3. Preuss LE and Profio A.E, Appl. Opt. **28**, 2207 (1989)
4. Cheong W F, Prahl SA and Welch A J, IEEE J. Quantum Electron. **26**, 2166 (1990)
5. Ishimaru A Appl. Opt. **28**, 2210(1989)
6. Kienle and M.S.Patterson, J. Opt. Soc. Am. A. **14**, 246 (1997)
7. R. C. Haskell, L.O.Svaasand, T.T.Tsay et al, J. Opt.Soc.Am.A.**11**, 2727(1994).
8. R. Marchesini, A. Bertoni, S. Andreola, E. Melloni et al, Opt. **28**, 2318 (1989)
9. T. Xu, C. P. Zhang, X.Y. Wang et al, Appl. Opt. **42**, 5777 (2003)

DETECTION AND ANALYSIS OF FLUORESCENCE IMAGE FOR PESTICIDE RESIDUES OF TYPICAL VEGETABLE

XIAOLEI YU, ZHIMIN ZHAO

Nanjing University of Aeronautics and Astronautics, Nanjing, P.R. China

Detection of the Pesticide Residues of vegetable is important for human life, which has attracted increasing attention in recent year. Feebleness light signal examining instrument with high sensitivity was adopted to capture the images of pesticide residues of typical vegetable and store them into computer. Spatial distribution of fluorescence property for related substance could be acquired based on the spectral data and image detection. Software-defined performance function of image processing was used to analyze experimental data deeply. Through data processing and analysis, pesticide residues of typical vegetable could be detected nondestructively. Results could be displayed intuitively. Through feature analysis of the fluorescence property for pesticide residues of typical vegetable, the experimental results and analysis were given. Some useful data could not be acquired by the traditional method. However, the new method can give the further analysis and experimental results. The technique of feature analysis for fluorescence of typical vegetable has wide use in scientific research, which also lays a foundation for advanced research.

Key word: Typical vegetable; Pesticide residues; Fluorescence property; Image detection

1. Introduction

The so-called pesticide residue refers to the pesticide residue of food after spraying pesticide, including the pesticide itself, the metabolism thing of the pesticide, and the poisonous impurity. It will cause poisonous function in human body when people eat some food remaining the pesticide, and this is called poisoning by the pesticide residue. Therefore, people attach importance to the problem of the pesticide residue when the pesticide promotes to the development the agriculture, wood, and livestock husbandry [1-3].

Although there are some mature detection methods of pesticide residue with high precision, their pre-processing of the samples is very complex [4-5]. These methods all need to take a series of processing of the samples, which not only needs much time and work, but also damages the samples in some way. So these methods cannot be suitable in our country and be used to quickly real-time detection of pesticide residue. If there is a method can be used to examine the pesticide residue of vegetable directly without damage, the method is very useful and valuable. So we should do research to get a quick and easy detection method of the pesticide residue urgently.

Based on the domestic and international research of the pesticide residue, this paper discusses the technique of fluorescence analysis [6-8], which is used to detect the pesticide residue of vegetable without damage through analyzing characteristics of the fluorescent image of pesticide. This is a quick and lossless new detection method.

2. Basic Theory of Fluorescence Analysis

When the material molecules absorb the photon of characteristics frequency, they will transit original ground state to vary and vibrating electron exciting state. Excited molecules consume partial energy at the time of impacting with the molecules surrounding them, and rapidly decline to the minimum vibrating level of the first electron exciting state, and seconds later, at about 10^{-9}, they continue to decline to vary and vibrating electronic original state after releasing superfluous energy through the direct form of light. This light is just fluorescence.

Generally, we can use the following parameters to describe the characteristics of the fluorescent spectrum: the distribution of fluorescent spectrum, the peak intensity of fluorescent spectrum, the width of fluorescent spectrum and so on. The characteristics of fluorescence is different under the same exciting condition because of the vary energy configuration of different substances. Through measuring the parameters of the fluorescent characteristics, we can distinguish between different substances.

Intensity of fluorescence that is usually used in the fluorescence analysis is related with the fluorescent intensity which is emitted by the substantial solution after absorbing light, the absorption coefficient and content of fluorescent substance, and the fluorescent efficiency.

The quantum generating rate of fluorescence is defined that the ratio of the fluorescent photons emitted by substance after absorption to the fluorescent photons of exciting light absorbed by substance. We use Y_F to show the rate, which is always less than 1. The bigger the rate is, the stronger the substance fluorescence.

Set that K is the instrumental constant of detection system, C is the consistence of the fluorescent substance, b is the thickness of sample, and ε is the mol absorption coefficient. Assume that a light which density is I_0 in one second per square meter irradiates to a sample that its absorption section is A, and we just consider the absorption light and emitting fluorescence of a very thin folium dx that x away from the surface. Before entering the folium, the intensity of the incident light is $AI_0e^{-\varepsilon Cx}$, and the intensity becomes to $AI_0e^{-\varepsilon C(x+dx)}$ after passing the folium. As εCdx is very small, through the integral of thickness b, and we can get the fluorescent intensity per square meter I_F:

$$I_F = KY_FI_o(1-e^{-\varepsilon Cx}) \qquad (1)$$

From the equation (1), we can see that the fluorescent intensity is related with the equipment. The value of the same sample measured by different equipment or even by the same equipment with different testing conditions is always distinguishing. We can also see that the fluorescent intensity is in direct ratio of the quantum generating rate of the sample itself and the intensity of exciting light. Quantum generating rate that means the ability of emitting fluorescence of the sample is the ratio of the number of photons emission and absorption, and the more quantum generating rate is, the stronger fluorescence is. The intensity of exciting light is related with the intensity of fluorescence because that it affects the power of light directly.

3. Test Results and Analysis

We use a high sensitivity testing instrument of weak light to do experiment. The fluorescence exciting and intensity testing of the sample system is showed in Fig. 1. This system is mainly composed of optical lens, image collection system, accumulation control processor and so on. The relative aperture of the optical lens used to collect fluorescence, that is, D/F=1/1.4,and its wavelength response range is 300-3000nm. The level resolving power of equipment is 600TV, and its SNR is more than 48db. So this machine is suitable to test the weak light such as this kind of fluorescence.

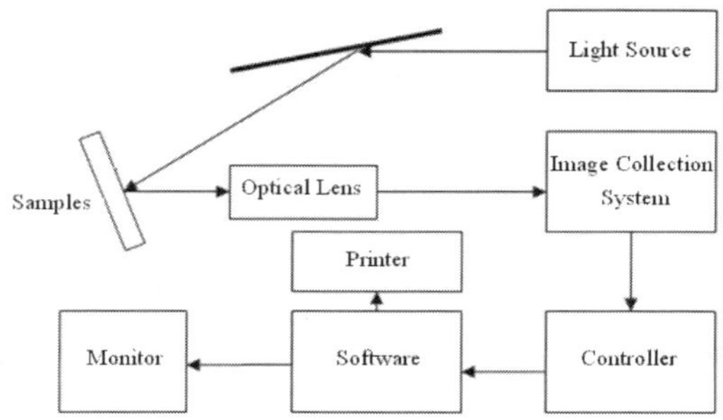

Fig. 1. Structure of the testing system

After excited by light, the sample emits fluorescence which has been filtrated by filter, focused and took photographs by optical lens. Now it will be transmitted to the accumulation control processor (controller), and the monitor will display the spot of the fluorescence emitted by the sample after the computer deals with the fluorescent signal detected. The intensity of fluorescence is related with the lightness of the spot.

We choose the representative fresh vegetables used in daily life such as cabbage, white Cabbage, and cucumber, which is marked sample 1, sample 2, sample 3 as the experimental sample. We select the beta-cypermethrin, bacillus thuringiensis and imidacloprid as the pesticide.

4. Research Method and Result Analysis

The on-line neps detection system based on machine vision has the function of high data collection, wide breadth recognition of neps characteristics, on-line and real-time processing (counting the number of neps), display, alarm and so on.

The combined light source can increase the intensity of illumination. Firstly, we used the combination of several linear CCD to partly collect the parallel image of the wide breadth surface in high speed. Secondly, we recognized and located neps through high speed image processing software. Thirdly, we counted the number of neps which were displayed time-sharing. The method overcomes the shortcoming such as low speed processing and narrow imaging breadth in previous linear CCD. The experiment shows that the system can on-line detect neps at the linear speed of 3m/s, the breadth of detection can reach 1m, and the area of detection can reach 0.5mm × 0.5mm. The on-line image processing software which has friendly interface and powerful function is used in the neps recognition, location, counting, and on-line monitoring the quality of products.

We can do further research in the system so that it can on-line detect the surface disfigurement of broad and moving object (such as steel, grass and prints). The system is of momentous practical significance and broad application in the field of automation devices and detection technology.

4.1 *Research Method*

From the relative report, we can see that the researcher usually use ready-made spectral equipment to find the distinct spectral peak value of the sample which distinguishes from other substances when we use fluorescence analysis to research the pesticide residues. But this method can not be used to see the spatial distribution of fluorescent substance, especially in measuring the pesticide residue of the surface of vegetable. We use a high sensitivity detector of weak light, and it transmits the fluorescent image that detected by the image collection system to the computer, then we can get some important data that can not be acquired through a normal spectrum equipment describing the spatial distribution of fluorescent substance.

In order to analyze the experimental data, we adopt the software programmed by ourselves, and define the relative functions of image processing which can directly display the needful results.

4.2 *The Pesticide Residue Analysis of the Vegetable Samples*

We find that the characteristics of the same sample, including the intensity of fluorescence and the rule changing of fluorescence after affiliating pesticide, is the same in spring and summer. But it is different in autumn and summer. So we mainly discuss the sample analysis in summer and autumn. It can be found that the acquired fluorescence of sample is very weak when the wave of exciting light is more than 450nm, such as 480nm, 500nm, 550nm, 600nm and 650nm. The fluorescent peak value of vegetable exists both in 430nm and 660nm when choosing the 365nm wavelength of the exciting light, but the acquisition value of the 660nm is very weak. So we choose 365nm as the wavelength of exciting light, and acquire fluorescent image at the wavelength of 430nm.

We adopt the method of histogram to analyze the intensity of fluorescence, which is a simple and useful tool in image processing. The histogram describes the gray-level content of an image. Each histogram includes the objective information, and completely describes some certain images. The function of gray-level value in histogram describes the number of pixels of histogram in the image in which abscissa represents the gray-level of pixels and longitudinal coordinates represents the frequency of this gray-level arising (numbers of pixels).

We spray pesticide to the selected samples, and process the images before and after sprayed. We will get s minus histogram of the image before and after sprayed. The minus histogram mainly records that the gray-level distribution of pesticide in vegetable. Figure 2(a), 2(b) and 2(c) are histograms of sample 1, sample 2 and sample 3 after sprayed by beta-cypermethrin.

The gray level in figure 2(a) is concentrated in (109,134) (184,255); the gray level in figure 2(b) is concentrated in (87,137) (195,255); the gray level in figure 2(c) is concentrated in (105,129) (190,255).

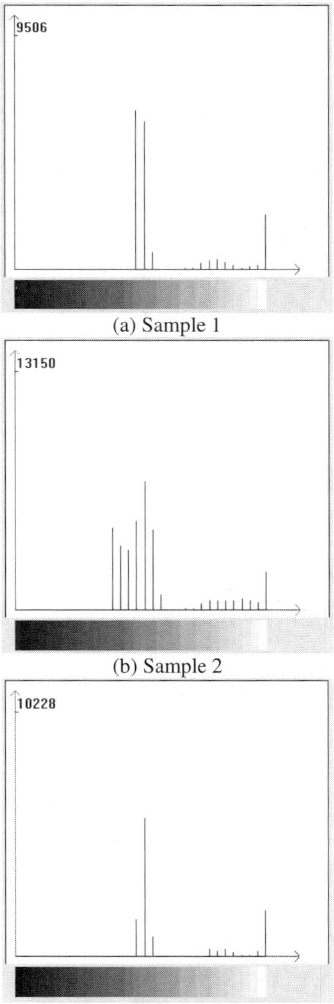

(a) Sample 1

(b) Sample 2

(c) Sample 3

Fig. 2. The histogram of sample after sprayed by beta-cypermethrin

Figures 3(a), 3(b) and 3(c) are the histograms of sample 1, 2 and 3 sprayed by bacillus thuringiensis. The gray-level in figure 3(a) is concentrated in (59,105); the gray-level in figure 3(b) is concentrated in (54,164); the gray-level in figure 3(c) is concentrated in (50,103).

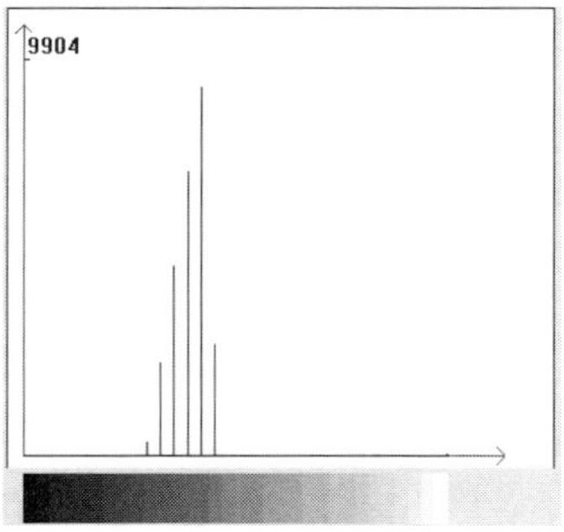

(a) Sample 1

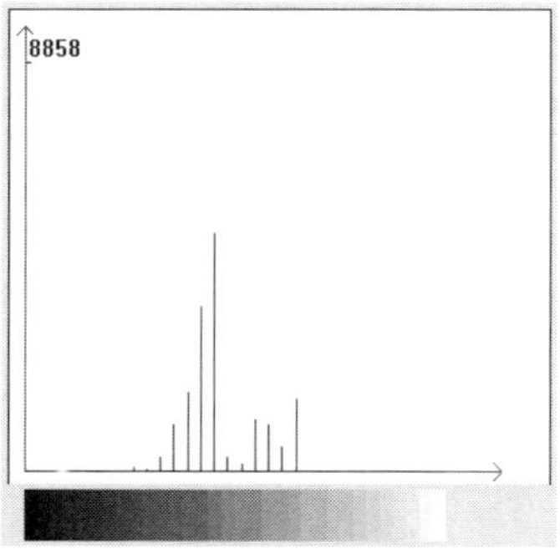

(b) Sample 2

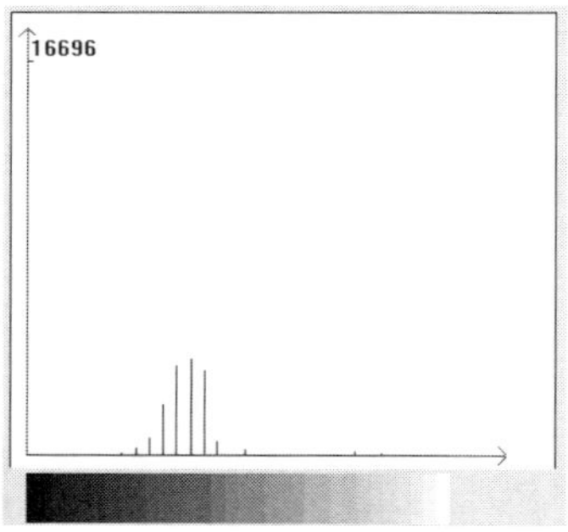

(c) Sample 3

Fig. 3. The histogram of sample after sprayed by bacillus thuringiensis

Figures 4(a), 4(b) and 4(c) are the histograms of sample 1, 2 and 3 sprayed by imidacloprid. The gray-level in figure 4(a) is concentrated in (77,105), (199,255); the gray-level in figure 4(b) is concentrated in (61, 65), (233,240); the gray-level in figure 4(c) is concentrated in (71, 75), (231,239).

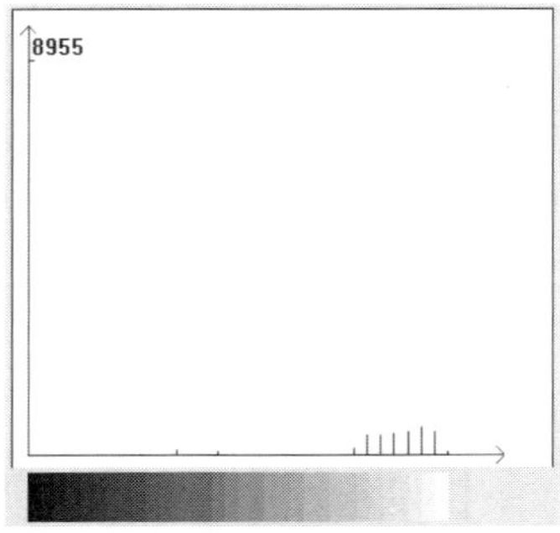

(a) Sample 1

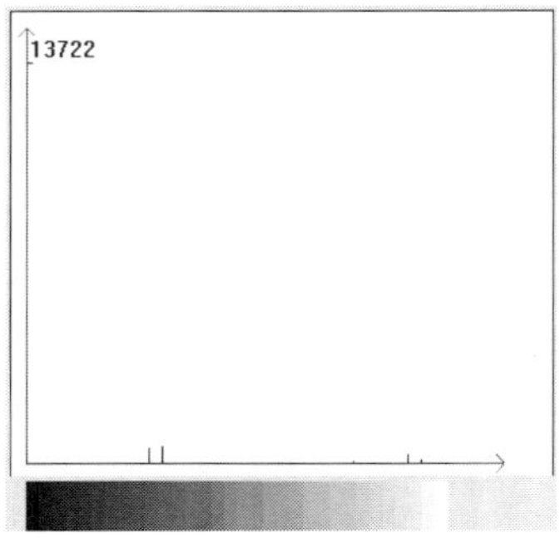

(b) Sample 2

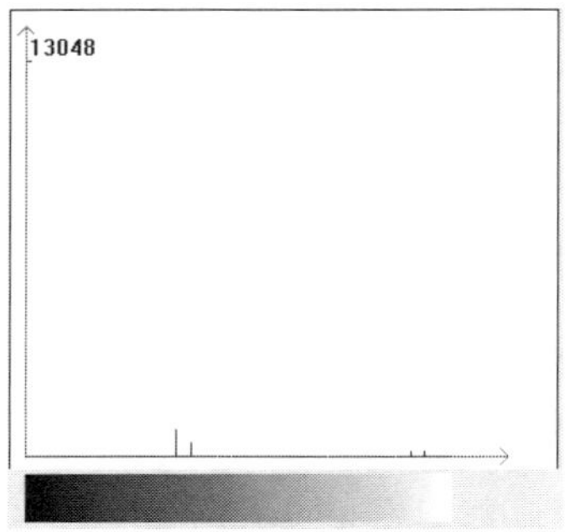

(c) Sample 3

Fig. 4. The histogram of sample after sprayed by imidacloprid

From the above figures we can see that the gray-level distribution of histogram is related with characteristics of the pesticide. There are two sections in the gray-level distribution of the vegetable sprayed by beta-cypermethrin and pixel in the gray-level of 255 is much more than others in which we can know the fluorescence is much stronger. Gray-level distribution is not continuous, and this discontinuous gray-level is big. There are more pixels in high and middle gray-level, so this pesticide should contain at least two fluorescent groups which emit different bright fluorescence. The gray-level distribution of bacillus thuringiensis is continuous while imidacloprid gray-level distribution is discontinuous. The fluorescence of the vegetables sprayed by these two kinds of pesticide is very weak because there are little pixels in the gray-level of 255 both of them.

5. Conclusions

From the above experimental results, we can find that: the selected pesticides are all have fluorescence, and the fluorescence of pesticide beta-cypermethrin is the strongest; the samples themselves also have fluorescence, which is related with their characteristics. For examples, the color and the shape of samples and the thickness of samples skin decide the fluorescence of samples together. So it is very important to select suitable samples.

The problem of measuring the pesticide residue of vegetables is related to people's life, and it attracts more attention in recent years. Through the relative study, we have got some initial useful result. We use image processing software programmed by ourselves in order to deal with the acquired information, and measure the pesticide residue of the vegetable surface without damage. The above work lays the foundation of further research.

References

1. Wang Mei, Duan Jin-sheng, Sun Ming-na, et al. Research progress on techniques of reducing pesticide residues in agro-product [J]. *Pesticides*, **46(7)**: 442-446 (2007).
2. Yang Xiu-li, Yan Guang-fan. Research on current situation and limit standards comments of pesticide resident in Chinese medicine [J]. *Journal of Chongqing University of Posts and Telecommunications*, **19(B06)**: 159-161 (2007).
3. He Juan, Wan Zheng-kai, Kang Chang-an, et al. Analysis of 16 pesticides residues in vegetable by SIM-GC-MS [J]. *Modern Agrochemicals*, **6(3)**: 27-29, 39 (2007).
4. Wang Xiang-hong, Cui Xiao-jun, Li Xin, et al. Summarized of organophosphorus pesticides analysis method in food [J]. *Food Research and Development*, **27(11)**: 190-194 (2006).
5. Zhang Hui-ting, Ye Gui-biao, Li Wen-ming, et al. The application of molecularly imprinted polymer sensor in pesticide detection [J]. *Chinese Journal of Pesticide Science*, **8(1)**: 8-13 (2006).
6. Zhao Zhi-min, Guo Lin-feng, Yu Xiao-lei, et al. Analysis and application of excitation fluorescence intensity of blood [J]. *Spectroscopy and Spectral Analysis*, **25(12)**: 2056-2060 (2005).
7. Zhao Zhi-min, Chen Yu-ming, Yu Xiao-lei. Characteristic analysis and application of blood fluorescence spectra [J]. *Spectroscopy and Spectral Analysis*, **23(5)**: 922-925 (2003).
8. Wang Le-xin, Zhao Zhi-min, Xin Yu-jun, et al. Experimentation research on excitation fluorescence intension of normal and abnormal cow blood [J]. *Applied Laser*, **26(5):** 339-340 (2006).

Topic II

Optical Molecular Imaging

LABEL-FREE DETECTION OF PROTEIN MICROARRAY WITH HIGH THROUGHPUT SURFACE PLASMON RESONANCE IMAGING (SPRI)

SHUKUAN XU [A, B], GUOLIANG HUANG [A, B*], CHENG DENG [A, B], JIANG ZHU [A, B], CHAO HAN [A, B], XIAOYONG YANG [B]

A. Medical Systems Biology Research Center, Tsinghua University School of Medicine, Beijing, China
B. National Engineering Research Center for Beijing Biochip Technology, Beijing, China
**Corresponding Author: Guoliang Huang, E-mail: tshgl@mail.tsinghua.edu.cn*

A surface plasmon resonance imaging (SPRI) system was developed for the discrimination of proteins on a gold surface. As a label-free and high-throughput technique, SPRI enables simultaneously monitoring of the biomolecular interactions at low concentrations. We used SPRI as a label-free and parallel method to detect different proteins based on protein microarray. *Bovine Serum Albumin (BSA)*, *Casein* and *Immunoglobulin G (IgG)* were immobilized onto the Au surface of a gold-coated glass chip as spots forming a 6×6 matrix. These proteins can be discriminated directly by changing the incident angle of light. Excellent reproducibility for label-free detection of protein molecules was achieved. This SPRI platform represents a simple and robust method for performing high-sensitivity detection of protein microarray.

1. Introduction

Surface plasmon resonance (SPR) is a surface sensitive optical detection technique based on total reflection. It measures changes of the refractive index onto a sensing surface coated with a thin layer of gold or silver film.[1-2] The optimum thickness of metallic film is usually about 50nm in order to provide better SPR coupling efficiency.[2] When resonance occurs, a fraction of the energy of incident light at a sharply defined angle can interact with the delocalized electrons in the metal film and reduce the reflected light intensity [3-4] Since its possible use as a biosensor was demonstrated in 1983 by Lidberg *et al.*,[3] SPR has been developed into a label-free and versatile analytical tool and used to detect a number of different biomolecular interactions in real-time[2,5-7] and to follow the kinetics of antibody-antigen reactions.[8]

Surface plasmon resonance imaging (SPRI), which is based on the traditional surface plasmon resonance principle,[1] is a label-free and high-throughput technique to study biomolecular interactions.[9-11] It involves measurement of the difference in reflected light intensity near the resonance response at a fixed-angle, due to differences in refractive index (RI), on the metal surface by using a charge-coupled device (CCD) camera through an imaging lens.[12] The advantages of SPRI furnish us with a tool for simultaneously monitoring of the biomolecular interactions at extremely low concentrations.

In this paper, we used SPRI as a label-free and parallel method to detect different protein molecules. *Bovine Serum Albumin (BSA)*, *Casein* and *Immunoglobulin G (IgG)* were immobilized as a 6×6 matrix of spots on the Au surface of a gold-coated glass chip. Our self-built multifunctional SPR system allowed the measurement of biomolecular interactions in three modes, angular interrogation, wavelength interrogation and intensity measurement. Three kinds of protein molecules can be discriminated directly by various reflected intensity or changing the incident angle of light. Instrument sensitivity was also demonstrated. Excellent reproducibility for label-free detection of protein molecules was achieved.

2. Materials and Methods

2.1. *Materials*

Bovine Serum Albumin (BSA, MW 66,200), *Casein* (MW 24,000), and *Immunoglobulin G* (IgG, MW 150,000) were received from Sigma. 11-mercaptoundecanoic acid (MUA) and 1,1'-carbonyldiimidazole (CDI) were purchased from Aldrich (Milwaukee, USA). Other chemicals used in this study were obtained commercially as the reagent grade, and unless mentioned otherwise, were used as received. Aqueous solutions were prepared in doubly distilled water. All experiments were performed at room temperature unless stated otherwise.

2.2. *Surface Chemistry of Protein Microarray*

Figure 1(a) shows the procedure of surface chemistry. Thin gold films (47 nm) with a 3-nm underlayer of chromium were deposited onto BK7 glass slides. The gold chips were then immersed in 1 mM ethanolic MUA solution overnight to form a well-packed self-assembled monolayer. The MUA monolayer was then reacted overnight with CDI in acetone solution (10 g/L).

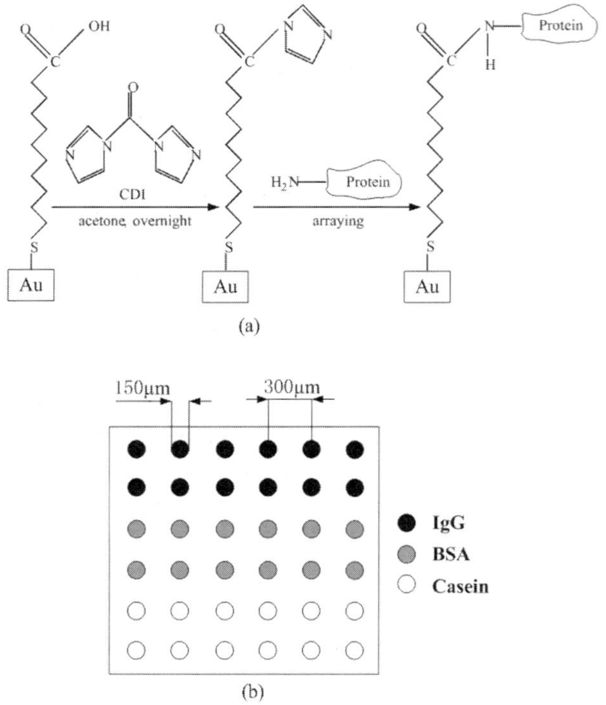

Figure 1. (a) A schematic showing the covalent attachment of protein onto gold surface. The reaction of a MUA monolayer with CDI forms an imidazolide monolayer that couples with a lysine residue on a protein to create an amide linkage to the surface. (b) Protein microarray pattern. IgG, BSA and casein forms a 6×6 matrix. The concentration of each of the proteins was 1mg/mL.

Immediately prior to arraying, the CDI-modified chips were rinsed in acetone and dried under a nitrogen stream. The chips were arrayed using a Microarrayer (CapitalBio, Beijing, China), equipped with a humidity control apparatus and ultrasonic wash station. During the array fabrication, the humidity was controlled at 40-55% and the platen temperature was set at 25 °C. Following antibody immobilization, the CDI-modified background was inactivated by washing thoroughly with water or buffer for 1 h before drying under a N_2 stream and storing at room temperature overnight.

2.3. *SPRI Setup and Procedure of Imaging*

A self-built multifunctional surface plasmon resonance system based on the Kretschmann geometry of the attenuated total reflection (ATR) method[13] was constructed to study biomolecular interaction. Different from former SPR sensors, this platform could undertake three modes of measurement, angular interrogation, wavelength interrogation and intensity measurement (i.e. SPRI). In this paper, we only used the intensity measurement mode, i.e. SPRI, to study protein microarray.

The incident light from a light emission diode was expanded, collimated, and then polarized through a polarizer. After being reflected from the gold film via the glass prism, it was imaged by optical lens and collected by CCD camera. The collected image was treated and saved by self-developed software. The polarizer (GCL-050002, China Daheng Group, Inc.) was adjusted to produce p-polarized wave only. The glass prism was made of BK7 with n=1.517. The slide with gold film was attached to the prism through index matching oil (n=1.51). A 0.9mm thick silicone gasket, with a 5 mm×20mm rectangular hollow in the middle, was mechanically pressed between polypropylene and the slide to form flow cell. During the experiment, the SPRI setup was placed in darkroom to avoid ambient light. All SPRI experiments of protein microarray were performed under equilibrium conditions.

3. Results and Discussion

As the molecular structures and molecular weights of IgG, BSA and casein are different, under identical conditions the reflectivity of the three different proteins should be different. Each of the proteins was spotted and immobilized on the same sensitive surface (pre-coated with MUA monolayer and CDI monolayer). While changing incident angle of light, the reflected light intensity of proteins would be very weak and accordingly should exhibit dark spots when the spotted proteins are in the state of surface plasmon resonance.

As shown in Figure 2(a), most of the incident light falling onto the protein spots was absorbed, so each of the protein spots appeared dark, indicating they were in the state of resonance. However, the degree of darkness was different for each protein. Casein exhibited much darker protein spots as it absorbed light more intensively, whereas the least absorption (or strongest reflected intensity) occurred on the BSA spots. The

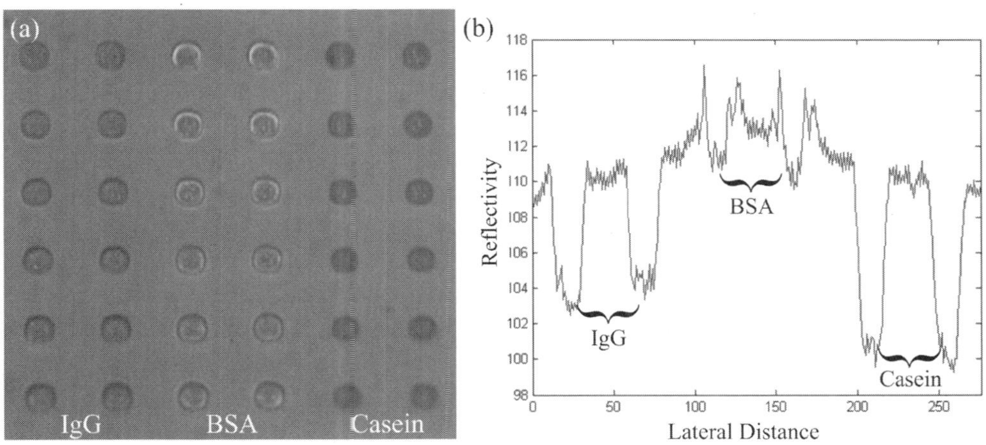

Figure 2. *In situ* SPRI of protein microarray at resonance angle (a) and the corresponding line profile (b). In (a), from left to right, IgG (line 1-2), BSA (line 3-4) and casein (line 5-6). In (b), the absorption peak represents average reflected intensity of each indicated protein.

intensity of light absorption by IgG was intermediate. Figure 2(b) shows the line profile that illustrates the quantitative relationship of reflected light among these three kinds of protein spots. We can see that the reflected intensities of two rows of same protein were similar, while the reflected intensity differed slightly with various proteins on the same sensitive surface, and thus made label-free and high-throughput discrimination of protein realizable.

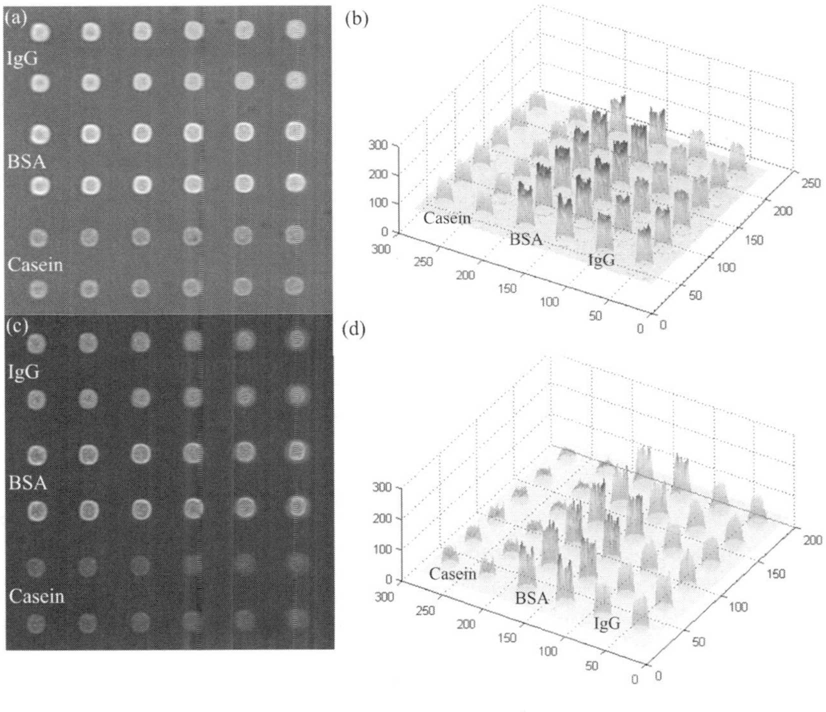

Figure 3. *In situ* SPRI images of protein microarray at different incident angle (a), (c) and their corresponding three-dimension SPRI images (b), (d). The microarray consists of IgG, BSA, casein, and inactivated CDI-modified background elements with the same pattern shown in Figure 1(b). In Figure 3 (b) and Figure 3 (d), the height of spots represents the degree of reflected intensity.

As noted above, the proteins appeared in dark spots due to the degree of surface plasmon resonance. However, the resonance signals can also be displayed by adjusting the position of the incident angle of light, causing the resonance condition be completely reversed. Thus under certain condition, the background would be in the state of surface plasmon resonance, while the spotted proteins would appear as bright spots because of stronger light reflectivity than the background.

Figure 3(a) and Figure 3(c) show each of spotted proteins as bright spots, changing from absorption peak (Figure 2) to pulse peak (Figure 3). Here it can be seen that BSA exhibits the strongest reflected intensity and the casein the least. The incident angles of Figure 3(a) and Figure 3(c) were 45 degrees and 47 degrees, respectively. Though the change in angle is small, the reflected intensity of different proteins increased significantly. The reflected intensities of IgG and casein increased greater than BSA, demonstrating a good sensitivity. When incident angle is 47 degrees (Figure 3(c)), the background is darker than at 45 degrees (Figure 3(a)) because of a better SPR condition of background, and the signal values of two angles are close, so the signal-to-noise at 47 degrees is better than at 45 degrees. This is an advantage of SPRI, the background of which could be kept at much lower values than traditional microarray based on glass or plastic.

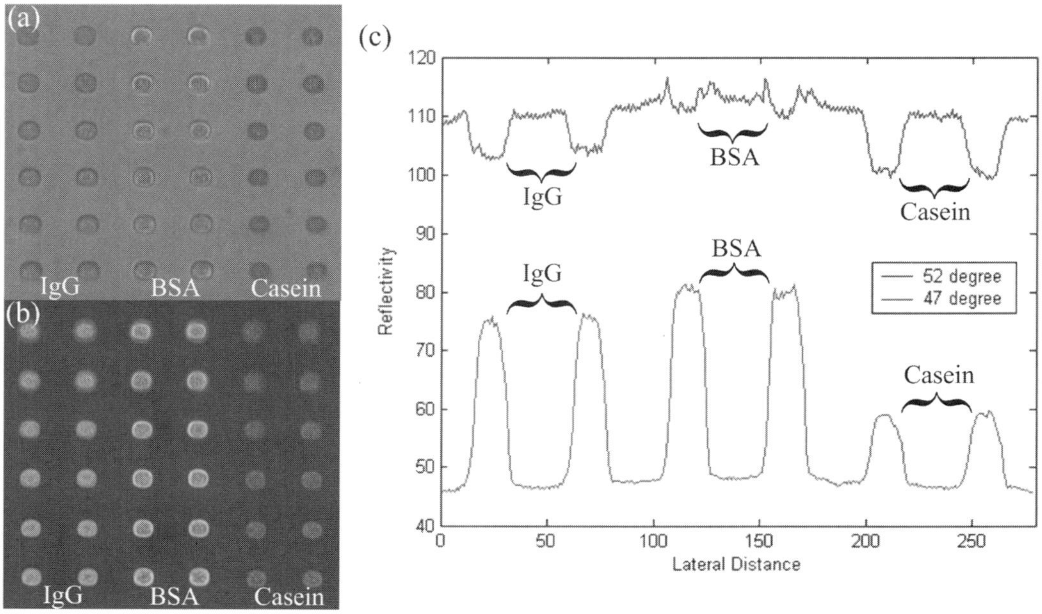

Figure 4. Background is at resonant condition (a) and proteins are at resonant condition (b). (c) Corresponding profile under two conditions. The incident angle of (a) is 52 degrees, and (b) is 47 degrees.

Figure 4 compares the signal-to-noise ratio that background is at resonant condition and protein is at resonant condition. We can see that the signal-to-noise ratio that background is at resonant is much larger than that protein is at resonant condition. However, the conditions shown in Fig. 4(a) are that reflected light of all proteins are absorbed, and one particular protein may be not in the optimal resonant situation. Thus, if only one kind of protein is target, we can expect a better signal-to-noise ratio at a given angle. In addition, the signal difference under diverse situations is also not the same. As shown in Figure 4(c), when incident angle is 47 degrees, the difference between IgG and BSA is smaller than when incident angle is 52 degrees, while the signal of casein is much smaller at 47 degrees than at 52 degrees.

4. Conclusion

We have demonstrated a simple method for label-free and high-throughput detection of protein microarray by using SPRI. Discrimination of three kinds of proteins was achieved by changing the incident angle of light.

The relevance between different proteins and incident angular positions was established. Excellent reproducibility for label-free detection of protein molecules was achieved. This SPRI platform represents a simple and robust method for performing label-free, high-sensitivity, and high-throughput detection of protein microarray. Other kinds of molecules could also be discriminated using this technique.

Acknowledgments

Supported by the National Foundation of High Technology of China (2006AA020701 and 2006AA020803), National Program on Key Basic Research Projects 973 of China (2006CB705700), the Nature Science Foundation of Zhejiang Province (2006C21G3210005) and Tsinghua-Yuyuan Medicine Foundation (40000510B).

References

1. E. Kretschmann and H. Raether, *Z. Naturforsch.* ***A*23**, 2135 (1968).
2. J. Homola, S. S. Yee, and G. Gauglitz, *Sensor Actuat.* ***B*54**, 3 (1999).
3. B. Liedberg, C. Nylander, and I. Lundstrom, *Sensor Actuat.* ***B*4**, 299 (1983).
4. A. Subramanian, J. Irudayaraj, and T. Ryan, *Biosens. Bioelectron.* **21**, 998 (2006).
5. R. Karlsson, *J. Mol. Recognit.* **17**, 151 (2004).
6. W. M. Mullett, E. PC Lai, and J. M. Yeung, *Methods* **22**, 77 (2000).
7. Q. M. Yu, S. F. Chen, and A. D. Taylor *et al.*, *Sensor Actuat.* ***B*107**, 193 (2005).
8. H. J. Lee, A. W. Wark, and R. M. Corn, *Langmuir* **22**, 5241 (2006).
9. H. J. Lee, D. Nedelkov, and R. M. Corn, *Anal. Chem.* **78**, 6504 (2006).
10. J. S. Shumaker-Parry, R. Aebersold, and C. T. Campbell, *Anal. Chem.* **76**, 2071 (2004).
11. T. Wilkop, Z. Z. Wang, and Q. Cheng, *Langmuir* **20**, 11141 (2004).
12. B. Rothenhausler and W. Knoll, *Nature* **332**, 615 (1988).
13. E. Kretschmann, *Z. Naturforsch.* ***A*241**, 313-314 (1971).

MEASURING DYNAMICS OF BID ACTIVATION IN LIVING CELLS DURING NPE6-PDT-INDUCED APOPTOSIS

LEI LIU[1], DA XING[1*], and QINGLING WAN

[1] *MOE Key Laboratory of Laser Life Science & Institute of Laser Life Science, South China Normal University, Guangzhou 510631, China*

WEI R. CHEN[1,2]

[2] *Biomedical Engineering Program, Department of Engineering and Physics, College of Mathematics and Science, University of Central Oklahoma, Edmond, Oklahoma 73034, USA*

Photodynamic therapy (PDT) employing photosensiter N-aspartyl chlorin e6 (NPe6) can induce lysosome disruption and initiate the intrinsic apoptotic pathway. Bid, a member of the Bcl-2 family of proteins, is an essential regulator of apoptosis. Bid is normally found in the cytosol of healthy cells in an inactive form, activation of Bid depends on the cleavage of intact Bid into its truncated form of tBid. The resulting tBid then translocates into mitochondria and leads to disruption of organelles and release of apoptogenic molecules such as cytochrome *c* in response to many apoptotic stimuli. In this study, we have investigated the molecular mechanism in NPe6-PDT-induced apoptosis in human lung adenocarcinoma cells (ASTC-a-1). A recombinant fluorescence resonance energy transfer (FRET) Bid probe was utilized to determine the kinetics of Bid cleavage. The cells were also transfected with Bid-CFP to dynamically detect tBid translocation. The results show that cleavage of the Bid-FRET probe occurred 150±5 min after NPe6-PDT treatment, and this process lasted for 45±5 min. The Bid cleavage coincided with a translocation of tBid from cytoplasm to mitochondria. Using real-time single-cell analysis, we first observed the kinetics of Bid cleavage and translocation to mitochondria in living cells during NPe6-PDT-induced apoptosis

1. Introduction

Photodynamic therapy (PDT) is a novel treatment modality for certain cancer and other indications. PDT employs a photosensitizer and visible light to generate singlet oxygen and other reactive oxygen species (ROS), which ultimately cause tumor destruction [1-4]. The subcellular location of a photosensitizer critically influences the kinetics and the regulatory pathway activated following a PDT treatment [5-6]. N-aspartyl chlorin e6 (NPe6), a powerful and natural occurring photosensitizer, is endowed with remarkable tumor localizing properties and *in vivo* antineoplastic activity upon irradiation, suggesting its clinical application as a promising PDT agent [7-8]. Irradiation of murine hapatoma 1c1c7 cells preloaded with NPe6 caused lysosomal disruption, the activation of pro-caspase-3 and apoptosis [9-11]. However, the precise cellular signal transduction mechanism in NPe6-PDT-induced apoptosis still remains poorly understood.

Bid, a member of BH3-only subgroup of Bcl-2 family, is a unique pro-apoptotic protein [12-13]. It plays an essential role in apoptotic signaling, by inducing the proapoptotic functionality of Bak and Bax, leading to cytochrome c release. The BH3 domain of Bid is required for its interaction with Bax or Bak [14]. It is also the target of the Bcl-2 or Bcl-xl [15]. Bid is normally present in its full-length inactive form in the cytosol. Activation of Bid depends on the cleavage of intact Bid into its truncated form of tBid. The activation of Bid can be conducted by several proteases such as caspases [15-16] and calpains [17]. Recently, Reiners, Jr *et al* reported that lysosomal extracts could cleave Bid in NPe6-PDT protocols [9]. However, the role of Bid in apoptosis induced by lysosomal disruption is not well known.

Fluorescence resonance energy transfer (FRET) is a nonradiative transfer of energy between two fluorophores that are placed in close vicinity and in a proper relative angular orientation. The emission spectrum of the donor molecule overlaps with the absorption spectrum of the acceptor molecule. When these two fluorophores are spatially close enough, there is energy transfer between the donor and acceptor molecules. This results in a reduction in donor fluorescence emission and, at the same

* Correspondence author: Da Xing; Tel: +86-20-85210089; Fax: +86-20-85216052; E-mail: xingda@scnu.edu.cn

time, an increase in acceptor fluorescence emission. In contrast, when the acceptor and the donor are separated by proteolytic cleavage of the linker, FRET cannot be observed [18-19]. Thus FRET is a powerful tool for investigation of molecular events in living cells. Recently, FRET has been used to study protein-protein interactions in living cells with improved spatial and temporal resolutions, spatial range, and sensitivity [20-23]. In our current study, a fusion protein pFRET-Bid (pYFP-Bid-CFP), constructed by connecting a yellow fluorescent protein (YFP) and a cyan fluorescent protein (CFP) to the N terminus and the C terminus of Bid, respectively [20], has been used to observe the activation of Bid in NPe6-PDT-induced apoptosis.

In this study, we used a confocal microscopy-based, real-time single-cell analysis approach to observe apoptosis in human lung adenocarcinoma cells (ASTC-a-1) induced by the application of NPe6-PDT. Bid cleavage and translocation were dynamically observed.

2. Materials and methods

2.1 *Materials*

Dulbecco's modified Eagle medium (DMEM) was purchased from GIBCO (Grand Island, NY, USA). Lipofectamine™ Reagent, mitochondrion-specific dye (Mito Tracker Red) and lysosome integrity (Acridine Orange, AO) were purchased from Invitrogen (Carlsbad, CA, USA). The photosensitizer NPe6 was a generous gift from Dr. C. Julie (Light Science Corporation, USA). Z-IETD-fmk (caspase-8 inhibitor) was purchased from BioVision (Mountain View, USA). Bid-FRET and Bid-CFP were kindly supplied by Dr. K.Taira (National Institute of Advanced Industrial Science and Technology, Tsukuba, Japan [20].

2.2. *Cell Culture, Transfection and PDT Treatment*

The human lung adenocarcinoma cell line (ASTC-a-1) was obtained from the Department of Medicine, Jinan University (Guangzhou, China) and cultured in DMEM supplemented with 10% fetal calf serum (FCS), penicillin (100 units/ml), and streptomycin (100 μg/ml) in 5% CO_2 at 37^{0}C in humidified incubator. Transfection was performed with Lipofectamine™ reagent according to the manufacturer's protocol. The medium was replaced with fresh culture medium after 5 hours. Cells were examined 36 hours after transfection. Cells were co-incubated with 33 μM NPe6 in the dark at 37^{0}C for 10 hours. After removing the culture medium, the cells were rinsed with phosphate buffered saline (PBS) and subsequently irradiated with a semi-conductor laser (NL-FBA-2.0-635, Shanghai, China) at a fluence of 4 J/cm^2 (fluence rate: 10 mW/cm^2).

2.3. *Confocal Microscopy*

CFP, Mito Tracker Red, NPe6 and AO emissions were monitored confocally using a commercial laser scanning microscope combination system (LSM510/ConfoCor 2, Zeiss, Jena, Germany) with a Plan-Neofluar 40×/1.3 NA Oil objective. Excitation wavelength and detection filter settings for each of the fluorescent indicators were as follows: CFP fluorescence was excited at 458 nm with an Ar-Ion laser and emission was recorded through a 470-500 nm band-pass filter. Mito Tracker Red and NPe6 fluorescence was excited at 633 nm with a He-Ne laser and emitted light was recorded through a 650 nm long-pass filter. AO fluorescence from the cells was excited at 458 nm with an Ar-Ion laser and the emission was recorded through a 650 nm long-pass filter. For time-course imaging, culture dishes were mounted onto the microscope stage equipped with a temperature-controlled chamber (Tempcontrol 37-2 digital, Zeiss).

2.4. *FRET Image Acquisition and Data Analysis*

FRET was also performed on LSM 510 confocal microscope. To monitor the fluorescence of FRET-Bid reporter, the 458 nm line of an Ar-Ion Laser was used to excite CFP, reflected by a dichroic mirror (main beam splitter HFT458/514). The emission fluorescence was split by a second dichroic mirror (secondary beam splitter NFT515) into separate CFP (470-500 nm band-pass) and YFP (530 nm long-pass) channels, respectively. The quantitative analysis of the fluorescence images was performed using Zeiss Rel 3.2 image processing software (Zeiss, Germany) by drawing regions around individual cells and obtaining the average fluorescence intensity of YFP, CFP and YFP/CFP emission ratio in each region for each image.

2.5. *Image Processing and Statistical Analysis*

Real-time single cell analysis was performed to study the kinetics of Bid distribution, the cells were labeled with Bid-CFP and Mito Tracker Red. In each experiment, 100-300 confocal images were recorded in a time series. The images were then analyzed using Zeiss Rel3.2 image processing software. The Bid-CFP localization in mitochondria was determined based on the overlap of Bid-CFP and Mito Tracker Red fluorescence images. The cells exhibiting strong punctate staining of Bid-CFP, which overlapped with the distribution of Mito Tracker Red, were counted as the cells with mitochondrially localized Bid. Untreated ASTC-a-1 cells usually had very low levels of diffuse Bid-CFP staining in the cytoplasm exhibiting only low levels of mitochondrial localization. At the selected time points, the percentage of cells showing Bid distribution to mitochondria was assessed by counting the number of cells exhibiting mitochondrial Bid. Date were collected from n=150-200 cells per treatment in 10-15 randomly selected image frames from three independent experiments. Data are represented as mean ± SEM.

3. Results

3.1. *Characterization of YFP-Bid-CFP in Living ASTC-a-1 Cells*

Activation of Bid was monitored by FRET technique following the methods of our previous study using a fusion protein YFP-Bid-CFP [23], which was constructed by connecting a yellow fluorescent protein (YFP) and a cyan fluorescent protein (CFP) to the N terminus and the C terminus of Bid, respectively [20]. At normal condition, CFP and YFP are covalently linked together, energy can be transferred directly from CFP to YFP, so fluorescence emitted from YFP can be detected when CFP is excited. Cleavage of full length Bid results in a separation of YFP and CFP, reducing the efficiency of resonance energy transfer. Acceptor bleaching experiments were carried out to assess the sensitivity of the FRET probe in our experimental system. The acceptor fluorophore YFP was selectively bleached in a single cell area by repeated scanning with a 514 nm laser (Fig. 1A, YFP). Upon 458 nm laser excitation, the level of CFP and YFP fluorescence emission was measured from the photobleached cell and plotted as a function of time (Fig. 1B). Upon bleaching there was a marked decrease in the acceptor fluorescence (YFP), which coincided with an increase in the donor fluorescence (CFP) because of an inability of the acceptor to accept energy from the donor after bleaching. Therefore the increase of CFP fluorescence upon YFP bleaching confirmed that FRET exists between the two fluorescent proteins in the Bid-FRET probe in living cells.

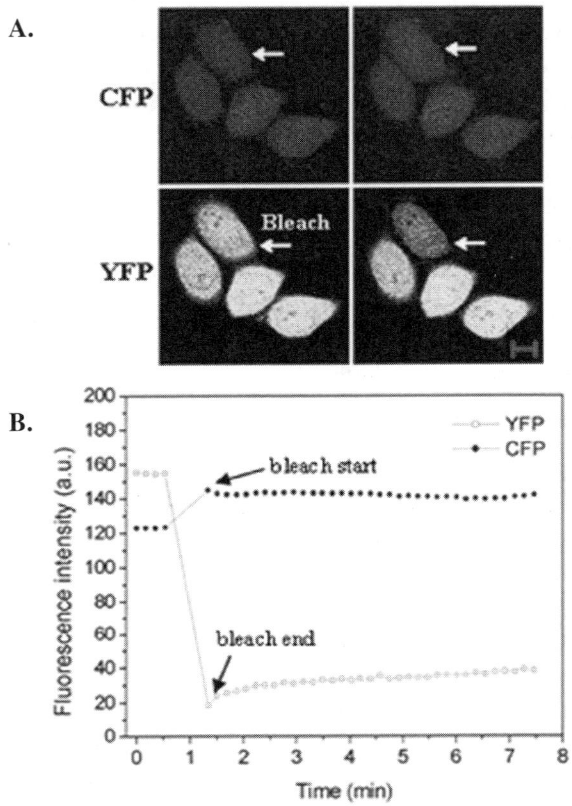

Figure 1. Characterization of Bid FRET probe in living ASTC-a-1 cells. FRET between CFP and YFP was confirmed by acceptor bleaching technique. (A) YFP was selectively bleached in a single cell by repeated scanning of the cell area at high laser power at 514 nm. Bar, 10 μm. The experiment was repeated twice. (B) Quantitative analysis of CFP and YFP fluorescence following the bleaching of YFP.

3.2. *FRET Analysis of Bid Cleavage Kinetics in NPe6-PDT-induced Apoptosis*

FRET was used to monitor the dynamics of Bid cleavage in ASTC-a-1 cells expressing Bid-FRET after NPe6-PDT treatment. Fig. 2A shows the FRET, CFP fluorescence emissions and FRET/CFP ratio. The intensity of CFP increased with time while that of FRET decreased after PDT treatment. The ratio of FRET to CFP fluorescence emission started decreasing about 150 min after treatment, as shown in Fig. 2B and Fig. 2C. The decrease of the FRET/CFP ratio indicated that Bid was cleaved.

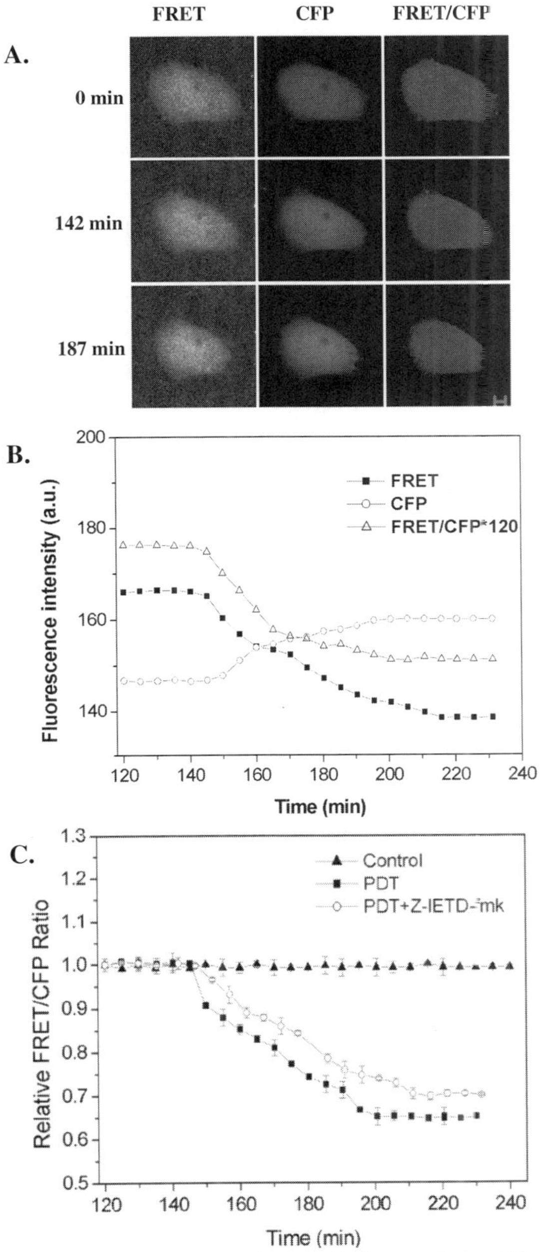

Figure 2. Dynamics detection of Bid cleavage during NPe 6-PDT induced apoptosis in living cells. (A) Fluorescence image series of FRET and CFP emission in ASTC-a-1 cells transfected with FRET-Bid after NPe6-PDT treatment. The panels of FRET, CFP and FRET/CFP are shown separately. Scale bar = 10 μm. (B) Dynamics of FRET, CFP emission intensities and FRET/CFP ratio after NPe6-PDT treatment. (C) Dynamics of FRET/CFP ratio after different treatments as indicated. The data represent the mean±SEM of three independent experiments

To determine whether Bid activation induced by NPe6-PDT was a caspase 8-dependent event, cells were pretreated with caspase-8 selective inhibitor Z-IETD-fmk for one hour before PDT treatment. In the presence of Z-IETD-fmk, the ratio FRET/CFP was the same as that of NPe6-PDT treated cells, indicating that Z-IETD-fmk did not block Bid activation (Fig. 2C). The results indicate that the Bid cleavage in our experimental model was caspase 8-independent.

3.3. *Real-time Detection of Bid Redistribution Induced by NPe6-PDT*

To monitor the dynamics of Bid redistribution after being cleaved by PDT treatment in ASTC-a-1 cells, Bid-CFP fluorescence image was used to monitor the Bid migration. Cells were loaded with Mito Tracker Red to label the mitochondria. The typical images of real-time distribution of Bid-CFP in control non-apoptotic cells are shown in Fig. 3A.

A.

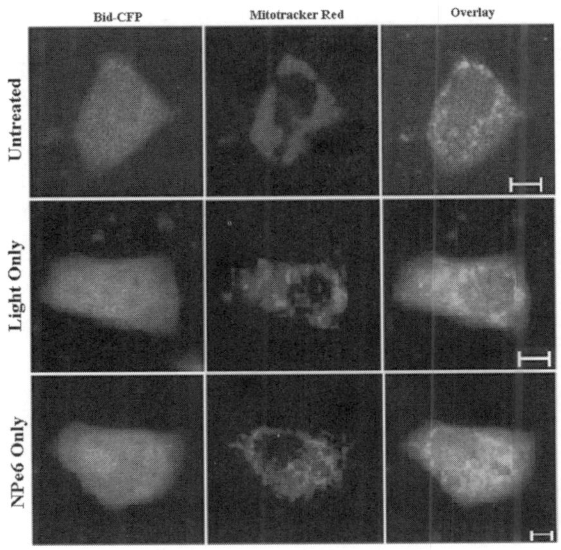

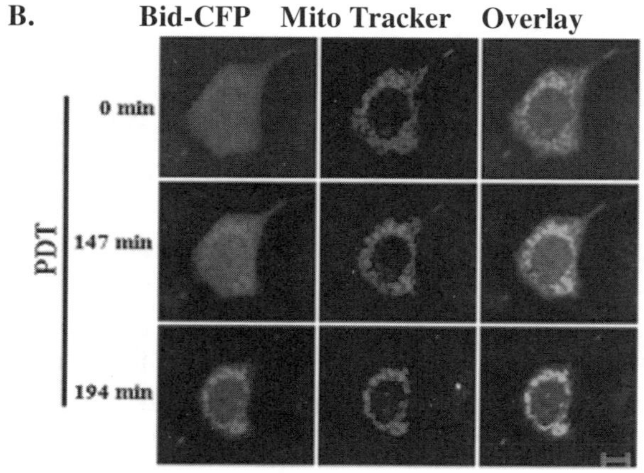

Figure 3. Dynamics of Bid translocation to mitochondria during NPe6-PDT-induced apoptosis in single living cell. Bid-CFP localization at mitochondria was determined based on the overlap of Bid-CFP and MitoTracker Red fluorescence images. (A) Control cells only treated with light or NPe6 without Bid translocation over time. (B) Time-lapse images of Bid-CFP translocation after NPe6-PDT treatment. Scale bar = 10μm.

Bid-CFP has a diffuse distribution throughout the cytoplasm, with little or no evident association with mitochondria. Fig. 3B shows typical spatial and temporal relationships of Bid-CFP and Mito Tracker Red after a PDT treatment. Bid-CFP translocated to mitochondria from cytoplasm as revealed by the overlaps of the Bid-CFP and Mito Tracker Red fluorescence images. Quantitative analysis of time-dependent redistribution of Bid-CFP confirmed these findings (Fig. 4).

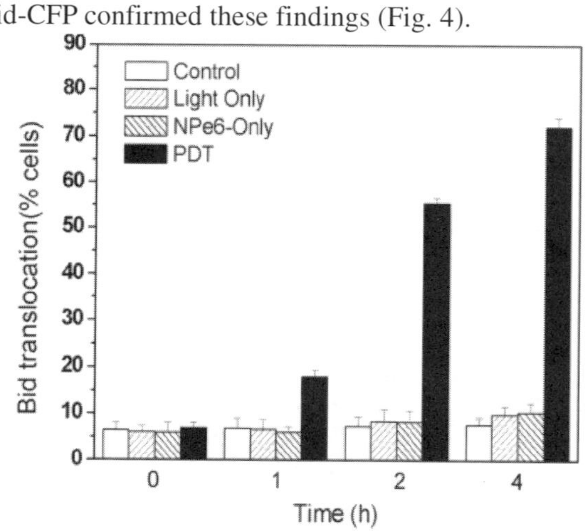

Figure 4. Quantification of cells showing mitochondrial Bid-CFP. At indicated time points, the percentage of cells showing Bid translocation to mitochondria was assessed by counting the number of cells exhibiting mitochondrial Bid. Date was collected from 150-200 cells per treatment in 10-15 randomly selected image frames from three independent experiments. Date represent the mean ±SEM.

4. Conclusions

In recent years, our laboratory has been focusing on developing real-time single-cell analysis for studying molecular events in living cells. In this study, using real-time single-cell analysis, we have investigated the kinetics of Bid activation during NPe6-PDT-induced apoptosis in human lung adenocarcinoma cells (ASTC-a-1).

In order to determine the kinetics of Bid cleavage, we utilized a recombinant fluorescence resonance energy transfer (FRET) Bid probe, pFRET-Bid, to monitor the whole process of Bid cleavage dynamically during NPe6-PDT-induced apoptosis. Our data clearly showed the cleavage of the Bid-FRET probe occurred 150±5 min after NPe6-PDT treatment, and this process lasted for 45±5 min (Figs. 2A and B). Then we investigated the kinetics of Bid distribution during NPe6-PDT-induced apoptosis. The cells were transfected with Bid-CFP to dynamically detect Bid translocation. Bid translocated from cytoplasm to mitochondria, occurring at about 150 min after treated with NPe6-PDT, and lasted about 50 min (Fig. 3B and Fig. 4). Using real-time single-cell analysis, we first observed the kinetics of Bid cleavage and translocation to mitochondria in living cells during NPe6-PDT-induced apoptosis.

Acknowledgments

We thank Dr. C. Julie of Light Science Corporation for providing photosensitizer NPe6. We thank Dr. K. Taira (National Institute of Advanced Industrial Science and Technology, Tsukuba, Japan) for providing FRET-Bid and Bid-CFP plasmids.

This research is supported by the National Natural Science Foundation of China (30600128; 30470494), and the Natural Science Foundation of Guangdong Province (7117864).

References

1. T. J. Dougherty, C. J. Gomer, B. W. Henderson, G. Jori, D. Kessel and M, Korbelik, *J. Natl. Cancer Inst.* **90,** 889 (1998).
2. B. W. Henderson and T. J. Dougherty, *Photochem. photobiol.* **55,** 145 (1992).

3. M. L. Agarwal, M. E. Clay, E. J. Harvey, H. H. Evans, A. R. Antunez and N. L. Oleinick, *Cancer Res.* **51**, 5993 (1991).
4. N. L. Oleinick, R. L. Morris and I. Belichenko, *Photochem. Photobiol. Sci.* **1**, 1 (2002).
5. D. Kessel, Y. Luo, Y. Deng and C. K. Chang, *Photochem. Photobiol.* **65**, 422 (1997).
6. N. L. Oleinick and H. H. Evans, *Radiat Res.* **150**, 146 (1998).
7. J. Usuda, H. Kato, T. Okunaka, K. Furukawa, H. Tsutsui, K. Yamada, Y. Suga, H. Honda, Y. Nagatsuka, T. Ohira, M. Tsuboi and T. Hirana, *J. Thorac. Oncol.* **5**, 489 (2006).
8. T. Kaneko, H. Chiba, T. Yasuda and K. Kusama, *Oral Oncol.* **40**, 787 (2004).
9. Jr. J. J. Reiners, J. A. Caruso, P. Mathieu, B. Chelladurai, X. M. Yin and D. Kessel, *Cell Death Differ.* **9**, 934 (2002).
10. J. A. Caruso, P. A. Mathieu, A. Joiakim, B. Leeson, D. Kessel, B. F. Sloane and J. J. Jr. Reiners, *Mol. Pharmacol.* **65**, 1016 (2004).
11. J. A. Caruso, P. A. Mathieu and J. J. Jr. Reiners, *Biochem. J.* **392**, 325 (2005).
12. K. Wang, X. M. Yin, D. T. Chao, C. L. Milliman and S. J. Korsmeyer, *Genes. Dev.* **10,** 2859 (1996).
13. J. Zha, S. Weiler, K. J. Oh, M. C. Wei and S. J. Korsmeyer, *Science* **290**, 761 (2000).
14. S. J. Korsmeyer, M. C. Wei, M. Saito, S. Weiler, K. J. Oh and P. H. Schlesinger, *Cell Death Differ.* **7,** 1166 (2000).
15. A. Gross, X. M. Yin, K. Wang, M. C. Wei, J. Jockel, C. Milliman, and H. E. Bromage, *J. Biol. Chem.* **274**, 1156 (1999).
16. H. Li, H. Zhu, C. J. Xu and J. Yuan, *Cell* **94**, 491 (1998).
17. M. Chen, H. He, S. Zhan, S. Krajewski, J.C. Reed and R.A. Gottlieb, *J. Biol. Chem.* **276**, 30724 (2001).
18. S. Fields and O. K. Song, *Nature*, **340**, 245 (1989).
19. R. B. Sekar and A. Periasamy, *J. Cell. Biol.* **160**, 629 (2003).
20. R. Onuki, A. Nagasaki, H. Kawasaki, T. Baba, T. Q. P. Uyeda and K. Taira, *Proc. Natl. Acad. Sci. USA* **99**, 14716 (2002).
21. J. Zhang, R. E. Campbell, A. Y. Ting and R. Y. Tsien, *Nat. Rev. Mol. Cell Biol.* **3**, 906 (2002).
22. X. Gao, T. Chen, D. Xing, F. Wang, Y. Pei and X. Wei, *J. Cell Physiol.* **206**, 441 (2006).
23. Y. Pei, D. Xing, X. Gao, L. Liu and T. Chen, *Apoptosis* **12,** 1681 (2007).

MICROTUBULE-DEPENDENT TRAFFICKING OF H-RAS TO CELL SURFACE AFTER LOW-POWER LASER IRRADIATION

XUEJUAN GAO, DA XING*, LEI LIU and YONGHONG TANG

MOE Key Laboratory of Laser Life Science & Institute of Laser Life Science, South China Normal University, Guangzhou 510631, China

Low-power laser irradiation (LPLI) has been shown to exhibit many biological effects including cell proliferation and cell apoptosis in various cell types. Ras/Raf signaling pathway is recently suggested to participate in the regulation of the biological effects of LPLI. H-Ras is one isoform of four Ras GTPases that function as molecular switches. In this study, we focus on the dynamic trafficking of H-Ras induced by LPLI, using GFP-H-Ras by performing confocal fluorescence microscope. Our results showed that H-Ras trafficked to cell surface induced by LPLI through near circular structure and that this structure was very important carriers in H-Ras trafficking. The large near circular structure was firstly characterized in our study as near-circular fluorescent profile continually emerging from juxtanuclear Golgi complex and moving toward cell surface. After disrupting microtubules with colchicines and nocodazole, we found that the trafficking of the near circular structure containing H-Ras toward cell surface induced by LPLI is microtubule-dependent. Taken together, the new insights into the regulation of H-Ras trafficking induced by LPLI contribute to clarify the mechanism of LPLI-induced biological effects.

1. Introduction

Low-power laser irradiation (LPLI) has been shown to stimulate cell functions. This is called "photobiostimulation" and has been commonly used in human medicine for treatment of various pathological conditions, including soft tissue injuries, severe wounds, chronic pain, and more, but the mechanisms of photobiostimulative processes are still poorly understood [1]. The signaling pathways involved in the laser irradiation have been investigated extensively. The activation of PKCs by LPLI was demonstrated in our previous study [2] and the activation of Ras has been suggested in the cell proliferation induced by LPLI [3].

Ras proteins coordinate a variety of cellular responses to extracellular stimuli and regulate cell growth, proliferation and differentiation [4-6]. Several isoforms of Ras: N-Ras, H-Ras and K-Ras, exist in human cells. It has been shown that activated H-Ras after EGF treatment redistributes to ruffled regions of the plasma membrane (PM) and to endosome-like structures [7]. However, detailed knowledge about the redistribution of H-Ras induced by LPLI is little.

Microtubules (MTs) are polarized filaments that enable transport of vesicles and organelles [8]. In mammalian cells the position of the Golgi complex correlates with that of the microtubule (MT)-organizing center or centrosome [9, 10]. Disruption of normal MTs cytoskeleton dynamics with agents such as colchicine and nocodazole that cause MTs depolymerization, results in reversible fragmentation and dispersion of the Golgi apparatus into smaller units or "mini-stacks" spreading throughout the cell, many in close proximity to the cell surface[9, 11-13]. Golgi apparatus stacks are known to remain functionally and cytologically intact after nocodazole- or colchicine-induced microtubules depolymerization [14]. Since H-Ras can traffic to the plasma membrane by both the classic secretory pathway and nonclassical secretory transport pathway [15], it is not surprising that cytoskeletal element also have been implicated in the redistribution of H-Ras. To date, however, the role of the MTs in the trafficking of H-Ras after LPLI treatment has not been explored.

In this study, we used confocal fluorescence microscope combined with green fluorescent protein (GFP)-tagged H-Ras to investigate the trafficking of H-Ras induced by LPLI. Our results showed that H-Ras trafficking to cell surface after LPLI treatment through near circular trafficking structure. The detail analysis of these trafficking structures induced by LPLI will be discussed in the context.

* Correspondence author: Da Xing; Tel: +86-20-85210089; Fax: +86-20-85216052; E-mail: xingda@scnu.edu.cn

2. Materials and Methods

2.1. *Materials*

Dulbecco's modified Eagle medium (DMEM) was purchased from GIBCO (Grand Island, NY). Colchicine and nocodazole were purchased from Sigma (St. Louis, MO, USA). Lipofectamin 2000 was from Invitrogen (Carlsbad, USA).

2.2. *Cell Culture and Transfection*

COS-7 cells were cultured in DMEM medium supplemented with 10% fetal calf serum (FCS), and cells were maintained at 370C in a humidified atmosphere (95% air and 5% CO_2). For microscopy purposes the cells were cultured on a coverslip-bottomed small chamber. Cell cultures at a confluency of 40-60% were used for transfection with Lipofectamin 2000.

2.3. *Laser Scanning Microscopes (LSM)*

Imaging experiments were performed on Laser Scanning Microscopes (LSM510/ConfoCor2) combination system (Zeiss, Jena, Germany) equipped with a Zeiss Plan-Neofluar 40×/1.3 NA Oil Dic objective or 100×/1.4 NA Oil Dic objective as described previously [2, 16, 17]. To monitor the fluorescence of GFP protein, the 488 nm line of Ar-Ion Laser was used to excite GFP, reflected by a dichroic mirror (main beam splitter HFT488). The 500-550 nm bandpass filter was used to record the emission fluorescence for GFP channel. During fluorescence imaging, the exciting power of 488 nm laser was reduced to the minimal level (0.5%-1%) to prevent the possible effects of exciting light.

For time-lapse imaging, the coverslip-bottomed small chambers were mounted onto the microscope stage equipped with a temperature-controlled chamber (Tempcontrol 37-2 digital, Zeiss). Digital image analysis for quantitative evaluation was performed using Zeiss Rel3.2 image processing software (Zeiss, Germany).

2.4. *LPLI Stimulation and Imaging with LSM*

COS-7 cells transfected with GFP-H-Ras reporter or GFP vector were cultured on a coverslip-bottomed small chamber. At least 24 hours after serum deprivation, the small chamber was mounted onto the stage of the LSM microscope. For laser irradiation, cells were exposed to 633 nm He-Ne laser from the epi-illumination port of the inverted objective of LSM microscope in the dark. The power measured at the epi-illumination port of the objective was 5 μW for 633 nm He-Ne laser. The irradiation time varied between 3.6 seconds and 15 seconds according to the individual cell area. The total fluence of every irradiation was 0.8 J/cm2. The corresponding images before and after LPLI were obtained by performing LSM microscope. During these processes, cells were maintained at 37^0C using the temperature regulator.

2.5. *Colchicine or Nocodazole Treatment*

COS-7 cells transfected with GFP-H-Ras or GFP-tubule reporter were cultured on a coverslip-bottomed small chamber. After serum deprivation, the cells were pretreated with colchicines (20 μg/ml) dissolved in 0.9 % NaCl for 4.5 hours, nocodazole (25 μg/ml) dissolved in dimethyl sulfoxide (DMSO) for 6.5 hours. Then the cells were used to do the following experiments. The corresponding images were obtained by performing LSM microscope.

3. Results

3.1. *Dynamics of Large Near Circular Structure Moving toward Cell Surface after LPLI*

To investigate the trafficking of H-Ras after LPLI treatment, COS-7 cells transfected with GFP-H-Ras probe were serum starvation for 36 hours and irradiated with 633 nm He-Ne lasers from the epi-illumination port of the inverted microscope objective as the methods described. In Fig. 1, before LPLI the corresponding fluorescence image of the cell indicated that a fraction of H-Ras was bound to Golgi complex adjacent to nucleus in stationary cell. The time-lapse images of the cell about 65 minutes after LPLI treatment were shown in Fig. 1. The large near-circular fluorescent profile (ranging from several to tens of microns in diameter) began to emerge from the juxtanuclear Golgi complex as the arrow shown about 30 minutes after LPLI treatment and gradually moved toward cell surface. This structure approached the surface after 65 minutes of treatment with LPLI and at this time point the second near circular structure has already originated from the same site as the arrow shown. Then this second near circular structure also moved toward surface along the similar tracks (data not shown). On the other hand, the results of that the control cells expressing GFP-H-Ras were not treated with LPLI were shown in Fig. 1B. No obvious trafficking of GFP-H-Ras and no near circular structure was observed in control nontreated cell. Thus, these results demonstrated that, after LPLI stimulation, the near circular structures containing H-Ras continually emerged from juxtanuclear Golgi complex and moved toward cell surface. We proposed that the emerging of near circular structure was the active response to LPLI stimulation.

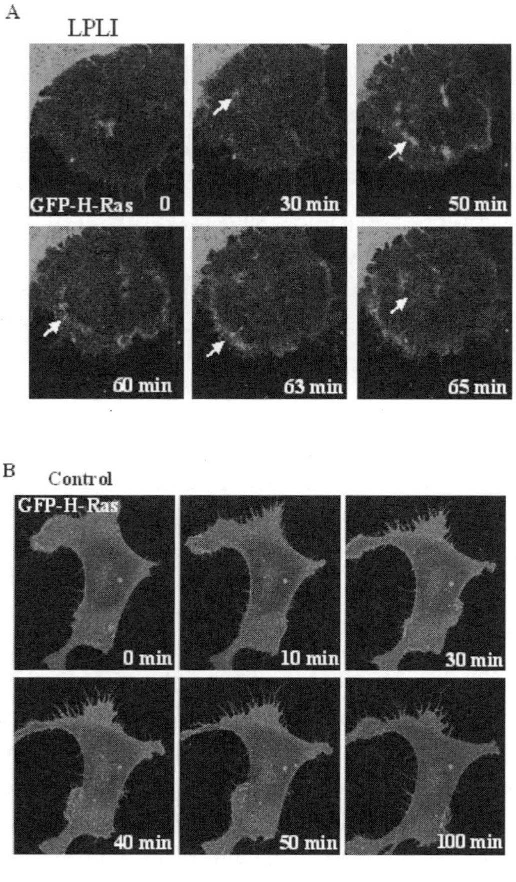

Fig. 1. Dynamics of large near circular structure moved toward cell surface after LPLI. (A) The time-lapse images ranging from 0 to 65 minutes after LPLI treatment. COS-7 cells transfected with GFP-H-Ras probe were serum starvation for 24 hours and irradiated with 633 nm He-Ne laser. (B) The time-lapse images ranging from 0 to 100 minutes of the control nonirradiated cell.

In addition, we found that near circular structure of H-Ras trafficking was not observed in every experimental cell. Statistics analysis of this structure emerged in 39 experimental cells were made. Percentage of near circular structure emerging is 56%. These results indicated that near circular structure played an important role in H-Ras trafficking induced by LPLI. One explanation for this finding was that the cells observed were not in the same physiological conditions.

3.2. Irregular trafficking of GFP after LPLI treatment

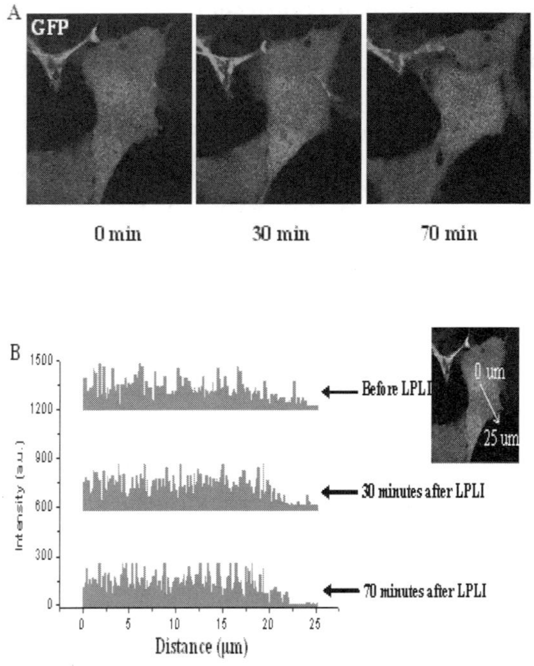

Fig. 2. Irregular trafficking of GFP after LPLI treatment. (A) Time-lapse images of GFP ranging from 0 to 70 minutes after LPLI. COS-7 cells were transfected with GFP vector and irradiated with 633 nm He-Ne laser using the same method. (B) Quantification of fluorescence intensities of GFP along the arrow from perinuclear region to cell surface.

To determine the fact that the above trafficking structures are the characterization of H-Ras trafficking, rather GFP vector, COS-7 cells were transfected with GFP vector and irradiated with 633 nm He-Ne laser using the same method. Time-lapse images of GFP after LPLI were shown in Fig. 2A. GFP trafficking was irregular and there were not any structures of GFP trafficking observed in 70 minutes after LPLI treatment. The quantified analysis of GFP intensities showed that there were not intensity peak observed along the arrow (Fig. 2B). Thus, these results suggested that the different trafficking structures were based on the movement of H-Ras protein rather GFP vector.

3.3. *Microtubule-dependent Trafficking of H-Ras after LPLI Treatment*

Since that H-Ras can traffic to the plasma membrane by the classic secretory pathway [15], we next tested whether microtubules, enabling transport of vesicles [18], were involved in the trafficking of H-Ras induced by LPLI. For this purpose, COS-7 cells expressing GFP-H-Ras after serum starvation were pretreated with colchicine (20 μg/ml) for 4.5 hours, a microtubule-disrupting drug, and then irradiated with 633 nm He-Ne laser. Before LPLI, a number of vesicles containing H-Ras emerged in the cytoplasm (Fig. 3A) and this phenomenon was observed in the majority of the cells after the treatment of microtubule-disrupting drugs (data not shown). This distribution of H-Ras was very different from that in the stationary cells with obvious binding to Golgi complex (Fig. 1). It has been reported that in colchicines- or nocodazole-treated cells, the Golgi apparatus rapidly fragmented into smaller units or "mini-stacks" spreading throughout the cell, many in close proximity to the cell surface[9, 11-13, 19]. It is likely that the different distribution of H-Ras in colchicines-treated cells is due to the fragmentation of Golgi complex. After LPLI, the results showed that near circular structures did not emerge. This suggested that the transport of H-Ras trafficking structure was blocked by colchicines pretreatment (Fig. 3A).

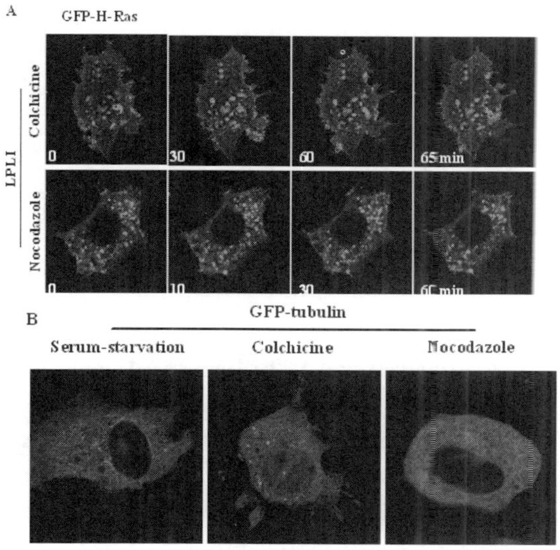

Fig. 3. Microtubule-dependent trafficking structures of H-Ras after LPLI treatment. **(A)** Time-lapse images of GFP-H-Ras of the cell pretreated with colchicine (20 μg/ml) for 4.5 hours or nocodazole (25μg/ml) for 6.5 hours, and then irradiated with 633 nm He-Ne laser. **(B)** The fluorescence images of GFP-tubulin under several conditions. COS-7 cells transfected with GFP-tubulin probe were observed by performing LSM microscope. Serum-starvation represents that the cells were deprived of serum for 24 hours. Colchicine means that the cells were serum starved and treatment with colchicine (20 μg/ml) for 4.5 hours. Nocodazole means that the cells were serum starved and treatment with nocodazole (25 μg/ml) for 1.5 hours.

To further confirm the role of MTs in H-Ras trafficking, we pretreated the cells with nocodazole (25μg/ml), another microtubule-depolymerizing drug, blocking microtubule-dependent translocation of membrane transport intermediates [9, 20]. The time-lapse images of the cell after LPLI irradiation were shown in Fig. 3A. The results showed that the transports of H-Ras trafficking structure were blocked by nocodazole pretreatment.

To confirm the effects of colchicine and nocodazole on MTs disruption, we transfected COS-7 cells with GFP-tubulin probe to observe the morphological properties of MTs. The fluorescence images of GFP-tubulin were collected by performing LSM microscope. In the cell serum-starved for 24 hours, MTs silkiness distributed in the cytoplasm and was somewhat sparse (Fig. 3B). After 4.5 hours of treatment with colchicines (20 μg/ml), all the MTs were disrupted and all GFP-tubulin uniformly distributed through the cytoplasm. Treatment with nocodazole (25 μg/ml) for 1.5 hours, gave quite similar result that all the MTs were disrupted. Taken together, the morphological effects of colchicine and nocodazole on MTs were consistent with the effects of colchicine and nocodazole on GFP-H-Ras trafficking induced by LPLI. These results demonstrated that MTs were responsible for the directly trafficking of H-Ras after LPLI treatment and that H-Ras vesicle probably moved along MTs tracks towards the cell surface.

4. Discussion

LPLI has been shown to stimulate many biological effects of cell [1], but the mechanisms of photobiostimulative processes are still poorly understood. In this study, we focused on the trafficking of H-Ras induced by LPLI. The near circular structure containing H-Ras induced by LPLI was visualized in our experiments (Fig. 1). The large near circular structure was firstly characterized as near-circular fluorescent profile (ranging from several to tens of microns in diameter) continually emerging from juxtanuclear Golgi complex and moving toward cell surface (Fig. 1A). These results of larger structures challenged the classical view that only small vesicles were involved in mediating late exocytic traffic [21]. Recently, with the advent of GFP technology, larger tubulovesicular structures have been identified and shown to play a major role in trafficking of newly synthesized proteins from TGN to PM [19-23]. However, there are no reports regarding near circular structure. Subsequently, to further confirm this phenomenon, we made the statistical analysis of the emergence of this structure induced by LPLI in 39 cells. The results showed that the percentage of near circular structure emerging was 56%. This suggested that the emergence of near circular structure was not occasional phenomenon and that this structure was the very important carrier in H-Ras trafficking. On the other hand, VSVG-GFP was commonly used as a reporter to study TGN-to-PM trafficking of secretory membrane protein [21, 23, 24]. Small vesicles and larger tubulovesicular structures were identified to traffic this transmembrane protein [21, 23, 24]. The differences in trafficking structures suggested various possibilities: lipid-anchored protein GFP-H-Ras and transmembrane protein VSVG-GFP may sort into different TCs or the different properties of these two proteins determined the different phenomenon observed.

Our results that MTs have been involved in the trafficking of H-Ras to cell surface induced by LPLI were consistent with the previous reports that exocytic cargo from Golgi complex moved along MTs tracks to the cell surface [21, 25]. When MTs were completely depolymerized by colchicines and nocodazole, all trafficking structures were inhibited (Fig. 3A). Since our results showed that the near circular structure was consisting of small vesicles and tubules (data not shown), it was easy to understand that the trafficking of near circular structure was mediated by MTs.

5. Conclusion

In conclusion, we for the first time investigated the trafficking of H-Ras induced by LPLI. Our results showed that H-Ras trafficked to cell surface after LPLI treatment through near circular structure (Fig. 1). The large near circular structure was firstly characterized in our study and this structure initiated from juxtanuclear Golgi complex trafficking H-Ras toward cell surface. Furthermore, the trafficking of this structure was MTs-dependent (Fig. 3). The new sights into the regulation of H-Ras trafficking induced by LPLI offer much new information for the mechanism of biological effects of LPLI.

Acknowledgments

The authors thank Prof. Piero Crespo of Unidad de Biomedicina de la Universidad de Cantabria-CSIC in Spain for providing GFP-H-Ras plasmid and Dr. Angela I. M. Barth of Stanford University in American for providing GFP-tubule plasmid.

This research is supported by the National Natural Science Foundation of China (30600128; 30470494), and the Natural Science Foundation of Guangdong Province (7117864).

References

1. R. Lubart, M. Eichler, R. Lavi, H. Friedman and A. Shainberg, *Photomed. Laser Surg.* **23**, 3-9 (2005).
2. X. Gao, T. Chen, D. Xing, F. Wang, Y. Pei and X. Wei, *J. Cell. Physiol.* **206**, 441–448 (2006).
3. G. Shefer, I. Barash, U. Oron and O. Halevy, *Biochim. Biophys. Acta*, **1593**, 131-139 (2003).
4. A. Apolloni, I. A. Prior, M. Lindsay, R. G. Parton and J. F. Hancock, *Mol. Cell. Biol.* **20**, 2475-2487 (2000).
5. T. G. Bivona and M. R. Philips, *Curr. Opin. Cell Biol.* **15**, 136–142 (2003).
6. S. Kupzig, S. A. Walker and P. J. Cullen, *Proc. Natl. Acad. Sci. USA.* **102**, 7577–7582 (2005).
7. X. Jiang and A. Sorkin, *Mol. Biol. Cell* **13**, 1522–1535 (2002).
8. S. L. Rogers and V. I. Gelfand, *Curr. Opin. Cell Biol.* **12**, 57-62 (2000).
9. A. A. Rogalski and S. J. Singer, *J. Cell Biol.* **99**, 1092- 1100 (1984).
10. A. Kupfer, D. Louvard and S. J. Singer, *Proc. Natl. Acad. Sci. USA.* **79**, 2603-2607 (1982).
11. A. A. Rogalski, J. E. Bergmann and S. J. Singer, *J. Cell Biol.* **99**, 1101-1109 (1984).
12. N. B. Cole, N. Sciaky, A. Marotta, J. Song and J. Lippincott-Schwartz, *Mol. Biol. Cell* **7**, 631-650 (1996).
13. B. Storrie, J. White, S. Rottger, E. H. Stelzer, T. Suganuma and T. Nilsson, *J. Cell Biol.* **143**, 1505-1521 (1998).
14. J. Thyberg and S. Moskalewski, *Exp. Cell Res.* **159**, 1-16 (1985).
15. J. S. Goodwin, K. R. Drake, C. Rogers, L. Wright,

J. Lippincott-Schwartz, M. R. Philips and A. K. Kenworthy, *J. Cell Biol.* **170**, 261-272 (2005).
16. Y. Wu, D. Xing and W. R. Chen, *Cell Cycle* **5**, 729-734 (2006).
17. F. Wang, T. S. Chen, D. Xing, J. J. Wang and Y. X. Wu, *Lasers Surg. Med.* **36**, 2-7 (2005).
18. A. Mu¨ sch, *Traffic* **5**, 1-9 (2004).
19. S. Waguri, F. Dewitte, R. L. Borgne, Y. Rouille´, Y. Uchiyama, J. F. Dubremetz and B. Hoflack, *Mol. Biol. Cell* **14**, 142-155 (2003).
20. J. F. Presley, N. B. Cole, T. A. Schroer, K. Hirschberg, K. J. Zaal and J. Lippincott-Schwartz, *Nature* **389**, 81-85 (1997).
21. D. Toomre, P. Keller, J. White, J. C. Olivo and K. Simons, *J. Cell Sci.* **112**, 21-33 (1999).
22. E. V. Polishchuk, A. D. Pentima, A. Luini and R. S. Polishchuk, *Mol. Biol. Cell* **14**, 4470-4485 (2003).
23. K. Hirschberg, C. M. Miller, J. Ellenberg, J. F. Presley, E. D. Siggia, R. D. Phair and J. Lippincott-Schwartz, *J. Cell Biol.* **143**, 1485-1503 (1998).
24. J. Schmoranzer, M. Goulian, D. Axelrod and S. M. Simon, *J. Cell Biol.* **149**, 23-32 (2000).
25. J. Schmoranzer and S. M. Simon, *Mol. Biol. Cell* **14**, 1558-1569 (2003).

QUANTITY RESEARCH ON VIRUS IN SEAWATER IN QINGDAO

JUAN LI[1,2], XIANG YONG Li[1,2], JIE HUANG[1*]
1 Yellow Sea Fisheries Institute, Qingdao, 266071
2 Dezhou University, Dezhou, 253023
E-mail: seamewlee@163.com

Viruses played a very important role in marine ecosystem. They have a significant impact on the occurrence of marine organism diseases. It made pelagic viruses have a very special meaning in the research of ocean ecosystem dynamics and marine organism disease. To have a concept of the count and distributing of the virus in seawater, we use fluorescent microscope counting stained with SYBR Green I to research the free virus quantity in seawater in Qingdao. Quantity analyzing by the photo analyze software shows that, virus concentration in seawater have great difference in different time and different place. At the same time, it shows that fluorescent microscope counting stained with SYBR Green I is a simple and high sensitive method. It can be used in the virus' dynamic quantity research in the water.

Keywords: count and distributing; fluorescent microscope counting

1. Introduction

It is well known that viruses are important in marine ecosystem. More and more researchers are concerntened into research of marine viruses. As a result, many significative works have been reported in marine virus research in recent years. For instance, study reports in quantity research technique and virus abundance in different sea area are mainly from Europe, North America, and Japan. Here we used fluorescent microscope counting stained with SYBR Green I to research the free virus quantity in seawater from Qingdao and Weifang, Shandong province, China respectively. Quantity analyzing by the photo analyze software BIO-RAD shows that, virus concentration in seawater have great difference in different time and different place.

Based on the Electron Microscope observation to the virus in seawater, we use fluorescent microscope counting technique to research the free virus quantity in seawater in Qingdao. It is mainly based on the virus characteristic in seawater. There are a large variety of virus in seawater. We cannot classify, filtrate and separate it out of the seawater at the present time. The only way we can use is microscope counting method. Virus thickness in free water is very low. For this kinds of quantity research in lesser sample, we chose a new kinds of nucleic acid fluorescence dyestuff SYBR Green I for use. It's a kind of fluorescent element of dissymmetry nitrile. Which was extensively be used as a kind of sensitive DNA fluorescence dyestuff recently, it can non- particularity match with the small ditch of the DNA double chain in its spiral structure. The fluorescence will strengthen 1000 remainings doubly after SYBR combining with DNA, SYBR has high affinity with DNA double chain, 20 pg DNA can be checked out at least, 25-100% higher than in EB dye method. SYBR Green I presents green fluorescence when it coalescence with DNA double chain under UV transmission. The color will be orange if it has single chain DNA in the sample, not green. It makes us can distinguish a live material nucleic acid from resolved plankton nucleic acid fragment very easily, Thus we can counting out of the amount of virus earsyly and smart under the fluorescence microscope. The SYBR Green I has certainly cooperate in person to glass and non-polypropylene material. So we all used polypropylene container in the process of the dilution, store save, dye etc.

2. Materials

(1) Sample, September 2004 to May 2005, draw from Qingdao and Weifang near the shore. (PRESERVED with 1% 0.02 um filtered formalin).

(2) Formaldehyde: Filtrate through 0.02μm filters film, the density is 37% (W/W);

(3) SYBR Green I: make a stock solution by diluting 1:10 of the original concentration with 0.02 um filtered deionized water, and stored frozen for a week or so. From this stock solution, make a 2.5% working solution (final 2.5×10^{-3} dilution of stock) just before use - preferably on the dish as described below.

Antifade mounting solution: 50% glycerol, 50% PBS (120 mM NaCl, 10 mM NaH_2PO_4 pH 7.5),

0.1% p-phenylenediamine (made fresh DAILY from the glycerol/PBS + frozen 10% stock of the phenylenediamine).

(4) 0.02 um pore size, 25 mm diameter Anodisc filters (from Whatman; made of aluminum oxide) 0.45 um, 25 mm diam. AA Millipore mixed-ester membrane filters.

(5 Glass 25 mm diameter filter holder with 15 ml funnel, Millipore-type plastic Petri dish, pipettes and tips.

(6) Microscope: The E800 type of Nikon ECLIPSE.

3. Methods (Done Away from Bright Light):

3.1 Sampling

Take 50 m L water in sample bottle, the sample bottle has already be sterilized and washed by distilled water which had be filtrated through 0.02μm filters film. Join 2.5 m L 2% L formaldehyde to the fixed sample (W/W). Place on the ice and avoid light, take back to the laboratory for analysis immediately.

3.2 Colander Assemble

Place the Anodisc filter over a pre-wetted 0.45 um Millipore filter (reusable many times, as long as it is completely intact and flat), in the glass filter unit. Apply the vacuum just after placing the Anodisc on the moist Millipore filter, and the Anodisc should moisten and stick in place completely flat and smooth with no air bubbles beneath.

3.3 Working Solution Prepare

Prepare a 100 ul drop of SYBR (2.5 ul of 1:10 diluted stock + 97.5 ul water) on the bottom of a plastic Petri dish.

3.4 Dye

Seawater samples typically 1 ml. Filter the preserved seawater sample through the Anodisc filter at approximately 20 kPa. The filter funnel should be removed shortly after the last liquid passes through, and the vacuum left on while removing the filter. Any seawater on the back of the Anodisc filters or top plastic rim are blotted with a Kimwipe or paper tissue, so that the filter is uniformly dry and looks opaque when held up to the light for examination. Best done by rubbing the bottom surface very gently on a dry Kimwipe for at least 2 minutes. The Anodisc filters are laid sample side up on the drops of the staining solution for at least 15 minutes.

3.5 Film Produce

Place the completely dried filter on a glass slide, then put a 30 ul drop of antifade mounting solution on a 25 mm cover slip and invert it over the filter. Push the cover slip to be sure the mounting solution fills the square space under the cover slip.

3.6 View and Counting

View with blue excitation. Count at least 200 each viruses and bacteria in at least 10 fields. The field size for the viruses can be much smaller than for bacteria, perhaps 3 small squares. To calculate total abundance, consider that the viruses appear to be filtered through the entire 20 mm diameter area enclosed by the plastic rim.

Calculation Formula: $VN=Na \cdot S / Sf \cdot (1 - 0.05) \cdot V$

Among them:

VN: The sample contains a virus number

Na: Each visual field average virus number

S: Anodisc filters filtrate real area(mm2)

Sf: Microscope view area(mm2)

V: Sample filtrate quantity(m L)

attention: 0.05 in the modle is the 37-40% formaldehyde proportion in fastness sample volume.

4. Results

4.1 Fluorescent Microscope Observation

Observed by microscope in natural light, the virus looks like bright green point, very small, and like needle groups. Having some big bright and green grain in the visual field either. They are the sea aquatic germ cells and unicellular planktons that have been dyed, shown in. figure1.

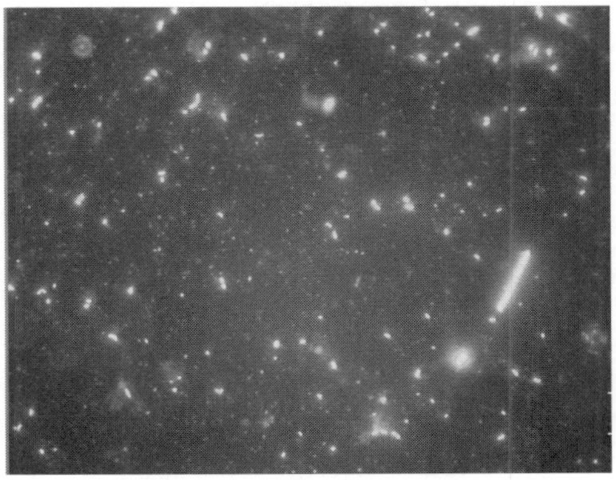

Fig 1 Fluorescence micrograph of seawater, Qingdao

4.2 Picture Analysis

Analysised the sea water fluorescence microphone by software BIO-RAD, we counting viruses, bacterias and algas etc. respectively according to the grain size. To the same sea water sample dyeeby fluorescence we selected three visual fields to make to count covariance, taking an average value towards statistising result, we get the fluorescent microscope counting result of the biomass in the circumlittoral seawater of the Qingdao downtown, shown in Table 1:

Table 1 Fluorescence micrograph counting of biomass in seawater, Qingdao

Sampling time(year)	Sampling time(month)	virus (10^6virus/ m L)	Germ(10^6germ/ m L)	Alga(10^6alga/ m L)
2004	9	3.36	1.06	1.95
	10	10.50	1.57	0.68
	11	0.86	0.055	0
	12	1.26	0.20	0.33
2005	1	1.32	2.02	0.48
	2	0.42	0.91	0.20
	3	3.05	2.63	0.88
	4	3.88	2.28	0.085
	5	2.68	2.38	0.132

4.3 Variety Relativity Analysis of Biomass in Seawater

Take the biomass's variety from September, 2004 to May, 2005 in the seawater for analyzing, just as figure 2 shown, algas' seasonal change is very obviously, larger in spring and autumn, few in winter. This fit with the phenomenon of marine biology's seasonal move and the water bloom in spring and autumn mutually. The bacteria density in seawater has a pinnacle in 2004 autumn, get down at winter, start a rally again in the next year spring, attain the pinnacle in May at the srage of the investigating. This probably have something to do with the mankind activity enlarges influence much to the seawater fluid matter in this period; after the hot weather turning into autumn coolly in September in Qingdao, the virus density in seawater attains a high peak in October, keeping with a more stable low peak between November to February after getting into winter. And start a rally again after weather become warm in the next year March. The Virus amount's variety and the bacteria in each period trend mutually consistently and shows large relativity, but have no such obvious relativity with the variety of alga amount.

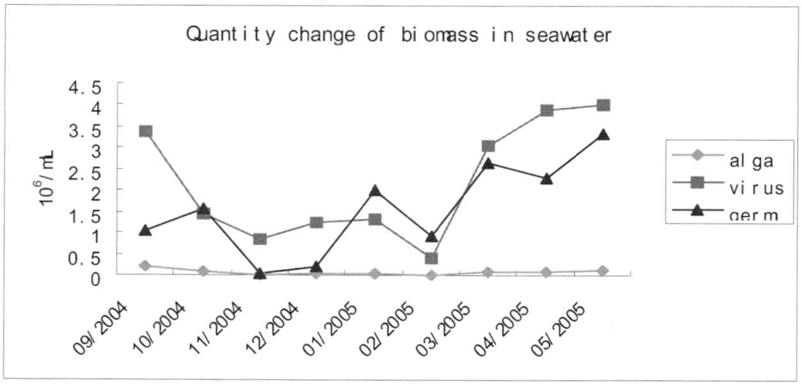

Fig 2 Quantity change of biomass in estuarine water in different time, Qingdao

We made a comparison with the covariance results of the biomass variety from April 27 to May 12, 2005 in the seawater near the shore in Weifang and Qingdao, just as figure 3 shown. At the same season, the virus density shows stronger positive relativity to the microorganisms density. At the same time, both of them shows stronger positive relativity to the plankton biomass change.

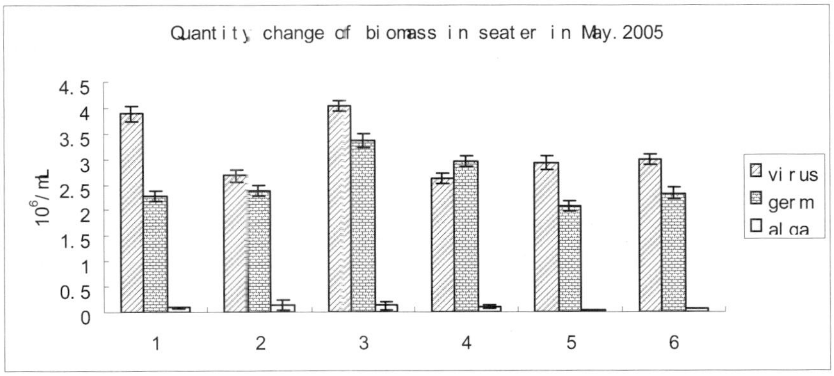

Fig 3 Quantity change of biomass in seawater during the same time, in difference place

4.4 Other Factors in Sampling Place in the Same Period

We made a accidence covariance to the research result of the water's fluid and farming circumstance in the sample collected location before and after May, 2005, just as table 2 shown. It shows that, the temperature and PH of estuarine seawater in same period in Qingdao have no much difference, the fluid matter and the algas' quantity have large difference in seawater, and farming surroundings is dissimilarity. There have more plankton in farming area, but less in no farming area. At the bathing beach with more mankind's activity, the fluid matter is more turbid.

Table 2 Water quality about water sample collection place, 05/2005

sample collects loaction	Temperature(°C)	pH	fluid	plankton	Surroundings
Taiping Cape, Qingdao	16	8.07	Pure	much	The deep water area has alga farmings
frist bathing beach, Qingdao	14	7.66	A little muddy	Less	No farming area surround, less mankind's activity
Diaosu park, Qingdao	14	7.25	Turbid	Much	Have a lot of sea cucumbers, abalone farming pond
Taiwan road, Qingdao	14	7.55	Pure	A great deal of algas	Have a lot of sea cucumbers, abalone farming pond
Shilaoren, Qingdao	15	7.?2	Fure	Have not seen by eye	No farming area surround, less mankind's activity
Fangzi, Weifang	12	7.68	Pure	Have not seen by eye	Saltworks and farming waste water inflow in

5. Discussion

5.1 The Virus Counts Method in Water

Research methods to virus counts in water are mainly lies in PFU (Plaque Forming Unit Numbering System), TEM (Transmission Electron Microscope), FIM (Fluorescence Imaging Microscope) and FCM (flow cytometry method) etc. PFU is the most traditional technique.

It is observes to the plaque forming units amount after virus infection in the foundation of sensitive cell culture, acquiring a amount information thus. That method is very accurate, but needs to notice the false positive results differentiation due to the cell toxic material in sample; TEM apply to virus quantify, is direct observation and counting under the scanning electron microscope after virus having been dealed with sample preparation technique such as redyeing. TEM can be used to virus appearance observation in the meantime, directly accurate. FIM make use of nucleic acid fluorescence dyestuff to dye to the sample, then observate and counting under a fluorescent microscope. This counting method is a precision and easy operation, and can make precision counting directly to very low virus density sample like seawater etc., and avoid of usage the expensive, big physical volume instrument. But it's accurate degree is slightly low to FCM when the higher density virus counting; FCM is a very directly and sensitive way to measurement the virus amount, single size of virus, and assistance to bio-chemical and physiology characteristic measurement. All sample's circulate and test carried on in the flow cytometry, for higher density virus sample test it's result is the most accurate one, but it can't test sparse virus sample.

5.2 Seawater FIM Counting Research

The virus FIM counting result to the near shore seawater sample in Qingdao shows, the virus density in the sea aquatic is variety. The germs and planktons like alga amount are also different in different time in seawater. The virus amount in seawater in Qingdao is slightly low than the report in Europe, North America and Japanese, this probably because of the ecosystem factors difference like nourishment salinity thickness and microorganism density. FIM counting result to the germ amount in seawater is, $0.055\text{-}3.35\times10^6$ germs/mL in Qingdao, the plankton is $0\sim1.95\times10^6$ algas/mL. With different sampling time and location, the content of living creature have season and region difference.

From September in 2004 to May in 2005 in Qingdao sea area, the virus density have a peak in October, this may be due to the temperature descend in September for the autumn coming, which is good to the survival of the free virus. After turning into winter the temperature goes down suddenly, the plankton moves down. Counting result shows that the algas and the germs amount descends significantly. This made the virus survival ecosystem factors change, the virus density gets into a low valley immediately. After the weather become warm in the next year spring, planktons moves up gradually, the virus amount also start it's increasing, the virus amount starts a rally again in March. Moreover, the spring water bloom of the plankton bringing stability breeds ecosystem environment for virus. The virus shows a prosperity prospects at this time. We can see this point in our research in the virus amount representation. It incarnate well in the relativity of time and the amount between the plankton, microorganism and the virus. To the variety of virus category, is what we will pay more attention to in the following research. To the variety regulation of the virus distribution in same time and same waters, and the distribute status in other times, we will also continue to do some experiments and research to collections more data.

The mensurate result of different area in the same time shows that, the content of various living creature also have certain difference in seawater. The difference of the various living creature measures have certain relativity. The virus amount and the germ content in seawater shows stronger positive relation particularly. Why like this? It's maybe because that phages occupy a bigger quota in the virus community, and the microorganism is the important host that the virus lives on. But germs acts as the role of disintegrator in the food chain, so, plankton likes algas measures variety influenced the variety of the germ amount in seawater directly. Moreover, we see under obvious difference farming circumstance and mankind's activity in sample collection area, the different colony's amount is also different to each other. Place having farming areas obvious have many algas. Virus and germ amount in these area are larger than other locations. But in the Bathing beach of Qingdao with more mankind's activity, the alga amount isn't very large, the virus and the germ amount there are very large. It shows that mankind activity have great influence to the virus and the germ's diffuse and multiplication in the seawater.

To the ecology research of virus in the sea water, to the virus research in plentiful, category in different waters', and the investigate on the depending relationship between virus and environment factors, will build a fundation to the research in the next step, such as in develop a high flux analysis system of the ocean virus gene chip, and in provide valid supervision measure to the key virus disease prevention and curing.

Reference

1. M. Barranco and J. R. Buchler, *Phys. Rev.* **Cf22**, 1729 (1980).
2. Tu Lihong, Li Chunlin, *Journal of Capital Normal University*. **78**, 17 (1996).
3. Fuhrman J A, *Nature*. **541**, 399(1999).
4. Zhao Yijun, et al., *Progress in Natural Science*. **923**, 12 (2002).
5. Zhao Yijun, et al., *Virologica Sinica*. **100**, 14 (1999).
6. Liu Yanming, Zhang Qiya, Yan Xiuping, *Acta Hydrobiologica Sinica*. **1**, 29 (2005).
7. Noble, R.T., and J.A. Fuhrman, *Aquat. Microb. Ecol.* **113**, 14(1998).
8. Xu Bing, Xu Huaishu, *Microbiology*. **357**, 20 (1993).
9. Cheng Kai et al., *Acta Hydrobiologica Sinica*. **535**, 27 (2003).
10. Li Juan, Hang Jie, Tang Xuexi, *Marine Fisheries Research*. **68**, 26 (2005).

INHIBITION EFFECT OF CITRUS CNHX1 GENE TO YEAST PROGRAMMED CELL DEATH INDUCED BY NaCL

WEI ZHEN-LIN †

Biology Science Department, DeZhou University, DeZhou, China, 253023

LI XIANG-YONG

Biology Science Department, DeZhou University, DeZhou, China, 253023

JIAO CHUAN-ZHEN

Biology Science Department, DeZhou University, DeZhou, China, 253023

LI JUAN

Biology Science Department, DeZhou University, DeZhou, China, 253023

RUI YU-KUI

Laboratory of Food Technology of China Agricultural University, Beijing, China, 100094

SHU HUAI-RUI*

College of Horticultural Science and Engineering, Shandong Agricultural University, Taian, China, 271018

The citrus *cNHX1* gene was inserted into yeast expressive vectors pYX212, which was then transformed into yeast strain AXT3K by electroporation to express cNHX1 protein. The inhibition effect of programmed cell death by cNHX1 protein was studied through DAPI dyeing and other methods. Fluorescence image analyzing showed that expression of *cNHX1* gene refrained nuclei fragment induced by 1.5M NaCl, delayed the process of cytolysis and decreased fragmented nuclei ratio. Clonigenic survival assay results demonstrated that cNHX1 protein expressed in AXT3K cells increased CFU value under 1.5M NaCl stress conditions. DNA contents detected by FACS method decreased more slowly in AXT3K cells expressed *cNHX1* gene than that in wild AXT3K cells. Sub-G^0/G^1 peak ratio were 12.68%, 30.82% and 50.41% at 1, 2 and 3h under 1.5M NaCl stress conditions, respectively, and responding value of control AXT3K cells were 19.42%, 41.96% and 67.07%.

1. Introduction

Salt and drought stress is the most important factor for plant growth to be limited, resulting in reduction of crop yield over 40% [2][12]. In China, the salinized soil adds up to approximately 3.3 × 10^7 ha which seriously hinders the production and development of agriculture. Recently, some researchers found that the Na^+/H^+ antiporter gene was one of important factors determining the salt-tolerant capacity of plants, which encoding plasma and vacuole membrane Na^+/H^+ antiporter, playing an important role in ion compartmentalization and regulation of cell pH [13]. The Na^+/H^+ transporting activity significantly rised in vacuole membrane vesicles isolated from salt-resisting plants such as sugar beet, barley, hoary plantain and ice plant which grew under the salt-stress conditions, proving that the low Na^+/H^+ transporting activity is a character of salt-sensitive plants [9].

Constitutive-expressing different plant Na^+/H^+ antiporter genes in *Arabidopsis, Oryza, Brassica*, and *Lycopersicon* showed a significant increase of salt tolerance [14][15][1][9][4]. The above-mentioned facts suggested that using such single gene might get salt-tolerance markedly increased crops [13]. Porat *et al.* (2002) isolated two segments of *cNHX1* gene from *Citrus paradise* [11], and these two fragments were fused to one complete gene by SOEing method in our lab recently (unpublished data).

Saccharomyces cerevisiae owns many similarities to plants in salt tolerant mechanism and easily to be

* Corresponding author: wzl19741028@163.com

operated, so widely used in cloning of plant salt-tolerant gene and analysis of gene function. In addition, *Saccharomyces cerevisiae* is a good tool for study on programmed cell death [8]. The yeast cytological phenomenon induced by high-concentration salt stress was very alike to that of lysigenous PCD, and the nucleus DNA fragment phenomenon was more clear in the *cnb1*△ yeast cell and in the *sos1* cell of *Arabidopsis thaliana* insufficient compartmentalization function [5]. The Bc1-2 gene relieved the yeast PCD process caused by NaCl, and this laxation depended on exist of ENA1 function, which demonstrated that lowering the Na^+ concentration in cytoplasm helped to relieve the yeast cell PCD process [5]. But little has been published concerning the plant *NHX* gene depressive programmed cell death encouraged by high-concentration salt. Making use of the widely adopted *scnhx1* mutanted yeast strain, we analyzed the inhibitory action of *cNHX1* gene upon the yeast programmed cell death led by high-concentration salt.

2. Materials and Methods

2.1. *Materials*

The *Saccharomyces cerevisiae* strain W3031B (MAT α △ura3-1 leu 2-3, 112 his 3-11, 15 trp 1-1 ade 2-1 can 1-100) and AXT3K (enal:: HIS 3:: ena 4, △nhal:: LEU2, △nhx 1:: Kan MX4) sended by Prof. Pardo J M (Consejo Superior de Investigaciones Cientificas, Sevilla 41080, Spain), the yeast express vector pYX212 (URA) provided by Prof. Hohmann S (Gteborg University, Sweden). The other materials were purchased as follows: the EX Taq enzyme, dNTP, and restriction enzyme from TaKaRa Co.; the agarose and agarose gel recovering kit from Shanghai Sangon Co.; ligase from Promega Co.; and DAPI from Sigma Co.

2.2. *Yeast Expressive Vector Construction*

The vectors Y15 contained completely *cNHX1* gene and pYX212 were double-digested by using the enzymes BamHI/SacI, and recovered the segments 1.63kb and 10kb, respectively. Then the two segments were connected and used for transformation to *E. coli* JM109, then the pYXNHX vector was gained finally.

2.3. *Yeast Transformation*

The pYXNHX plasmid DNA at 1 μ g purified with agarose gel recovering kit and pulsing parameters (1.5kv, 25uF and 200 Ω) were used in transformation protocol [3]

2.4. *Growth Curve*

The effect of NaCl on growth curve of yeast cells was determined by using the method of Huh *et al.*[5]with a little amendment. The overnight cultured AKT3X yeast cell mass was collected and diluted into freshly prepared YPD medium with 50mM NaCl, with the OD_{600} being adjusted to 0.05, shake-cultured at 28℃ for 20h, and them the OD_{600} value was measured.

2.5. *Experiment of PCD Depression Effect*

The experiment was carried out by using DAPI staining method[5]. The overnight-cultured yeast cell mass was adjusted with liquid YPD medium to OD_{600}=0.5, with the NaCl added at 1.5M of final concentration, then treated for 0, 1, 2 and 3h, respectively. The cell mass collected by centrifugation was fixed in 70% ethanol and dyeing with 1 μ $g.L^{-1}$ DAPI in PBS buffer solution for 10 min. Finally, make observation under an OLYMPUS BX51 microscope at 359nm wavelength and fluorescent photographs were taken. Fragmented nuclei ratio were calculate by counting cells contained fragmented nuclei in all statistical cells (200 cells) under all treated times.

2.6. *Clonigenic Survival Assay*

The assay was done following Huh *et al.*[5]. The AXT3K yeast cell, with or without *cNHX1* gene was put on SC medium to grow overnight, the culture was adjusted with fluid YPD medium to OD_{600}=0.5, treated with 1.5M NaCl for 0,1, 2 and 3h, respectively, and washed twice with sterile water.

The cell mass collected by centrifugation was diluted with the liquid YPD medium to about 150 cells per 100 μl, then coated on the solid YPD medium at 100 μl per 90mm plate, after incubation at 28℃ for 60h, the yeast colonies of each plate were counted. Taking the untreated AXT3K as the control, the clonigenic survival unit(CFU) was calculated: CFU=colonies per plate of different treatments/colonies per untreated plate (three repeats per treatment).

2.7. *FACS Analysis*

DNA contents of NaCl treated yeast cells were detected by FCM method as described by Herskowitz *et al.* (http://www.fhcrc.org/labs/gottschling/yeast/facs.html). About 10^7 cells of each treatment were stained with PI, sorted by BECTON DICKINSON FACScan machine and calculated by ModFit LT software.

3. Results

3.1. *Effect of Citrus cNHX1 Gene on Yeast Cell Growth Curve*

Under the 50mM NaCl stress conditions, the yeast growth was inhibited in different degrees, elongating the cell logarithmic growth stage and decreasing cell growth rate. Under the same conditions, the yeast strain AXT3K with expressive *cNHX1* gene got a higher growth rate than that of the control, retarding the time to entry the platform stage (Fig.1), which suggested that the *cNHX1* gene increased cell division speed and cell growth rate of yeast under the salt stress conditions.

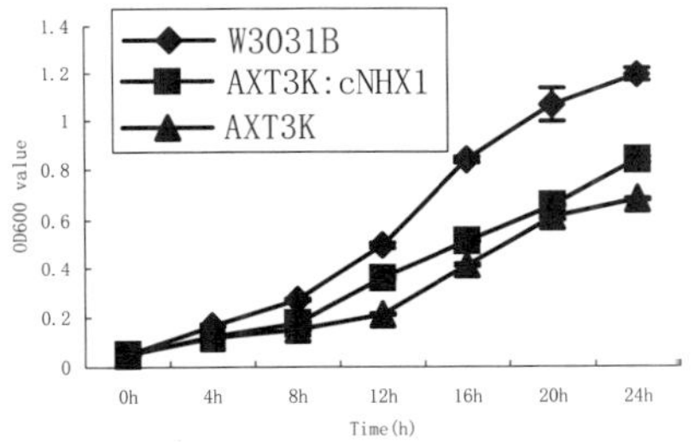

Fig. 1. Effect of citrus *cNHX1* gene on yeast cell growth curve under 50mM NaCl conditions.

3.2. *Fluorescence Image Analyzing*

From Fig.2 can be seen that the expression of *cNHX1* gene reduced the 1.5M NaCl inducing programmed cell death scale of AKT3K cell. In untreated cell, no matter with or without *cNHX1* gene, a huge blue zone was displayed by DAPI staining, which represented intact nucleus. In the yeast cell treated with 1.5M NaCl for 1h, in comparison with control, all their nuclei had a shrinking appearance, a part of AKT3K yeast without expressive *cNHX1* gene exhibited multiple DAPI dyeing zones, suggesting there has happened the nucleus fragment phenomenon, the typical symbol of programmed cell death, meanwhile, in the AKT3K yeast cell containing the cNHX1 gene has not happened this phenomenon but displayed nuclei shrinkage phenomenon. After 2h treatment with 1.5M NaCl, a part of AKT3K yeast cell without expressive *cNHX1* gene had no distinct nuclei, and some individuals of the yeast cell were in lysis. We yet can see the clear nuclei in the AKT3K cell with expressive *cNHX1* gene, but there happened the fragment phenomenon in the nuclei of some cells, too. After 3h treatment, the AKT3K cell having no expressive *cNHX1* gene are on the whole in lysis, however, single or fragmented nuclei yet can be seen in the AKT3K cell with expressive *cNHX1* gene, and a few cell have been lysed. These results basically agreed with the cytological phenomenon discovered by ultra-thin section [5]

The color photos could be downloaded from http://www1.dzu.edu.cn /dzxyswx /sjjx.html.

In this experiment, fragmented nuclei ratio was analyzed also, results showed (Fig.2) AXT3K cell expression of *cNHX1* gene have lower fragmented nuclei ratio after treated 1 or 2 h by 1.5M NaCl, which consist with its higher clonigenic survival unit (Fig.2); When these two cells were treated 3h, the fragmented ratio of AXT3K cell not contain cNHX1 gene was lower than that of cells containing *cNHX1* gene, because much more cells were in the period of lysis result in no visible nuclei stained by DAPI can be observed.

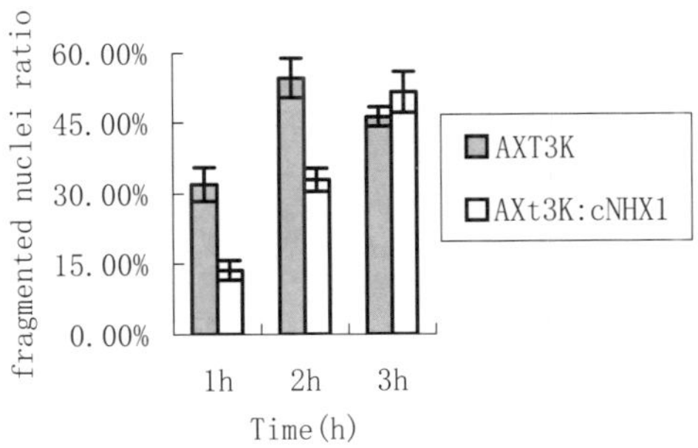

Fig. 2. Fragmented nuclei ratio.

3.3. *Clonigenic Survival Assay*

The clonigenic survival assay and cell death restraining effect of expression of Bcl-2 protein can be used to confirm whether or not has happened the PCD process [5]. Since the yeast strains adopted by us in the experiment was the *enal* mutants, so we used the clonigenic survival assay to further affirm the role of *cNHX1* gene in the inhibition of yeast programmed cell death induced by high-level salt. The results indicated that the clonigenic survival units of the yeast AXT3K cell with expressive *cNHX1* gene after different treatment durations (no matter 1, 2, and 3h) were higher than that without expressive *cNHX1* gene (Fig.3), demonstrating the expression of *cNHX1* gene increased the CFU in AXT3K strains under the 1.5M NaCl stress conditions.

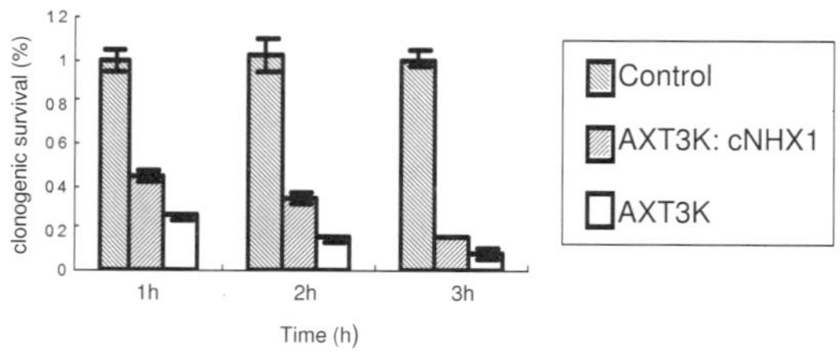

Fig. 3. Results of clonigenic survival assay of yeast cell with or without citrus cNHX1 gene.

3.4. *FACS analyzing*

DNA contents variation was a rapid and reliable indicator of the PCD phenomena appearance (Madeo F, personal communication). At present experiment, FACS assay results showed expressive *cNHX1* gene restrained DNA degradation of transformed AXT3K cells at all treating times, subG^0/G^1 peak ratio (an index scaling undergone PCD phenomena cells) were 12.68%, 30.49% and 50.41%, respectively, while the responding value of wild AXT3K were 19.42%, 41.96% and 67.07%, respectively (Table.1). This results proved further that cNHX1 protein restrained high concentration NaCl induced PCD phenomena in yeast cells.

Table.1 FACS assay results of yeast cells at 1,2 and 3h treated by 1.5M NaCl

Different cells	Sub G^0/G^1 peak ratio treated for 1h	Sub G^0/G^1 peak ratio treated for2h	Sub G^0/G^1 peak ratio treated for 3h
AXT3K	19.42%	41.96%	67.07%
Transformed AXT3K	12.68%	30.82%	50.41%

4. Discussion

In this experiment, expressive *cNHX1* gene alleviated restrain effect to growth curve under 50mM NaCl stress conditions, in our previous work, cNHX1 protein was confirmed to located in vacuolar membrane increased Na^+ contents under NaCl stress conditions, enhanced Li^+, K^+ and Hyg tolerance of AXT3K cells (unpublished data), showed *cNHX1* gene was a functional homologue to *AtNHX1* gene.

Several reports showed NaCl stress and SFK1 protein induced programmed cell death phenomena in yeast cells which related to ion disequilibrium under stress conditions [5][16].

Our experiments revealed that the cNHX1 protein inhibited high-concentration NaCl inducing PCD process in AXT3K yeast cells. It also showed decreasing of Na^+ concentration in cytoplasm by compartment excessive Na^+ to vacuolar was an efficient NaCl stress resistance strategy. Peinado *et al.* (personal communication) showed high osmotic stress provoked yeast cells PCD process, so, it was likely that cNHX1 protein performed negative effect to PCD process by its antiport activity and decreased osmotic imbalance between external medium contained 1.5M NaCl to internal solutions of cells as the result of Na^+ influx into vacuolar.

Plants and yeasts showed a similar PCD form [5], for example, in barley roots treated with NaCl, the typical symptoms of PCD including nucleus fragment and DNA cleavage were observed [6][7]. In plants, the Bcl-2 protein expressed in tobacco cell also enhanced resistance to salt stress, and the resistance enhance also depended on the function improvement of ion compartmentalization [10].

References

1. M.P. Apse, G.S. Aharon, W.A. Sneddon, E.Blumwald. *Science*. **285**, 1256(1999).
2. M.P. Apse, E. Blumwald. *Current Opinion in Biotechnology*. **13**, 146(2002).
3. F.M. Ausbel, R. Brent, R. E. Kingston, D. D. Moore, J. G. Seidman, J. A. Smith, K. Struhl. *Current Protocols in Molecular Biology*. Press John Wiley and Sons. **512**(1995).
4. A. Fukuda, A. Nakamura, A. Tagiri, H. Tanaka, A. Miyao, H. Hirochika, Y. Tanaka. *Plant and Cell Physiology*. **45**, 146(2004).
5. G. H. Huh, B. Damsz, T. K. Matsumoto, M. P. Reddy, A. M. Rus, J. I. Ibeas, M. L. Narasimhan, R. A. Bressan, P. M. Hasegawa. *Plant J.* **29**, 649(2002).
6. M. Katsuhara. *Plant Cell Physiol.* **38**, 1091(1997).
7. M. Katsuhara, T. Kawasaki. *Plant Cell Physiol.* **38**, 169(1996).

8. S. Matsuyama, S. Nouraini, J. C. Reed. *Current Opinion in Microbiology*. **2**, 618(1999).
9. M. Ohta, Y. Hayashia, A. Nakashimaa, A. Hamada, A. Tanaka, T. Nakamura, T. Hayatawa. *FEBS letters*. **532**, 279(2002).
10. J. B. Qiao, I Mitsuhara, Y. Yazaki, K. Sakano, Y. Gotoh, M. Miura, Y. Ohashi. *Plant and Cell Physiology*. **43**, 992(2002).
11. R. Porat, D. Pavoncello, B. H. Gozal, S. Lurie. *Plant Science*.**162**, 957(2002).
12. J. D. Rhoades, J. Loveday. *American Society of Agronomists*. **1089**(1990).
13. M. Tester, R. Davenport. *Annals of Botany*. **91**, 503(2003).
14. H. X. Zhang, E. Bumwald. *Nature Biotechnolog*. **19**, 765(2001a).
15. H.X. Zhang, J.N. Hodson, J. P. Williams, E. Blumwald. *Proc Nat Acad Sci USA*. **98**, 12832 (2001b).
16. I. Wadskog, C. Maldene, A. Proksch, F. Madeo, L. Adler. *Mol. Biol. Cell*. **15**, 1436(2004).

INVESTIGATING THE DISPERSION AND STABILITY OF AL_2O_3 NANO-SUSPENSIONS*

XIAN-JU WANG[A,B]

[A]*Key Lab of Enhanced Heat Transfer and Energy Conservation, Ministry of Education, School of Chemical and Energy Engineering, South China University of Technology, Guangzhou 510641, China*

[B]*Navy Arms of Services Command Academy, Guangzhou 510431, China*

DONG-SHENG ZHU [A], XIN-FANG LI [A]

This paper presents a procedure for preparing a nanofluid which is solid–liquid composite material consisting of solid nanoparticles with sizes typically of 1–100 nm suspended in liquid. By means of illustrating the transmission electron microscope (TEM) photograph, sediment photographs and particle size distribution, the stability and evenness of Al_2O_3-H_2O suspension with dispersant were investigated. Meanwhile, aiming at the dispersion of nano-Al_2O_3 is regarded as the guide of heat transfer enhancement, the dispersion behavior of Al_2O_3 nanoparticles in water were studied under different pH value. The experimental results show that zeta potential has good corresponding relation with absorbency, the higher magnitude of zeta potential and the absorbency are, the better dispersion and stability the Al_2O_3-H_2O suspension system are, and the optimizing pH of the suspension corresponding to the highest magnitude of zeta potential and absorbency are at pH 9.0, which has the best disperse results.

1. Introduction

In order to manage the growing demand from different industries such as electronic, automotive and aerospace industries, heat exchanger devices have to be small in size, light in weight and of high performance. Low thermal conductivity of conventional heat transfer fluids such as water, oil, and ethylene glycol mixture is a serious limitation in improving the performance and compactness of these engineering equipments [1-4]. Development of the nanomaterial technology makes it possible to structure a new type of heat-transfer fluid by suspending nanoparticles in conventional liquids. Choi [5] used the term nanofluid to refer to this class of fluids. Compared to conventional heat transfer fluids, nanofluids show a superior potential for increasing heat transfer rates [6-10]. Researchers have demonstrated that oxide ceramic nanofluids consisting of Al_2O_3 nanoparticles in water or ethylene glycol exhibit enhanced thermal conductivity [11]. Because of high surface area, nanoparticles form aggregates or agglomerates due to the action of van der Waals or other attractive forces. On the other hand, their stability is promoted through the electrostatic or electrosteric repulsion. In most of applications, a high stable dispersion is required. Therefore, controlling the coagulation of nanoparticles in the nanofluid becomes a common factor in the current technological limitations for their potential benefits and applications [6].

Note that physical or chemical treatment such as the addition of surfactant, surface modification of the suspended particles or applying strong force on the clusters of the suspended particles can change the suspension stability [9, 10]. Donggeun Lee [7] studied the stability of nanofluids by changing pH of the solution systematically to control surface charge density and surface potential, they found that the surface charge states directly affected the suspension stability and presented the strong correlation between hydrodynamic size of particles and stability coefficient. UV–vis spectrophotometric measurements have been used to quantitatively characterize colloidal stability of the dispersions [10].

In this paper a procedure for preparing a nanofluid was presented. By means of illustrating the transmission electron microscope (TEM) photograph, sediment photographs and particle size , the stability and evenness of Al_2O_3-H_2O suspension with dispersant were investigated. Meanwhile, the dispersion behavior of Al_2O_3 nanoparticles in water is studied under different pH value.

* The corresponding author is Xianju Wang. E-mail: xwqwxj@tom.com

2. Materials and Methods

2.1. *Chemical*

In this work, Al_2O_3 particles were purchased from a commercial company (Nanophase Technologies Corporation, USA) which is chosen as a source material for a nanofluid. The nanoparticles were used as received, which were produced by a physical vapour deposition technique. Figure 1a shows the TEM photograph of Al_2O_3 powder. It can be seen that the primary Al_2O_3 nanoparticles are spherical and their size is widely distributed in a range of 10–50 nm. The BET surface area of the powder is found to be (30-50) m^2/g.

The dispersant used in the investigation was SDBS (sodium dodecylbenzene-sulfonate) in chemical grade. The water used was distilled and deionized. Analytical grade hydrochloric acid and sodium hydroxide were used to modify pH value of the system.

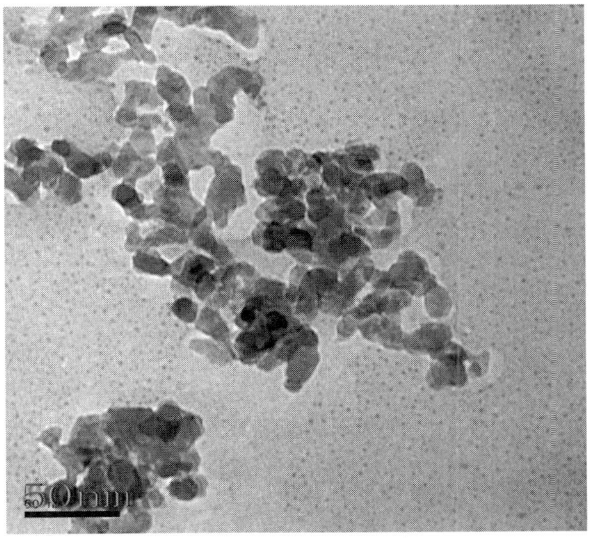

Figure 1. TEM micrograph of Al_2O_3 powder

2.2. *Preparation of Nanofluids*

Two-steps method was selected to prepare the nanofluids, which is shown in Fig.2. Preparation of nanofluids is the first key step in applying nanophase particles to changing the heat transfer performance of conventional fluids. The nanofluid does not simply refer to a liquid-solid mixture. Some special requirements are necessary, such as even suspension, stable suspension, durable suspension, low agglomeration of particles and no chemical change of the fluid. In general, these are effective methods used for preparation of suspensions: (1) to change the pH value of suspensions; (2) to use surface activators and/or dispersants; (3) to use ultrasonic vibration. Selection of the suitable dispersants mainly depends upon the properties of solutions and particles. In this study, the nanostructured material of Al_2O_3 particles of about 25 nm is used to form water-particle suspensions. Al_2O_3 nanoparticles are uniformly dispersed in the water by using SDBS as a dispersant. The Al_2O_3-H_2O nanofluids suspension was vibrated for 1 h in an ultrasonic vibrator (KQ2200DE Ultrasonic Cleanser, 100W, Kunshan of Jiangsu Equipment Company, China). At the same time, the Al_2O_3 suspension without SDBS dispersant was also vibrated for 1 h in the ultrasonic vibrator for comparison. Their sedimentation photographs were shown in Figure 3, which shows that the stabilization of the suspension with dispersant can last about 1 week in the stationary state and no sediment was found, while the suspension without dispersant exhibited weaker dispersion and aggregatation quickly occurring.

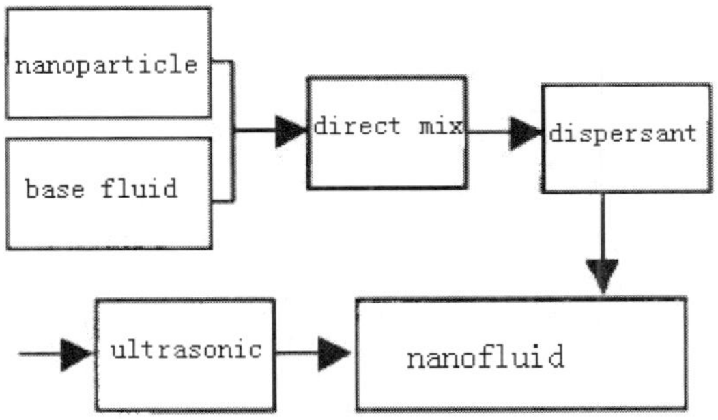

Figure 2. The preparation process of nanofluid

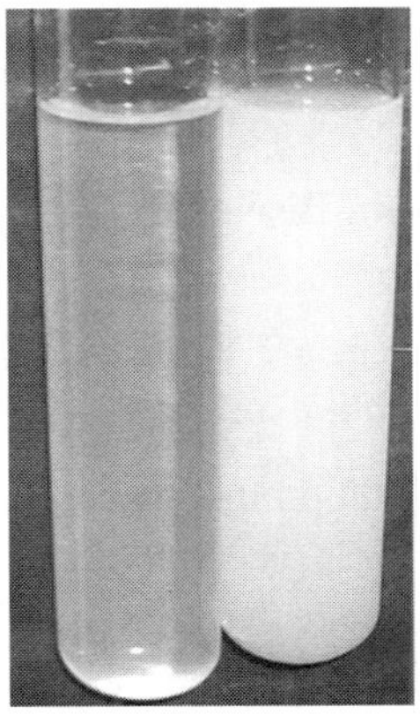

Figure 3. Photographs of Al_2O_3-H_2O suspensions in the absence (left) and in the presence (right) of the dispersant for a week: Concentration of Al_2O_3 is 0.1%.

2.3. *Evaluation of Dispersion by Sedimentation Method and Absorbency Measurement*

Evaluation of dispersion characteristics of the Al_2O_3 suspension was carried out using the conventional sedimentation method and measurement of absorbency in the test tubes. Different concentrations of the dispersant and different mass faction of the Al_2O_3 nonoparticles were added to the suspensions, which were stirred thoroughly and ultrasonicated for at least 1 hour. The pH value of system was adjusted with HCl and NaOH solution by precise pH Meter (PHS-25, China). Then 10 mL of each suspension was then poured into test tubes and the samples were allowed to deposit for several days. Then the above solution for certain height of the suspensions at different pH value was taken and the absorbency of was measured on a U-3010 spectrophotometer (Hitachi Instrument Inc., Japan) after depositing for 24 h. Meanwhile, the states of sedimentation were observed every 24 h, and the sediment photographs were then taken directly from the test tubes.

2.4. *Measurement of Zeta Potential and Hydrodynamic Size*

The experiments were conducted using dilute different mass fraction Al_2O_3 nano-suspensions, and the suspensions were stirred thoroughly and ultrasonicated for at least 30 minutes. Then the zeta potential and hydrodynamic size of the particles were measured at 25 ℃ in a Malvern Zeta sizer 3000 particle analyzer (Malvern Instrument Inc., London, UK). Each parameter was measured 10 times per sample. The pH value of system was also adjusted with HCl and NaOH solution by precise pH Meter.

3. Results and Discussion

3.1. *Influence of pH Value on Zeta Potential and Hydrodynamic Size*

In aqueous medium, the interaction between the colloid particles is influenced by the local electric field developing around the particles. The relation between the local electric field and the stability of the lyophobic colloids is described in the DLVO theory [12-14]. The theory considers the London–van der Waals attractive forces and the electrostatic repulsion between two charged particles due to the overlapping of electric double layers. Since solution pH is an important factor controlling the stability of metal hydroxides precipitated in aqueous media, by use of the acid form of the anionic surfactants affords an opportunity to tune to the pH. Therefore, the study of the electrophoretic behavior through measurement of the zeta potential and hydrodynamic size at different pH becomes important for understanding the dispersion behavior of nano-particles in a liquid medium. Fig. 4 shows the zeta potential and hydrodynamic size of alumina powder as a function of pH with SDBS. It can be seen that the surface of alumina powder is negatively charged by adding SDBS dispersant. It may be due to dissociation of SDBS in solution producing phenyl sulfonic group adsorbed on the alumina particle, which in turn increases the net negative charge of powder surface and also increases the repulsive forces. At pH below 9.0, with pH increase the absolute zeta potential increase and hydrodynamic size decrease, and above pH 9.0 with pH increase the magnitude of zeta potential decrease but hydrodynamic size increase. This result suggests that zeta potential has very corresponding relation with hydrodynamic size, the higher magnitude of the zeta potential and the smaller hydrodynamic size of the nano-Al_2O_3 and the better dispersive stability of the nano-suspension is, and that an optimizing pH value of the nano-suspension is best for good dispersion stability of alumina powder at about pH 9.0.

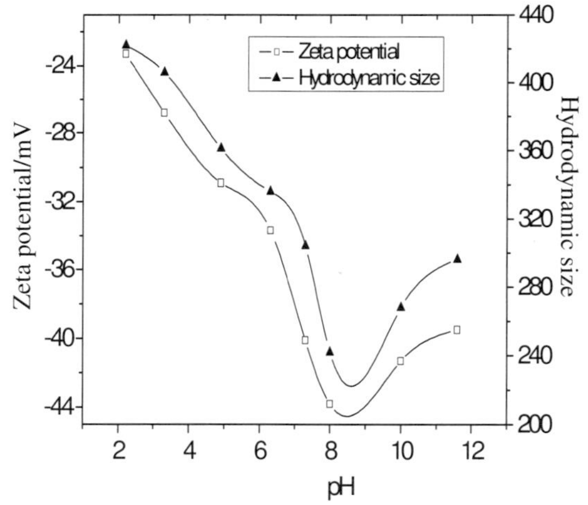

Figure 4. The effect of pH on the zeta potential and hydrodynamic size

3.2. *Influence of pH Value on on Zeta Potential and Absorbency of Al_2O_3-water Suspension*

Figure 5 shows the changes of zeta potential and absorbency of Al_2O_3-water nanofluids as a function of pH with 0.1% SDBS dispersant in the case of Al_2O_3 suspension with dispersant, all the zeta potential values were negative in the range of pH (from 2 to 11). At lower pH values (from 2 to 6), the absolute values of zeta potential and the values of absorbency almost increase with pH values increasing, then, both the magnitude of zeta potential and absorbency reach the maximum with the pH≈9. 0, afterwards, in the region of 9.0<pH<11, the magnitude of zeta potential and absorbency all tend to decrease with the increase of pH value. These results suggest that zeta potential has very corresponding relation with absorbency at any value of pH.

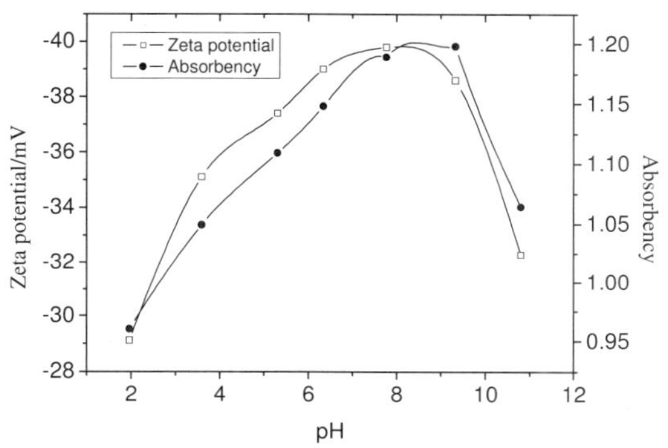

Figure 5. The effect of pH on the zeta potential and absorbency

3.3. *Influence of pH Value on Stability of Al_2O_3-water Suspension*

The above experimental results showed that: as pH increases, the magnitude of zeta potential on the particle surface increases, the electrostatic repulsion force between particles becomes stronger and sufficient to prevent attraction and collision, on the other hand, greater electrostatic force can also lead to more free particles by increasing particle–particle distance so that the distance exceeds the hydrogen bonding range between particles and further reduces the probability of particle coagulation and settling; when the pH increase to certain degree, the value of zeta potential arrive the maximum and the electrostatic repulsion force between particles become the strongest, and the coagulated particles can be redispersed through mechanical force, so the dispersion stability of Al_2O_3 nanoparticles is the best at this case; but as pH continues to further increase, the concentration of the pH adjustment reagent (NaOH) in the system increases, which causes the compression of electrical double-layer, so lowering the magnitude of zeta potential and electrostatic repulsion force on the particle surface, at this case the repulsion barrier become lower and lower, the attraction between particles become stronger and stronger, which means that suspension exhibits a poorer dispersion, larger hydrodynamic diameter and less stability. The relation between hydrodynamic diameter and stability obtained by numerical integration following reference [7] is shown in figure 6. In figure 6 shows clearly that the stability (W) is inversely proportional to the hydrodynamic diameter (D_h) at all pHs which is in good agreement with showing in figure 4. The above analysis shows that the dispersion stability behavior of Al_2O_3 nanoparticles can be affected by pH, and the best dispersion and stability corresponding to an optimization pH value.

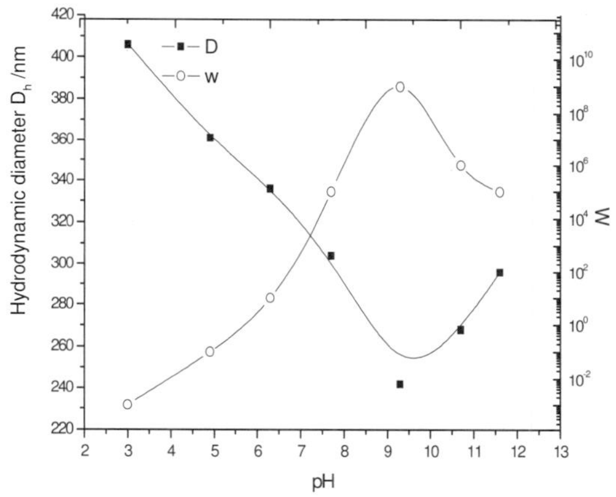

Figure 6. The effect of pH on hydrodynamic diameter and stability

4. Conclusion

By measurement of the zeta potential, absorbency and hydrodynamic diameter, the dispersive behavior of Al_2O_3 nanoparticles in water were investigated under different pH values. Based on the work done in this research, we can get the following general conclusions: at first, Al_2O_3-H_2O nanofluids by two-step method were prepared and the hydrodynamic diameter distribution show better dispersion behavior in the suspension with the addition of dispersant. In addition, it is found that the dispersion and stability behavior of Al_2O_3 nanoparticles can be affected by pH, and the best dispersion and stability corresponding to an optimization pH value. That is to say, among the zeta potential, absorbency and hydrodynamic diameter there are a good correlation: the higher the absolute value of zeta potential and absorbency are, the smaller the hydrodynamic diameter is, and the better the stability of the system is. Furthermore, the experimental results show that the optimizing pH of the suspension corresponding to the highest magnitude of zeta potential and absorbency are at pH 9.0, which has the best disperse results.

Acknowledgments

This work was supported by the National Natural Science Foundation of China (Grant No. 20346001), and Post-doctor Foundation of China (Grant No. 20060400219) Specialized Research Fund for the Doctoral Program of Higher Education (Grant No. 20050561017).

References

1. J.A. Eastman, S.U.S. Choi, S. Li, W. Yu, L.J. Thompson, Appl. Phys. Lett. **78**, 718 (2001).
2. H.Q. Xie, J.C. Wang, T.G. Xi, Y. Liu, F. Ai, J. Appl. Phys. **91**, 4568 (2002).
3. S.U.S. Choi, Z.G. Zhang, W. Yu, F.E. Lockwood, E.A. Grulke, Appl. Phys. Lett. 79, 2252 (2001).
4. A. Akbarinia, A. Behzadmehr, Applied Thermal Engineering **27**, 1327 (2007).
5. S.U.S. Choi, ASME, FED, **231**, 99 (1995).
6. X. F. Li, D.S. Zhu, X.J. Wang, J. Colloid Interface Sci. **310**,456(2007).
7. L. Donggeun, Jae-Won Kim, Bog G. Kim, J. Phys. Chem. B **110**, 4323(2006).
8. Y. Yang, Z.G. Zhang, E.A. Grulke, W.B. Anderson, G. Wu, Int. J. Heat Mass Trans. **48**, 1107(2005).
9. Y.L. Ding, H. Alias, D.S. Wen, R.A. Williams, Int. J. Heat Mass Trans. **49**, 240(2006).
10. S. Cong, L. Fu, X. H. Ning, L.Y. Ping, Materials Letters **59** 3494 (2005).
11. S. Lee, S.U.S. Choi, S. Li, J.A. Eastman, ASME J. Heat Transfer, **121**, 280(1999).
12. R.J. Hunter, Foundations of Colloid Science, Vol. 1, Clarendon Press, Oxford, 1987.
13. D.J. Shaw, Introduction to Colloid and Surface Chemistry, Butterworths, London, 1966.
14. M. Elimelech, J. Gregory, X. Jia, R.A. Williams, Particle Deposition and Aggregation: Measurement, Modeling and Simulation, Butterworths, Oxford, 1995.

MEASUREMENT OF INTRACELLULAR pH IN SINGLE CA46 CELL USING NANOSENSOR

NA FANG, YISHEN QIU†, ZHIHAO CHEN, QINMIAO CHEN, QINGHONG LU, HUIYUE QIU, GAOMING LI

Key Laboratory of Optoelectronic Science and Technology for Medicine of Ministry of Education, Fujian Normal University, Fuzhou 350007, China

A method was developed to determine the intracellular pH (pH_i) of single CA46 cell using nanosensor and pH-dependent fluorescent dye. To correct for changes in intracellular dye concentration, the excitation wavelength was changed between the pH-sensitive wavelength and pH-insensitive wavelength, then a ratio was computed between fluorescence intensities at these two wavelengths. Calibration of the intracellular fluorescence spectra is obtained by equilibration of the internal and external pH with Nigericin in K+ buffers. Using this method, the intracellular pH (pH_i) of single CA46 cell was found. This technique appears adequate to monitor pH_i in Single CA46 cell, offer the advantages of good time resolution, inexpensive, simplify, high sensitivity, and little influence on cell functions and will allow to study regulatory mechanism in single living cell.

1. Introduction

The intracellular pH of cells is important in the understanding of their physiology. Intracellular pH (pH_i) is expected to be tightly regulated process of paramount importance of the normal function, including cell growth/division, cell differentiation, glycolysis, DNA synthesis, cell motility/contractility, exocytosis/ endocytosis and so on. Most of these function are sensitive to small change in pH_i, maintain their pH_i within narrow range for optimum activity. On the contrary, metabolisms also affect on Intracellular pH of cells. Intracellular pH (pH_i) is an indicator of cell function and metabolism. for these reasons, Various of methods have been developed to measure the intracellular pH like labeled weak acids[1,2], Flow-Cytometry[3,4], confocal laser microscope[5,6], fiber-optic sensor[7-10], microelectrodes[11,12], nuclear magnetic resonance spectroscopy[13,14] and so on.

Since most of these methods are tedious and elaborate, or require extensive manipulations of cell, or require complex expensive equipment. In this paper we developed a new method by which intracellular pH in single living cell could be determine using nanosensor and pH-dependent fluorescent dye 5(6)-Carboxyeosin diacetate (CFDA). Compared to other techniques, this technique appears adequate to monitor pH_i in Single living CA46 cell, offer the advantages of good time resolution, inexpensive, simplify, high sensitivity, and little influence on cell functions.

† Corresponding author, E-mail: ysqiu@fjnu.edu.cn

2. Materials and Methods

2.1. *Chemical Reagent*

5(6)-Carboxyeosin diacetate (CFDA), Nigericin, HEPES, RPMI 1640, Dimethyl sulfoxide (DMSO), Fetal Bovine Serum were obtained from Sigma. NaCl, KCl, KH_2PO_4, $Na_2HPO_4 \cdot 7H_2O_4$, glucose, $CaCl_2$, HCl, NaOH and $MgSO_4$ were purchased from Guanghua chemical Factory.

2.2. *Cell Line*

CA46 cells were kindly provided by Fujian Institute of Hematology. CA46 cells were grown in RPMI 1640 medium supplemented with 10% Fetal Bovine Serum. Cell line was cultured in the growth medium in standard flasks. The flasks were incubated at 37°C, 5% CO_2 and 86% humidity. Cell growth was monitored daily by microscopic.

2.3. *Nanosensor Fabricate*

The apparatus used for fabricate the nanosensor was FDCW-2000(s)-CH2-H (Fusotek, shanghai) fiber–optic puller. The method of fabricate nanosensor has been described previously [7, 8]. Only a brief description is presented here. The jacket of multimode silica fiber–optic with its core diameter is 125um was removed, then the fiber-optic was pulled down to 300-400nm at the tip. Finally coated with approximately 50nm of aluminum to prevent light leakage and maximize the delivery of light to the distal end, in essence creating quantum confinement. The distal end of the fiber should be free from aluminum for delivering excitation light and collecting fluorescence.

2.4. *Loading Cell with CFDA*

Harvested cells were first incubated for 45 min with 10 mmol l^{-1} CFDA. Then washed twice with phosphate-buffered saline (PBS; containing, in mmol l^{-1}, 137 NaCl, 2.68 KCl, 1.47 KH_2PO_4, 8.06 $Na_2HPO_4{\cdot}7H_2O$, 10 glucose, 1 $CaCl_2$ and 1 $MgSO_4$; pH7.4) to remove excess dye. Measurements were started approximately 20 min later so that pH_i was not influenced by CFDA incubation; In order to exclude other disturbing influences such as autofluorescence of the cells or of the chemicals used for measurements, both cells and media were tested before incubate with CFDA.

2.5. *Fluorescence Measurement System*

Fig. 1 shows a schematic representation of fluorescence measurement system. The 458nm and 488nm laser from Argon ion-laser was focused onto the fiber which is terminated to a 2×1 fiber connector, then coupled into the nanosensor which was mounted on Nikon NF88-V_3 three-dimensional manipulator holder. By modulate this three-dimensional manipulator the nanosensor can move. The emission light from the cell was collected with this nanosensor. The emission light from the nanosensor passed though a band pass filter (CWL 520nm, FWHM 10nm, Tpk 50%) and coupled into the single photon detection then to a personal computer with custom-designed software for date treatment and process. Nikon TE2000-U Invert fluorescence microscope, CCD and another personal computer was used for image collection.

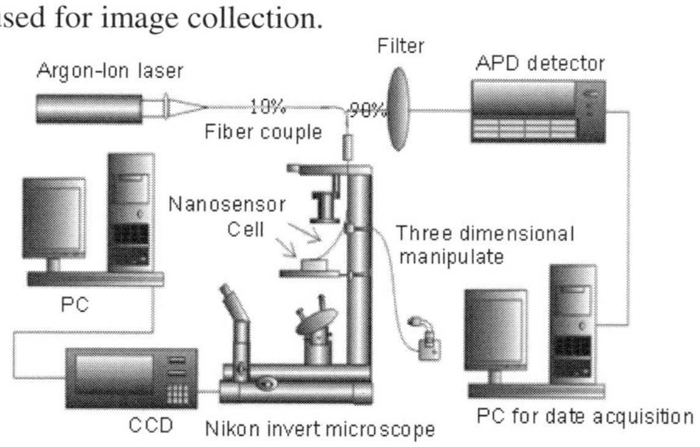

Fig. 1. Shows the components of the laser-induced fluorescence measurement system.

2.6. *Find the Optimal Excitation Wavelength*

To find the optimal excitation wavelength we employed FLS920 combined fluorescence lifetime& steady state spectrometer (Edinburgh Instruments Ltd, UK).

Using CFDA, all reading followed an excitation scan between 400nm and 500nm with emission at 520nm (bandwidths 10nm).

2.7. *Measurement of Intracellular pH Using Nanosensor*

By modulate this three-dimensional manipulator the nanosensor can move to the surface of cell (see fig. 2). then turn invert fluorescence microscope off and turn on Argon ion laser, the monochromatic light from Argon ion laser was focused on single cell, the total fluorescence can be gained from the date. In order to obtained the fluorescence of single cell. The background fluorescence must be subtracted from the total fluorescence. To acquire the background fluorescence we move the nanosensor away from of cell. The excitation wavelength was changed from 488nm to 458 nm, the 488nm and 458nm ratio was calculated as follows:

Ratio= (488total-488background)/
(458total-458background)

Calibration curves for CA46 cell were determined in buffer with pH values ranging from 6.2 to 7.9. The buffer was prepared from HEPES (10mM), KCl (45mM), Nigericin (10mM). The pH of the buffer was adjusted with either NaOH or HCL. The pH_i and pH_{out} were equilibration by addition of Nigericin and the ratio was determined as described previously. Calibration curves were constructed by plotting the ratio of fluorescence intensities (emission wavelength 520 nm) at the excitation wavelengths of 488 (pH-dependent point) and 458 nm (pH-independent point) as a function of pH. Intracellular pH was calculated from this calibration curve.

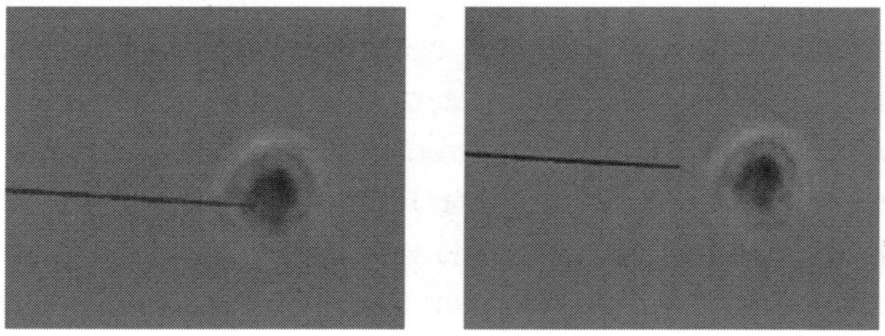

(a) (b)

Fig. 2. (a) CCD image of fiber–optic nanoprobe just outside of a single CA46 cell illustration how the total fluorescence measurement was performed. (b) CCD image of fiber–optic nanoprobe away from the single CA46 cell illustration how the background fluorescence measurement was performed.

3. Result

3.1. *Result of Nanosensor Fabrication*

After pulling the multimode fiber, the end structure tapered from the original fiber 125um to a nanoscale tip with a flat surface. When argon laser is coupled into the nanosensor, a spot at the tip could be seen under the invert fluorescence microscope. This probe delivers light very efficiently to the aperture since all the radiation remains bound to the core until a few nano-sizes from the tip.

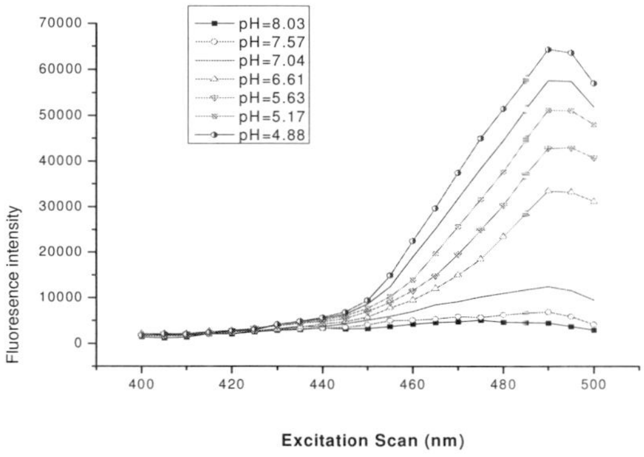

Fig. 3 The pH-dependent excitation spectra of CFDA in Nigericin calibration solution in the presence of permeabilised cells of CA45.

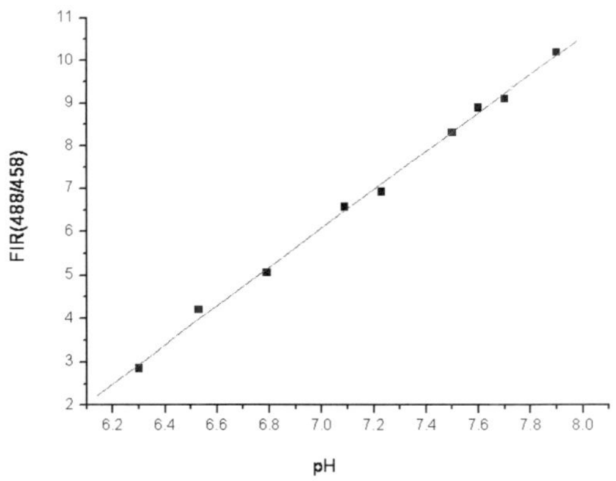

Fig. 4. Fluorescence intensity ratio (488 nm/453 nm) vs. pH calibration curve of CFDA with permeability cell of single CA46 cell. The fitted equation best describing the calibration curve is also shown.

3.2. *Intracellular pH Calibration Curves and Intracellular pH Determination*

The excitation spectrum of CFDA (emission 520) between 400nm and 500nm as a function of pH is show in Fig. 3. From this date the pH-dependent nature of the excitation spectrum of the probe is clearly seen. The fluorescence intensity of CFDA is independent of proton concentration at an excitation wavelength of 458nm, but changes according to pH when excited at 488 nm.

Intracellular pH value was therefore quantified by calculating the ratio of fluorescence recorded at these two excitation wavelengths. The ratio of fluorescence intensity can be calculated and plotted as a function of pH to construct a calibration curves. The calibration curves of ratio values of fluorescence intensity of CFDA (488/458nm) plotted against pH is show in Fig. 4. A poly nominal function can be fitted to calculate $pH_i(y)$ from ratios(x). The fitted equitation is

$$Y=-25.27826+4.48026X \quad (P<0.0001)$$

the regression (*R* value) for this function was 0.99851.

Discussions

We have explored a ratiometric approach to determine the intracellular pH on the single-cell level. The ratio is dependent on the pH_i only. Ratio measurements have the advantage of eliminating a number of variables that lead to inaccuracies including probe bleaching and variations in loading intensity between cells as well as variations in machine performance or biomass. Other advantages of this optic ratio technique include increased sensitivity and specificity for pH, minimal perturbation of the cell, and improved temporal and spatial resolution.

A number of techniques have been used for the measurement of pH, in a variety of cell types. Advantages and limitations of each have been reviewed by others. These techniques involve measurements on either single cells or cell populations. Typical techniques include 1) ^{31}P nuclear magnetic resonance (NMR), 2) weak acid/base distribution, and 3) pH-sensitive microelectrodes and so on.

NMR provides the only approach that is totally noninvasive, although non-invasive; a universal problem with this technique has been the need to use high cell densities meaning that determination of pH single cell is not possible. The technique also suffers from limited time resolution and the need for complex, expensive equipment.

Another method used to measure pH is the equilibrium distribution of radio labeled weak acids. Although widely used, the method has limited application as an in vivo method because measurements are made in buffers following centrifugation steps.

Other potential difficulties with this method include the inhibitory effects of weak acids on cell physiology, compartmentalization of weak acids the cell and the estimation of an unchanging cell volume which has been shown to alter significantly upon environmental change. it is not very suitable for following rapid changes in intracellular pH. Moreover, the intracellular pH value obtained represents a statistical average for all cell compartments. Not suitable for studying living cell.

Measurement of pH can also be carried out using pH-sensitive microelectrodes but this technique is destructive and clearly, only suitable for rather large cell.

The technique described allowed us determine the intracellular pH (pH_i) of single CA46 cell. Compared to other techniques, this technique appears adequate to monitor pH_i in Single living CA46 cell, offer the advantages of good time resolution, inexpensive, simplify, high sensitivity, and little influence on cell functions.

References

1. W. J. Waddell and T. C. Butler, *J. Clin Invest.* **38**, 720 (1959).
2. A. Kotyk, *Folia Microbiol.* **8**, 27 (1963).
3. G. Valet, A. Raffael, L. Moroder, E. Wünsch and G. Ruhenstroth-Bauer, *Naturwissenschaften.* **68**, 265 (1981).
4. E. D. Wieder, H.Y. Hang and M.H. Fox, *Cytometry.* **14**, 916 (1993).
5. M. Weinlich, G. Capasso and R. K. H. Kinne, *Pflügers Archiv European Journal of Physiology.* **422**, 523 (1993).
6. S. J. Cope, S. Hibberd, J. Whetstone, R. J. MacRae and C. D. Melia, *Pharmaceutical Research.* **19**, 1554 (2002).
7. W Tan, ZY Shi, S Smith, D Birnbaum, and R Kopelman, *Science.* **258**, 778 (1992).
8. W.H. Tan, Z.Y. Shi and R. Koplman, *Anal Chem.* **64**, 2985 (1992).
9. J. Samuel, A. Strinkovski, S. Shalom, K. Lieberman, M. Ottolenghi, D. Avnir and A. Lewis, *Materials Lett.* **21**, 431 (1994).
10. A. Song, S. Parus and R. Kopelman, *Anal Chem.* **69**, 863 (1997).
11. C. C. Aickin and R. C. Thomas, *J. Physiol.* **245**, 348 (1977).
12. C. C. Aickin and R. C. Thomas, *J. Physiol.* **252,** 803 (1975).
13. O. Jardetzky and G. C. K Boberts, *NMR in Molecular Biology* (Academic Press, New York, 1981).
14. G. Navon, R. G. Schulman, T. Yamane, T. R. Eccleshall, K. B. Lam, J. J. Baronofsky and J. Marmur, *Biochemistry.* **18**, 4487 (1979).

A HIGHLY SELECTIVE PHOTOOXIDATION OF 2,3,5-TRIMETHYLPHENOL TO 2,3,5-TRIMETHYLBENZOQUINONE USING HYPOCRELLINS AS CATALYSTS †

PEI ZHANG

Faculty of Science, Honghe University, Mengzi 661100, P. R. China
Faculty of Technology, Honghe University, Mengzi 66110, P. R. China

YING LI, WEI LIU‡, CANBANG ZHANG, JUCHENG ZHANG, LIHUA YAO, YONG MIN, JIN YANG

Faculty of Science, Honghe University, Mengzi 661100, P. R. China
Key Laboratory of Natural Pharmaceutical & Chemical Biology of Yunnan Province, Honghe University, Mengzi, 661100, P. R. China

The photooxidation of 2,3,5-trimethylphenol by hypocrellins in organic solvent has been studied under various operating conditions for the purpose of learning more about the reactions involved and the extent of the oxidation process. This study realized nearly 100% selectivity for 2,3,5-trimethylbenzoquinone at 2.5% conversion of 2,3,5 trimethylphenol under visible light irradiation at aerobic atmosphere. The photooxidation reaction has been studied by different parameters like the effect of solvents and the influence of catalyst concentration. Compared to the conventional methods, this method could be more eco-friendly.

1. Introduction

Recently, a great deal of attention has been devoted in the last years to photocatalytic oxidation processes both in the homogeneous phase and in heterogeneous systems [1]. At the same time, development of the efficient oxidation catalyst for hydrocarbon is a challenging goal for chemists [2]. Photocatlysts have drawn much attention, since they are capable of activating the unreactive C-H bonds to functionalize hydrocarbon [3]. Homogeneous oxidation under irradiation catalyzed by polyoxotungstate [4], Fe(III)-prophyrins [5], transition metal chlories [6] have been reported, however, in these methods products would be contaminated by metal ions. Hypocrellins, including hypocrellin A and hypocrellin B (Fig.1), are a type photosensitizers naturally present in a fungus named Hypocrella bambusae grown mainly in Yunnan province, China. It was known that their high photodynamic activity but very low dark-toxicity [7-8]. However, to the best of our knowledge, there is no reported that hypocrellins could be used to catalytic oxidation of 2,3,5-trimethylphenol (TMP) to 2,3,5-trimethylbenzoquinone (TMQ). TMQ, as an important intermediate, is used to synthesize Vitamin E, and it stems from the oxidation of TMP (Fig. 2) or 2,3,6-trimethylphenol. In industry, a current method of TMQ production is para-sulfonation of TMP, followed by oxidation with manganese dioxide. But this method is not suitable for currently industrial manufacture because of the contamination of inorganic salts and of waste water. Previously, various oxidation methods for trimethylphenol have been reported [9-11]. These different methods, however, show varying degrees of success as well as limitations due to the need for equimolar catalysts, high reaction temperatures, and the use of large quantities of toxic organic solvents with the possibility of explosion. Recently, we reported that $CuCo_2O_4$ as catalyst showed high activity and selectivity in the oxidation of TMP to TMQ with 30% aqueous hydrogen peroxide at room temperature. However, the excessive use of H_2O_2 is unrealistic in the industrial process [12].

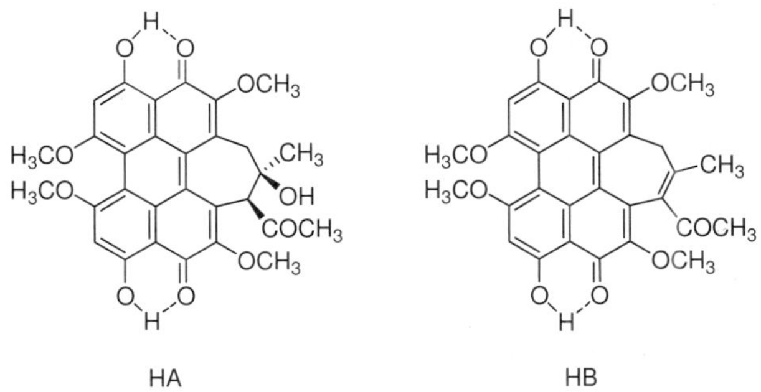

Figure 1. The molecular structure of HA (left) and HB (right). The dash links for denoting the intramolecular hydrogen bonds.

* This work is supported by Science Fund from Government of Yunnan Province (2005C0078M).

‡ Corresponding author: E-mail:liuwei4728@163.com, Tel: +86-873-3699979, Fax: +86-873-3699979

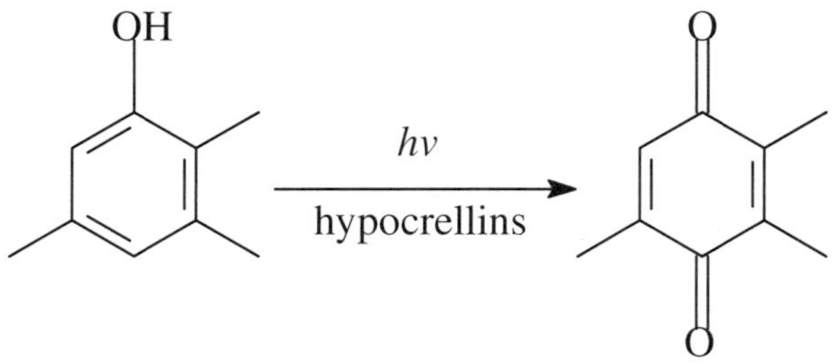

TMP TMQ

Figure 2. The oxidation of 2,3,5-trimethylphenol to 2,3,5-trimethylbenzoquinone.

Herein, we report in this work a highly selective oxidation of 2, 3, 5-trimethylphenol to 2, 3, 5-trimethylbenzoquinone using hypocrellins as catalyst, developed under mild conditions. The reaction can take place under visible light irradiation at aerobic atmosphere. On the other hand, the products are not contaminated with metal ion, since metal catalyst was not used in this photocatalysis system.

2. Experiments

2.1 Materials

2,3,5-trimethylphenol (TMP) and 2,3,5-trimethylbenzoquinone (TMQ) were donated by J&K. Hypocrellins was derived as described previously [13]. All organic solvents were analytical grade

2.2 General Procedure for the Oxidation of TMP

The photo-catalytic oxidation reactions of TMP were carried out using a 100 ml three-necked flask equipped with a reflux condenser, magnetic stirrer. In a typical reaction, TMP (1mmol), hypocrellins (0.0005mmol, 0.05 mol% of TMP) and ethanol (50 ml) were placed in the flask, and then the solution was radiated by visible light 12h. The selectivities and conversions were determined with by GC and Uv-Vis using authentic samples for comparison. And the products were determined by GC using authentic samples for comparison.

2.3 Physical Measurements

GC analyses were performed using a gas chromatograph (GC-2010, Shimadzu) equipped with a flame ionization detector and a quartz capillary column (25 m ×0.3 mm) filled with Carbowax 20M. The UV–vis spectra of hypocrellins were performed with a TU-1901 UV–vis spectrophotometer using a quartz cell.

3. Results and Discussion

3.1 Effect of Solvents on the TMP

In order to choose the best oxidation solvent in the presence of hypocrellins as catalyst, the oxidation reactions were carried out in the different solvents using TMP as the representing substrate. The results are presented in Table 1. We can see that the conversion of TMP decreased in the order ethanol > acetonetrile > methanol > chloroform > acetone ≈ no solvent. The conversion of TMP was not occurred when no solvent was used because of unmixture of TMP and hypocrellins. In the acetone medium, at the same time, TMP was not oxidized under visible light irradiation at aerobic atmosphere. From table1, however, it was also found that ethanol provides the best oxidation medium for high TMP conversion. The observed differences from methanol to acetone may be attributed to the difference in polarities of solvents [14]. Futhermore, changes of catalyst in the reaction medium were investigated and we could found that catalyst was hardly changed in the ethanol (Fig.3). This implied that ethanol was a suitably innocuous solvent in this photooxidation system.

Table 1. The effect of solvents

Solvents (ml)	Conversion of TMP (mol%) [a]	TON [b]
No solvent	0.00	0
Methanol (50)	0.02	3
Ethanol (50)	0.1	17
Acetone (50)	0.00	0
Chloroform (50)	0.01	2
Acetonetrile (50)	0.03	5

Reaction conditions: substrate, 1mmol; Catalyst, 0.0005mmol; reaction time, 12h; other conditions, visible light irradiation at aerobic atmosphere.
[a] The selectivity of TMQ is nearly 100%
[b] TON = (converision of TMP) / (catalyst * reaction time), e.g. TON (methanol) = 0.02/(0.0005 *12) ≈ 3

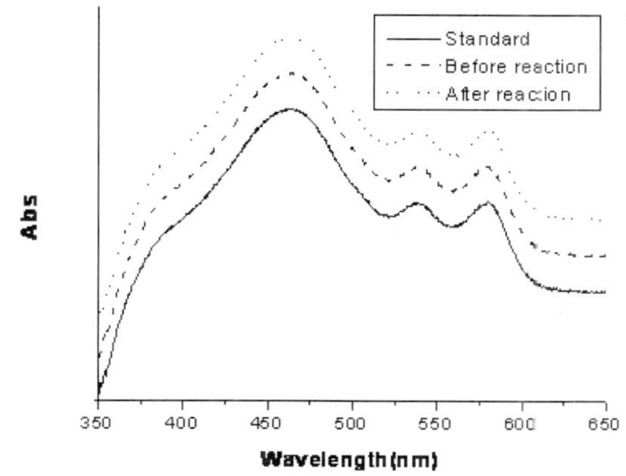

Figure 3. Uv-Vis absorption spectra of standard, before reaction, and after reaction.

3.2 Influence of hypocrellins concentration

The influence of catalyst (hypocrellins) concentration on TMP oxidation reaction was also investigated and depicted in Table 2. The different ratios of catalyst to TMP were obtained by varying the amount of the catalyst and keeping the concentration of TMP constant. An increase in conversion of TMP was observed when the amount of the catalyst was incerased up to 0.005 mmol. While the conversion of TMP was obviously reduced when the amount hypocrellins was exceeded 0.005mmol. according to general knowledge, the conversion of substrate would increased along with the increase of the amount of catalyst. However, the result is reversed in this reaction system. The reason would be treated further studies. Therefore, 0.005mmol would be the optimum catalyst amount used for the selectively producing TMQ under this conditions.

Table 2. The influence of hypocrellins concentration

hypocrellins(mmol)	Conversion of TMP (mol%) [a]	TON [b]
0.0005	0.1	17
0.005	2.5	42
0.0125	0.06	<1
0.025	0.07	<1

Reaction conditions: substrate, 1mmol; reaction time, 12h; other conditions, visible light irradiation at aerobic atmosphere.
[a] The selectivity of TMQ is nearly 100%
[b] TON = (converision of TMP) / (catalyst * reaction time), e.g. TON (0.005) = 2.5/(0.005 *12) ≈ 42

4. Conclusions

In conclusion, The photooxidation of 2,3,5-trimethylphenol by hypocrellins in organic solvent has been studied under various operating conditions. This study realized nearly 100% selectivity for 2,3,5-trimethylbenzoquinone at 2.5% conversion of 2,3,5 trimethylphenol under visible light irradiation at aerobic atmosphere. The photooxidation reaction has been stuied by different parameters like the effect of solvents and the influence of catalyst concentration. Compared to the conventional methods, this method could be more eco-friendly. Thorough research and its applications in the selective oxidation of other simliar compounds are underway.

Appendix

TMP: 2,3,5-trimethylphenol
TMQ: 2,3,5-trimethylbenzoquinone
TON: turnover numbers

References

1. A. Maldotti, A. Molinari, R. Amadelli, *Chem. Rev.*, **102**, 3811 (2002).
2. M. Yamaguchi, M. Tomizawa, K. Takagaki, M. Shimo, D. Masui, T. Yamagishi, *Catal. Today*, **117**, 105 (2006).
3. M. Yoshizawa, S. Miyagi, M. Kawani, K. Ishiguro, M. Fujita, *J. Am. Chem. Soc.*, **124**, 9172 (2004).
4. R. F. Renneke, C. L. Hill, *J. Am. Chem. Soc.*, **110**, 5461 (1988).
5. A. Maldoti, C. Bartocci, R. Amadelli, P. Battioni, D. Mansuy, *J. Chem. Soc. Chem. Commun.*, 1487 (1991).
6. K. Takaki, J. Yamamoto, K. Komeyama, T. Kawabata, K. Takehira, *Bull. Chem. Soc. Jpn.*, **77**, 2251 (2004).
7. G.A. Kraus, W.J. Zhang, M.J. Febr, J.W. Petrich, Y. Wannemuehler, S. Carpenter, *Chem. Rev.*, **96**, 523 (1996).
8. J. Ma, J. Zhao, L. Jiang, *Photochem. Photobio.*, **74**, 143 (2001).
9. G. S. Chernikova, N. G. Baranova, U. M. Azizov, I. B. Afanasëv, *Pharm.Chem. J.*, **22**, 788 (1988).
10. V. A. Bushmelev, T. A. Kondratëva, M. A. Lipkin A. V. Kondratëv, *Pharm. Chem. J.*, **25**, 340(1991).
11. H. Sun, K. Harms, J. Sundermeyer, *J. Am. Chem. Soc.*, **126**, 9550 (2004).
12. Y. Li, W. Liu, M. Wu, Z. Yi, J. Zhang, *J. Mol. Catal. A Chem.*, **261**, 73 (2007).
13. Y.Y. He, J.Y. An, L.J. Jiang, *Dyes Pigments*, **41**, 93 (1999).
14. S. Gontier, A. Tuel, *Appl. Catal.*, **118**, 173 (1994).

SPECTROSCOPIC STUDIES OF THE INTERACTION BETWEEN HYPOCRELLIN B AND GELATIN [†]

PEI ZHANG

Faculty of Science, Honghe University, Mengzi 661100, P. R. China
Faculty of Technology, Honghe University, Mengzi 66110, P. R. China

YING LI, WEI LIU[‡], CANBANG ZHANG, ZHAOLONG HUANG, YONG MIN, JUCHENG ZHANG

Faculty of Science, Honghe University, Mengzi 661100, P. R. China
Key Laboratory of Natural Pharmaceutical & Chemical Biology of Yunnan Province, Honghe University, Mengzi, 661100, P. R. China

Interaction of hypocrellin B (HB) with two types of gelatin was studied by the spectroscopic responses. Interestingly, it was found that, bound to type □ gelatin at pH 7.4, the absorption peak of HB was greatly red-shifted from 470nm to 514nm with appearance of an isobestic point at around 480nm and a new absorption peak at 280 nm. Furthermore, the fluorescence was almost all silenced. The spectral feature suggested that a chemical reaction, instead of weak interactions, took place.

1. Introduction

Hypocrellin A (HA) and hypocrellin B (HB) (Fig. 1), as a type of photosensitizers, are naturally present in a fungus named Hypocrella bambusae, which is grown mainly in Yunnan province, China. They have high photodynamic activity and very low dark-toxicity [1-4], and it was known that the hypocrellin fluorescence was strengthened in an organic solvent. Recently, it was proved that hypocrellin fluorescence was originated from the intramolecular excited-state proton or hydrogen atom transfer, in turn related to the intramolecular hydrogen bonds [5-7]. Therefore, the spectral properties of hypocrellins should be environmentally sensitive to be used as a probe to detect the microenvironment in a biological macromolecule or a supramolecule system. The pH-sensitive fluorescence properties were also observed for HB in a PBS solution, for which a fluorescence minimum would appear at pH 7. At the same time, they were observed in liposome, for which a fluorescence maximum at pH 7.4 [8]. On the other hand, gelatins are widely-used denatured proteins derived from collagens, especially for drug preparations, such as HB-gelatin nanoparticles [9]. To observe how HB molecules would bind to gelatin, therefore, would be meaningful. Furthermore, as a kind of denatured proteins, gelatins may undergo some additional reactions with HB, compared to the natured proteins, which may be used to distinguish a natured protein from a fully or partly denatured one.

In our work, the interaction of HB with two types of gelatins (including Type I and Type II) was studied. We found that Type II gelatin, containing a cysteine-rich domain could chemically react with HB to produce a covalently combined derivative, which is very uncommon for HB binding to other biological molecules. On the other hand, interaction of HB with the type I gelatin reproduced the general feature, strengthening the fluorescence, for HB binding to the hydrophobic area in biological molecules, such as human serum albumin, some polysaccharides and also liposome. The chemical derivative was proved to be an HB-cysteine compound formed by reaction of HB with one-terminal free cysteine residue in the gelatin, perhaps produced via breaking of the disulfide bond during denaturation of a protein.

[*] This work is supported by Science Fund from Government of Yunnan Province (2005C0078M).
[‡] Corresponding author: E-mail:liuwei4728@163.com, Tel: +86-873-3699979, Fax: +86-873-3699979

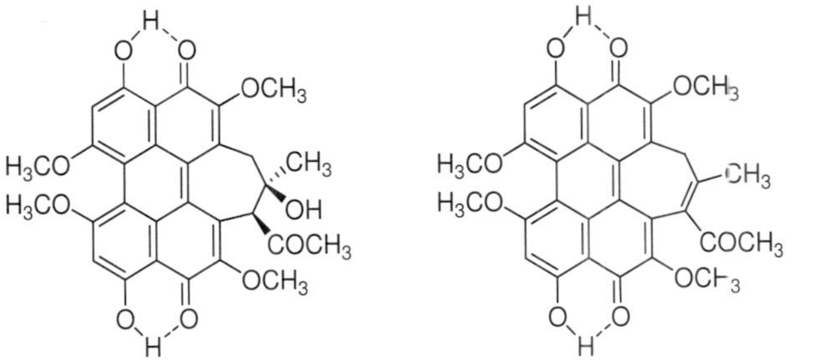

Figure 1. The molecular structure of HA (left) and HB (right). The dash links for denoting the intramolecular hydrogen bonds.

2. Experiments

2.1 Materials

HB was derived as described previously [10]. Type I gelatin was purchased from Sigma Chemical Co, and the Type II gelatin was provided by Institute of Biophysics, Chinese Academy of Sciences. Phosphate buffered saline (PBS) of pH 7.4 was prepared by the use of 10 mM KH_2PO_4 and 10 mM K_2HPO_4. The working solutions were prepared immediately before use.

2.2 Preparation of Sample

Previously, it was reported that that 8 μM of HB could be well soluble in aqueous PBS solution [11]. Therefore, 8 μM of HB was used for all of the experiments in this work. Furthermore, for preparation of the complexes of HB with gelatins, a concentrated solution of HB in dimethylsulfoxide (DMSO) (2mg/ml) was added slowly into the aqueous solution of the gelatin in microliter quantities and mixed quickly by shaking until reaction completion.

2.3 Spectral Measurements

Absorbance spectra of HB were obtained on a Shimadzu UV-1601 UV-Vis spectrophotometer and fluorescence spectra on an F-4500 spectrofluorimeter (Hitachi, Japan) at room temperature. The fluorescence spectra of HB were measured with selective excitation of HB at 470 nm.

3. Results and Discussion

As mentioned previously, it was observed that both of the absorption and fluorescence spectra would be strengthened as long as HB binding to a bio-molecule. However, the spectral feature for HB binding to type II gelatin is very distinctive, as demonstrated by the time-dependent absorption and fluorescence spectra for HB in an aqueous solution (pH 7.4) of type II gelatin of 12.5g/l which was the "saturated" concentration for a complete binding of HB.

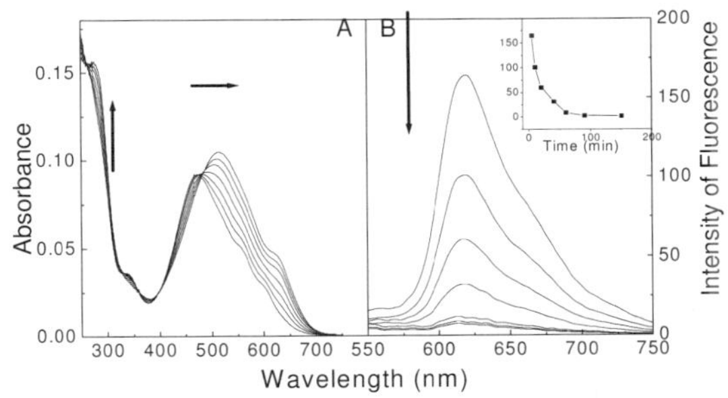

Figure 2. The absorption (A) and fluorescence (B) spectra of HB (8 μM) in aqueous solution (pH 7.4) of type II gelatin (12.5g/l) at different time (5, 10, 20, 40, 60, 90, 150 and 180min). The insets are the plots of the fluorescence intensities (■) to the reaction time.

Binding to type II gelatin, the absorption spectra of HB were red-shifted about 50 nm with a isobestic point at 480 nm accompanied by appearance of new absorption peak at 280 nm (Fig. 2A), while type II gelatin without HB did not show any spectral changes under the same condition. At the same time, the fluorescence was seriously quenched until almost silent finally (Fig. 2B). Furthermore, HB could not be extracted by chloroform from the HB-type II gelatin complex. All of these observations suggest that a chemical reaction, instead of weak interaction, took place between HB and type II gelatin. Besides, the absorption and fluorescence spectra of HB-type II gelatin complex reproduce the basic feature of those of 5, 8-2-cysteine-HB reported previously [16], as shown in figure 3.

From Fig. 3, it can be seen that the absorption and fluorescence spectra of HB-type II gelatin complex are comparable to those of 5,8-2-cysteine-HB but apparently different from those of HB. The vanished fluorescence is also the feature of HB-cysteine derivative, low fluorescence quantum yield.

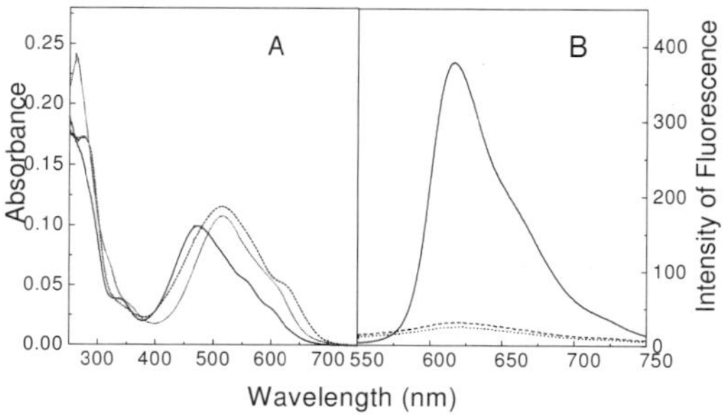

Figure 3. A: Absorption spectra of HB (solid line), 5,8-2-cysteine-HB (dashed line) and HB (dotted line) in type II gelatin in aqueous solution. B: fluorescence spectra of HB (solid line), 5,8-2-cysteine-HB (dashed line) and HB (dotted line) in type II gelatin in aqueous solution.

Conclusion

In conclusion, the interaction of HB with two types of gelatins was studied. And it was found that Type II gelatin was very uncommon for HB binding to other biological molecules. On the other hand, interaction of HB with the type I gelatin reproduced the general feature, strengthening the fluorescence. The chemical derivative was proved to be an HB-cysteine compound formed by reaction of HB with one-terminal free cysteine residue in the gelatin, perhaps produced via breaking of the disulfide bond during denaturation of a protein. Farther research is underway.

Appendix

HA: hypocrellin A
HB: hypocrellin B

References

1. G.A. Kraus, W.J. Zhang, M.J. Febr, J.W. Petrich, Y. Wannemuehler, S. Carpenter, *Chem. Rev.*, **96**, 523 (1996).
2. M.J. Fehr, S.L. Carpenter, Y. Wannemuehler, J.W. Petrich, *Biochem.*, **34**, 15845 (1995).
3. J.B. Hulson, J. Zhou, J. Chen, L.Harris, L. Yip and G.H.N. Towers, *Photochem. Photobiol.*, **60**, 253 (1994).
4. J. Ma, J. Zhao, L. Jiang, *Photochem. Photobiol.*, **74**, 143 (2001).
5. S.J. Xu, S. Chen, M.H. Zhang, T. Shen, *J. Org. Chem.*, **68**, 2048 (2003).
6. K. Das, D.S.English, M.J.Fehr, J.W.Petrich, *J. Am. Chem. Soc.*, **119**, 2763(1997).
7. K. Das, A.V. Smirnov, J. Wen, P. Miskovsky, J.W. Petrich, *Photochem. Photobiol.*, **69**, 633 (1999).
8. L. Song, B. Zhao, J. Xie, J. Zhao, *Biochim. Biophys. Acta–General.*, **1760**, 333 (2006).
9. M. C. Ryan, L. J. Sandell., *J. Biol. Chem.*, **265**, 10334(1990).
10. Y.Y. He, J.Y. An, L.J. Jiang, *Dyes and Pigments*, **41**, 93 (1999).
11. B. Zhao, J. Xie, J. Zhao, *Biochim. Biophys. Acta*, **1670**, 113 (2004).

NEAR-INFRARED FLUORESCENCE LABELING AND *IN VIVO* MONITORING OF DYNAMICS OF INSULIN IN MOUSE MODEL

CHUNSHENG FANG, YUEQING GU

School of Life Science and Technology, China Pharmaceutical University, Nanjing 210038 China

In vivo evaluation of drug properties becomes a crucial in the drug development. Near infrared techniques contributed considerably to the development of protein polypeptide drugs by performing non-invasive *in vivo* real time measurements on experimental subjects. In this study, a NIR monitoring system comprised of NIR spectroscopy and imaging system for *in vivo* real time monitoring of the dynamic of protein polypeptide drugs on mouse models was designed. Insulin acted as polypeptide drug model was covalently labeled with NIR fluorescence probe cypate (Ex = 780nm, Em = 830nm). The labeling route was optimized with different factors to obtain the maximum labeling efficiency, which was evaluated by RP-HPLC. The labeling result show that the labeling efficiency was approximated to 95%. Pharmacodynamics study showed that insulin labeled with cypate did not affect its function of hypoglycemic. The half-time of insulin—cypate complex at arteria cruralis blood was 17min,and the dynamic distribution in mouse model indicated that the labeled insulin firstly accumulated in liver and clearance from the enteron system. The NIR monitoring system combined with NIR probe showed a promising way for *in vivo* monitoring the dynamics of protein polypeptide drugs in small animals.

***Keywords*:** NIR monitoring system, cypate, insulin, *In vivo*, labeling

1. Introduction

Characterized with the virtue of high physiological activity, specificity, and low toxicity, protein polypeptide drugs are becoming mainstream therapeutic agents and constitute a substantial portion of the compounds under preclinical and clinical development in the global pharmaceutical industry. And the *in vivo* evaluation of drug properties, from animal models to human subjects, becomes a crucial in the protein polypeptide drug development. It involves drug delivery, drug bio-distribution, pharmacokinetics, pharmacodynamics and so on. Each of these processes requires quantitative monitoring of drug concentrations in the specific organ/tissue or qualitative imaging of drug distribution in whole subjects. Thus, a number of analytical techniques were exploited to analyze the protein polypeptide drug process within the experimental subjects. At present the most commonly used methods for quantification of macromolecule drug in biological fluid include immunoassay, bioassay, and isotope labeling and liquid chromatography mass spectrometry [1~4]. However, all these techniques, except isotope labeling, are unable to perform the *in vivo* measurement of protein polypeptide drug in biological matrices. They require the tedious and intricate process of sample preparation such as sample collection and purification. Furthermore, the clearance of protein polypeptide drugs from the specific tissue/organ is quick and the above analytical techniques cannot catch up with its dynamic changes. Therefore, a non-invasive *in vivo* real time monitoring modality for protein polypeptide drugs in biological matrices will timely meet the needs and greatly expedite the step of protein drug development.

Near infrared (NIR) light (700 ~ 900 nm) possesses the capability of penetrating living tissues several centimeters due to the low absorbance of tissue intrinsic chromophores such as oxy- and deoxy-hemoglobin (the main absorber of visible light), melanin, water, and lipid (the principal absorber of infrared light) [5~6]. Featured with the deeper tissue penetration, non-ionizing and non-radioactive, NIR light attracted extensive attentions on the development of non-invasive techniques for *in vivo* real time monitoring of biological signals in living tissues. Recent studies have reported the capability of the near infrared spectroscopy to quantitative accurately optically active drugs in mouse tissues noninvasively *in vivo* [7~9]. Kanick et al [10] use the optical pharmacokinetic system (OPS) to measure concentrations of the anticancer agent motexafin gadolinium (MGd) in mouse tissues noninvasively and nondestructively using elastic-scattering spectroscopy. But due to the fact that most of the drug lacks the optical properties within the near infrared range, At the same time the absorption spectra difficult to achieve relatively low detection limit. All these reasons make the extrinsic NIR fluorescent labeling for drug is very necessary. Protein polypeptide drug was an ideal choice because which usually have a larger molecular weight than fluorescence dyes, labeled with dye will not change its delivery properties. Near-infrared imaging is currently the typical method for accessing the distribution of fluorescent marker in biological matrices [11~12]. Together with biological marker, NIR reflectance imaging offers additional insights into the drug targeting. Ramjiawan et al [13] used a fluorescent dye (Cy5.5.18) labeling a novel human anti-tumor antibody fragment, NovoMab-G2-scFv, and monitored its distribution in node mice bearing subcutaneous human melanoma tumor cells *in vivo* by near-infrared image system. Images show that the antibody-dye complex was distributed mainly in the tumors, kidney and liver. Although they obtained the half-time of the antibody at the tumor site by fitting an exponential decay to fluorescence data extracted from the images, The algorithmic complexity and the result is not very accurate, So it has fundamental limitations for quantitative measurement.

In this study, we designed a NIR monitoring system comprised of NIR spectroscopy and imaging system for *in vivo* real time monitoring of the dynamic of protein polypeptide drugs on mouse models. In general, NIR spectroscopy offers a quantitative way to measure the drug concentrations in specific tissue/organs, while NIR imaging can image the distribution of drug in animal models. Insulin acted as polypeptide drug model was covalently labeled with NIR fluorescence probe cypate (Ex = 780nm, Em = 830nm,). Insulin is the most important regulatory hormone in the control of glucose homeostasis consisting of 51 amino acids shared between two intramolecular chains and with a molecular weight of 5800 g/mol. The amino of Lysine and the N-terminal of insulin can covalent binding with the carboxyl of the cypate. In the process of cypate conjugation to insulin, many factors including catalyzing system, pH condition, temperature, reacting duration, feed ratio of different components, were intensively investigated to achieve an optimal labeling result, which was evaluated by RP-HPLC before administer to mouse. The function of hypoglycemic of the labeled insulin is tested in diabetes mode mouse by glucose oxidase method. Finally, the insulin-cypate complex was administrated through tail vein into the blood stream of mouse, the drug concentration in the thigh femoral artery was quantitative measured by NIR spectroscopy system, and the distribution of the complex was monitoring by NIR imaging system.

2. Materials and Methods

2.1 Materials

Cypate (C41H41N2O2) was offered by Dr. Achilefu from Washington University School of Medicine. Insulin was purchased from JiangSu WanBang Bio-pharmaceutical Co., Ltd.insulin was directly dissolved in buffer and stored in 0-4°C, two catalysts, N, N'-Dicyclohexyl- carbodiimide (DCC, purchased from GL Biochem) and N-hydroxyl succinimide (NHS, purchased from Qiude Biochemical Engineering Co., Ltd), was used to catalyze the covalent binding of insulin and cypate with an optimal pH, temperature, reacting duration, feed ratio of different components. Streptozotocin was purchased from Sigma. Blood glucose kit (glucose oxidase method) was purchased from ShangHai Rongsheng Biotech Co.; Ltd. Acetonitrile was purchased from JiangSu HanBon sci & tech. Co., Ltd.All chemicals were of analytical reagent grade unless otherwise mentioned.

2.2 Methods

2.2.1 HPLC methods

The JASCO HPLC Programmable Gradient System consisted of a model 2089-plus gradient solvent delivery module Pump, a model 2055-plus Autosampler and a model 2075-plus UV detector module. The Marked mixture was separated on a Diamonsil Lichrospher C18 (5μm, 150mm×4.5mm, I.D.) column (Dikma Technologies) fitted with a Diamonsil Lichrospher C18 (5μm, 3.0mm×4.0mm, I.D.) guard column (Dikma Technologies). A model 2067-plus temperature controller colemn heater (JASCO) was used to maintain the column temperature at 300°C. HPLC solvents consist of water containing 0.05% TFA (solvent A) and acetonitrile containing 0.05% TFA (solvent B). The elution protocol for analytical HPLC starts with 70% an immediately by a linear gradient to 67% an over 9 min, continued by a linear gradient to 10% an over 11min, and held at 10% A for 10 min. The elution profile was monitored by UV absorbance at 214 nm.

2.2.2 Assessment of hypoglycemic function of insulin-cypate complex in diabetic mouse

The hypoglycemic function of insulin-cypate complex were tested on acute diabetic mouse model induced by chemical drug streptozotocin (STZ), the mouse obtained from Laboratory Animal Center of China Pharmaceutical University (Nanjing, China). Adding 70mg of STZ in 10mL 0.1mol/L citric acid and citrate sodium (pH4.2) in the ice bath, after fully dissolved then filtration and disinfection. The mouse first to adapt pre-keeping one week, before the establishment of model mouse prohibited food (supply water) for 12 hour, then weighing the body weight and measuring the blood glucose concentration. Measurement methods: about 0.2ml blood sample via orbital vein were collected in heparinized 2.0 ml polythene tubes, and the blood samples were immediately centrifuged at 4000 rpm for 10 min. A 10 μl volume of plasm was used for Measuring the blood glucose concentration by blood glucose test kit (According to the operation manual). 10 mouse with the blood glucose concentration in the range of 3.5 to 5.5 mmol/l were selected and administered STZ by i.p. injection at a dose of 190 mg/kg. After 72 h the mouse blood glucose concentration was measured and 5 success diabetic model mice (blood glucose concentration≥11.1mmol/L) were selected for test. The mouse were administered insulin-cypate complex by i.v. injection at an insulin dose of 67μg (The molar ratio of insulin and cypate was about 1:1), and about 0.2mL blood sample were collected at 0, 10, 20, 30min for the blood glucose concentration measurement.

2.2.3 NIR spectroscopy system

A schematic of instrument is shown in Fig. 1. A laser diode (λ=765.9nm, NL-FC-2.0-763) with a wavelength of 760 nm was used to excite the cypate to emit fluorescence at around 820 nm. Which was detected by the photomultiplier (PMT,) with a high pass filter (> 800 nm) placed in front of PMT to block the incident light. A 5% reference measurement of light intensity from a spectroscope was made once at the beginning of a procedure prior to all other measurements. Light is delivered to the tissue and collected from the tissue using silica optical fibers, and the delivering and collecting fiber bundles were positioned in a perpendicular geometry in the leg area that contains abundant vasculature. All the detected light transform to data by Analog to Digital converters which displayed in the computer.

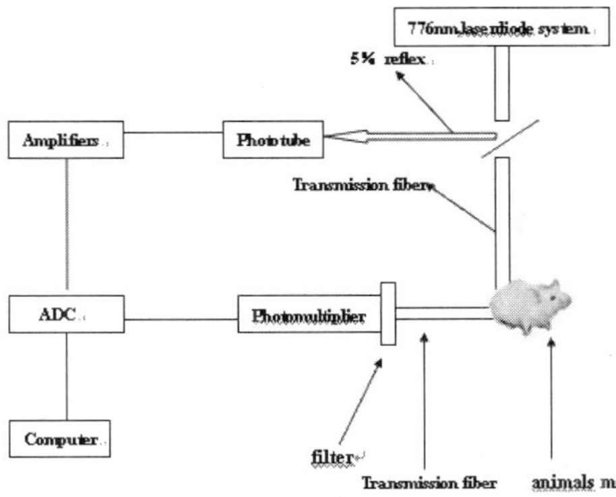

Figure 1. Scheme for the near-infrared spectroscopy system.

2.2.4 NIR image system

A NIR reflectance imaging system shown in Fig.2, equipped with the excitation laser (λ=765.9nm, NL-FC-2.0-763), a high sensitivity CCD camera (PIXIS 512B, Princeton Instrument), and an 800nm long pass filter for capturing the emitted fluorescence from the tissue. This system was used to real time image the distribution of cypate-labeled protein drug. Besides, another laser (λ=808nm, HLU32F400-808, LIMO Co. Ltd) is supplied as background light to obtain the profile of the subjected animal (Otherwise, the imaging was completely black except the fluorescence spot).

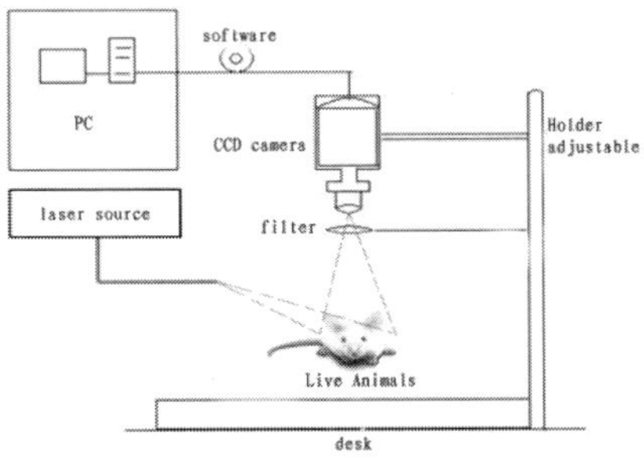

Figure 2. Scheme for the near-infrared image system.

2.2.5 Measurement method

Mouse was anaesthetized with 20% urethane and fixes on the stent.Each mouse was injected with 67 μg insulin-cypate complex via the tain vein. The blood dynamic concentrations of insulin in the thigh femoral artery were *in vivo* monitoring by the NIR spectroscopy system.Fixed the equipment when the fluorescence intensity reach the largest by regulating the fibers head.

Here, only fluorescence was directly recorded to quantify the cypate in thigh femoral artery blood, and the concentrations of insulin was obtained by applying a pre-calibrated curve. Images acquired 0.5, 1, 2, 11, 90, 660min post-injection by NIR image system were used to trace the distribution of insulin-cypate complex in the mouse body. The imaging camera was positioned 15cm above the mouse, the room lights were turned off and the laser source was activated to stimulate fluorescence and positioned such that the mouse was evenly illuminated. Images were acquired at 820nm.

3 Results and Discussion

3.1 The Labeling Efficiency Evaluated by RP-HPLC

As show in Fig. 3, the better chromatographic separation of cypate, insulin and insulin-cypate complex, was achieved using a C18 reversed phase column with a gradient water and acetonitrile mobile phase, which was adjusted to the final conditions and allowed a good resolution in a run time of 30 min. The retention time of insulin, cypate and insulin-cypate complex was respectively 4.4min, 24.8min and 18.8min. In order to support that the peak at retention time of 18.8min is represent the insulin-capote complex, we collected the elate of this time and done fluorescence detection by NIR image system (as show in fig.4.) And protein measurement by Folln-phenol reagent method, the result show that this kind of material is the complex. And the HPLC result also show that we access a high label efficiency of about 95% under our optimum conditions.

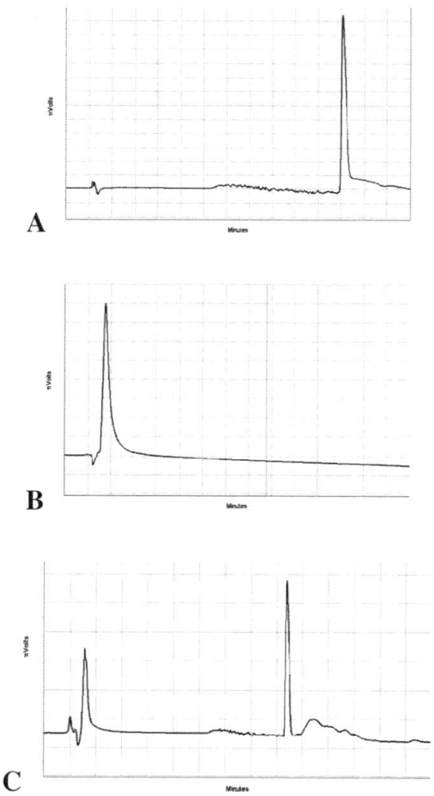

Figure 3. The HPLC chromatograms of cypate (A, dissolved in acetonitrile), insulin (B, dissolved in 0.01 mol/L hydrochloric acid) and insulin-cypate complex (C, directed injection after dialysis).

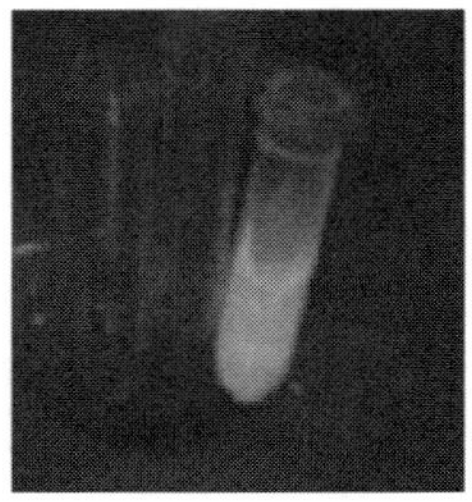

Figure 4. Fluorescence imaging of the eluate of 18.8 min peak by NIR image system.

3.2 Hypoglycemic Function of Insulin-cypate Complex in Diabetic Mouse

After inducing diabetes with streptozotocin at 3 day, STZ mouse showed a significantly decreased weight of 15±2g compared with an increased blood glucose concentration of 22 ±3 mmol/L (> 11.1 mmol/L). The diabetic mouse were administered insulin-cypate complex by i.v.injection and measured blood glucose concentration at 0, 10, 20, 30min. Fig. 5 show that the insulin-cypate complex has obvious hypoglycemic effect. So labeled with cypate does not affect hypoglycemic function of insulin.

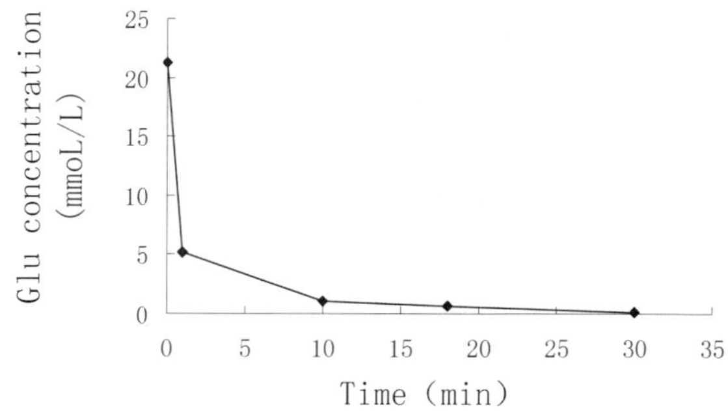

Figure 5. Mean plasma concentration–time profile of glucose after the i.v. administration of 67μg insulin-cypate complex to diabetic mode mouse. Each point represented the mean of the concentration obtained from five mouse.

3.3 *The Blood Dynamic Concentrations of Insulin in the Thigh Femoral Artery of Mouse*

The optimal conjugation of cypate-labeled insulin, with higher labeling efficiency and hypoglycemic activity, was administrated through tail vein into the blood stream of mouse. The blood dynamic concentrations of insulin were *in vivo* monitored by a NIR spectroscopy system. Here, only fluorescence was directly recorded to quantify the cypate in blood, and the concentrations of insulin were obtained by applying a pre-calibrated curve. The detecting sensitivity on the dye-labeled drug is much higher and the detectability can reach 0.1 μg/mL (data not show). Fig. 6 shows the time profile of relative insulin concentration according to the pre-calibrated curve in thigh femoral artery blood stream. The half-time of insulin-cypate complex at arteria cruralis blood to be calculated, and a value of 17min were obtained.

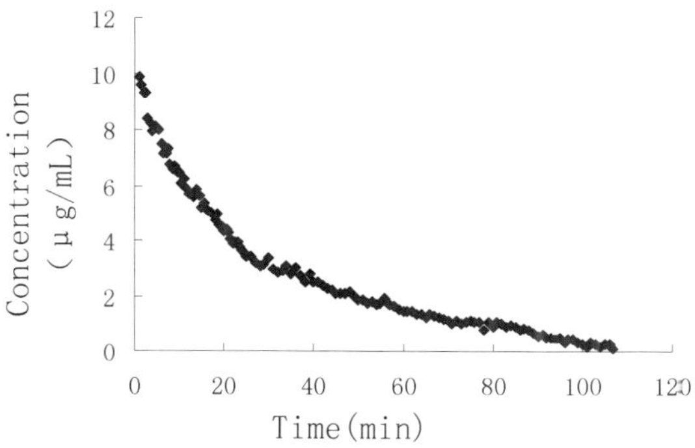

Figure 6. Time profile of relative insulin concentration according to a pre-calibrated curve in thigh femoral artery blood stream.

3.4 *The Distribution of Insulin-cypate Complex in Mouse*

We utilized the NIR imaging system to successfully trace the route of insulin-cypate complex in mouse, as shown in Fig. 7. The optimal conjugation of insulin-cypate complex was administrated through tail vein into the blood stream of mouse. The distributions of insulin-cypate complex were imaged at certain time intervals by NIR imaging system. As demonstrated in the six selected map, insulin-cypate complex was quickly accumulated in liver (30s minutes after administration) and reached the maximum concentration in liver at 2 min post-injection. At about 11min post-injection, the complex began to excrete to the intestine through biliary. The back and forth moving of fluorescence in duodenum and intestines was observed during the experiment, these fluorescence substances may be separate dye or dye-labeled drug complex. There is no fluorescence can detected on mouse at about 11h post-injection, note that the complex has been completely removed from the body.

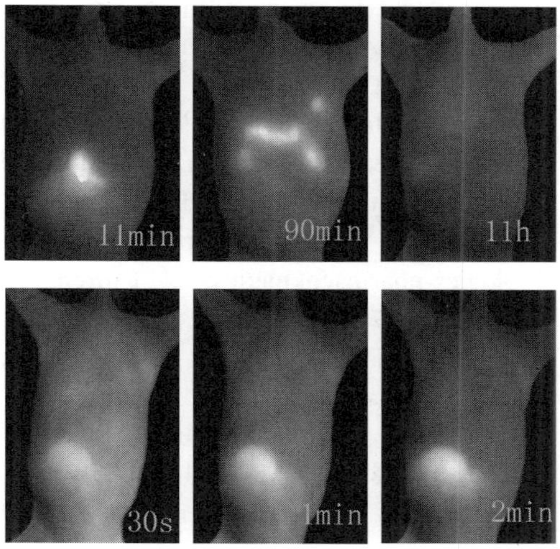

Figure 7. Dynamic route of insulin-cypate complex in mouse monitored by NIR imaging system.

4 Conclusion

The emergence of NIR techniques offers great potential to expedite drug discovery and development. In NIR techniques based on the intensity profile of the incident light, spectroscopy and imaging are the most widely implemented systems for non-invasive *in vivo* real time monitoring of optical signals in biological matrices. Usually, NIR spectroscopy offers a quantitative way to determine the endogenous and exogenous chromophores in living body, while the NIR imaging provides a qualitative method to image the distributions of biomarkers in whole animal body. The results obtained from NIR spectroscopy and imaging in our experiments and from literatures demonstrated that NIR technique is a practical and reliable monitoring modality for *in vivo* tracing of drug concentration and distribution in small animal models. Although there are still many problems to be solved such as low quantum yield, inconvenience to manipulate the wavelength and requiring high Marking conditions of near-infrared dye, verification by a Conventional.

Parallel method, the widespread application of NIR technique will significantly expedite the step of drug development.

References

1. Tang L., Persky AM., Hochhaus G, Meibohm B, Pharmacokinetic aspects of biotechnology products, Journal of pharmaceutical Science, **93(9)**: 2184-2204 (2004).
2. Lobo E.D., Hansen R.J., Balthasar J.P., Antibody pharmacokinetics and pharmacodynamics, Journal of Pharmaceutical Science, **93(11)**: 2645-68 (2004).
3. Baumann A., Early development of therapeutic biologics-pharmacokinetics, Current Drug Metabolism, **7:** 15-2 (2006).
4. Mahmood I and Green MD. Pharmacokinetic and Pharmacodynamic Considerations in the Development of Therapeutic Proteins. Clin Pharmacokinet, **44(4)**: 331-347 (2005).
5. Ntziachristos V, Bremer C, Weissleder R.Fluorescence imaging with near- infrared light: new technological advances that enable *in vivo* molecular imaging [J]. Eur Radiol, **13(1)**: 195-208 (2003).
6. Rao JH, Dragulescu-Andrasi A, Yao H Q. Fluorescence imaging *in vivo*: recent advances [J]. Curr Opin Biotechnol. **18(1)**: 17-25 (2007).
7. J.R. Mourant, T.M. Johnson, G. Los, I.J. Bigio, Non-invasive measurement of chemotherapy drug concentrations in tissue: preliminary demonstrations of *in vivo* measurements, Phys. Med. Biol. **44**: 1397–1417 (1999).
8. M. Canpolat, J.R. Mourant, Monitoring photosensitizer concentration by use of a fiber-optic probe with small source–detector separation, Appl. Opt. **39**: 6508–6514 (2000).
9. I.J. Bigio, S.G. Bown, Spectroscopic sensing of cancer and cancer therapy, Cancer Biol. Ther. **3**: 259–267 (2004).
10. Stephen C. Kanick, Julie L. Eiseman, Erin Joseph et al. Noninvasive and nondestructive optical spectroscopic measurement of motexafin gadolinium in mouse tissues: Comparison to high-performance liquid chromatography. Journal of Photochemistry and Photobiology B: Biology **88**: 90-104 (2007).
11. Kim DE, Schellingerhout D, Jaffer FA, Weissleder R, Tung CH, Near infrared fluorescent imaging of cerebral thrombi and blood brain barrier disruption in a mouse model of cerebral venous sinus thrombosis, J. Cereb Blood Flow Metab, **25(2)**: 226-33 (2005).
12. Shah K, Weissleder R, Molecular optical imaging: applications leading to the development of present day therapeutics, NeuroRx. **2(2)**: 215-25 (2005).
13. Ramjiawan B, Maiti P, Kaplan H, et al. *In vivo* monitoring of the distribution of a monoclonal antibody fragment [J]. Vibrational Spectroscopy, **28**: 177-188 (2002).

CULTURE OF GABAERGIC NEURONS FROM TRANSGENIC MICE ON MULTI-ELECTRODE ARRAY*

XIANGNING LI, GENG ZHU, WEI ZHOU, SHAOQUN ZENG, QINGMING LUO

Britton Chance Center for Biomedical Photonics, Wuhan National Laboratory for Optoelectronics, Huazhong University of Science and Technology
Wuhan, Hubei, 430074, China

GABAergic neurons, considered as the chief inhibitory neurons in the mammalian hippocampus, comprise a myriad of diverse neuronal subtypes. However, due to difficulties in the identification of GABAergic neurons in living tissues, characteristics of GABAergic neurons involved in the balance controll of neural network remains unclear. To facilitate the study of these neurons, neurons that express enhanced green flourscent protein (GFP) from neonatal transgenic mice were isolated and purified to examine its biological characters in vitro. The multi-electrode array and confocal microscopy is ideal for repeatly recording over hours or days. Combining neural activities recorded simultaneously from 60 channels in a multi-electrode array with morphological changes imaged with the confocal microscopy technique, we have succeeded in analysing the electrophysiological and morphological characteristics of GABAergic neurons in cultured hippocampal networks. The result showed that morphological changes might relate with signal processing in neural network. These suggest that the cultured transgenic neural networks of hippocampal neurons grown on microelectrode arrays could be a useful model system for future studies on properties of neuron ensembles and roles of GABAergic neurons in living neuronal network.

1. Introduction

The proper functioning of the hippocampus requires neural networks produced by projection neurons and interneurons that primarily use the neurotransmitters glutamate and GABA, respectively [1-3] GABAergic interneuron constitutes a diverse population of neurons which precisely control the timing of action potential generation in principal neurons. GABAergic interneuron constitutes a diverse population of neurons which precisely control the timing of action potential generation in principal neurons. The spatial segregation of interneuron terminals on pyramidal neurons suggests functional specialization among GABAergic interneurons. While excitatory interactions between principal cells mediate information processing and storage in the hippocampus, the pattern of activity across distinct interneuronal subtypes are critical in determining the computational functions performed by the network [3, 4].

The spatiotemporal patterns of neuronal firing during network activities, and the mechanisms of information processing and storage, would ideally be explored *in vivo*. However, in order to physiologically and pharmacologically dissect effects of pathway across the network, it is expedient to use a relevant *in vitro* model to test hypotheses for the basic mechanisms involved[5].

While excitatory interactions between principal cells mediate information processing, the pattern of activity across distinct interneuronal subtypes are critical in determining the computational functions performed by the network. The development of an alternative approach relying on optical signal would prove most useful. Such an attempt was made with the autofluorescent serotonin analogue. However, one of the major disadvantages of this method is the rapid photobleaching of the signal during fluorescence imaging [6]. More problematic is the fact that it has been found to be highly unreliable to identify GABAergic neurons. Although GABAergic neurons are the most interneuron in hippocampus, it only labels a small proportion of GABAergic neurons and is quite time-consuming [7]. Due to difficulties in the identification of GABAergic neurons in living tissues, their characteristics involved in the balance control of neural network remains unclear.

To study the morphological dynamics of a living neuronal network, while simultaneously monitoring its electrical activity, dissociated brain cell cultures (including neurons and glias) were grown on glass dishes with arrays of embedded electrodes (multi-electrode arrays, MEAs). These electrodes are capable

* This work is supported by the National Natural Science Foundation of China (Grant No. 30700215) and China Postdoctoral Science Foundation (Grant No. 20070410284).

of recording and stimulating the neurons extracellularly and non-destructively.

2. Materials and Methods

2.1. *Neuronal Culture Preparation*

Hippocampal neurons from mice (P1~3, 67-kDa isoform of glutamic acid decarboxylase-green fluorescence protein (GAD67-GFP) (Δneo) knock-in (GAD67GFP/+)) were prepared. Cells were plated on poly-D-lysine precoated MEA at a concentration of $5*10^6$ cells/mL and maintained in DMEM supplemented with 10% fetal calf serum, 10% horse serum and 2% glutamax. Half of the medium was changed twice a week. Neuronal cultures were kept in an incubator providing a controlled level of CO_2 (5%), temperature (37°C).

Bicuculline was purchased from Sigma As the blocker of GABAA receptor, Bicuculline (10 μM) was added to the cultured medium, and was washed away by changing the medium three times after recording.

2.2. *Electrical Recordings*

The MEA system used for electrophysiology was supplied by Multi Channel Systems (MCS, Gmbh, Germany). MEA dishes had 60 TiN electrodes with the interelectrode spacing of 200 μm and each electrode had a diameter of 30 μm. In order to keep the desired environmental conditions during the electrical recordings, the dish was sealed with a FEP cap in order to eliminate evaporation and contamination.

3. Results

3.1. *GFP Expression in Hippocampal Neurons in vitro*

Hippocampal neurons began neuritic outgrowth within several hours after plating on multi-electrode array plate. With the fluorescent microscope, cultured hippocampus neurons labeled with GFP could survive for more than two weeks and formed a neuronal network within a few days. The processes of the neurons were thick and the neurons connected one another to form the network through their processes. And then the network appeared to be a good status in two weeks after being plated. Spontaneous activities became obvious as early as 5 days in vitro and could be recorded over 24 weeks in vitro.

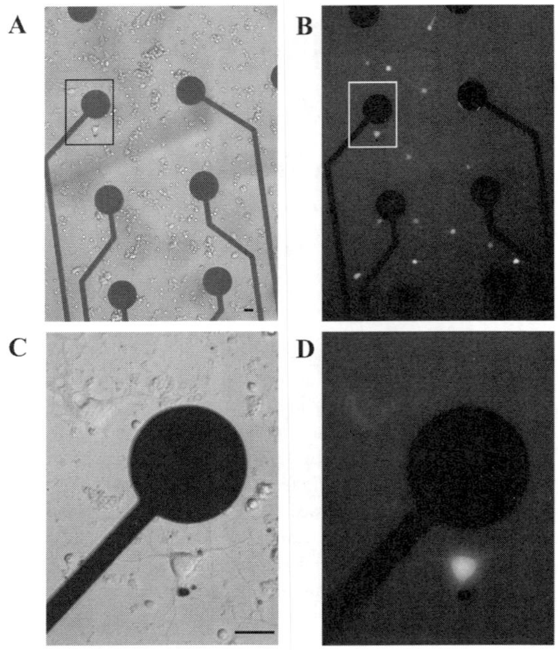

Figure 1. Living dissociated neurons from the hippocampus of GFP mice in phase contrast and during fluorescence on the multi-electrode array. The dark squares are platinized substrate electrodes. The scale bar, 10 μm.

3.2. *Firing Pattern of Disinhibited Networks*

To access the changes in firing patterns in cultured neuronal networks during bicuculline treatment, we compared spontaneous activities with activities during pharmacological treatment. Removal of GABAergic inhibition with bicuculline led to synchronized bursts throughout all active sites on the multi-electrode array. Comparison to the timing of the bursts from different sites suggested that there was a very small variability in the burst frequency value among the electrodes treated with bicuculline.

4. Discussion

GABAergic interneurons are necessary for the emergence of hippocampal network oscillations, during which they play a key role in the synchronization of firing[8]. However, it remains to be resolved how interneuron contribute to low-frequency synchronized oscillations during the spontaneous development, in what way the spatiotemporal pattern of interneuronal firing affects principal cell activity, and by which mechanisms the interneurones themselves are synchronized[9]. In order to physiologically and pharmacologically dissect effects of pathway across the network, it is expedient to use a relevant *in vitro* model

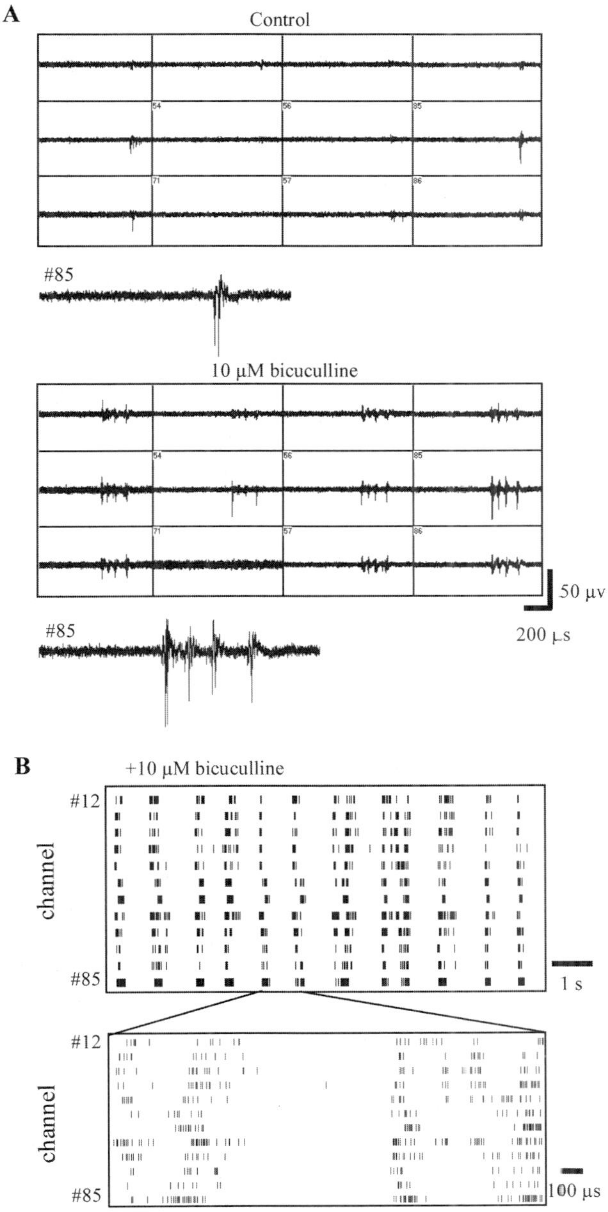

Figure 2. The effect of disinhibition on the cultured neuronal networks. A, a typical example of an MEA recording from a hippocampal network before and during being exposed to bicuculline. B, C, raster plot of the activities from neuronal networks exposed to bicuculline. Simultaneous recordings from 12 separate electrodes are shown in each panel.

to test hypotheses for the basic mechanisms involved[7, 10, 11].

We are developing a useful tool aimed at studying the morphological dynamics of a living neuronal network during the development and the signal process, while simultaneously monitoring its electrical activity. We grow dissociated hippocampal cell cultures (including neurons and glia) on glass dishes with arrays of embedded electrodes. These electrodes are capable of recording and stimulating the neurons extracellularly and non-destructively. Combining multi-electrode array technology with transgenic neurons allows us to follow the electrical activity and morphological changes in hundreds of neurons for weeks to months *in vitro*. Compared to *in vivo* investigation, cultured networks offer unparalleled access for long-term electrical and pharmacological recording; every neuron in the network can be imaged at the submicron with the electrical changes in microsecond. This will enable us to determine the causal interrelationships between morphology and network activity.

Acknowledgments

We gratefully acknowledge the financial support by the National Natural Science Foundation of China (Grant No. 30700215) and China Postdoctoral Science Foundation (Grant No. 20070410284).

References

1. M.M. Lindroos, S.L. Soini, T.K. Kukko-Lukjanov, E.R. Korpi, D. Lovinger and I.E. Holopainen, *Int J Dev Neurosci.* 23, 65 (2005).
2. V. Nimmrich, N. Maier, D. Schmitz and A. Draguhn, *J Physiol.* 563, 663 (2005).
3. Y.J. Karten, M.A. Jones, S.I. Jeurling and H.A. Cameron, *Hippocampus.* 16, 312 (2006).
4. A.R. Woodruff, H. Monyer and P. Sah, *J Neurosci.* 26, 11881 (2006).
5. S.W. Hughes and V. Crunelli, *Int J Psychophysiol.* 64, 3 (2007).
6. F. Minneci, M. Janahmadi, M. Migliore, N. Dragicevic, D. Avossa and E. Cherubini, *Hippocampus.* 17, 538 (2007).
7. M. Uusisaari, K. Obata and T. Knopfel, *Journal of neurophysiology.* 97, 901 (2007).
8. G. Horcholle-Bossavit, B. Quenet and O. Foucart, *Bio Systems.* 89, 244 (2007).
9. L. Sirovich, A. Omurtag and K. Lubliner, *Network (Bristol, England).* 17, 3 (2006).
10. B. Chattopadhyaya, G. Di Cristo, C.Z. Wu, G. Knott, S. Kuhlman, Y. Fu, R.D. Palmiter and Z.J. Huang, *Neuron.* 54, 889 (2007).
11. M. Miura, S. Saino-Saito, M. Masuda, K. Kobayashi and T. Aosaki, *J Neurosci.* 27, 9721 (2007).

DETECTION OF BACE CLEAVAGE IN THE SECRETARY PATHWAY

JINLING LU, JIE YANG, SHAOQUN ZENG, ZHIHONG ZHANG*

Britton Chance Center for Biomedical Photonics, Wuhan National Laboratory for Optoelectronics, Huazhong University of Science and Technology, Wuhan 430074, PR China

Abnormal production and accumulation of amyloid beta-peptide (Aβ) play major roles in the pathogenesis of Alzheimer's disease (AD). Aβ is the product of sequential cleavage of amyloid precursor protein (APP) by β-secretase (BACE) and γ-secretase. To measure BACE activity in living cells, we have constructed a genetically encoded fluorescence resonance energy transfer (FRET) biosensor, dYβC, which is directed to the secretory pathway for monitoring BACE activity in vivo. It has been demonstrated that this FRET sensor can be cleaved by BACE both in vitro and in living cells in our previous report. To further identify the cellular compartments in which the cleavage occurs, FRET measurements with spectral scanning and acceptor photobleaching was used to follow the BACE cleavage of the probes in the endoplasmic reticulum (ER) and Golgi apparatus. In our experiments, the cells expressing dYβC showed significant higher FRET efficiency comparing to cells co-expressing dYβC and BACE both in ER and Golgi apparatus, indicating that the FRET probes are cleaved in present of BACE. The results provide in vivo evidence that the β-secretase cleavage occurs within the secretory pathway, as early as the ER.

1. Introduction

Alzheimer's disease (AD) is characterized pathologically by senile plaques containing the amyloid β-peptide (Aβ), the product derived from sequential cleavage of amyloid precursor protein (APP) by β-secretase and γ-secretase [1].

Although it is known that APP undergoes sequential cleavage to produce Aβ, the subcellular compartments in which Aβ is produced are not fully determined in living cells yet. Previous studies emphasized the acidic pH optimum of BACE [2-3] and provided evidences that Aβ generated in acidic compartments, such as trans-Golgi network (TGN) and endosomes [4-5]. However, retention of APP in the endoplasmic reticulum/intermediate compartment did not alter intracellular Aβ (1-42) synthesis, which suggests that Aβ can be generated from the ER [6]. BACE-APP interactions also occur at the cell surface [7] and inhibition of receptor-mediated endocytosis demonstrates that the BACE cleavage can occur directly at the plasma membrane [8].

Until now, however, experimental evidence about the multi-subcellular loci of BACE cleavage has been limited to the studies using immunocyto-chemistry methods, which could not provide in vivo evidence of BACE cleavage in living cells. Using our new developed dYβC FRET biosensor [9], together with time-lapse confocal imaging, we investigate the BACE cleavage within discrete regions of the endoplasmic reticulum (ER), Golgi apparatus and plasma membrane in living cells.

2. Materials and Methods

2.1. *Plasmids*

The plasmids of pDis-YβC and pBACE-mCherry were constructed as described in previous report [9]. The pEYFP-ER and pEYFP-Golgi plasmids were kindly provided by Dr. Davis (Albany Medical College, NY, USA). pEYFP-ER encodes a fusion protein (EYFP-ER) containing an enhanced yellow fluorescence protein (EYFP) with the ER targeting sequence of calreticulin at the 5' end and a KDEL retention sequence at the 3' end. The pEYFP-Golgi plasmid encodes an EYFP fusion protein (EYFP-Golgi) that specifically targets the trans-medial region of the Golgi apparatus. EYFP-ER and EYFP-Golgi were used as markers to visualize ER and Golgi in living cells, respectively.

2.2. *Cell Culture and Transient Transfection*

Human cervical cancer cells, HeLa cells, were cultured in Dulbecco's modified Eagle's medium (DMEM, Gibco) supplemented with 10% fetal calf serum, 100 units/ml penicillin and 100 μg/ml streptomycin sulfate at 37°C in a humidified atmosphere of 5% CO_2/95% air.

* To whom correspondence should be addressed. E-mail: czyzzh@mail.hust.edu.cn, Tel: +86 27 8779 2033; Fax: +86 27 8779 2034

Transfections were performed using Lipofectamin 2000 (Invitrogen) according to the manufacturer's protocol.

2.3. *Confocal Imaging*

All images were acquired on an Olympus FV1000 confocal imaging system (Olympus corp., JP) with a 40 mW argon laser and a 60 × 1.35 NA oil immersion objective at 3 × zoom. The laser was tuned to lines at 458 and 515 nm. Images of mCerulean were obtained through photomultiplier 1 (PMT1, bandpass filter 470-500 nm) with 458 nm excitation, which is termed as the 'CFP channel' here. The 'YFP channel' refers to the combination of 515 nm laser-line excitation and PMT2 (bandpass filter 530-580 nm) detection, and it was used to get images of mCitrine or EYFP. The BACE-mCherry was imaged by 'RFP channel' with multi-photon excitation at 800 nm and PMT2 (bandpass filter 600-640 nm) detection. All transfected cells were imaged live at 37°C in phenol red-free DMEM (Sigma).

2.4. *FRET Measurement*

FRET was measured using the acceptor photobleaching method [10], in which the energy transfer is detected as an increase in donor (mCerulean) fluorescence after complete photobleaching of the acceptor molecules (mCitrine). Acceptor photobleaching protocol was performed according to reference 19, with the following modifications. Before performing phtotobleaching, the prebleach mCerulean and mCitrine images of transfected cells were sequentially collected through the CFP channel (3% intensity) and the YFP channel (0.2% intensity) as baseline, respectively. The cells were then bleached by scanning a selected region of interest (ROI) 20 times for 10 s using the 515 argon laser line at 10% intensity. Post-bleach mCerulean and mCitrine images were collected immediately following photobleaching using the same parameters as those used for acquiring prebleach images. FRET efficiency (E_F) was calculated as a percentage increase in mCerulean fluorescence intensity following the mCitrine photobleaching using the formula $E_F = (I'-I) \times 100 / I'$, where I and I' were the mCerulean fluorescence intensities before and after acceptor photobleaching, respectively, taking into account the background noise in each image.

Single cell spectrum scanning was performed on a FV1000 confocal imaging system by using lambda scan with 458nm excitation (10% power, 1nm bandwidth) and a series of images of emission were collected from 460 to 600nm (1nm steps, 5nm bandwidth). The emission spectra data were processed using FV10-ASW1.4 software (Olympus corp., JP).

3. Results

3.1. *Real-time Imaging of dYβC Trafficking from the ER to the Golgi and Plasma Membrane*

The EYFP-ER and EYFP-Golgi were used as markers to visualize ER and Golgi apparatus, respectively.

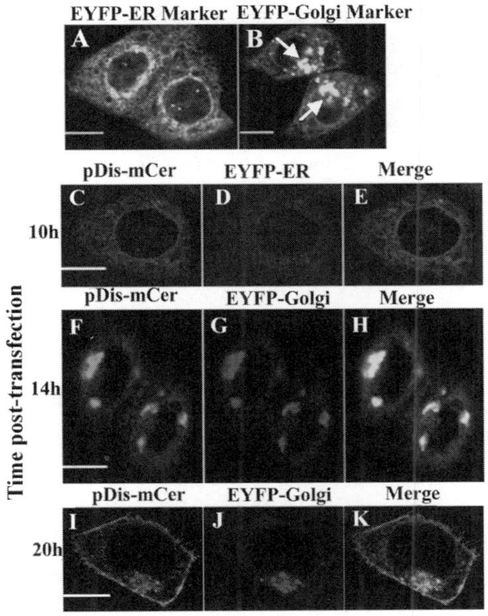

Figure 1. Confocal fluorescence imaging of ER and Golgi apparatus. The white scale bar represents 10 μm. (A) HeLa cells 24 h post-transfection with the pEYFP-ER plasmid. (B) HeLa cells 24 h post-transfection with pEYFP-Golgi. The Golgi apparatus is indicated by white arrows. (C-E) Living HeLa cells co-transfected with pDis-mCer (green) and the ER-EYFP marker (red) at 10 h after transfection. (F-H) Living HeLa cells co-transfected with pDis-mCer (green) and the EYFP-Golgi marker (red) at 14 h after transfection. (I-K) Living HeLa cells co-transfected with pDis-mCer (green) and the Golgi-EYFP marker (red) at 20 h after transfection.

HeLa cells transfected with pER-EYFP or pGolgi-EYFP 24 h post-transfection are shown in Fig. 1A and B. The ER is a diffuse reticular network in the living cell (Fig. 1A) and the Golgi apparatus appears as a dense area of intense fluorescence confined to the perinuclear region (Fig. 1B).

As a plasmid constructed based on pDisplay vector as well as pDis-YβC, the pDis-mCer was used as a substitute for pDis-YβC to visualize the protein expressed from pDisplay vector when co-transfected with pEYFP-ER or pEYFP-Golgi markers. HeLa cells co-transfected with pDis-mCer and pEYFP-ER showed co-localization of mCerulean (green) and EYFP (red) fluorescence in the ER at 10 hours after transfection (Fig. 1C-E). Cells co-transfected with pDis-mCer and pEYFP-Golgi showed co-localization of mCerulean (green) and EYFP (red) fluorescence in the Golgi at 14 hours after transfection (Fig. 1F-H). At twenty hours after transfection, mCerulean fluorescence was mainly on the plasma membrane and the EYFP-Golgi marker remained in the Golgi (Fig. 1I-K).

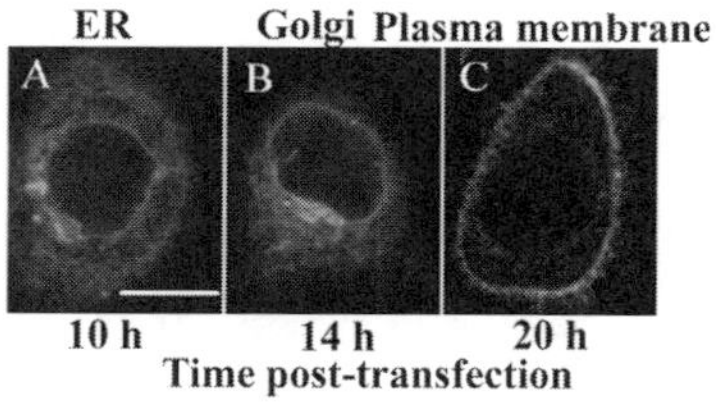

Figure 2. Real-time imaging the dYβC proteins trafficking from the ER to the Golgi and plasma membrane. The white scale bar represents 10 μm.

The cells transfected with pDis-YβC were imaged at various times post-transfection. The time-lapse confocal imaging revealed that the dYβC fusion protein in many cells was located in the ER from 10 to 14 hours post-transfection (Fig. 2A), and mainly in Golgi apparatus from 14 to 18 hours (Fig. 2B). At 20 hours after transfection, the expression of the FRET probe reached a steady-state level, indicating that majority of the dYβC probes were on the plasma membrane with little intracellular expression (Fig. 2C). This phenomenon has also been observed in HEK 293 cells expressing CFP or YFP tagged serotonin 5-HT_{2C} receptors [11]. Although the reason for this phenomenon is not clear, it has been observed in all of our transient transfections with pDis-mCer, pDis-YβC, pDis-YcC and pDis-mCit. Based on these results, it is possible to follow the β-cleavage of dYβC probes through the secretory pathway from the ER to Golgi according to the above time windows.

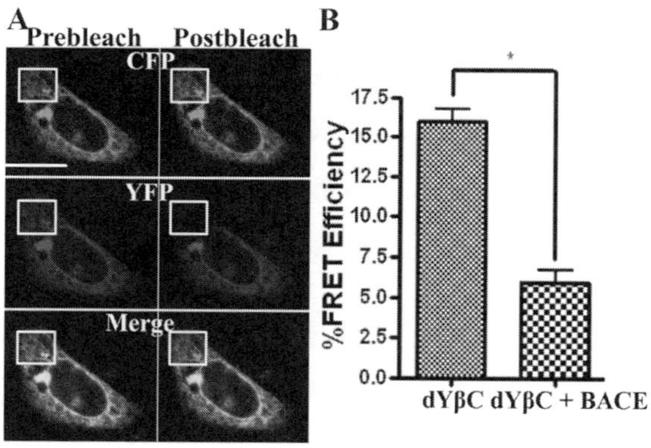

Figure 3. Detection of β-secretase activity in the ER by acceptor photobleaching. (A) A region of ER (marked by the white rectangle) was selectively irradiated using the 515 nm laser line. The white scale bar represents 10 μm. (B) The efficiency of FRET (E_F) in the ER was detected by acceptor photobleaching. A significant difference (* P<0.001) of FRET efficiency was observed between the groups of dYβC and dYβC+BACE. dYβC, transfected with pDis-YβC; dYβC +BACE, co-transfected with pDis-YβC and pBACE-mCherry. All experiments were performed 10-14 hours post-transfection.

3.2. *BACE cleavage of the dYβC probe is initiated in the ER*

Time-lapse confocal imaging and acceptor photobleaching were used to visualize and quantify FRET of dYβC probes in the ER and Golgi apparatus of living cells.

To determine if the dYβC probes were cleaved by BACE in the ER, FRET measurements were performed 10-14 hours post-transfection when dYβC probe expression was supposed to occur in the ER according to the above results. When a region of the ER (marked by the white rectangle in Fig. 3A) was selectively irradiated using the 515 nm laser line, an increase in mCerulean fluorescence was observed following the mCitrine photobleaching. The FRET efficiency of dYβC in HeLa cells transfected with pDis-YβC was 15.98±0.82 (mean±SEM, n=41). For the cells co-transfected with both pDis-YβC and pBACE-mCherry, the FRET efficiency was 5.95±0.79 (mean±SEM, n=54). Significant differences in FRET efficiency (*P<0.001) were observed between the two groups (Fig. 3B), which suggests that β-cleavage occurred in the ER.

To investigate whether the dYβC probe is cleaved in the Golgi apparatus, FRET measurements were performed 14-18 hours after transfection when the dYβC protein expression was supposed to occur in the Golgi. When a region of Golgi (marked by the white rectangle in Fig. 4A) was irradiated using the 515 nm laser line, the mCerulean fluorescence was increased following mCitrine photobleaching. The FRET efficiency of dYβC in HeLa cells transfected with pDis-YβC was 18.99±1.08 (mean±SEM, n=39), and that in the cells co-transfected with pDis-YβC and pBACE-mCherry was 6.97±1.07 (mean±SEM, n=26). Significant differences in FRET efficiency (*P<0.001) were observed between the two groups (Fig. 4B), which suggests that the β-cleavage also occurred in the Golgi apparatus.

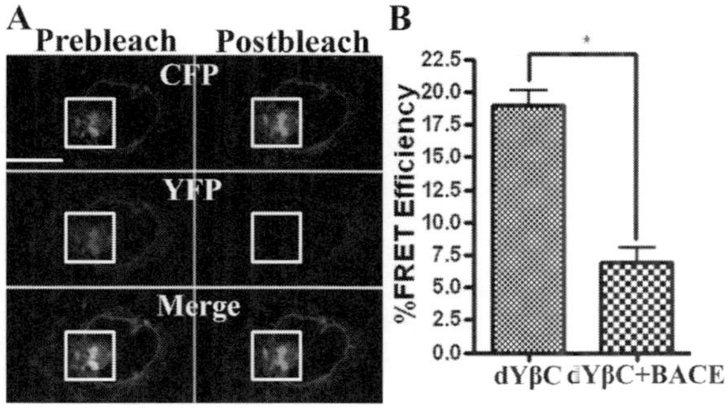

Figure 4. Detection of β-secretase activity by acceptor photobleaching in the Golgi. (A) A region of the Golgi (marked by the white rectangle) was selectively irradiated using the 515 nm laser line. The white scale bar represents 10 μm. (B) The efficiency of FRET (E_F) in Golgi was detected by acceptor photobleaching. A significant difference (* $P<0.001$) in FRET efficiency was observed between the groups of dYβC and dYβC+BACE. dYβC, transfected with pDis-YβC; dYβC + BACE, co-transfected with pDis-YβC and pBACE-mCherry. All experiments were performed 14-18 hours post-transfection.

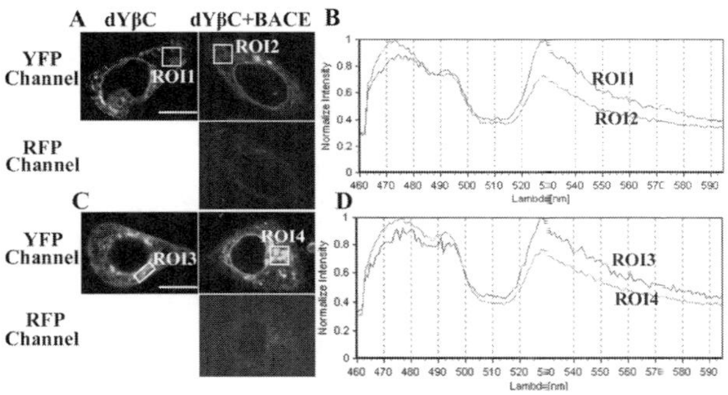

Figure 5. Visualization of the cleavage of dYβC by BACE in the ER and Golgi apparatus in living cells by spectrum scanning. The white scale bar represents 10 μm. (A) The HeLa cells were imaged 10 h after transfection with pDis-YβC (left) or co-transfection with pDis-YβC and pBACE-mCherry (right). The dYβC proteins were mainly in the ER. (B) The emission spectrum of ROI 1 and ROI 2 in A with 458 nm excitation. (C) The HeLa cells were imaged 14 h after transfection with pDis-dYβC (left) or co-transfection with pDis-YβC and pBACE-mCherry (right). The majority of dYβC proteins were expressed in the Golgi. (D) The emission spectrum of ROI 3 and ROI 4 in C with 458 nm excitation. dYβC, transfection with pDis-YβC; dYβC+BACE, co-transfection with pDis-YβC and pBACE-mCherry.

The acceptor photobleaching results were confirmed by single cell spectrum scanning. As shown in Fig. 5A and B, the fluorescence intensity at the peak wavelength of mCitrine (528 nm) was markedly decreased in the cells co-transfected with pDis-YβC and BACE-mCherry (Fig. 5B ROI 2), compared to that in the cells transfected with pDis-YβC alone (Fig. 5B ROI 1). The similar result was found in Golgi apparatus at 14 hours post-transfection (Fig. 5 C and D).

Discussion

Aβ plays a central role in AD pathogenesis, and therefore, understanding the generation of Aβ is critical importance for AD therapeutic strategy. The present study was performed to identify the cellular compartments in which β-secretase cleavage occurs in vivo. Although immunocyto-chemical evidence has suggested that β-secretase cleavage occurs in multi-subcellular loci, there is no in vivo evidence of β-secretase cleavage in live cells. To address this issue, we constructed a genetically encoded FRET biosensor and detected cleavage of the FRET sensor by BACE in living cells using the time-lapse confocal imaging in combination with acceptor photobleaching FRET technique, which allows the determination of β-secretase cleavage within different intracellular compartments, such as the ER, Golgi and plasma membrane, in real-time and in living cells. Our results suggest that β-secretase cleavage was initiated in the secretory pathway, as early as in the ER.

In present study, β-secretase cleavage in different cellular compartments was determined by monitoring the FRET efficiency of the probe. There are many methods to measure FRET efficiency. The most widely used approach is intensity-based FRET technique that requires excitation of the donor and detection of acceptor emission [12]. This approach suffers from contamination of the FRET images caused by bleed-through of CFP fluorescence into the YFP channel as well as excitation of YFP during CFP excitation [11]. In this study, we chose another increasingly popular approach called acceptor photobleaching [11] to detect FRET. The principle of acceptor photobleaching is that energy transfer is reduced or eliminated following acceptor photobleaching, resulting in an increase in donor fluorescence. Its combination with laser scanning confocal microscope enables faster and targeted bleaching to be confined to very small intracellular regions, which makes the technique suitable for live cell imaging.

Acknowledgments

This work is supported by by the National High Technology Research and Development Program of China (Grant No. 2006AA020801) and National Natural Science Foundation of China (Grant No. 90508003).

References

1. D.J. Selkoe, *Physiol. Rev.* **81**, 741 (2001).
2. J. Ermolieff, J.A. Loy, G. Koelsch, J. Tang, *Biochemistry* **39**, 12450 (2000).
3. M. Shoji, T.E. Golde, J. Ghiso, J. Ghiso, T.T. Cheung, S. Estus, L.M. Shaffer, X-D. Cai, D.M. McKay, R. Tintner, B. Frangione, S.G. Younkin, *Science*, **258**, 126 (1992).
4. G. Thinakaran, D.B. Teplow, R. Simani, B. Greenbergi, S.S. Sisodia, J. Biol. Chem. **271**, 9390 (1996).
5. E.H. Koo, and S.L. Squazzo, *J. Biol. Chem.* **269**, 17386 (1994).
6. D.G. Cook, M.S. Forman, J.C. Sung, S. Leight, D.L. Kolson, T. Iwatsubo, V.M.-Y. Lee, R.W. Doms, *Nat. Med.* **3**, 1021 (1997).
7. A. Kinoshita, H. Fukumoto, T. Shah, C.M. Whelan, M.C. Irizarry, B.T. Hyman, *J. Cell. Sci.* **116**, 3339 (2003).
8. J.H. Chyung, D.J. Selkoe, *J. Biol. Chem.* **278**, 51035 (2003).
9. J. Lu, Z. Zhang, J. Yang, J. Chu, P. Li, S. Zeng, Q. Luo, *Biochem. Biophys. Res. Commun.* **362**, 25, 2007.
10. A.K. Kenworthy, *Methods*, 24, 289 (2004).
11. K. Herrick-Davis, B.A. Weaver, E. Grinde, J.E. Mazurkiewicz, *J. Biol. Chem.* **281**, 27109 (2006).
12. M. Zaccolo, *Circulation Res.* **94**, 866 (2004).

MEASUREMENT OF OPTICAL PATH LENGTH VARIATION IN NANOMETER LEVEL BY USE OF PHASE INFORMATION IN SPECTRAL-DOMAIN INTERFEROMETRY *

JIANGANG WANG, QING XIAO, SHAOQUN ZENG, QINGMING LUO

Key Laboratory of Biomedical Photonics of Ministry of Education, Wuhan National Laboratory for Optoelectronics, Huazhong University of Science and Technology, Wuhan 430074, China

The Michelson type spectral-domain interferometer is setup with a broadband light source and a spectrometer. The central wavelength of the broadband light source is 1316nm and its bandwidth is 65nm. Using the phase information directly from the results of the Fourier transform of interferograms in spectral-domain, the optical path length variation in nanometer level can be extracted. The common path setup in interferometer is important for stability of phase information and suppression of vibrational and environmental influence. Extraction and import of phase information are discussed. The experimental results show that the standard deviation (std) of phase valuation with time is ±0.007 radian, i.e. the std of the optical path variation for 1316nm is ±1.5nm. This paper also points out that the standard deviation of phase values of much deeper sample is much larger than that of phase values of shallower sample and a way to reduce the std of phase values in deeper sample.

1. Introduction

Recently people have great interests in obtaining phase information about biological samples. The first reason is most biological samples have large phase contrast and small intensity contrast; the second reason is phase information can give high sensitivity image and measurement in some cases, even nanometer-level optical path variation; the third reason is phase-sensitive detection can observe unstained biological samples. Traditional interferometric techniques such as digital holographic microscopy [1], Fourier phase microscopy [2], Hilbert microscopy [3] and phase shifting interferometry [4] have been applied to measurement of optical path length variation in nanometer level. The coherent gate inherent in optical coherence tomography (OCT) is used to extract depth-dependent phase information [5]; spectral-domain phase microscopy takes advantage of phase information directly from Fourier transforming spectral-domain time serial interferograms to study cellular dynamics and cytoskeletal rheology[6,7,8].

This paper discusses the setup of the common-path spectral-domain interferometry. The measurement of the optical path length variation in nanometer level is realized by this common-path interferometric setup. This paper also points out the standard deviation of phase values of much deeper sample is much larger than that of phase values of shallower sample. Section 2 gives the detailed illustration about the experimental setup and brief discussion on the principle, Section 3 gives the experimental results and some talks on the results, finally one simple summary is given in Section 4.

2. Experimental Setup

The basic difference between time domain and spectral domain interferometry is that in time domain the reference mirror is a moving one but in spectral domain the reference mirror is static. One of the mirrors can be replaced by an observed sample. For phase information, suppression of vibrational and environmental influence and phase stability of the light source is the key to achieve nanometer level measurement of optical path length variation. The common-path interferometer is better for this purpose. The common-path spectral domain interferometer based on Michelson interferometer is shown in Fig.1. For the common-path setup, the mask plate is used to block one arm of the interferometer. The reflective light from the first surface of the coverslip is taken as the reference light, and the light reflected from the surface of the sample as the object light. Except for the realization of the common-path setup in the interferometer, another advantage is that the fringe contrast becomes better due to the decrease of the reference light intensity. In the setup, the central wavelength of the SLED broadband light source λ_0 = 1316nm, its bandwidth = 65nm. The spectrometer is used to detect the spectral distribution of the interference light coupled into the spectrometer by a fiber.

The basic principle is described in detail by Reference [6]. In spectral-domain OCT, the spectral

* This work is supported by the National Natural Science Foundation of China (Grant No.60708025).

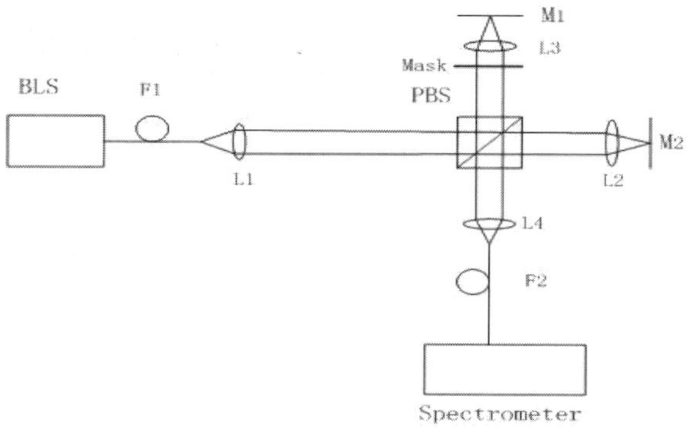

Fig.1. Schematic of the experimental setup. BLS, broadband light source; PBS, polarization beam splitter; Mask, mask plate; F1, F2, fibers; L1, L2, L3, L4, lenses; M1, M2, mirrors.

distribution of interferogram is Fourier transformed to get the amplitude and position of the reflective surfaces of the sample. Also the phase information is associated with this position directly from the Fourier transformed complex value. If the phase noise is small and the phase stability of the light source is high, then the phase variation at this position with time reflects the subcoherence [6] optical path length variation with time. Because the light source distribution is not symmetrical about the central wavenumber k_0, its Fourier transform can introduce the extra phase information from the light source; this makes it difficult to give the absolute phase value from the reflective sample surface. In most cases the relative phase values are important enough to give interesting results.

3. Results and Discussion

At present, the spectrometer is based on scanning due to point detection, its speed is lower than one based on CCD array detection. As shown in Fig.2, there are four peaks in the half side of the z coordinates; the mirror image of these four peaks locates in the other half side. These four peaks stand for four surfaces of two coverslips. Because the reference light is from the top surface of the first coverslip, the position of the first peak in the middle of the z coordinates locates about the zero optical path length position. We can see that the two coverslips are separated 3mm and the thickness of each coverslip is about 230μm and the axial resolution of OCT is about 15μm as shown in Fig.3, which is the enlarged part of Fig.2. The phase stability of the system is measured through many measurements at different times. The measurement process lasts for about one hour. The phase values associated with the second peak is shown as in Fig.4. We can see that the standard deviation of the phase value for the second peak is 0.007 radian, which means the optical path length variation is 1.5nm by the following equation

$$\varphi = 2n\delta d k_0 \qquad (1)$$

where φ represents the phase value, n is the refractive index, δd is the variation of the length, k_0 is the central wavenumber.

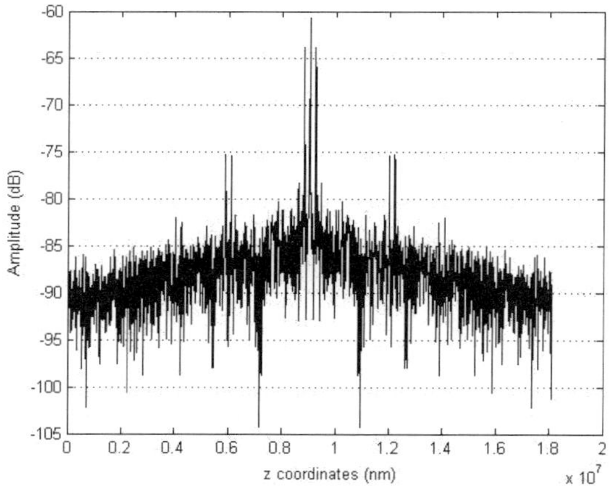

Fig.2. Amplitude of Fourier transform of one spectral-domain interferogram.

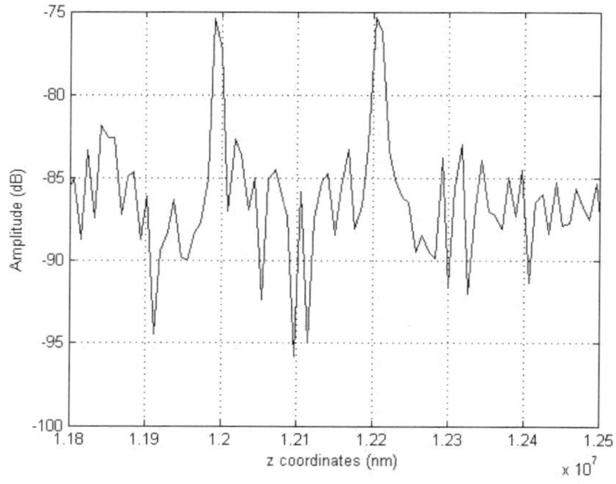

Fig.3. The enlarged part of Fig.2.

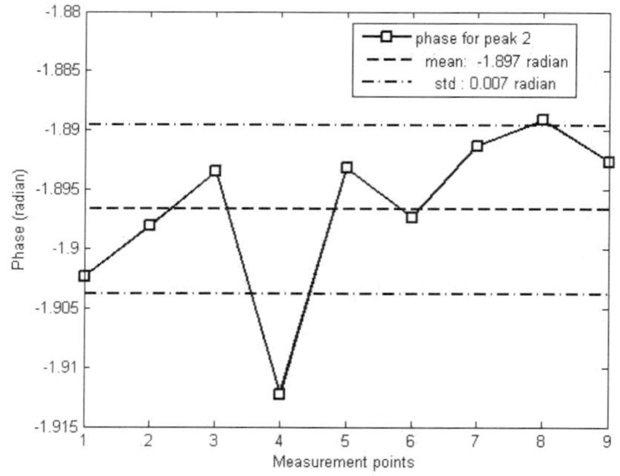

Fig.4. The phase measurement with time for the second peak.

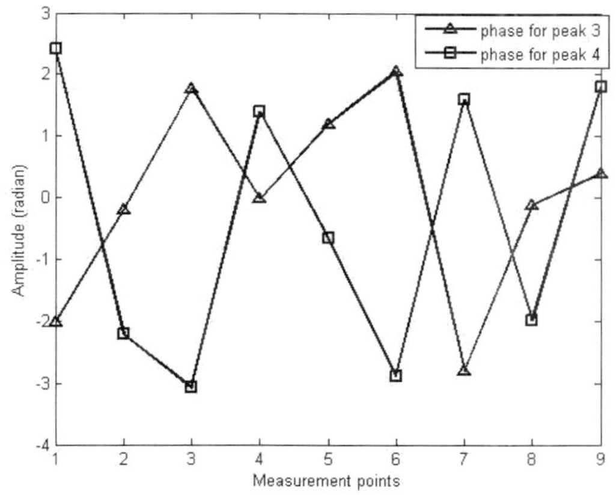

Fig.5. The phase measurements with time for the third and fourth peaks.

But for the third and fourth peaks, their standard deviations are respectively 338.5nm and 463.0nm, which are large compared with the standard deviation for the second peak. The first possible reason is that the sensitivity of phase measurement is dependent inversely on the signal noise ratio (SNR) [6], the SNR for these two peaks are less than that for the second peak; the second possible reason is that the longer the light transverses, the worse the influence of the optical path on the phase value of the sample position due to the phase noise and the decrease of the quality of interferogram; the third reason is the phase error from phase wrapping.

The position for the second peak can be chosen as the reference position for the phase measurement; i.e. both of the change of refractive index and the variation of the position for the second peak lead to the phase change of the second peak that reflects the optical path length variation within subcoherence length. The heated water is put into the middle part of the two coverslips. The time serial interferograms are recorded. By Fourier transforming these interferograms, the phase values of the second peak with time can be given directly as shown in Fig.6. Clearly the coverslip contracts about 11nm during the measurement process. Now we can not measure the temperature change during the measurement process exactly. The expansion coefficient can not be given at present but the estimated value is the same amount order with that given in Reference [6].

Phase wrapping is always a problem in phase measurement. For the third and fourth peaks, the phase values have negative and positive values. The phase range [-π, π] can be changed to [0, 2π] by adding π to the negative phase values. In this way the standard deviations of the third and fourth peaks become

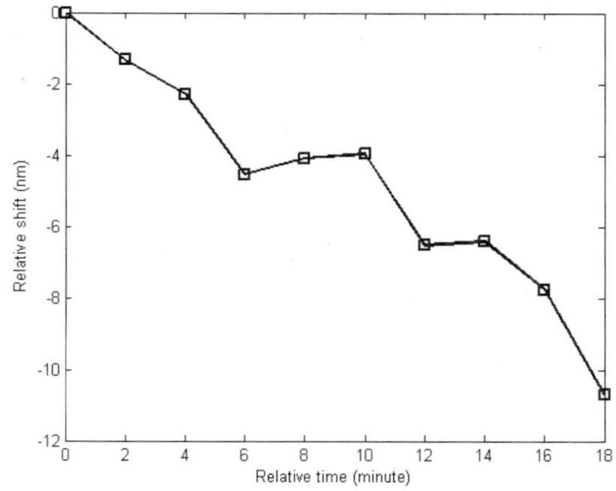

Fig.6. Relative shift for the position of the second peak with time when cooling the heated water.

228.5nm and 177.5nm respectively. Comparison with the previous results shows that this method can largely improve the measurement of phase values.

4. Summary

Spectral-domain interferometry can measure the optical path length variation in nanometer level. The coherent gate of the broadband light source can give us deep-dependent phase information. The common-path setup and phase stability and small phase noise of the light source are the key to achieve this power. The experimental results show that this kind of the common-path spectral-domain interferometry using SLED broadband light source can measure the optical path length variation with time in nanometer level. For deeper samples, the standard deviation of measured phase values becomes larger than one for shallower samples. Transforming the phase range [-π, π] to [0, 2π] by adding π to the negative phase values could improve the standard deviation of measured phase values.

References

1. P. Marquet, B. Rappaz, P. J. Magistretti, E. Cuche, Y. Emery, T. Colomb, and C. Depeursinge, Opt. Lett. **30**, 468 (2005).
2. N. Lue, W. Choi, G. Popescu, T. Ikeda, R. R. Dasari, K. Badizadegan, and M. S. Feld, Appl. Opt. **46**, 1836 (2007).
3. T. Ikeda, G. Popescu, R. R. Dasari, and M. S. Feld, Opt. Lett. **30**, 1165 (2005).
4. Christopher Fang-Yen, Seungeun Oh, Yongkeun Park, Wonshik Choi, Sen Song, H. Sebastian Seung, Ramachandra R. Dasari, and Michael S. Feld, Opt. Lett. **32,** 1572 (2007).

5. C. Yang, A. Wax, M. S. Hahn, K. Badizadegan, R. R. Dasari, and M. S. Feld, Opt. Lett. **26**, 1271 (2001).
6. Michael A. Choma, Audrey K. Ellerbee, and Changhuei Yang, Opt. Lett. **30**, 1162 (2005).
7. Emily J. McDowell, Audrey K. Ellerbee, Michael A. Choma, Brian E. Applegate, Joseph A. Izatt, Journal of Biomedical Optics **12**(4), 044008 (2007).
8. Audrey K. Ellerbee and Joseph A. Izatt, Opt. Lett. **32**, 388 (2007).

THEORETICAL ANALYSIS OF PHASE-STEPPING ALGORITHMS IN LOW COHERENCE INTERFEROMETRY

QING XIAO, JIANGANG WANG, SHAOQUN ZENG, QINGMING LUO

Key Laboratory of Biomedical Photonics of Ministry of Education

Huazhong University of Science and Technology, Wuhan 430074, China

Several phase-stepping algorithms are presented, including three-stepping algorithm, Carré algorithm, Hariharan algorithm and compensating five-stepping algorithm. And respective detailed error analyses are provided in this paper. By comparing the influence of PZT step error and noise error on phase and amplitude measurement between these algorithms, Carré algorithm should be the first choice, but when the phase shifting is far away from 90 degrees, compensating five-stepping algorithm will be a better choice, and when larger noise exist in imaging system, Hariharan algorithm will show better result, and simultaneously a fact was revealed that the distance in which we could reconstruct the phase signal will be narrower as the noise increase, which is the reason why the mismatch of two beams of interference is important for phase reconstruction in low coherence interferometry.

1. Introduction

Phase shifting interferometry(PSI) provide a method to measure the optical phase accurately through several intensity patterns, and PSI has been performed in almost every interferometric imaging system by using different algorithms[1][2][7]. Phase-stepping algorithms are derived from PSI. In phase-stepping algorithms, phase is stepped by special or equal amount between each intensity measurement[3]. There are many correlative algorithms are proposed, but in this paper, we just emphasize the analysis of three kinds of algorithms in common use, including carré algorithm (four-stepping algorithm), Hariharan algorithm (five-stepping algorithm), compensating five-stepping algorithm, and introduce three-stepping algorithm briefly. Afterward respectively detailed step error analysis (usually from PZT) and the signal analysis when adding noise (usually from CCD) are provided in this paper.

The principle of low coherence interferometry is given out firstly. All the works have been done by a numerical simulation in low coherence interferometry, considering as the object a slightly tilted mirror located within the maximal coherence distance, and the spectrum of light source is in Gaussian shape. By comparing the influence of step error and noise error on phase and amplitude measurement between these algorithms, we could find that carré algorithm, Hariharan algorithm, and compensating five-stepping algorithm are all the very mature algorithms, but they are suitable for different conditions. And as we known, the imaging distance between two interference beams should be equal or almost equal. If any deviation occurred, the quality of imaging will degrade greatly, because the amplitude information is limited by the Gaussian autocorrelation fringe. But for the phase measurement, it should not have any infection. Through comparing of the signal by adding different amount noise of each algorithm, the fact why the equal light path is so important in low coherence interferometry is show out.

2. Principle of Low Coherence Interferometry

The electric field E could be expressed as a complex exponential[4]:

$$E(\omega, t) = s(\omega)\exp[-i(\omega t + kz)] \tag{1}$$

The electric fields of incidence, sample beam, reference beam and output beam could be expressed as E_{in}, E_s, E_r, E_{out} respectively:

$$E_{in}(\nu, t) = s(\nu)e^{-i2\pi\nu t} \tag{2}$$

$$E_s(\nu, t) = (T_r T_s)^{1/2} E_{in}(\nu, t) H(\nu) \tag{3}$$

$$E_r(\nu, t, \tau) = (T_r T_s)^{1/2} E_{in}(\nu, t) e^{-i\phi(\tau)} \tag{4}$$

$$E_{out}(\nu, t, \tau) = E_s(\nu, t, \tau) + E_r(\nu, t) \tag{5}$$

The intensity of interference signal recorded by detector is:

$$\begin{aligned} I(\nu, \Delta z) &= \left\langle E_{out}(\nu, t, \tau) E^*_{out}(\nu, t, \tau) \right\rangle \\ &= \lim_{T\to\infty} \frac{1}{2T} \int_{-T}^{T} E_{out}(\nu, t, \tau) E^*_{out}(\nu, t, \tau) dt \end{aligned} \tag{6}$$

$$I(\nu,\tau) = T_r T_s |S(\nu)H(\nu)|^2 + T_r T_s |S(\nu)|^2 + 2T_r T_s R\{S(\nu)^* S(\nu) H(\nu) e^{-i\phi(\tau)}\} \quad (7)$$

In this paper, we choose a mirror as sample simply, so assumed $H(\nu)=1$.

The spectrum of light source S(v) has a Gaussian shape that[5]

$$S(\nu) = \frac{2\sqrt{\ln 2/\pi}}{\Delta\nu}\exp[-4\ln 2(\nu-\bar{\nu})^2/\Delta\nu^2] \quad (8)$$

The intensity of interference signal is then

$$I(\nu,\tau) = I_S + I_R + I_T \exp[-\left(\frac{\pi\Delta\nu\tau}{2\sqrt{\ln 2}}\right)^2]\cos(2\pi\bar{\nu}\tau) \quad (9)$$

Considering the practical condition, we assume the center wavelength of light source is 600nm, and the FWHM is 300nm. The following analyses are all based on this principle of low coherence interferometry. The object is considered as a slightly tilted mirror (the phase change corresponding to the continuously variety of points on mirror is consecutive), and the interference fringe is present in Fig1.. For the phase measurement, we could express the intensity of interference signal simply as[6]:

$$I = I_0 + A\cos(\phi+\varphi) \quad (10)$$

cos(.) represents the interference pattern, ϕ is the original phase difference between the two beams of the interferometry, φ is the additional phase shift.

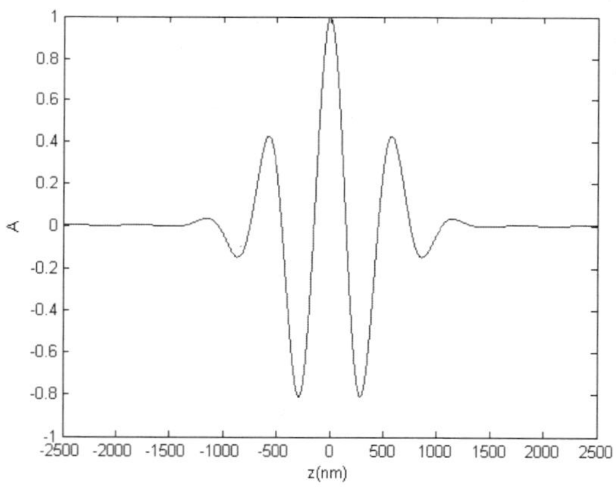

Fig1. The interferometric image of tilted mirror. Assuming the center wavelength of light source is 600nm, and the FWHM is 300nm.

3. Phase-stepping Algorithms:

To retrieve the phase and amplitude reflected by object hidden in interference fringes, expressed as $\cos(\phi)$ and A respectively, several interferograms have to be taken by digital imaging system.

3.1 Three-stepping Algorithm[2]

$$A = 1/2[(I_3 - I_2)^2 + (I_2 - I_1)^2]^{1/2} \quad (11)$$

This algorithm is correct only for exact 90degree phase steps. Because no angle error compensation is provided, if there is any deviation from the 90degree, big error will occur. To solve the problem, phase-shift angle should be known in advance, and then more phase-shift images have to be acquired. So this paper emphasizes the analysis of four-, five-stepping and compensating five-stepping algorithm.

3.2 Four-stepping Algorithms (Carré Algorithm)

This algorithm requires four measurements of the intensity at each point corresponding to three equal phase steps, needn't has to be the special phase angle 90degree. So we consider it's exactly for angle compensation[2][6].

$$\begin{cases} I_1 = I_0 + A\cos[\phi - 3\varphi] \\ I_2 = I_0 + A\cos[\phi - \varphi] \\ I_3 = I_0 + A\cos[\phi + \varphi] \\ I_4 = I_0 + A\cos[\phi + 3\varphi] \end{cases} \quad (12)$$

$$A = [(\frac{I_1 - I_4 + I_2 - I_3}{8\sin\varphi\cos^2\varphi})^2 + (\frac{I_1 + I_4 - I_2 - I_3}{8\cos\varphi\sin^2\varphi})^2]^{1/2} \quad (13)$$

$$tg\varphi = [\frac{3(I_2 - I_3) - (I_1 - I_4)}{I_1 - I_4 + I_2 - I_3}]^{1/2} \quad (14)$$

$$tg\phi = (I_1 + I_2 - I_3 - I_4)/(I_1 - I_2 - I_3 + I_4) * tg\varphi \quad (15)$$

But if the original phase difference between the two beams is closed to mπ, where m is an integer, the numerator and denominator of Eqs. (14) and (15) tend to be zero. Five-stepping algorithm could avoids this problem.

3.3 Five-stepping Algorithm (Hariharan Algorithm)

Five intensity measurements are needed for this algorithm. Formulas are exact when phase shift is 90degree, when shift region near 90degree, the error both in ϕ and A are small, because the first-order error terms is canceled. [2][3]

$$\begin{cases} I_1 = I_0 + A\cos[\phi - 2\varphi] \\ I_2 = I_0 + A\cos[\phi - \varphi] \\ I_3 = I_0 + A\cos[\phi] \\ I_4 = I_0 + A\cos[\phi + \varphi] \\ I_5 = I_0 + A\cos[\phi + 2\varphi] \end{cases} \quad (16)$$

$$\tan\phi = \frac{2[I_2 - I_4]}{2I_3 - I_1 - I_5} \quad (17)$$

$$A = 1/4[4(I_2 - I_4)^2 + (2I_3 - I_1 - I_5)^2]^{1/2} \quad (18)$$

3.4 Compensating Five-stepping Algorithm

Comparing to the five-stepping algorithm, this algorithm introduce a term $\sin\varphi$. So that the formulas are exact when the phase shifts needn't have to be the 90degree. The essential background to this exact compensating five stepping algorithm is present in the paper of Hariharan et al.[3] firstly, but it is not shown explicitly[2].

$$\sin\varphi = \frac{[4(I_2 - I_4)^2 - (I_1 - I_5)^2]^{1/2}}{2(I_2 - I_4)} \quad (19)$$

$$\tan\phi' = \frac{2[I_2 - I_4]}{2I_3 - I_1 - I_5} * \sin\varphi \quad (20)$$

$$A = 1/4[4((I_2 - I_4)/\sin\varphi)^2 + ((2I_3 - I_1 - I_5)/\sin^2\varphi)^2]^{1/2} \quad (21)$$

Supposing phase shifting angle has a deviation to 90degree, comparing to the algorithm above, the compensating five stepping algorithm has exact error compensation, but Hariharan algorithm only has partial (first order) error compensation[3]:

$$\varphi = (\pi/2) + \varepsilon \quad (22)$$

$$\tan\phi' = [1 - (\varepsilon^2/2)]\tan\phi \quad (23)$$

4. Result of Numerical Simulation

A simple way to offer phase shifts in an interferometric system is to mount one of the mirrors of the interferometry on a piezoelectric transducer (PZT) and supply suitable voltages to the PZT. But the nonlinear response of PZT always gives a series of phase-shift that deviate from what we set, and sometimes we could not sure if the voltage is exactly fit to the value we want[3]. So it is necessary to make the analysis of amplitude measurement and phase measurement in three algorithms for step error, corresponding to the step error of PZT appears in experiment. According to presentation of the three algorithm, the amplitude range of five-stepping algorithm at 90 degree step size and four-compensating five-stepping algorithms at arbitrary degree will be all right, they are shown in Fig2., and the reconstruction of z (corresponding to the continuously variation points on mirror) are shown in Fig3.

The influences of equal step size deviation from 90degree are shown in Fig4, 5, and the parameter rms (root mean square) are also given as an assistant criterion.

$$rms = \sqrt{\frac{\sum\Delta^2}{N}} \quad (24)$$

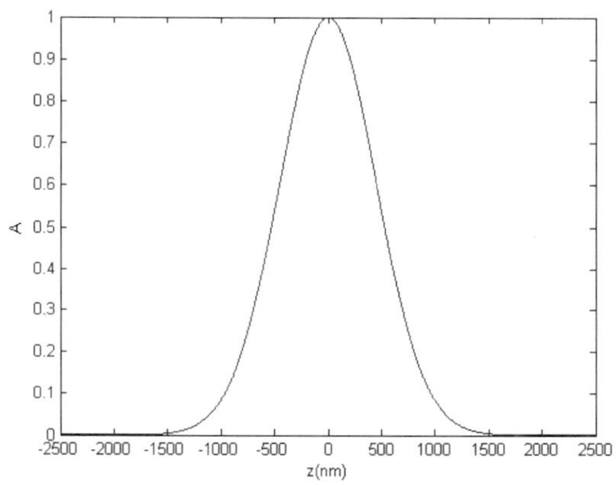

Fig2. Amplitude measurement of Gauss shape.

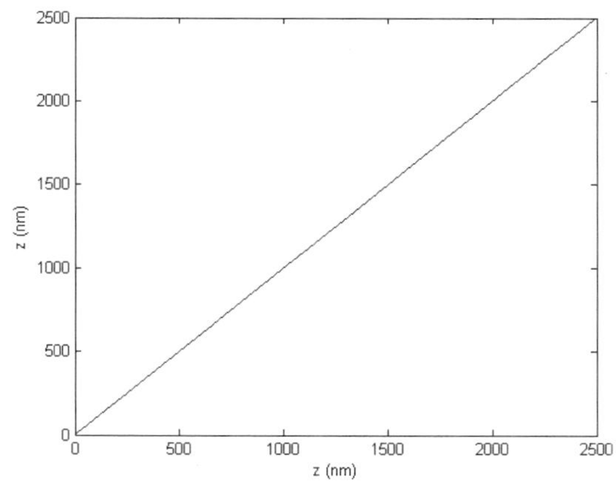

Fig3. Reconstruction of z corresponding to the continuously points on the mirror.

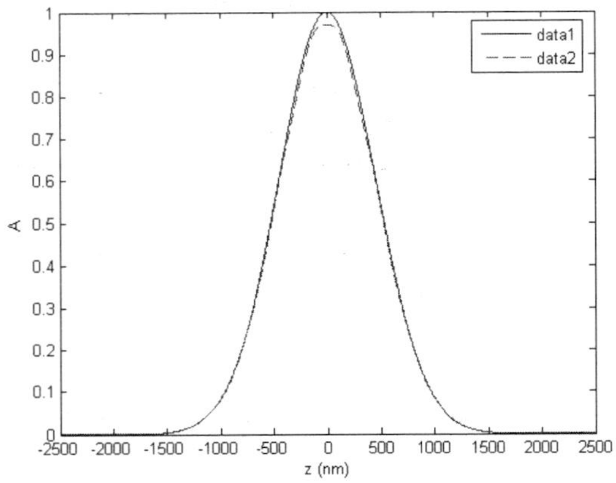

Fig4. the amplitude calculated by exactly compensating algorithms is shown as date1. Amplitude calculated by five-stepping algorithm at100^0 step size is presented as date2.

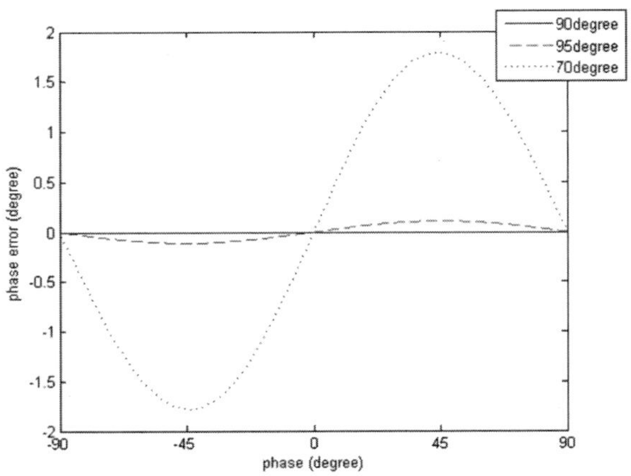

Fig5. Phase error against the phase difference in the phase step of 90^0(rms=0), 95^0(rms=0.077), 70^0(rms=1.2598) respectively.

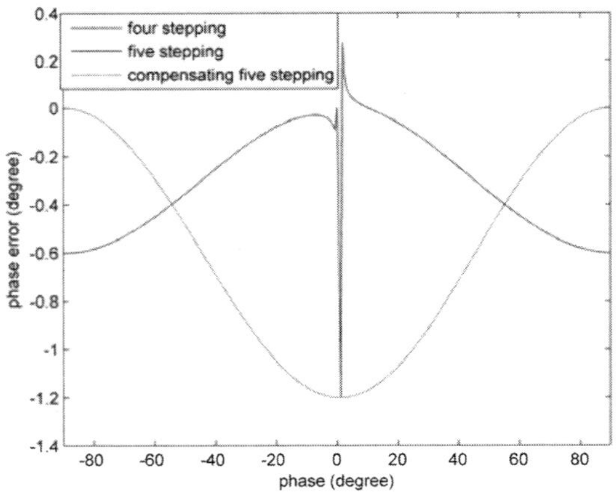

Fig6. Phase error against the phase difference for a PZT with deviations from linearity of 1.2degree at nominal phase shifts of 900. Real line show the result by four-stepping algorithm, rms=0.376. Broken line show the result by five-stepping algorithm, and compensating five-stepping algorithm (they are the same), rms=0.735.

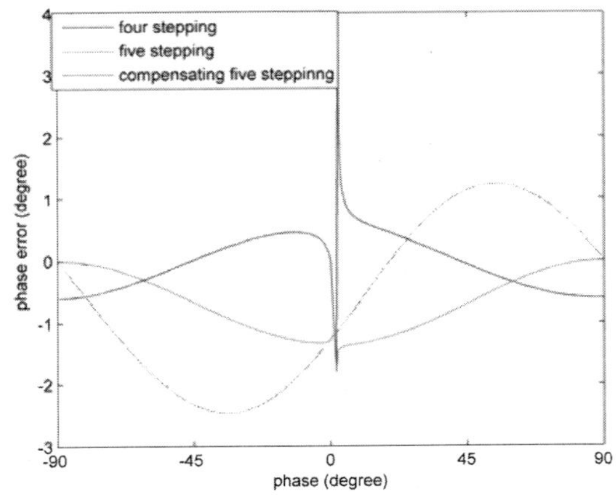

Fig7. Phase error against the phase difference for a PZT with deviations from linearity of 1.2degree at nominal phase shifts of 70^0. Real line show the result by four-stepping algorithm, rms=0.504. Point line show the result by 5 stepping algorithm, rms=1.440. Broken line show the result by compensating 5stepping algorithm, rms=0.832.

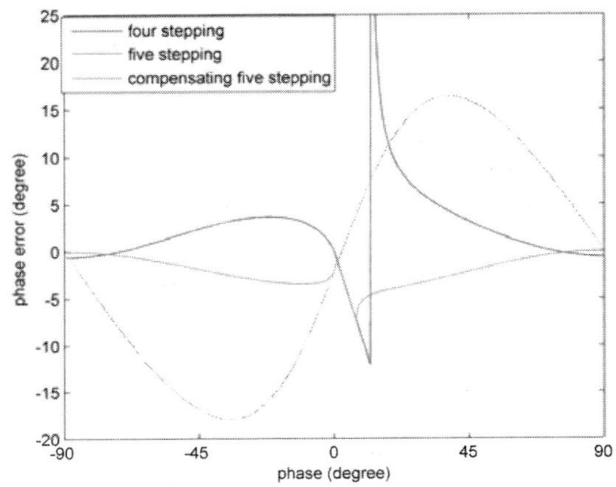

Fig8. Phase error against the phase difference for a PZT with deviations from linearity of 1.2degree at nominal phase shifts of 30^0. Real line show the result by four-stepping algorithm, rms=6.796. Point line show the result by five-stepping algorithm, rms=12.111. c. show the result by compensating five-stepping algorithm, rms=2.577.

For amplitude and phase measurement, when phase shifting is not sure to be 90^0, compensating five-stepping algorithm and four-stepping algorithm will be exactly correct, the rms=0. But the error of five-stepping algorithm could be small when the shifting range near 90^0.

The phase error against phase difference with deviation of 1.2^0 differences between each phase shifts by three algorithms are shown in Fig6, 7, 8 (for example: in five-stepping algorithm, if the phase shift is set to 90^0, the actual phase shifts are 88.8^0, 91.2^0, 88.8^0, 91.2^0 respectively).

By comparing the Fig6, 7, 8, we could think that the errors because of different distance between each step are smaller when phase shifting is near 90 degree. Four-stepping algorithm should be the first choice (but it is complex). When the phase shifting is far away from 90^0, compensating 5 stepping algorithm will be the better choice, and then 4 stepping algorithm second.

Otherwise noise would exist in every imaging system, usually from the CCD. So adding noise to each intensity measurement to analyze the noise influence to the reconstruction of amplitude and phase is necessary. In Fig9, 10, 11, we assuming the signal is -20dB, the phase error against the distance of light path mismatch by 4-, 5-, compensating 5 stepping-algorithms are simulated and the distance (d for short) in which we could reconstruct phase signal is recorded in every figure with S/N=90dB, 60dB, 30dB respectively.

Comparing the d and rms in Figure9, 10, 11 above, 5-stepping algorithm will show better result than the others. And we could easily find that the distance in which we could reconstruct the phase signal will be narrower as the noise increase, which is the reason why

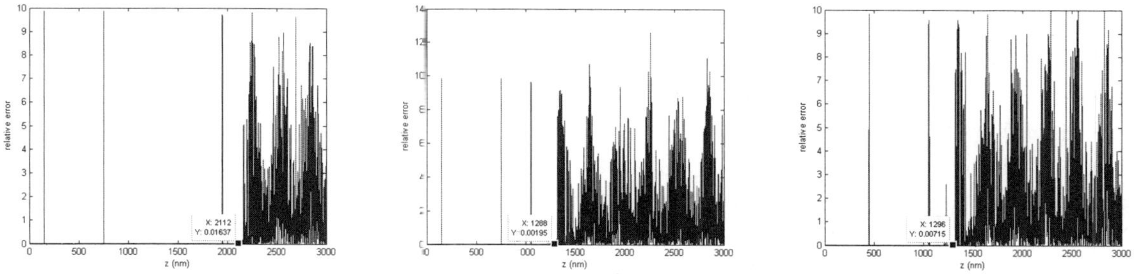

Figure9. a Figure9. b Figure9. c

Fig9. Calculated by 4-stepping algorithm. a. S/N=90dB, d=2.112nm, rms=0.693. b. S/N=60dB noise, d=1.739nm, rms=0.846. c. S/N=30dB, d=1.296nm, rms=0.966.

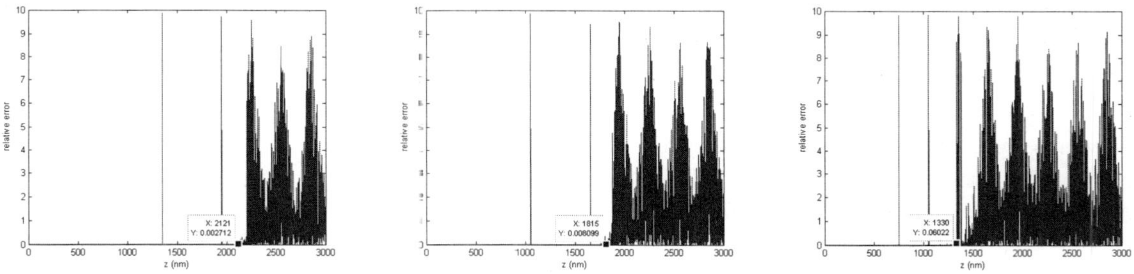

Figure10. a Figure10. b Figure10. c

Fig10. Calculated by 5-stepping algorithm. a. S/N=90dB, d=2.121nm, rms=0.661. b. S/N=60dB noise, d=1.815nm, rms=0.816. c. S/N=30dB, d=1.330nm, rms=0.947.

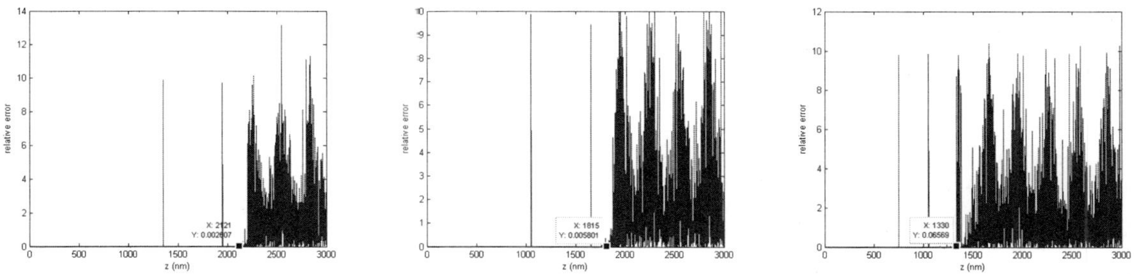

Figure11. a Figure11. b Figure11. c

Fig11. Calculated by compensating 5-stepping algorithm. a. S/N=90dB, d=2.121nm, rms=0.726. b. S/N=60dB noise, d=1.815nm, rms=0.816. c. S/N=30dB, d=1.330nm, rms=1.005.

the mismatch of two beams is so important for phase reconstruction in low coherence interferometry. This could be a reference point for experiment analysis.

5. Discussion and Conclusions

In this paper, we emphasize the analysis of three kinds of algorithms including carré algorithm, Hariharan algorithm and compensating five-stepping algorithm. They are all mature algorithms and in common use, but by comparing the influence of step error and noise error on phase and amplitude measurement between these algorithms, we also could find that they have a few differences. Step error usually from PZT, and noise error usually from CCD or system, we could choose suitable algorithm according to different condition. If biggish error, Carré algorithm should be the first choice, but when the phase shifting is far away from 90 degree, compensating five-stepping algorithm will be a better choice. When larger noise exist in imaging system, Hariharan algorithm will show better result. And as we known, the imaging distance between two interference beams should be equal or almost equal. If any deviation occurred, the quality of imaging will degrade greatly, because the amplitude information is limited by the Gaussian autocorrelation fringe. But for the phase measurement, it should not have any infection. Through comparing of the signal by adding different amount noise of each algorithm, a fact was revealed that the distance in which we could reconstruct the phase signal will be narrower as the noise increase. All the works have been done by a numerical simulation in low coherence interferometry, the principle of low coherence interferometry is given out in part 2. We hope this work would be helpful for experiment.

Acknowledgements

We gratefully acknowledge the financial support by the National Nature Science Foundation of China (grant No.30727002).

References

1. A. Dubois, J. Opt. Soc. Am. A. **18**, 1972 (2001).
2. K. G. Larkin, J. Opt. Soc. Am. A. **13**, 832 (1996).
3. P.Hariharan, B. F. O., and T.Eiju, Applied Optics. **26**, 2504 (1987).
4. P. H. Tomlins and R. K. Wang, Appl. Phys. **38**, 2519 (2005).
5. J. M. Schmitt, IEEE Journal of Selected Topics in Quantum Electronics, **5**, 1205 (1999).
6. Shoude Chang, Sherif Sherif and Costel Flueraru, Proceedings of SPIE **6343,** (2006).
7. Christopher Fang-Yen, Seungeun Oh, Yongkeun Park, et al., Optics Letters. **32**, 1572 (2007).

Topic III

Multiphoton Microscopy in Biomedical Sciences

NONLINEAR OPTICAL MICROSCOPY OF MOUSE ORAL MUCOSA

SHUANGMU ZHUO, JIANXIN CHEN*, XINGSHAN JIANG, ZUFANG HUANG AND SHUSEN XIE

Key Laboratory of Optoelectronic Science and Technology for Medicine (Fujian Normal University), Ministry of Education, Fuzhou 350007, China

Nonlinear optical microscopy (NOLM) is employed in this study to non-invasively elucidate the layered–resolved microstructure and spectroscopy of mouse oral mucosa. It was found that NOLM is effective for acquiring the layered–resolved microstructure and spectroscopy of mouse oral mucosa, providing quantitatively information on the biomorphology and biochemistry of tissue. Our results suggest that the NOLM technique has potential to provide more important and comprehensive information for early diagnosis of oral cancer.

1. Introduction

Most people are surprised to learn that one American dies every hour from oral cancer; a death rate that has remained virtually unchanged for more than 40 years [1, 2] In fact, recent statistics published by the American Cancer Society indicate that while the incidence and death rates for cancers overall has decreased, the incidence of oral cancer has increased by 5.5% and the death rate has increased by 1.5% [1, 2]. Oral cancer is far too often discovered in late stage development, the primary reason for the consistently high death rate. Oral cancer treatment often results in disfiguring effects on patients, and can seriously compromise their quality of life. Early detection and diagnosis can make a tremendous difference in life expectancy; oral cancer is 90% curable when found in its early stages [1, 2]. So, early detection of oral cancer and its curable precursors remains the best to ensure patient survival and quality life. During the development of oral precancer, not only epithelial cells undergo pathological transformation that give rise to increased metabolic activity, increased cell density and the variability in nuclear shape and size, and overall thickening of the epithelial layer, but also the stroma can be changed, such as collagen loss in malignancy [3-5]. So, obtaining the information of these morphological and biochemical changes in various layers is of great value for the diagnosis of oral cancer at an early stage.

A promising imaging modality is nonlinear optical microscopy (NOLM). Based on two-photon fluorescence (TPF) or second-harmonic generation (SHG), NOLM has been widely applied to diverse areas of the life sciences, including neurobiology, development biology, oncology, immunology, cell biology, and deep-tissue imaging [6-14]. NLOM is a non-destructive, laser scanning microscopy technique that renders images of living tissue with subcellular resolution and component specificity without use of exogenous stains or dyes. Using backreflected experimental geometry, NLOM is not limited by sample thickness. NLOM is very sensitive in defining cellular and extracellular structures as well as metabolic assessment [6-14].

Our primary goals in this study are to investigate layered–resolved microstructure and spectroscopy of mouse oral mucosa using the NOLM technique and to extract the comprehensive information for potential oral diagnostics.

2. Materials and Methods

2.1. *Preparation of Sample*

The fresh oral mucosa tissues of experimental mice were examined in this study. The fresh specimens were excised from live experimental mice, provided by the Animal Center of Fujian Medical University. Firstly, they were rinsed briefly with PBS solution (PH 7.4) to remove the residual blood on their surfaces. Then oral mucosa tissues sandwiched between the microscope slide and a piece of the cover glass with the keratinizing layer facing the objective. Moreover, to avoid dehydration or shrinkage during the whole imaging process, the specimen was sprinkled with PBS solution.

* Address all correspondence to Jianxin Chen, Tel: +86-591-22686078; Fax: +86-591-83465373; E-mail: chenjianxin@fjnu.edu.cn

2.2. *Nonlinear Optical Microscopy (NLOM)*

The nonlinear optical microscopy system used in this study has been described previously [8-13].The system is an Axiovert 200 microscope (Zeiss LSM 510 META) equipped with a mode-locked femtosecond Ti: sapphire laser (110fs, 76MHz), tunable from 700nm to 980nm (Coherent Mira 900-F). The polarization direction of the laser light is the horizontal polarization. An Acousto-Optic Modulator (AOM) is used to control the laser intensity attenuation. A galvanometer-driven optical scanner (Zeiss) is used to scan the beam through the objective and across the sample generating images. A Plan-Apochromat 63× (N.A.=1.4) oil immersion objective (Zeiss) was employed for focusing the excitation beam and for collecting of the backward signals. We detected all signals in the backward direction. The signals were directed by a main dichroic beam splitter (MDBS) to the META detector. The META detector consists of high-quality, reflective grating as a dispersive element and an optimized 32-channel PMT array to collects the dispersed output in 10.7nm samples across the visible spectrum (377-716nm). An IR beam block filters (Zeiss KP650), which is in front of META detector, was be used to ensure that excitation light was filtered out and only emission signals were recorded. An optional HRZ 200 fine focusing stage (HRZ 200 stage, Carl Zeiss) is used to change the focus position, allowing for imaging at various depths. The LSM 510 META laser scanning microscopy is controlled via a standard high-end Pentium PC and linked to the electronic control system via an ultrafast SCSi interface. This system has two especial modes: Channel mode and Lambda mode. These two modes use the same detector, the META detector. The Channel mode has many independent-channels (maximum up to 8) and each channel can selectively be set to detect emission signals within the random range from 377 to 716nm to achieve imaging. In our study, in keratinizing layer one channel (382-618nm) is used to image keratinocytes; in the epithelium one channel (425-618nm) is employed for image epithelial cells; in the stroma, one channel (414-436nm) is used to obtained SHG images. The Lambda mode can obtain emission spectra by plotting the mean intensity of all pixels with the image versus the center wavelength of each emission band. We used this mode to emission spectra of various layers. The obtained spectra have been corrected for the wavelength dependent instrument response and dark noise spectra are subtracted from the acquired sample spectra. Each spectrum is normalized to the maximal peak intensity. Using our setting, about 1s of exposure were necessary for the alternation of two modes. In this work, all images were 512×512 pixels and have a 12-bit pixel depth. The images were acquired at $2.56\mu s$ per pixel. The average laser power at the specimen was maintained $\leq 10mW$ and no photobleaching was observed at this low power level.

3. Results and Discussion

3.1. *Characterization of Keratinizing Layer*

To characterize microstructure and spectral features of the keratinizing layer, the multiphoton images and corresponding spectra excited at 810nm were obtained and shown in Fig.1(a, b). As can be seen, this layer is consist of some keratinocytes without nuclei (regular polygons), which is in excellent agreement with previous histological observations [15].

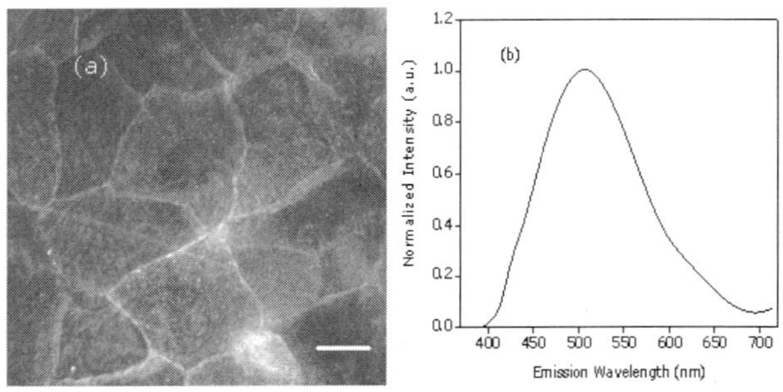

Figure 1. (a) High-resolution multiphoton image of keratinocytes at depth of $2\mu m$ for $\lambda_{ex} = 810nm$. (b) Emission spectra recorded from the keratinizing layer at depth of $2\mu m$ $\lambda_{ex} = 810nm$ .Scale bar $20\mu m$.

For emission signal, a 510nm peak was acquired. According to previous study [16], the 510nm peak corresponds to keratin emission spectra. Moreover, with the increasing of imaging depth, keratinocytes are gradually replaced by individual epithelial cells. Using the optional HRZ 200fine focusing stage, the images and signals in the transition zone at the depths of $10-15\mu m$ was observed. The demonstration of the keratinocytes in oral mucosa is valuable for analysis and comparison of the keratinization processes, providing an additional morphological substrate for the identification and classification of mucosal premalignant and malignant lesions.

3.2. *Characterization of Epithelium*

To demonstrate cellular microstructures of epithelial layer, multiphoton images were obtained at various depths, z, into the whole epithelial layer ($10-45\mu m$) for an excitation wavelength $\lambda_{ex}=730nm$, which is preferred for imaging cellular structures [17] In Fig.2 (a), the epithelia squamous cells can be identified through the fluorescent cytoplasm and dark non-fluorescent nuclei (arrow heads) without any artifact, which always occurs in the histological processing disturb the topography of cells. The cytoplasm is filled with the fluorescent granules stemming from the intrinsic mitochondrial NADH. Owing to the absence of suitable fluorophore at $\lambda_{ex}=730nm$, the nuclei surrounded by the fluorescent granules remain non-fluorescent and appear as dark regions in figures. The cells connect with each other through cellular processes, ranging from replication and energy conversion to molecule transport and the various signaling used in cellular communication. Moreover, the cellular nucleus area could be quantitatively analyzed. For example, the cellular nucleus area in $z=15\mu m$ is $\approx 61.9\mu m^2$, as was shown in Fig.2 (a).

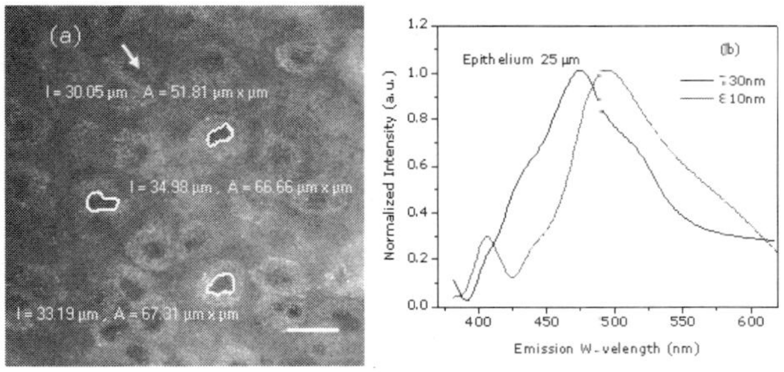

Figure 2. Subcellular-spatial-resolution multiphoton images of oral epithelium at depth of $15\mu m$ for excitation wavelength of $730nm$. The fluorescent granules of cytoplasmatic mitochondria and the non-fluorescence dark nuclei [arrow heads] can be here perfectly identified without any staining assistance. Scale bar $20\mu m$. (b) Epithelium emission spectra at $15\mu m$. The excitation wavelengths used were 730 and 810nm.

The corresponding spectra from the epithelium was acquired at $\lambda_{ex}=730, 810nm$ and shown in Fig.2 (b). As can be seen, an interesting result was also obtained, that is, a small SHG signal from the epithelium. In Fig.2 (b), for $\lambda_{ex}=730$, we could not acquire the whole SHG signal due to our system's limitations described in Materials and Methods, but the part of SHG signal at near 377nm could be seen; for $\lambda_{ex}=810nm$, an obvious SHG signal was observed. Because DNA in chromatin typically has large molecular anisotropy and second-order nonlinearlity [18], we believe that a small SHG signal is associated with the presence of DNA. To the best of our knowledge, SHG signal from DNA in the unstained thick tissue for the first time obtained in backscattering geometry.

3.3. *Characterization of Stroma*

To selectively study stromal collagen, SHG images and corresponding spectra were obtained at various depths, z, into the stroma ranging from 45 to 100 μm. Figures 3(a) illustrate SHG images of stroma at depths of 90 μm, for $\lambda_{ex}=850nm$, at which almost no TPA signal from collagen was detected and sample photodamage is minimal due to no molecular transition [5-7], depicted by arrow heads in Fig.3(b). As can be seen, the distribution and orientation of collagen lamellas are visible. In this work, the SHG signal showed no signs of bleaching during acquisition of repeated images from area, showing that no damage to the collagen structure was occurring. This is to be expected since SHG is coherent process and no energy is lost.

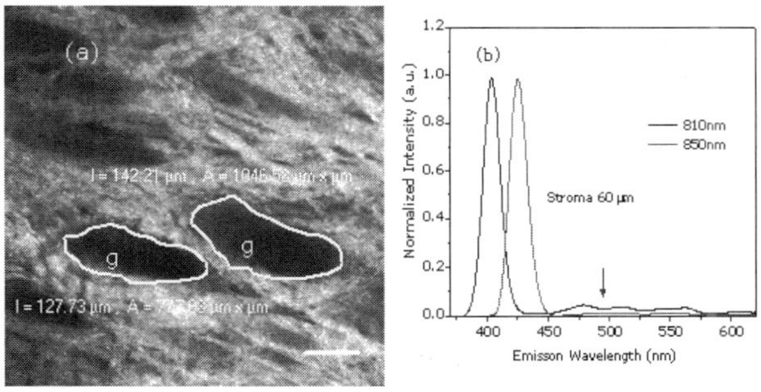

Figure 3. (a) SHG images from stroma at depths of 90 μm for $\lambda_{ex}=850nm$. (b) Emission spectral from stroma at depth of 90 μm for $\lambda_{ex}=810$, $850nm$ .Scale bar $20\mu m$.

The study of collagen is of crucial importance in the anatomy and pathology. NLOM has promise for imaging collagen organization, and provides higher resolution and more detailed images of collagen organization and distribution, and obtains quantitative collagen structural and biochemical information. The technique has the potential to become the most valuable tools in microscopy of collagen and could be also employed to inspect collagen-related changes such as epithelial disease, skin cancer, burn depth and severity, and extracellular matrix related disorders. Moreover, some minor salivary glands were observed. As can be seen, some minor salivary glands (depicted by "g" symbol) are surrounded collagen and display the morphology of concave polygon. Moreover, the area of concave polygon can be measured. For example, two minor salivary glands in Fig.3 (a) were outlined, and the areas are approximately $1046.58\mu m^2$ and $777.98\mu m^2$, respectively. Because pathological transformation of normal mucosal tissue often is accompanied by specific alterations in minor salivary glands microstructure and function [19], selective displaying of minor salivary glands is of substantial interest.

4. Conclusions

The nondestructive selective visualization of various microstructural components and the characterization of spectra features in oral mucosa play a critical role in the diagnosis of oral cancer at an early stage. In this study, we used the nonlinear optical microscopy to acquire layered-resolved microstructure and spectroscopy of mouse oral mucosa. Multiphoton images and spectra obtained from mucosa were shown to provide excellent microstructural and biochemical characterization of the oral tissue. With additional developments, we foresee potential applications of the nonlinear optical microscopy only with different excitation wavelengths in the diagnosis of oral cancer at an early stage.

Acknowledgments

The project was supported by the National Natural Science Foundation of China (No. 60508017, 60678054), the Scientific Program of the Educational Hall of Fujian Province (NO.JA05215), the Natural Science Foundation of Fujian Province of China (2007J0007, C0720001) and Program for New Century Excellent Talents in University (2007).

References

1. American Cancer Society. Cancer Facts and Figures, *Atlanta: American Cancer Society* (2005).
2. American Cancer Society. Cancer Facts and Figures, *Atlanta: American Cancer Society* (2006).
3. M. C. Skala, J. M. Squirrell, K. M. Vrotsos, J. C. Eickhoff, A. Gendron-Fitzpatrick, K.W. Eliceiri, and N. Ramanujam, *Cancer. Res.* **65**, 1180 (2005).
4. R. K. Rebecca and S. M. Eva, *Annu. Rev. Phys. Chem.* **47**, 555 (1996).
5. J. X. Chen, S. M. Zhuo, R. Chen, X. S. Jiang, S. S. Xie, and Q. L. Zou, *New. J. Phys.* **9**, 2121 (2007).
6. P. J Campagnola and L. M. Loew, *Nat. Biotechnol.* **21**, 1356 (2003).
7. A. Zoumi, A. Yeh, and B. J. Tromberg, *Proc. Natl. Acad. Sci. USA.* **99**, 11014 (2002).
8. S. M. Zhuo, J. X. Chen, X. S. Jiang, X. C. Cheng, and S. S. Xie, *Skin. Res. Techno.* **13**, 406 (2007).
9. S. M. Zhuo, J. X. Chen, X. S. Jiang, S. S. Xie, and Q. L. Zou, *Scanning.***29**, 219 (2007).
10. J. X. Chen, S. M. Zhuo, T. S. Luo, X. S. Jiang, J. J. Zhao, *Scanning.* **28**, 319 (2006).
11. J. X. Chen, S. M. Zhuo, T. S. Luo, and J. J. Zhao, *Chin. Opt. Lett.* **4**, 598 (2006).
12. S. M. Zhuo, J. X. Chen, X. S. Jiang, S. S. Xie, R. Chen, N. Cao, Q. L. Zou, and S. Y. Xiong, *Phys. Med. Biol.* **52**, 4967 (2007).
13. S. M. Zhuo, J. X. Chen, T. S. Luo, D. S. Zou, and J. J. Zhao, *Opt. Express.* **14**, 7810 (2006).
14. W. R. Zipfel, R. M. Williams, and W. W. Webb, *Nat. Biotechnol.* **21**, 1369 (2003).
15. A. Keszler, R. L. Cabrini, *J. Oral. Pathol.***12**, 330 (1983).
16. A. M. Pena, M. Strupler, T. Boulesteix, and M. C. Schanne-Klein, *Opt. Express.* **13**, 6268 (2005).
17. W. R. Zipfel, R. M. Williams, R. Christie, A. Y. Nikitin, B. T. Hyman, and W. W. Webb, *Proc. Natl. Acad. Sci. USA.* **100**, 7075 (2003).
18. R. Gauderon, P. B. Lukins, C. J. R. Sheppard, *Micron.* **32**, 685 (2001).
19. J. M. Crawford, M. A. Taubman, D. J. Smith, *science.* **190,** 1206 (1975).

Topic IV

Photonic Therapeutics, Diagnostics and Instrumentations

REAL-TIME MONITORING OF MITOCHONDRIAL FUNCTION AND CEREBRAL BLOOD FLOW FOLLOWING FOCAL ISCHEMIA IN RATS

LIVNAT AMIR, BARBIRO-MICHAELY EFRAT, MAYEVSKY AVRAHAM

The Mina & Everard Goodman Faculty of Life-Sciences and The Gonda Multidisciplinary Brain Research Center
Bar-Ilan University
Ramat-Gan, 52900, Israel
Email: amirtolaat@yahoo.com

Introduction: Focal ischemia due to reduction of cerebral blood flow (CBF) creates 2 zones of damage: the core area, which suffers severe damage, and penumbra area, which surrounds the core and suffers intermediate levels of injury. Objectives: A novel method is introduced, which evaluates mitochondrial function in the core and in the penumbra, during focal cerebral ischemia. Methods: Wistar rats underwent focal cerebral ischemia by middle cerebral artery occlusion (MCAO) for 60 minutes, followed by 60 minutes of reperfusion. Mitochondrial function was assessed by a unique Multi-Site – Multi-Parametric (MSMP) monitoring system, which measures mitochondrial NADH using flourometric technique, and CBF using Laser Doppler Flowmetry (LDF). Results: At the onset of occlusion, CBF dropped and NADH increased significantly only in the right hemisphere. CBF levels were significantly lower and NADH significantly higher in the core than in the penumbra. After reperfusion, CBF and NADH recovered correspondingly to the intensity of ischemia. Conclusion: Application of the MSMP system can add significant information for the understanding of the cerebral metabolic state under ischemic conditions, with an emphasis on mitochondrial function.

1. Introduction

Brain focal ischemia involves a shortage of cerebral blood flow (CBF).[4,12] Since the brain function is completely dependent on a continuous blood supply, CBF deterioration leads to a disruption of cerebral oxygen balance (supply versus demand) resulting in diverse changes in the tissue, creating a new microenvironment in the ischemic area.[6,9,14] These changes include the disruption of extra-cellular ion balance, increased Lactate levels, decrease in extra-cellular pH, Glutamate and free radicals accumulation and suppression of spontaneous electrical activity.[2-4,12,14] Oxygen shortage in the brain affects the intracellular respiratory chain activity, resulting in a decrease in adenosine triphosphate (ATP) synthesis, and an increase in mitochondrial NADH levels.[6,15] Due to its high sensitivity to intracellular oxygenation, mitochondrial NADH level can serve as an indicator of tissue pathological state.[6,8]

Brain focal ischemia entails gradual injury, spreading from severe injury in the lesion center (core) toward moderate damage in periphery regions (penumbra), due to collateral blood supply,[4] affecting the energy balance and leading to mitochondrial injury.[14,15] In reperfusion, mitochondrial recuperation is more clearly observed in the core than in the penumbra.[13,15]

The present study introduces a novel method, which combines the evaluation of mitochondrial function and cerebral blood flow in the core and in the penumbra, during and following focal cerebral ischemia of various severity.

2. Methods

The monitoring of the rat brain was performed by a unique Multi-Site–Multi-Parametric (MSMP) monitoring system, designed for real-time, continuous measurements.[1,10] The MSMP system includes 2 channels, each comprising a probe, with a bundle of optical fibers. Each probe combines NADH monitoring using the fluorometric technique, and cerebral blood flow (CBF) monitoring using Laser Doppler Flowmetry (LDF), as specified in previous articles.[1,7,10]

The experiments were performed on male Wistar rats (200-300 g.), according to the NIH guidelines for the care and use of laboratory animals, and approved by the Institutional Animal Care committee of Bar-Ilan University. Animals were anesthetized by an IP injection (0.3ml/100g) of Equithesin,[1] and additional injections were given every 30 minutes during monitoring (0.1ml/100g). A heating pad was placed under the rat to maintain body temperature at 37°C.

Each animal was placed in a ventral position, and the skull was exposed by a midline incision. Holes were drilled in the parietal bones for the cannula/s, in which the probes were placed. The cannulas were placed epidurally and fixated to the skull using dental acrylic cement, with additional screws for better fixation. Two models of cerebral measurement during and after MCAO were developed (fig. 1). In the bilateral model (n=8), 2 probes were placed in parallel locations on both hemispheres (each on a different hemisphere), 1mm posterior and 5 mm lateral to the Bregma. In the unilateral model (n=9), both probes were placed on the right hemisphere in medial and lateral positions, using a double cannula that was placed 1mm posterior and 1mm lateral to the Bregma. At this point, the rat was turned over on its back, and the MCAO surgery was performed on the right artery, according to the protocol of Longa *et al.*,[5] with several adaptations. A 4/0 nylon filament, with a heated bulb head covered with Poly-L-Lysine, was inserted through a puncture to the external carotid artery, onto the common carotid artery. The MSMP probes were then placed in their cannulas, and the monitoring began. After a short anoxia, the nylon filament was removed from the common carotid artery and inserted 21mm into the internal carotid artery, occluding the basis of the MCA. The animals were monitored before, during and after focal ischemia for 60 minutes. At the end of each experiment, the rats were sacrificed by N_2 inhalation.

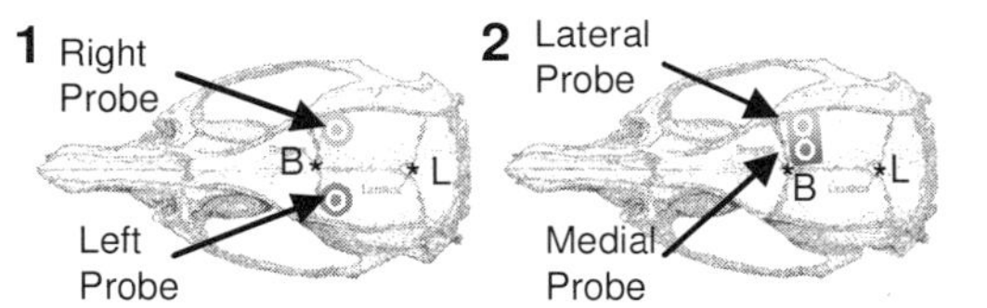

Fig. 1. Schematic representation of the location of the MSMP system fibers on the monitored sites in the bilateral model (1) and unilateral model (2). B-Bregma, L-Lambda.

Mean ± S.E. values were calculated for the following parameters: reflectance (Ref), fluorescence (Flu), NADH and CBF. T-test was used to determine significant differences between each parameter and its baseline, and between 2 areas in the brain.

3. Results

During the bilateral monitoring of the brain (fig. 2), CBF dropped to a level of 22±5% at the onset of the occlusion only in the right hemisphere. This decrease was significant during the whole ischemic period, as compared to basal levels ($p<0.001$) and to the left hemisphere ($p<0.001$). Fluorescence levels in the right hemisphere showed an increase to a level of 170±15% 5 minutes after the occlusion, with significant differences from the baseline ($p<0.05$) and from the contra-lateral hemisphere ($p<0.01$) during 30 minutes of the ischemic period. Reflectance levels showed an increase of 23±9%, which was significant as compared to the contra-lateral hemisphere for one minute ($p<0.05$), then gradually decreased until stabilizing. Due to these changes, NADH levels showed a 50±5% increase in the right hemisphere, which was significantly different from the contra-lateral hemisphere during 30 minutes after the occlusion ($p<0.001$), and then returned to the baseline. In reperfusion, all parameters returned to the baseline, except for a significant decrease in NADH levels to 75±7% ($p<0.05$). No significant changes were observed in the left hemisphere, during the entire experiment.

During the unilateral brain monitoring (fig. 3), CBF levels decreased to 20±7% in the lateral site ($p<0.001$, compared to the baseline) and to 40±10% in the medial site ($p<0.001$), exhibiting significant differences between the sites ($p<0.05$). Reflectance increased to a level of 170±19% in the lateral site 3 minutes after the occlusion onset; these values were significantly higher than the baseline ($p<0.001$) and when compared to the medial site (122±9%, $p<0.05$). Fluorescence reached peak levels of 213±22% in the lateral site ($p<0.001$, compared to the baseline) and 167±19% in the medial site ($p<0.01$), and then stabilized and remained higher than the baseline for the rest of the ischemic period, exhibiting a significant difference between the 2 sites ($p<0.05$). NADH increased to peak levels of 145±10% in both sites ($p<0.01$, compared to the baseline), followed by a decrease to the baseline in the lateral site and to 120±13% in the medial site.

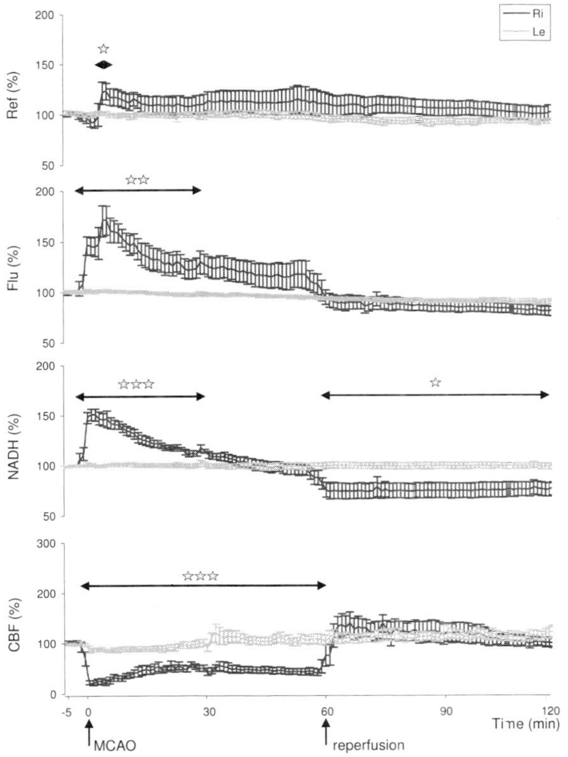

Fig. 2. Hemodynamic and metabolic changes in the bilateral monitoring model, following right MCAO for 60 minutes and reperfusion. Right hemisphere – Ri, Left hemisphere – Le. The results are presented as mean ± S.E. (n=8). Significant differences between the two hemispheres (in each minute) are marked by asterisks: (*) p<0.05 ; (**) p<0.01 ; (***) p<0.001.

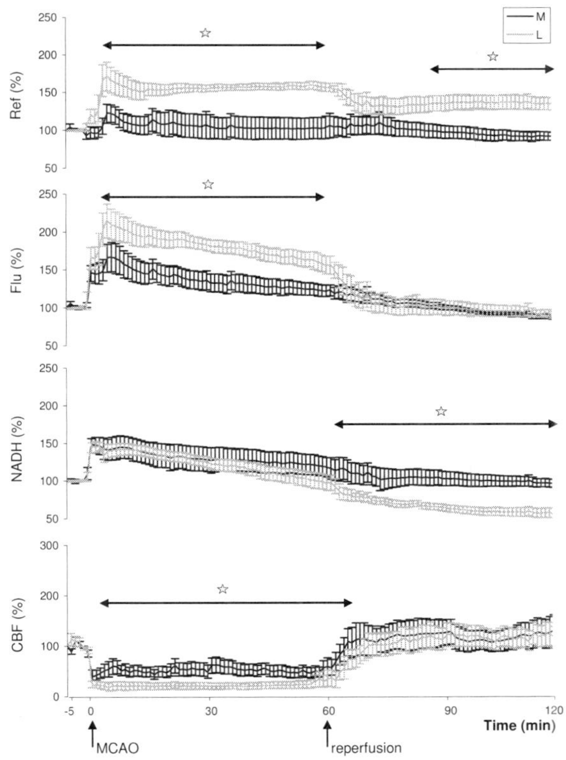

Fig. 3. Hemodynamic and metabolic changes in the unilateral monitoring model following right MCAO for 60 minutes and reperfusion. Lateral monitoring location – L, Medial monitoring location – M. The results are presented as mean ± S.E. (n=9). Significant differences between the two hemispheres are marked with asterisks (*) p<0.05

In reperfusion, CBF levels increased and returned to the baseline one hour following reperfusion. The recovery of CBF was more rapid in the medial area, as compared to the lateral area. Reflectance levels declined to the baseline in the medial site and to 135±9% in the lateral site ($p<0.01$), demonstrating significant differences between the 2 sites ($p<0.05$). Fluorescence decreased gradually until reaching baseline levels 30 minutes after reperfusion. NADH levels decreased significantly in the lateral site to 80±7% as compared to the baseline ($p<0.001$), while a gradual decrease toward the baseline was observed in the medial site, resulting in significant differences between the 2 sites ($p<0.05$).

4. Discussion

The monitoring of the brain during an ischemic stroke induced by MCAO, may contribute to the understanding of the events occurring in ischemia and reperfusion. The effect of right MCA occlusion on both brain hemispheres was tested to validate the model integrity, and indeed, our results showed that ischemic injury was restricted to the right hemisphere, as demonstrated by the decrease in blood supply and the increase of reduced NADH, while the left hemisphere showed no significant changes, as documented in the literature.[3,4,15] These findings verify the restricted and local nature of the damage in the cerebral tissue following MCAO, as well as the reliability of the MSMP monitoring system in detecting focal brain ischemia.

In the current study, the lateral site of the unilateral model represents the core and the medial site represents the penumbra of the focal ischemia, as can be seen from the differences in CBF between the two sites and as was reported in earlier works.[3,4] The higher levels of CBF in the penumbra are due to a partial blood supply by collateral blood vessels, resulting in a minor degree of damage compared to the core.[3,4] In addition, the recovery phase of CBF was delayed in the core, as compared to the penumbra, indicating that tissues in different degrees of ischemia have different oxygen demands.[4,9]

The ischemic core showed a secondary reflectance increase (SRI) 3 minutes after the occlusion onset, which was absent in the penumbra. This phenomenon, known as ischemic depolarization (ID),[14] is caused by an increase in extra-cellular potassium levels, due to oxygen shortage in the brain, and leads to the absence of cerebral electrical activity.[11] The high levels of reflectance for the entire ischemic period, lead to an artifact in NADH calculation, as previously observed in cases of ischemia and at death.[8,11] As a result, low NADH levels were observed during ischemia and at reperfusion. In such cases, the non-corrected fluorescence might serve as a better indicator of the mitochondrial state,[8,11] and indeed, fluorescence levels were higher in the core than in the penumbra, suggesting a more severe mitochondrial injury in the former area,[6,7] as seen in other studies.[4,15]

In conclusion, the application of the MSMP monitoring system can contribute additional novel information regarding the mitochondrial function and cerebral tissue state following focal ischemia, and may be potentially used in clinical practice.

References

1. Barbiro-Michaely E, Tolmasov M, Rinkevich-Shop S, Sonn J and Mayevsky A. Can the "brain-sparing effect" be detected in a small-animal model? *Med Sci Monit* 2007; **13:** BR211-BR219.
2. Gido G, Kristian T, Siesjo BK and Koehler RC. Extracellular Potassium in a Neocortical Core Area After Transient Focal Ischemia. *Stroke* 1997; **28:** 206-210.
3. Ginsberg MD. Adventures in the Pathophysiology of Brain Ischemia: Penumbra, Gene Expression, Neuroprotection: The 2002 Thomas Willis Lecture. *Stroke* 2003; **34:** 214-223.
4. Lipton P. Ischemic cell death in brain neurons. *Physiol Rev* 1999; **79:** 1431-1568.
5. Longa EZ, Weinstein PR, Carlson S and Cummins R. Reversible middle cerebral artery occlusion without craniectomy in rats. *Stroke* 1989; **20:** 84-91.
6. Mayevsky A. Brain NADH redox state monitored in vivo by fiber optic surface fluorometry. *Brain Res* 1984; **319:** 49-68.
7. Mayevsky A and Chance B. Intracellular oxidation reduction state measured *in situ* by a multichannel fiber-optic-surface fluorometer. *Science* 1982; **217:** 537-540.
8. Mayevsky A and Chance B. Oxidation-reduction states of NADH in vivo: From animals to clinical use. *Mitochondrion* 2007; **7:** 330-339.
9. Mayevsky A, Meilin S, Manor T, Ornstein E, Zarchin N and Sonn J. Multiparametric monitoring of brain oxygen balance under experimental and clinical conditions. *Neurol Res* 1998; **20 Suppl 1:S76-80.:** S76-S80.
10. Mayevsky A, Meilin S, Manor T, Zarchin N and Sonn J. Optical monitoring of NADH redox state and blood flow as indicators of brain energy balance. *Adv Exp Med Biol* 1999; **471:133-40:** 133-140.
11. Mayevsky A and Rogatsky G. Mitochondrial function in vivo evaluated by NADH fluorescence: from animal models to human studies. *Am J Physiol Cell Physiol* 2007; **292:** C615-C640.
12. Nagasawa H and Kogure K. Correlation between cerebral blood flow and histologic changes in a new rat model of middle cerebral artery occlusion. *Stroke* 1989; **20:** 1037-1043.
13. Pestalozza IF, Di LS, Calabresi M and Lenzi GL. Ischaemic penumbra: highlights. *Clin Exp Hypertens* 2002; **24:** 517-529.
14. Siesjo BK. Pathophysiology and treatment of focal cerebral ischemia. Part I: Pathophysiology. *J Neurosurg* 1992; **77:** 169-184.
15. Sims NR and Anderson MF. Mitochondrial contributions to tissue damage in stroke. *Neurochem Int* 2002; **40:** 511-526.

HYPERBARIC HYPEROXIA AND THE BRAIN IN VIVO: THE BALANCE BETWEEN THERAPY AND TOXICITY*

JUDITH SONN, ELHANAN MEIROVITHZ, AVRAHAM MAYEVSKY†

The Mina & Everard Goodman Faculty of Life Sciences and the Leslie and Susan Gonda Multidisciplinary Brain Research Center, Bar-Ilan University
Ramat-Gan 52900, Israel
Email:mayevsa@mail.biu.ac.il

Hyperbaric oxygenation (HBO) treatment protocols utilize low pressures up to 3ATA. Higher pressures may induce side effects such as convulsions due to brain toxicity. The optimal HBO pressure allowing for maximal therapy and minimal toxicity, is under controversy. However, it can be evaluated by monitoring oxygen delivery, saturation and consumption. In the present study, the monitoring system fixed on the rats' brain cortex included a time-sharing fluorometer-reflectometer for monitoring mitochondrial NADH and hemoglobin oxygenation (HbO_2) combined with laser Doppler flowmetry for blood flow monitoring. Rats were located in a hyperbaric chamber and exposed to different pressures. HBO pressure caused an increase in HbO_2 and a decrease in NADH in proportion to the increase in hyperbaric pressure, up to a nearly maximum effect at 2.5ATA. Above this pressure, at a variable time, the brain tissue showed an increase in blood volume and NADH, while tissue O_2 supply by hemoglobin remained stable. Oxygen pool includes oxygen dissolved in the plasma and HbO_2. Over 2.5ATA, hemoglobin is fully saturated and the oxygen pool nourishment derives only from the oxygen dissolved in the plasma, exceeding the physiological ability for auto-regulation; hence, homeostasis is disturbed and convulsions appear.

1. Introduction

The systemic circulation involves O_2 diffusion from the capillary bed to the mitochondria, that entirely depends on the oxygen gradient between the microcirculation and the mitochondria.[19] Thus, there is a direct connection between blood and tissue hemoglobin saturation, which enables metabolic mitochondrial activity and affects the oxidation-reduction state of the respiratory chain.[10] Under normal conditions, most of the oxygen carried in the blood is bound to hemoglobin, which is 98% saturated in the blood, leaving the heart at sea level pressure. However, some dissolved oxygen is carried in the plasma and, according to Henry's Law, its level is proportional to the increase in O_2 partial pressure, maximizing tissue oxygenation.[6] In our previous article[15] we hypothesized that brain hemoglobin oxygenation at the microcirculation level, is in the range of 50-60% and in order to achieve 100% Hb oxygenation in the microcirculation, the tissue has to be exposed to hyperbaric oxygenation (HBO). Then, the O_2 consumed by the cells will be supplied by high levels of dissolved O_2. Araki et al.[1] found that at 2ATA O_2 hemoglobin in the brain tissue is only 80% saturated. We showed in a previous study[15] that tissue HbO_2 reaches a maximal level at 2.5ATA. Higher pressures did not increase significantly the tissue hemoglobin saturation and all O_2 supply to the cells was provided by the O_2 dissolved in the plasma. At higher pressures, such as 4.5ATA and 6ATA, tissue HbO_2 remained unchanged even when convulsion periods were noted. These pressures are known to cause oxygen toxicity manifested by convulsion periods and even death.[2,3,13,14]

The aims of this study were to verify the effects of high HBO pressures on oxygen supply (CBF and HbO_2) and on tissue oxygen balance (mitochondrial NADH redox state) and to examine the interrelation between these parameters during the development of oxygen toxicity in relation to HBO therapy.

2. Materials and Methods

The hemodynamic and metabolic functions of the cerebral cortex, were monitored by the Tissue Vitality Monitoring System (TVMS) that includes two devices: a time-sharing fluorometer-reflectometer (TSFR) for mitochondrial NADH redox state and microcirculatory hemoglobin oxygen saturation (HbO_2) measurements combined with a laser Doppler flowmeter for CBF monitoring. Details of this monitoring system and methods have been presented in our previous articles.[10,12,15].

* This work is supported by etc, etc.
† Work partially supported by grant 2-4570.5 of the Swiss National Science Foundation.

2.1. *Animal Preparation*

All experiments were performed in accordance to the Guidelines of the Animal Care Committee of Bar-Ilan University. Male Wistar rats (n=28) weighing 220-320g, were used. The animals were anesthetized by IP injection of Equithesin, 0.3 mg/100g body weight. A black delrin light-guide holder combined with silver wires, for measuring bipolar electrocorticogram (ECoG), was placed epidurally on the left parietal brain and cemented to the skull by dental acrylic cement. More details can be found in our previous study.[15] Body rectal temperature was monitored continuously using a thermistor (Yellow Springs Instruments Co. Inc., type 402), and was kept constant at 36-37°C. The animal was placed in a Plexiglas cage, fitted to the rat's size, and was allowed to recover for about 30 minutes. The normal response to anoxia was checked by the NADH increase and HbO_2 decrease. Thereafter, the rat (in the Plexiglas cage) was placed in a 150 liter hyperbaric chamber, containing a special cage heated by water (Bethlehem Corp. FM-21-A).

2.2. *Experimental Procedure*

Four hours after anesthesia, in a fully awake state, the experiment started. The chamber was flushed with 100% oxygen, to establish normobaric hyperoxia (NH) for 15 minutes at atmospheric pressure (cleaning phase) in order to prevent acid-base disturbances and physiological CO_2 elevation in blood. Then, the oxygen pressure was elevated at a rate of 1 atmosphere per minute to a maximum pressure of the hyperbaric hyperoxia (HH). The respective maximum pressures were applied to 4 groups of animals. Group 1 (n=7) was exposed to 2.5ATA for 140 minutes. Ten minutes after decompression, the animals were exposed to pure N_2 inhalation until death. Group 2 (n=7) was subjected to 4.5ATA until the appearance of the first tonic-clonic convulsions, or until a maximum of 140 minutes, and then decompression was performed. Ten minutes later, the animals were exposed to pure N_2 inhalation until death. Group 3 (n=7) was pressurized to a maximum of 6ATA. The animals were left in the chamber until their death which occurred approximately 0.5 hours after the compression induction. An additional group of rats (n=7), the control group of normobaric normoxia (NN), underwent the same preparation and procedures. The rats were monitored at 0.2ATA air in the chamber for 2.5 hours without pressure and were exposed to pure N_2 inhalation at the end of the experiment.

2.3. *Data Collection, Processing and Statistical Analysis*

Since neither of the optical signals is calibrated in absolute units, we set all signals to read 100% after connecting the brain to the monitoring system in the normoxic state, and measured the percent change of the signal. Data were collected and stored in computer files, using LabView A/D hardware and software (National Instruments Inc. USA). Simultaneously, electrocorticogram (ECoG) and electrocardiogram (ECG) were recorded on a polygraph.

All statistical analyses and calculations were done by Matlab software (Ver. 2006b, The MathWorks, Inc.). The latency time for the first convulsion was measured from the experiment initiation until the appearance of the first convulsion, or until a maximum of 166 minutes (a maximum of 140 minutes under pressure). The distribution of the time until the first convulsion was found to be abnormal. Therefore, we used the Wilcoxon signed-rank test to calculate the changes in the appearance of the first convulsion between the pressure groups. The median amplitudes at 5, 20 and 30 minutes after experiment initiation (in the 2.5ATA and 6ATA groups) and 5, 10 and 15 minutes before the first convulsion (in the 6ATA group), were calculated. In the 2.5ATA group, the times until convulsions (for comparative calculations) were set at 61 minutes from the experiments initiation, since this is the median time to the first convulsion in the 6ATA group. Additionally, the differences between the control and 2.5ATA groups as well as between 6ATA and 2.5ATA groups at specific time points were tested by the Student t-test. A value of $p<0.05$ was considered significant.

3. Results

Four levels of pressure were selected: 1.75ATA and 2.5ATA (utilized for clinical treatments) and two high pressures, 4.5ATA and 6ATA, known for their toxic effects.[8,14,15] Figure 1 shows the effects of 2.5ATA compared to the control group at 0.2ATA, on CBF, reflectance, mitochondrial NADH fluorescence and HbO_2. The experiments started from normal pressure (0.2ATA), then, pressure was elevated gradually starting from (NH) at 100% O_2 and continued to hyperbaric pressure (2.5ATA). 2.5 ATA had no toxic effects during the experiments period. Starting with 100% O_2, HbO_2 and reflectance increased significantly ($p<0.01$, $p<0.001$ respectively), whereas NADH oxidized (decreased) significantly ($p<0.001$). Under elevated pressure (2.5ATA), a further augmentation in HbO_2 and reflectance was observed ($p<0.001$). HbO_2 reached a steady state level about 5 minutes after the compression started. In parallel to these changes, mitochondrial NADH oxidized and reached a steady state. No significant changes were observed in CBF. Figure 2 presents an illustrative experiment, showing the effect of high pressures (6ATA) on the measured parameters (left side) and on ECoG (right side). Trace D is zoomed to illustrate the epileptic convulsions and their effects on the measured parameters. High pressure (6ATA) initially yielded changes identical to 2.5ATA (Fig. 1) until the toxic effects started. At the 50th minute under 6ATA, HbO_2 reached its maximal amplitude and remained stable until the end of the experiment (even during the epileptic convulsions). Fifteen minutes after 6ATA induction, CBF and NADH started to increase whereas reflectance decreased and reached the maximum changes few minutes before the first convulsion was observed. The increase in pressure was followed by pre-convulsion ECoG spikes that became amplified and synchronized over time (Fig. 2, right). Fig. 2 (right) shows 5 ECoG traces of one rat exposed to 6 ATA, from the beginning of the experiment (control anesthesia), during HH (6 ATA), the convulsions period and until death. ECoG shows clear synchronized tonic-clonic convulsions (the last 2 right-hand traces), synchronization of high amplitude activation that terminated in a silent trace (the last trace).

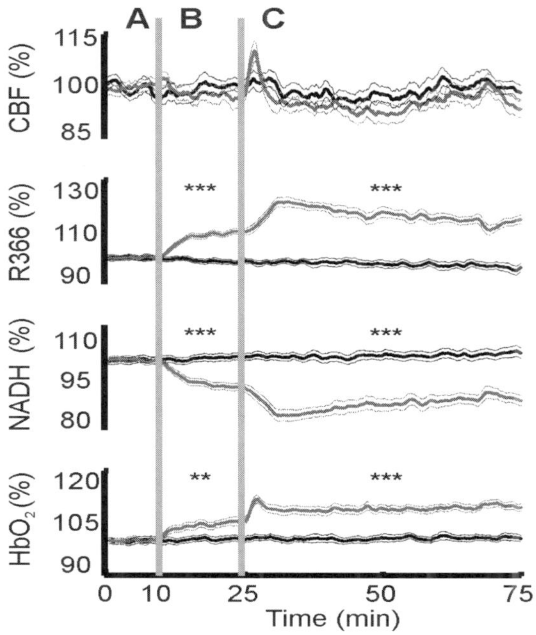

Fig. 1. The effect of 100% O_2 and 2.5ATA O_2 (gray line) as compared to normobaric normoxia (black line). A–normobaric normoxia; B-100% O_2; C–2.5ATA. Data are presented as Median ± Mad. CBF-cerebral blood flow; R366 - reflectance at 366nm; NADH - mitochondrial NADH redox state; HbO_2–brain tissue hemoglobin saturation. **=p<0.01, ***=p<0.001.

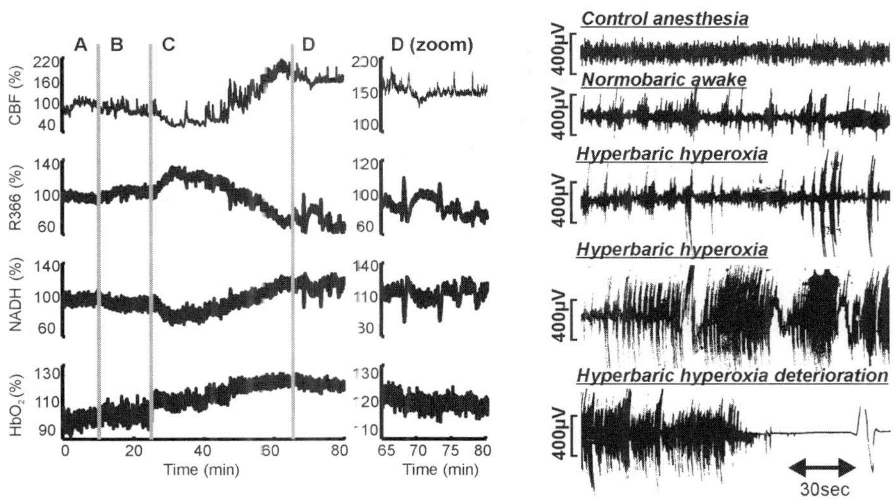

Fig. 2. Illustrative traces showing the effect of 6 ATA on: left side - the measured parameters; right side -the ECoG during the different phases of the experiment. Abbreviations are as described in Fig.1 including A - Control, B - 100% O_2, and C - high pressure (6 ATA); D - the first convulsion period during 6ATA; D (zoom) - enlargement of the convulsion periods

In the 2.5ATA group, no convulsions occurred; at 4.5ATA, 3 of 7 animals developed convulsions and 2 of the 3 died. In contrast, in the 6ATA group, all the animals developed convulsions and died during 166 minutes of measurement. Thus, the gradual decrease in latency is parallel to the pressure increase. The differences in the latency time to the first convulsion between 4.5ATA and 6ATA pressure groups were assessed by the Wilcoxon signed-rank test and were found to be significant ($p<0.016$). For the statistical calculations, animals in the 4.5 ATA group that did not develop convulsions during the 166 minutes period, were considered to develop the first convulsion at 166 minutes. Not only the convulsion latency (time) decreased with increasing pressure, but the time until oxygen toxicity death also diminished. In the 6ATA group, all the animals died during the recorded time (between 78-148 minutes of experiment), whereas in the 4.5ATA group, only two animals died very near to the end of the experiment (158 and 165 minutes within the 166 minutes of recording).

Table 1. Median amplitude values (%) ± Mad of the measured parameters, at selected time points before the first convulsion started. Abbreviations as described in fig. 1. min=minutes; *= $p<0.05$.

Parameter	Pressure (ATA)	Median % Values ± Mad. at time before the first convulsion		
		15 (min)	10 (min)	5 (min)
CBF	2.5	90± 6	91±6	96±4
(%)	6	106±14	113±27 *	101±45
Ref-366	2.5	118±4	118±5	115±4
(%)	6	113±20	108±19	94±14 *
NADH	2.5	87±3	87±3	88±3
(%)	6	87±17	93±15	107±13 *
HbO_2	2.5	107±2	108±2	109±2
(%)	6	111±19	115±23	116±37

Substantial points, after the experiments initiation, were chosen over the 2.5ATA and 6ATA time course, in order to find the pre-convulsion significant changes in the measured parameters (Table 1). Since during 2.5ATA application, no convulsions were observed, to find the parallel time for comparing the changes in the measured parameters between the two pressures, we used the median time to the first convulsion in the 6ATA group, which was 61 minutes into the experiment. The effects of 2.5ATA and 6ATA pressures on the measured parameters were similar from the beginning of the experiment and lasted until 10 minutes before the first convulsion. Ten and five minutes before the convulsions started, CBF and NADH increased significantly ($p<0.05$) while the reflectance significantly decreased ($p<0.05$). HbO_2 levels did not change at both 2.5 and 6ATA despite the convulsions at 6ATA (table 1).

4. Discussion

This study examines the relationship between microcirculatory blood flow, tissue reflectance, hemoglobin oxygenation and intracellular mitochondrial NADH redox state under normoxia (21% O_2), normobaric hyperoxia (100% O_2) and variable levels of hyperbaric hyperoxia (HH). The combination of these parameters may significantly contribute to the understanding of brain energy metabolism in vivo at various levels of oxygen supply above normoxia. We tested the hypothesis that the mitochondrial NADH is not fully oxidized in the normoxic brain in vivo and can be oxidized even further under HH conditions. The majority of studies investigating the effect of HBO focused on monitoring intracranial pressure (ICP), CBF, EEG and blood pressure.[16] Consequently, important information was to be gained on the relationship between brain tissue O_2 delivery, O_2 consumption and utilization (mitochondrial NADH redox state).

In previous works,[7,11,13,14,22] we showed that the reflectance changes occurring immediately after increasing oxygen levels are due to changes in the hemodynamic responses tending towards vasoconstriction (decrease in blood volume). This vasoconstrictive effect was shown by a reduction in CBF and a significant increase in reflectance, indicating a significant decrease of blood volume in the awake rat brain under various oxygen levels, immediately following an elevation in brain oxygenation (Fig. 1). A decrease in blood volume was also found in our previous studies.[11,15,22] In contrast, under high pressures, at the beginning of convulsions (at 4.5 and 6 ATA), the CBF increased, while the reflectance decreased showing an increase in blood volume (clearly seen at 6ATA in Fig. 2, left). Similar reductions in reflectance under high pressures were shown previously.[7,11,13-15,22] The effect of hyperbaric oxygenation on CBF was found to be highly dependent on the CBF monitoring method. A detailed discussion of this issue can be found in our previous article[15]. Studies using laser Doppler flowmetry [4,17,18,23] revealed no changes or only a small decrease or increase under exposure to 1-4ATA. At pressures above 4ATA, in the first few minutes, the CBF decreased and returned to the base line within a variable period. Several minutes before convolutions started, the CBF rose significantly (Fig. 2 left).

Van Hulst et al.[21] found that tissue pO_2 increases in a linear relation to the increase in the hyperbaric chamber O_2 pressure, due to the high plasma pO_2. We established that the tissue HbO_2 reached a maximal value at 2.5ATA. Pressures above 2.5ATA did not alter the tissue hemoglobin saturation.[15] In addition, mitochondrial NADH reaches a plateau concomitantly to tissue HbO_2. At 2.5ATA, NADH oxygenation is maximal and in linear correlation to hemoglobin oxygenation. When increasing the hyperbaric pressure above 2.5ATA, a reduced NADH molecule will be quickly transformed into an oxidized form[15] (see Figs. 1 and 2 left). A similar oxygenation effect was found in our previous studies during various pressures ranging 3-11ATA.[7,8,11,13,14,22] Pressures above 4ATA caused oxygen toxicity within a variable period after the pressure induction. Similar results have been obtained by others.[4,7,8,11,13,14] The results of the present study show that during pressures higher than 4ATA, the latency time to the first convulsion and death, decreases with increasing pressure. Thus, there is a gradual effect of oxygen pressure on the brain: the higher the pressure – the faster the convulsions' emergence and the shorter the time until oxygen toxicity and death.

In addition, under 6ATA, 10 minutes before the convulsions onset, mitochondrial NADH redox state increased (Table 1), even though CBF rose and reflectance decreased, representing an increase in the blood flow and volume. Similar results have been reported by other groups.[1,3,4,14,16,17,21] Furthermore, HbO_2 levels did not change during 6 ATA as compared to 2.5 ATA (Table 1) showing no decrease in oxygen supply. The increase in mitochondrial NADH may result from mitochondrial dysfunction due to oxygen toxicity triggered by the high pressures, since other works showed NADH oxidation during convulsion.[5,9,20] The convulsions period is very variable among animals, but in all the animals we found spreading depression cycles.[6,7,9,11,12] The post-convulsive phase starts with the stops of the convulsions and flattening of the ECoG until death (Fig. 2 right).

In conclusion, the present work shows the following. 1. The maximum saturation (100%) of hemoglobin in the tissue is achieved at 2.5 ATA and higher hyperbaric pressures produce no further hemoglobin oxygenation. 2. During pressures higher than 4ATA, increasing pressure shortens the time to oxygen toxicity. 3. Multi-parametric monitoring (CBF, mitochondrial NADH redox state and tissue HbO2) is important for detecting tissue oxygen toxicity during hyperbaric treatment, since monitoring only CBF and HbO2 without NADH, impedes the identification of tissue oxygen toxicity.

References

1. Araki R, Nashimoto I and Takano T. "The effect of hyperbaric oxygen on cerebral hemoglobin oxygenation and dissociation rate of carboxyhemoglobin in anesthetized rats: spectroscopic approach". *Adv Exp Med Biol* 1988, **222**: 375-381.
2. Bean JW, Lignell J and Coulson J. "Regional cerebral blood flow, O_2, and EEG in exposures to O_2 at high pressure". *J Appl Physiol* 1971, **31**: 235-242.
3. Bergo GW and Tyssebotn I. "Cerebral blood flow distribution during exposure to 5 bar oxygen in awake rats". *Undersea Biomed Res* 1992, **19**: 339-354.
4. Chavko M, Braisted JC, Outsa NJ and Harabin AL. "Role of cerebral blood flow in seizures from

hyperbaric oxygen exposure". *Brain Res* 1998, **791**: 75-82.

5. Dora E. "Glycolysis and epilepsy-induced changes in cerebrocortical NAD/NADH redox state". *J Neurochem* 1983, **41**: 1774-1777.
6. Gill AL and Bell CN. "Hyperbaric oxygen: its uses, mechanisms of action and outcomes". *QJM* 2004, **97**: 385-395.
7. Mayevsky A. "The effect of trimethadione on brain energy metabolism and EEG activity of the conscious rat exposed to HPO". *J Neurosci Res* 1975, **1**: 131-142.
8. Mayevsky A. "Multiparameter monitoring of the awake brain under hyperbaric oxygenation". *J Appl Physiol* 1983, **54**: 740-748.
9. Mayevsky A and Chance B. "Metabolic responses of the awake cerebral cortex to anoxia hypoxia spreading depression and epileptiform activity". *Brain Res* 1975, **98**: 149-165.
10. Mayevsky A and Chance B. "Oxidation-reduction states of NADH in vivo: from animals to clinical use". *Mitochondrion* 2007, **7**: 330-339.
11. Mayevsky A, Jamieson D and Chance B. "Oxygen poisoning in unanesthetized brain: correlation of pyridine nucleotide redox state and electrical activity". *Brain Res* 1974, **76**: 481-491.
12. Mayevsky A and Rogatsky GG. "Mitochondrial function in vivo evaluated by NADH fluorescence: from animal models to human studies". *Am J Physiol Cell Physiol* 2007, **292**: C615-C640.
13. Mayevsky A and Shaya B. "Factors affecting the development of hyperbaric oxygen toxicity in the awake rat brain". *J Appl Physiol* 1980, **49**: 700-707.
14. Mayevsky A, Wrobel-Kuhl K and Mela L. "High pressure oxygenation in unanesthetized brain: mitochondrial activity, pyridinenucleotide redox state, and electrical activity". *Neurol Res* 1980, **1**: 305-311.
15. Meirovithz E, Sonn J and Mayevsky A. "Effect of hyperbaric oxygenation on brain hemodynamics, hemoglobin oxygenation and mitochondrial NADH". *Brain Res Rev* 2007, **54**: 294-304.
16. Rogatsky GG, Shifrin EG and Mayevsky A. "Physiologic and biochemical monitoring during hyperbaric oxygenation: a review". *Undersea Hyperb Med* 1999, **26**: 111-122.
17. Sato T, Takeda Y, Hagioka S, Zhang S and Hirakawa M. "Changes in nitric oxide production and cerebral blood flow before development of hyperbaric oxygen-induced seizures in rats". *Brain Res* 2001, **918**: 131-140.
18. Thom SR, Fisher D, Zhang J, Bhopale VM, Ohnishi ST, Kotake Y, Ohnishi T and Buerk DG. "Stimulation of perivascular nitric oxide synthesis by oxygen". *Am J Physiol Heart Circ Physiol* 2003, **284**: H1230-H1239.
19. Tsai AG, Johnson PC and Intaglietta M. "Oxygen gradients in the microcirculation". *Physiol Rev* 2003, **83**: 933-963.
20. Van Buren JM, Lewis DV, Schuette WH, Whitehouse WC and Marsan CA. "Fluorometric monitoring of NADH levels in cerebral cortex: preliminary observations in human epilepsy". *Neurosurgery* 1978, **2**: 114-121.
21. Van Hulst RA, Haitsma JJ, Klein J and Lachmann B. "Oxygen tension under hyperbaric conditions in healthy pig brain". *Clin Physiol Funct Imaging* 2003, **23**: 143-148.
22. Yoles E, Zurovsky Y, Zarchin N and Mayevsky A. "The effect of hyperbaric hyperoxia on brain function in the newborn dog *in vivo*". *Neurol Res* 2000, **22**: 404-408.
23. Zhang J, Sam AD, Klitzman B and Piantadosi CA. "Inhibition of nitric oxide synthase on brain oxygenation in anesthetized rats exposed to hyperbaric oxygen". *Undersea Hyperb Med* 1995, **22**: 377-382.

EFFECTS OF SEVERE HEMORRHAGE ON IN VIVO BRAIN AND SMALL INTESTINE MITOCHONDRIAL NADH AND MICROCIRCULATORY BLOOD FLOW

MANDELBAUM MIRAM, BARBIRO-MICHAELY EFRAT, TOLMASOV MICHAEL, MAYEVSKY AVRAHAM [†]

The Mina & Everard Goodman Faculty of Life-Sciences and The Gonda Multidisciplinary Brain Research Center
Bar-Ilan University
Ramat-Gan, 52900, Israel
Email: mayevsa@mail.biu.ac.il

Severe body stress induced by hypoxemia and hypotension may lead to total body energy state deterioration. The perfusion of the most vital organs is maintained at the expense of "less vital" organs. In the present study, we used a multi-site multi-parametric (MSMP) monitoring system for real time evaluation of Tissue Blood Flow (TBF) and mitochondrial NADH fluorescence of the brain and the small intestine following hemorrhage. In Group 1 of rats, after hemorrhage, Mean Arterial Pressure (MAP) was decreased to 40 mmHg with no further perturbations for 30 min and after 2 hours of resuscitation. In Group 2, MAP was maintained at 40 mmHg for 30 min. During hemorrhage, in both groups, the intestinal TBF and NADH deteriorated, while the brain remained relatively well protected. In Group 1, all parameters partly recovered within the hemorrhage phase, while in Group 2 complete recovery occurred only after resuscitation. At the end of the experiment, both models showed a decrease in intestinal viability (TBF decreased, NADH increased), while the brain metabolic state in Group 2 slightly declined. Our unique multi-parametric monitoring device demonstrated that, under hemorrhage, the small intestine responded entirely differently from the brain. This may suggest the potential usefulness of the monitoring of less vital organs, as proxy organs, in critical conditions such as massive hemorrhage. The present study also highlights the importance of mitochondrial function monitoring in similar conditions in the clinical environment.

1. Introduction

Severe hemorrhage involves a depression of cardiac output, decrease of perfusion pressure and oxygen delivery (DO_2), decreased systemic blood pressure and a minor change in oxygen consumption (VO_2).[10,11,13] Severe and sustained decreases in DO_2 eventually overwhelm the micro-vascular response, until the mitochondria cannot sustain aerobic metabolism and VO_2 declines. Under these conditions, cellular function depends on the aerobic and anaerobic mechanisms providing ATP; with insufficient ATP levels, irreversible tissue damage ensues. Under severe hemorrhage, cardiac output and perfusion pressure increase as a compensatory mechanism.[4] However, under significant loss of intravascular volume, hemodynamic instability develops, tissue perfusion decreases, leading to cellular hypoxia, organ damage and finally death.[10,11,13] The mortality rate of patients following the first 24 hours after hemorrhagic shock is 50%, and survivors are at a high risk of developing infection and organ failure.[9] Therefore, the early detection of the patients' condition is clearly important.[2,12] In cases of severe stress, such as hemorrhagic shock, blood is redistributed to the most vital organs, such as the brain, heart and adrenal gland, whereas the "less vital" and peripheral organs and tissues, such as the G-I tract, muscles and skin undergo vasoconstriction and a decrease in blood flow and oxygen supply,[5,6] making them a useful, early indicator of the body condition.

In the present study, we developed a unique multiparametric monitoring device for real time measurements of the cerebral cortex simultaneously with the small intestine in various hemorrhage models in rats. We also tested the effects of controlled versus uncontrolled hypotension, on the responses of these two organs.

2. Methods

2.1. *The Multi-Site Multi-Parametric Monitoring System*

In the present study, the MSMP system enabled simultaneous monitoring of the brain and small intestine. The MSMP (Figure 1) monitors both NADH redox state using Fluorometry[7,8] and Tissue Blood Flow (TBF) using Laser Doppler Flowmetery (LDF) from the same location, simultaneously in each organ. The system specifications have been presented in our previous article.[1]

[†] This work was partly supported by grant 358/04-3 of "The Israeli Science Foundation".

2.2. *Animal Preparation*

All experiments were performed in accordance with the guidelines of the Animal Care Committee of Bar-Ilan University. Wister male rats (220–300g) were anesthetized by an IP injection of 0.3 ml/100g Equithesin. A heating pad was placed under the rat to maintain body temperature at 37°C and a rectal thermistor probe was inserted to measure body temperature. The cerebral cortex and the small intestine were exposed and the monitoring probes were placed on the two organs as detailed in our recent article.[1] The femoral arteries were cannulated for blood pressure monitoring and for bleeding, and the femoral vein was cannulated for the resuscitation phase. Hemorrhage was induced until blood pressure decreased to 40mmHg and was kept low for 30 minutes in the controlled hypotension group, whereas in the uncontrolled hypotension group, after reaching this blood pressure, no further perturbations were performed.

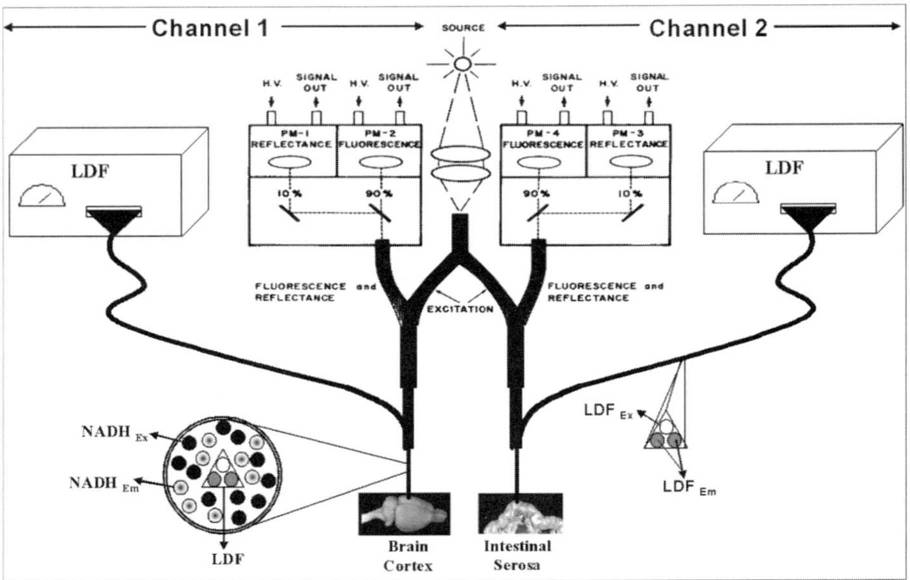

Figure 1: Schematic representation of the MSMP probe containing optical fibers for NADH monitoring (Ex - Excitation, Em - Emission, H.V.- High Voltage) and for Tissue Blood Flow (Laser Doppler Flowmetry - LDF).

3. Results

In order to evaluate the reliability of our continuous monitoring model, a group of 4 rats were monitored for 180 minutes with no further perturbation. The results indicated that throughout the entire experiment, there were no significant changes in any of the parameters in both organs (results not shown).

When uncontrolled hypotension was induced (Figure 2), MAP immediately decreased (44±5mmHg, $p<0.01$) and remained significantly low for 10 minutes, followed by a full recovery within the hemorrhage phase. The intestinal responses were greater and faster than those of the brain. Following bleeding, intestinal TBF immediately decreased by 21%±5 ($p<0.01$) and later almost fully recovered. In contrast, in the brain, TBF increased by 28%±13 ($p<0.05$) only several minutes after hypotension induction and remained high for the entire period. Intestinal reflection increased (8%±3, $p<0.05$) and then decreased back to its basal level, while cerebral reflection remained stable. Intestinal NADH immediately increased (13%±5, $p<0.05$) followed by full recovery, whereas cerebral NADH did not significantly change. During resuscitation and afterwards, MAP remained relatively stable. Intestinal TBF transiently recovered reaching high levels of 115% followed by a gradual decrease by 35.5%±10 to very low levels. In comparison, TBF in the brain slowly increased reaching a level of 147±0.12%. About 20 minutes after resuscitation, the intestinal reflection decreased by 11%±2 ($p<0.05$) and remained significantly low for an hour, while the cerebral reflection remained stable. However, NADH in both organs remained stable, with no significant changes.

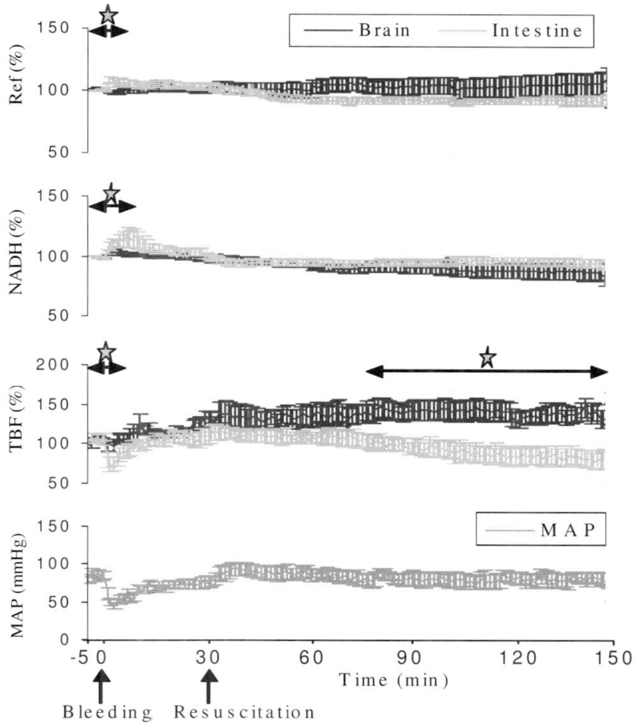

Figure 2: The responses of the brain, small intestine and MAP to uncontrolled hypotension for 30 minutes (n=9, mean±S.E). The arrows represent the period in which significant differences were found between the two organs in each minute. (*) $p<0.05$. REF - reflectance, NADH - mitochondrial NADH redox state, TBF - tissue blood flow, MAP - mean arterial pressure.

The comparison between organs showed significant differences in all of the parameters during the bleeding session ($p<0.05$). TBF showed major differences between the organs, during the bleeding session and an hour after resuscitation ($p<0.05$).

Upon controlled hypotension induction (Figure 3), MAP significantly decreased (62±1 mmHg, $p<0.001$). The intestine and the brain responded quite differently. Intestinal TBF significantly decreased (30%±7, $p<0.01$) while TBF in the brain remained relatively stable. Intestinal reflectance significantly increased (28%±7, $p<0.01$) while the cerebral reflection remained stable. Intestinal NADH had two phases, an increase of 32%±10 ($p<0.01$) followed by a slight decrease 10 minutes after bleeding began. Cerebral NADH increased significantly (9%±2, $p<0.01$) and was the only parameter in the brain that showed a significant response.

Following resuscitation, MAP levels fully recovered. However, about an hour after resuscitation, MAP decreased by 14±5 mmHg ($p<0.01$) until reaching significantly low levels. About 50 minutes after resuscitation began, intestinal TBF levels were again significantly low following a decrease of 39%±12 ($p<0.01$), while TBF in the brain increased. Following resuscitation, intestinal reflectance decreased and gradually returned to its basal level (with no significant change). Cerebral reflection also decreased gradually with no significant change. Intestinal NADH level showed a trend of decrease below the basal level, but then it increased toward the basal level and remained steady for the rest of the experiment. Cerebral NADH decreased gradually but without a significant alteration, except for resuscitation itself and several minutes afterwards where changes were significant ($p<0.01$). When comparing the responses of the intestine and the brain to the bleeding episode, significant differences can be seen in the reflection ($p<0.01$) and in NADH ($p<0.05$). During resuscitation, there were significant changes in the reflection ($p<0.01$) and in the TBF ($p<0.05$).

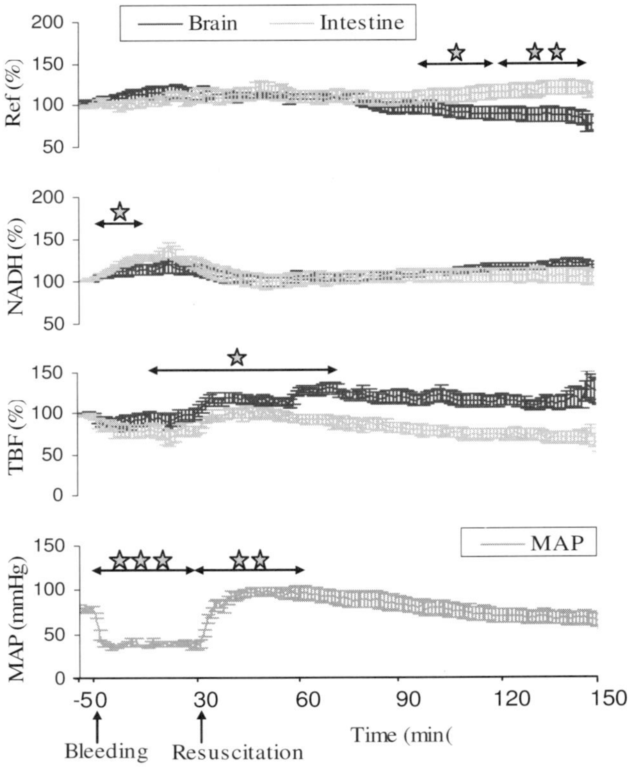

Figure 3: The responses of the brain, small intestine and MAP to controlled hypotension for 30 minutes (n=9, mean±S.E). The arrows represent the period in which significant differences were found between the two organs in each minute. (*) $p<0.05$, (**) $p<0.01$, (***) $p<0.001$. Abbreviations are as earlier presented.

4. Discussion

According to the global study of disease, 30-40% of trauma mortality is attributable to hemorrhage.[3] Moreover, the patients who survive are at a high risk of developing infection and organ failure.[9] Early diagnosis of hemodynamic catastrophe and early resuscitation are crucial in hemorrhagic shock,[2,12] therefore many experimental studies are conducted using various models of hemorrhage. In the present study, we tested two models: of uncontrolled and controlled hypotension. These two models actually differ in the time of hypotension, which was very short in the uncontrolled group (~2min) due to the activation of auto-regulation mechanisms,[4] and longer (30min) in the controlled hypotension group. The distinction between the organs is thus clearly emphasized. In the intestine, even a very short hypotension produced tissue state deterioration. In contrast, in the brain, the injury level appeared to be correlated to the duration of hypotension, namely, the longer the hypotension phase, the greater the ischemic reperfusion injury. In both models, during hemorrhage, the intestinal tissue deteriorated, as reflected in its hypo-perfusion state and the elevation of NADH level. These observations suggest the use of the small intestine as a proxy organ for monitoring in similar clinical conditions. Regarding the brain, although most parameters were not significantly changed, there was still an increase in NADH level, suggesting the importance of mitochondrial function monitoring for early detection of tissue deterioration, rather than monitoring tissue blood flow alone.

References

1. Barbiro-Michaely E, Tolmasov M, Rinkevich-Shop S, Sonn J and Mayevsky A. "Can the "brain-sparing effect" be detected in a small-animal model?". *Med Sci Monit* 2007, **13**: BR211-BR219.
2. Boldt J. "Clinical review: hemodynamic monitoring in the intensive care unit". *Crit Care* 2002, **6**: 52-59.
3. Grottke O, Henzler D and Rossaint R. "Use of blood and blood products in trauma". *Best*

Pract Res Clin Anaesthesiol 2007, **21**: 257-270.
4. Harbrecht BG, Alarcon LH and Peitzman AB. Managment of shock. In: Principles of Surgery, edited by Schwartz SI. New York: McGrow-Hill, p. 101-122, 1994
5. Jakob SM. "Clinical review: splanchnic ischaemia". *Crit Care* 2002, **6**: 306-312.
6. Kraut A, Barbiro-Michaely E and Mayevsky A. "Differential effects of norepinephrine on brain and other less vital organs detected by a multisite multiparametric monitoring system". *Med Sci Monit* 2004, **10**: BR215-BR220.
7. Mayevsky A and Chance B. "Intracellular oxidation reduction state measured *in situ* by a multichannel fiber-optic-surface fluorometer". *Science* 1982, **217**: 537-540.
8. Mayevsky A and Chance B. "Oxidation-reduction states of NADH in vivo: From animals to clinical use". *Mitochondrion* 2007, **7**: 330-339.
9. Peng TC, Liao KW, Lai HL, Chao YF, Chang FM, Harn HJ and Lee RP. "The physiological changes of cumulative hemorrhagic shock in conscious rats". *J Biomed Sci* 2006, **13**: 385-394.
10. Vajda K, Szabo A and Boros M "Heterogeneous microcirculation in the rat small intestine during hemorrhagic shock quantification of the effects of hypertonic-hyperoncotic resuscitation". *Eur Surg Res* 2004, **36**: 338-344.
11. Vatner SF. "Effects of hemorrhage on regional blood flow distribution in dogs and primates". *J Clin Invest* 1974, **54**: 225-235.
12. Yu ZY, Ono S, Spatz M and McCarron RM. "Effect of hemorrhagic shock on apoptosis and energy-dependent efflux system in the brain". *Neurochem Res* 2002, **27**: 1625-1632.
13. Zakaria eR, Ehringer WD, Tsakadze N, Li N and Garrison RN. "Direct energy delivery improves tissue perfusion after resuscitated shock". *Surgery* 2005, **138**: 195-203.

FIBER LENSES FOR ULTRA-SMALL PROBES USED IN OPTICAL COHERENT TOMOGRAPHY

YOUXIN MAO, SHOUDE CHANG, SHERIF SHERIF, COSTEL FLUERARU

Institute for Microstructural Sciences, National Research Council Canada,
1200 Montreal Rd, Ottawa, K1A 0R6, ON, Canada
Linda.mao@nrc-cnrc.gc.ca; phone 1 613 993-7331; fax 1 613 954 3338; http://ims-ism.nrc-cnrc.gc.ca/

Quality and parameters of probing optical beams are extremely important in biomedical imaging systems both for image quality and light coupling efficiency considerations. For example, the shape, size, focal position, and focal range of such beams could have a great impact on the lateral resolution, penetration depth, and signal-to-noise ratio of the image in optical coherence tomography. We present a design, construction and characterization of different variations of GRIN and ball fiber lenses, which were recently proposed for ultra-small biomedical imaging probes. Those fiber lens modules are made of a single mode fiber and a GRIN or ball fiber lens with or without a fiber spacer between them. The lens diameters are smaller than 0.3 mm. We discuss design methods, fabrication techniques, measured performance compared with modeling results, and present a swept-source OCT image measured by the fiber lens probe.

Keywords: Optical Fiber Probe, Optical Coherence Tomography, Bio-medical Imaging

1. Introduction

Optical biomedical imaging techniques, such as optical coherence tomography (OCT)[1] and Doppler OCT[2,3], are becoming increasingly important tools for both diagnosis and guided surgery because of their high image resolutions. OCT can provide images on the cellular level while Doppler OCT can detect blood flow with velocity sensitivities approaching a few micrometers per second[4,5]. However, in most optically nontransparent tissues, OCT has a typical imaging depth limitation of 1-3 mm. Similarly, Doppler OCT systems suffer from limitations where blood flow can rarely be detected beyond 1–2 mm from the tissue surface without *a priori* velocity profile information and digital extrapolation algorithms. As a result, the earliest *in vivo* OCT imaging of tissue microstructure and microvasculature was restricted to a few transparent or superficial organ sites, such as the retina[6,7] and skin[8,9]. To overcome this depth limitation, optical probes, such as endoscopes, catheters, and needles have been investigated for *in vivo* OCT imaging in mucosal layers of the gastrointestinal tract[10,11], deep organs and tissues[12,13], and inter-arterial and intra-vascular[14,15]. However, for the imaging of small lumen, narrow space, and deep tissue and organ of humans and small animals, a key concern is the possible damage from the mechanical insertion of the optical probe. Therefore it is critical to develop an ultra-small optical probe that is compatible with the current optical biomedical imaging systems, which results in minimum tissue damage.

In vivo optical imaging of internal tissues is generally performed using a fiber-optic probe, since an optical fiber can be easily and cheaply produced with a diameter of less than 150 μm. The key components of such optical fiber probe include a small lens and a beam director, where both provide a focused optical beam directing it to a location of interest through a guide-wire. Traditionally, this type of small optical probe has been implemented by attaching a small glass GRIN or SELFOC lens (0.25-1.0 mm) and a glass micro-prism to a single mode fiber (SMF) with optical adhesive or optical epoxy[12]. However, the gluing of a separate small lens and a tiny prism to a fiber is a complex fabrication process that results in a low quality optical interface. A new probe design that uses optical fiber lenses, e.g., fiber GRIN lens or fiber ball lens, has recently been proposed[16,17]. The main advantage of fiber lenses over conventional glass lenses are their small size, ability to auto-align to a fiber, thus creating a fusion-spliced interface with low loss, low back-reflection, and high mechanical integrity. In addition, a beam director can be easily attached to the fiber lenses by the fusion-splice of a polished fiber spacer and direct polish on the ball lens.

Beam quality of a fiber-optic probe is crucial for the imaging system. Ideal characteristics of a fiber-optic probe include a high Gaussian beam intensity profile, an appropriate intensity-distance shape, high flexibility, and low optical aberration and loss. To date, no paper has compared in detail the actual optical performance of a fiber-based optic probe with modeling results. Swanson *et al.* and Shishkov *et al.* proposed the fiber based optic probes design, but presented the variations of probe structure instead of the characteristics of their performance[16,17]. Reed *et al.* demonstrated the usage of such probes with emphasis on their insertion loss only[18]. Yang *et al.*[13], Jafri *et al.*[19], and Li *et al.*[20] reported OCT images without detailed characterization of the used fiber lens based probes. In this presentation, we discuss design methods and fabrication techniques of fiber lens based optic probes. We compare in detail measured performance with expected theoretical performance.

2. Method

For most optical imaging systems, light is guided through a SMF and focused on a sample using a lens. Back-reflected light carrying information about the sample is coupled back into the fiber, and then signal detection and data processing systems will collect the useful information for imaging. The optical probe is one of the crucial parts of such an optical imaging system. The beam profile, i.e. working distance (focal distance from lens surface), depth of field (two times the Raleigh range), and spot size (waist diameter) of the optical probe in the sample will directly determine properties of the image such as image location, depth, and resolution. Thus, image quality and light coupling efficiency from the sample will be directly influenced by the beam quality of the probe. For the best optical performance of a probing lens, its beam profile must be designed to be consistent with the light penetration depth in the sample. In most biomedical imaging systems, light from the probe will be directed into a turbid tissue. Based on interaction properties of light with turbid tissue[21], the range of penetration depth is from 0.5-3 mm at near infrared wavelengths. For example, the penetration depths are 0.7 mm and 3.0 mm in human skin and liver respectively at 1300 nm, a conventional wavelength used in OCT systems. Thus, for designing an optical probe, working distance should be in the range of 0.4 – 1.2 mm in the air that depends on the tissues to be tested. There is a tradeoff between the depth of field and beam spot size because the depth of field of a lens is positively related to the square of the spot size according to the theory of Gaussian beam. A large depth of field unavoidable results in a large spot size. Thus, the optimal depth of field is in the range of 0.8 - 1.5 mm in the air; this keeps the spot size in the range of 26 - 35 μm at the 1300 nm wavelength. For an ultra-small optical lens, it is not possible to achieve a large working distance by directly attaching a lens to a SMF because of the strong focus ability of lens and the small mode field diameter (MFD) of the SMF[22]. A fiber spacer with a homogeneous index of refraction has to be added between the SMF and the lens for beam expansion prior to focusing to obtain a larger working distance. Therefore, theoretical modeling becomes necessary to obtain a lens with optimized optical beam performance for imaging different tissues.

In this work, we used the commercially available numerical optical modeling software, ZEMAX (ZEMAX

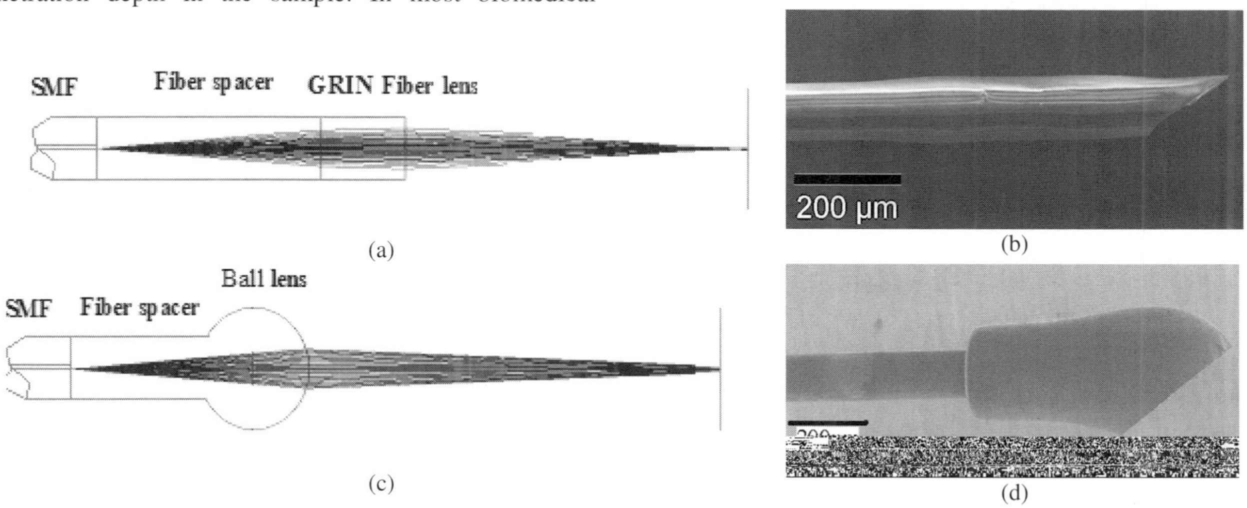

Figure 1: ZEMAX layouts of ray trace for the GRIN (a) and ball (c) fiber lens systems, respectively. Scanning electron micrographs of the GRIN (b) and ball (d) fiber lens tips attached with angle-polished beam directors.

Development Corp., WA, USA), to design both GRIN and ball fiber lenses by choosing an appropriate surface type and analysis method. For a GRIN fiber lens, the gradient surface type used in ZEMAX will depend on the profile of the refractive index. The refractive index profile of the GRIN fiber lens used in this work is a radial index gradient, which is very similar to that of a conventional GRIN (or SELFOC) lens. The index of refraction is highest in the center of the lens and decreases with radial distance from the axis. The following quadratic equation closely describes the refractive index of a GRIN fiber lens[23,24]:

$$n(r) = n_0 (1 - \frac{g^2}{2} r^2) . \quad (1)$$

where r is the radial position from the axis, n_0 is refractive index on the lens axis, g is the gradient constant. The pitch, $p = 2\pi/g$, is the spatial period of the ray trajectory. The Standard Surface type in ZEMAX is employed for modeling of the ball fiber. Figure 1 (a) and (c) show the ZEMAX ray trace layout of the GRIN and ball fiber lenses, respectively. The working distance, depth of field and spot size were calculated in ZEMAX using the Gaussian beam theory.

The GRIN and ball fiber lens probes were made from a standard Corning SMF-28 single mode fiber as the principal light guide, a fiber spacer and a GRIN or a ball fiber lens as the focusing lens. For the GRIN lens probe, a fiber spacer with same outer diameter (0.125 mm) as SMF-28 was fusion-spliced via arc welds to the Corning SMF-28 and then accurately cleaved to a theoretically-determined length. A GRIN fiber was then fusion-spliced to the cleaved fiber spacer and precisely cleaved at a pre-calculated length to generate a desired beam-distance profile (i.e., working distance, depth of field, and spot size). For a short working distance probe, the section of the fiber spacer was omitted resulting in a simple fabrication process. For the ball lens probe, a fiber spacer with 250 μm outer diameter was fusion-spliced via arc welds to the Corning SMF-28 and then accurately cleaved to a theoretically-determined length plus extra 200 μm. The tip of the fiber spacer then was fused via arc welds to a perfect ball shape by inputting an appropriate fusion setting. To ensure minimum back-reflection for both probes, the indexes of the fiber spacer and the center of GRIN fiber were matched to the core index of the SMF.

We used a conventional low cost off-the-shelf optical multi-mode GRIN fibers as the GRIN lens, which has 100 μm core size, 140 μm outer diameter, a core refractive index $n_0 = 1.487$, and a gradient constant $g = 3.76$ at 1300 nm (Prime Optical Fiber Corp., Taiwan). The fiber spacers (Prime Optical Fiber Corp., Taiwan) are made of pure silica without a core. Fusion-splicing was processed using an Ericsson FSU 995 fusion-splicer and an EFC11 fiber cleaver (3SAE Technologies, TN, USA). The spliced interfaces produced minimum back-reflections since the mechanical strength at the interface was similar to that of the untreated fiber. The desired focused beam profile was obtained by tailoring the length of the fiber spacer and parameters of fiber lenses (length of GRIN fiber and diameter of the ball) based on the theoretical results. We fabricated the different variations of the GRIN and ball fiber lens modules with the different length of the fiber spacer and the different lens parameter. All samples were listed in Table 1 with detailed information.

A beam profile measurement system (BeamView Analyzer, OR, USA) with an infrared camera (Electrophysics, NJ, USA) and a Super Luminous Diode source (Covega, MD, USA) with 60nm 3dB bandwidth at 1310 nm center wavelength was used to characterize the beam parameters of the lens system. A 40X JIS microscopic objective lens and a related objective tube were attached to the input window of the camera to increase the image resolution. The horizontal and vertical resolutions of the system were 1.0 μm and 1.1 μm, respectively. The distribution of light intensity at various distances along the direction of propagation after the lens was accurately measured by the beam profile system. Working distance, depth of focus, $1/e^2$ spot size, and Gaussian fitting were analyzed from the measured intensity distribution. The results demonstrated in this work are all in the air medium.

In addition, for the GRIN fiber probe, after characterization of the lens, another fiber spacer could be fusion-spliced to the lens tip as a beam director by polishing the end of the fiber spacer to a 45 degrees angle and coating the polished surface with a total reflection film. This then allowed the beam to be reflected at a 90 degrees angle creating a side-view probe. For the ball lens probe, the beam can be 90 degree reflected by a 45 degree polished face on the ball lens. Figure 1 (b) and (d) show the typical scan electron microscope (SEM) pictures of the GRIN and ball fiber lens tip with beam directors, respectively. The fiber lens tip together with a tubing system and a connected linearly scanning or 360 degrees rotated motor could be built as an endoscope or a needle probe using in the biomedical optical imaging system for *in situ* and *in vivo* minimally invasive diagnostic and/or guided surgery and treatment applications[13,19].

Table 1. Structures of the various samples with measured beam properties.

Samples	Length of Fiber Spacer (mm)	Fiber Lens		Measured Beam Properties		
		Type	Length/ Diameter (mm)	Working Distance (mm)	Depth of Field (mm)	Spot Size (μm)
1	0	GRIN100/140	0.60	0.18	0.16	13
2	0	GRIN100/140	0.55	0.20	0.30	16
3	0	GRIN100/140	0.52	0.28	0.50	22
4	0	GRIN100/140	0.50	0.38	0.60	23
5	0	GRIN100/140	0.48	0.41	0.85	25
6	0	GRIN100/140	0.46	0.40	1.30	30
7	0	GRIN100/140	0.45	0.38	1.45	32
8	0.48	GRIN100/140	0.17	1.00	0.95	28
9	0.48	GRIN100/140	0.16	1.10	1.5	35
10	0.48	GRIN100/140	0.145	1.20	1.8	41
11	0.48	GRIN100/140	0.14	1.05	2.0	45
12	0.52	Ball Lens	0.15	1.00	3.6	50
13	0.55	Ball Lens	0.15	1.4	2.1	45
14	0.62	Ball Lens	0.15	1.20	1.1	27
15	0.70	Ball Lens	0.15	1.00	0.5	20
16	0.75	Ball Lens	0.15	0.90	0.48	18

3. Results and Discussion

For each sample in this study, optical intensity distribution data on the radial (i.e. x and y) planes were collected along the beam propagation (i.e. optical axial z) direction from the plane of the first half peak intensity (beginning-plane), through the maximum intensity plane, i.e. focus plane (center-plane), to the second half peak intensity plane (end-plane). Beam properties including working distance, spot size, and depth of field were analyzed by measured intensity distribution data with distance from the lens surface to the focal plane, $1/e^2$ beam diameter at the focal plane, and the distance between the begin-plane and the end-plane, respectively. The measured results of the beam properties are listed in Table 1 along with detailed descriptions of the samples. The theoretical and experimental results of working distance, depth of focus, and spot size of different variations vs. length of GRIN fiber or diameter of the ball lens (bottom x-axis), and length of fiber spacer (top x-axis) are shown in Figure 2 (a), (b), and (c), respectively, where, lines represent the theoretical results from ZEMAX at 1300 nm, amount them, dark doted line represent GRIN fiber lens without a fiber spacer, dark and light solid lines represent GRIN fiber lens with a constant length of fiber spacer (0.48 mm) and a constant length of GRIN fiber (0.17 mm), respectively; dark and light dash lines represent ball fiber lens with a constant length of fiber spacer (0.62 mm) and a constant diameter of the ball (0.30 mm), respectively. The related experimental results were represented by points, amount them, triangle points represent GRIN fiber lens and square points represent the ball lens.

From the theoretical result shown in Figure 2, short working distance (<0.4 mm) could be obtained by the GRIN fiber lens without fiber spacer shown as the dark dotted lines. To obtain larger working distance, a fiber spacer has to be used. In these cases, the working distance varies sharply with the length of GRIN fiber for the GRIN fiber lens and with the diameter of the ball for the ball fiber lens, but it varies less sharply with the length of the fiber spacer. The working distances have saturated values for each case. By compensating the working distance with the depth of field and the spot size, the optimized parameters (i.e. 0.9 – 1.2 mm working distance, 0.9 – 1.1 mm depth of filed, and <30 μm spot size) are not at the position of the largest working distance, instead, the optimized positions are around 0.17 mm length of the

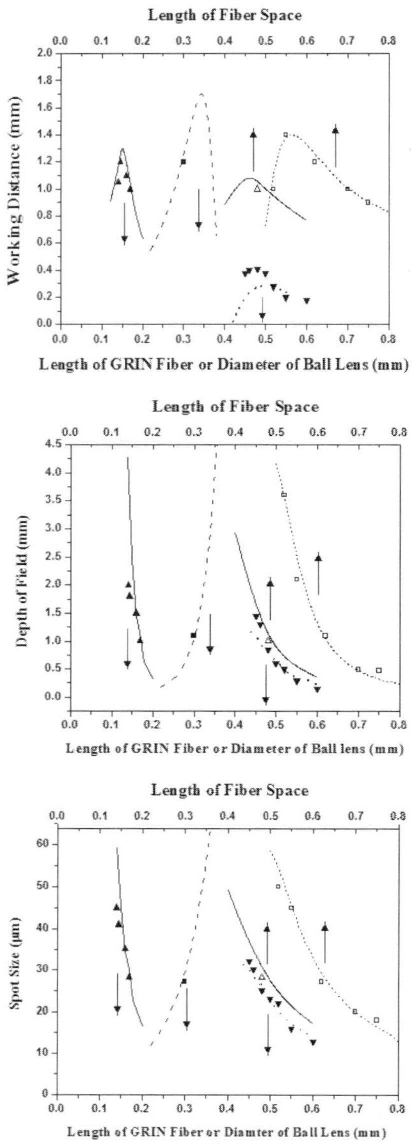

Figure 2: Theoretical and experimental results of working distance (a), depth of field (b), and spot size (c) vs. length of GRIN fiber or diameter of the ball lens (bottom x axis), and length of fiber spacer (top x axis), where, lines represent the theoretical results from ZEMAX at 1300 nm, amount them, dark doted line represent GRIN fiber lens without a fiber spacer, dark and light solid lines represent GRIN fiber lens with a constant length of fiber spacer (0.48 mm) and a constant length of GRIN fiber (0.17 mm), respectively; dark and light dash lines represent ball fiber lens with a constant length of fiber spacer (0.62 mm) and a constant diameter of the ball (0.30 mm), respectively. The related experimental results were represented by the points.

GRIN fiber with 0.48 mm length of fiber spacer for GRIN fiber lens and 0.3 mm diameter of ball and 0.62 mm length of fiber spacer for ball fiber lens. From the experimental results shown in Figure 2 and Table 1, we obtained the working distance of 1.0 mm, the depth of field of 0.95 mm, and the spot size of 28 µm from a GRIN fiber lens module (sample #8) and the working distance of 1.2 mm, the depth of field of 1.1 mm, and the spot size of 27 µm from a ball fiber lens module (sample #14). The results from the ZEMAX numerical optical design software were in a good agreement with the experimental results. Considering chromatic aberrations, from ZEMAX simulation for the ball fiber lens in the wavelength range of 1260 – 1370 nm, the relative variations of the working distance, depth of field and spot size were calculated all smaller than 4.0%. For the GRIN fiber lens, the range of the zero-dispersion wavelengths, $\{\lambda_0\}$ is 1297-1316 nm. The zero-dispersion slope, S_0, equal to or smaller than 0.101 ps/nm^2-km. Using the standard formula of fiber dispersion, $D(\lambda) = S_0[\lambda - \lambda_0^4 / \lambda^3]/4$ (ps/nm-km), we calculated the changes of refractive index in the 1260 – 1370 nm wavelength range. By using these values in ZEMAX, we calculated the relative changes of the working distance, depth of field and spot size were all smaller than 3%. Based on our results, the desired beam profile for the application of optical biomedical imaging systems can be obtained by the GRIN and ball fiber lens with or without fiber spacers. The technique described here possesses a high degree of flexibility for designing ultra-small optical probes with different beam shapes for the different tissue imaging.

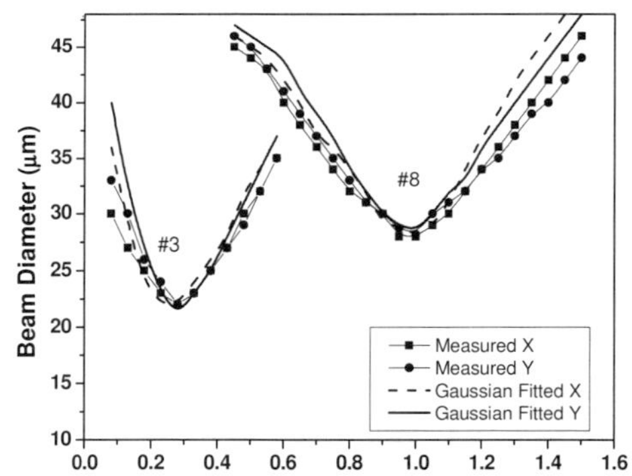

Figure 3: Measured and Gaussian-fitted 1/e^2 intensity beam diameters along the axial distance (zero is the position of the lens surface) at x (horizontal) and y (vertical) radial coordination in the distance range of depth of field of the samples #3 and #8, which was made from the GRIN fiber lens.

We found quality of the beam depends very much on the quality of the surface cleaving and the alignment of the fusion-splicing between the fiber spacer and the fiber lens. The high quality beam is easier to obtain for the probe with the ball fiber lens because the ball is made from the fiber spacer and there is no interface between the fiber spacer and the ball lens. By well controlling the cleaving and the fusion-splicing, we obtained high quality of beam for the probe of the GRIN fiber lens as well. Figure 3 shows measured and Gaussian-fitted $1/e^2$ intensity beam diameters along the axial distance z (zero is the position of the lens surface) at x (horizontal) and y (vertical) radial coordination in the distance range of depth of field for the samples #3 and #8 with the GRIN fiber lenses. In Figure 3, the smallest beam diameter value indicates spot size, x-coordinate value at the pole point indicates the working distance, and the distance range of the curve indicates the depth of field. The working distance of 0.28 mm, depth of field of 0.5 mm, and spot size of 22 μm were obtained for the sample #3 with 0.52 mm length of 100/140 GRIN fiber lens and without fiber spacer. The working distance of 1.0 mm, depth of field of 0.95 mm, and spot size of 28 μm were obtained for sample #8 with 0.17 mm length of 100/140 GRIN fiber lens and 0.48 mm fiber spacer. The x and y symmetry of the beam diameter is very good for both samples. The measured beam diameters are well matched to Gaussian-fitted values in the center (focused) regions, but have small deviations on either side of the center regions.

To further examine the beam quality of the GRIN fiber lens system, Table 2 shows measured beam profile images and measured normalized intensity distributions with Gaussian-fitted results at the x and y directions on the three typical planes (i.e. begin-plane, center-plane, and end-plane) of the sample #8. From the profile images and distributions shown in Table 2, the measured beam profiles match very well with Gaussian distributions at the beginning-plane and center-plane. On the end-planes, the measured and Gaussian-fitted intensity distributions

Table 2. Measured beam profile images and normalized intensity distributions with Gaussian-fittings at x (horizontal) and y (vertical) radial coordination for sample #8.

	Begin-Plane	Center-Plane	End-Plane
Images	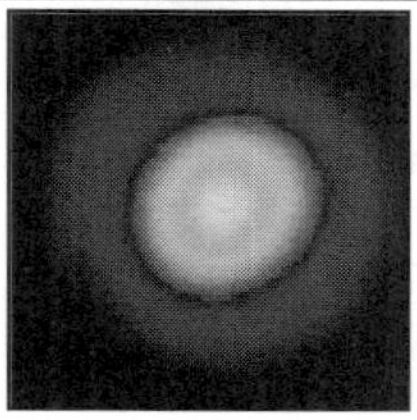		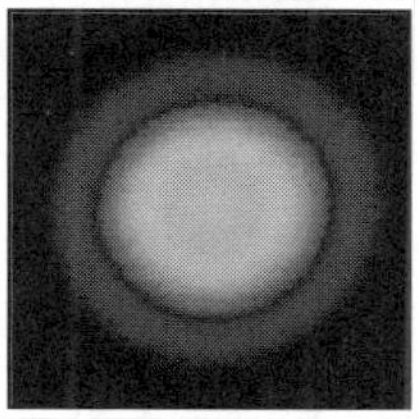
X[a]	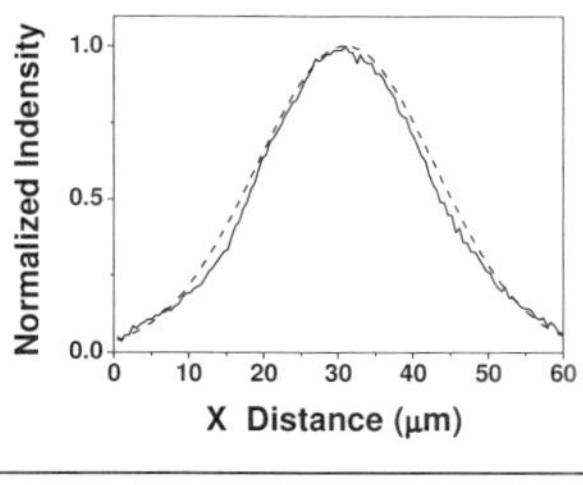	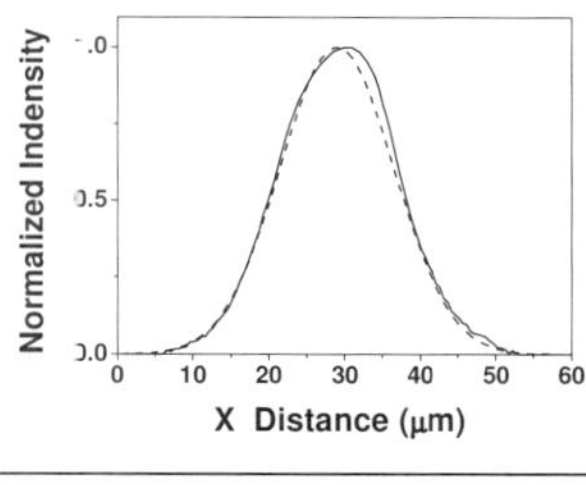	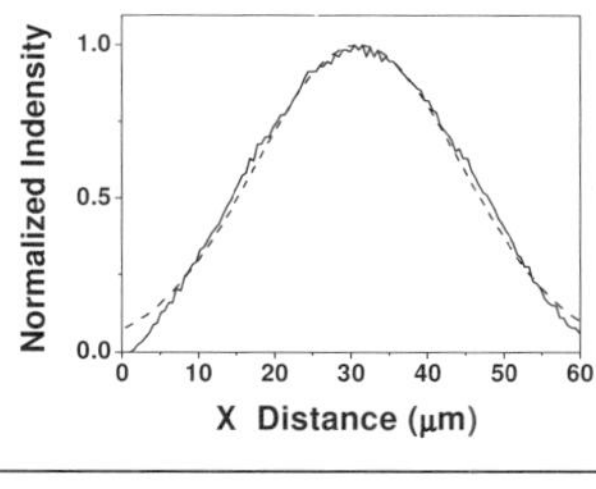
Y[a]	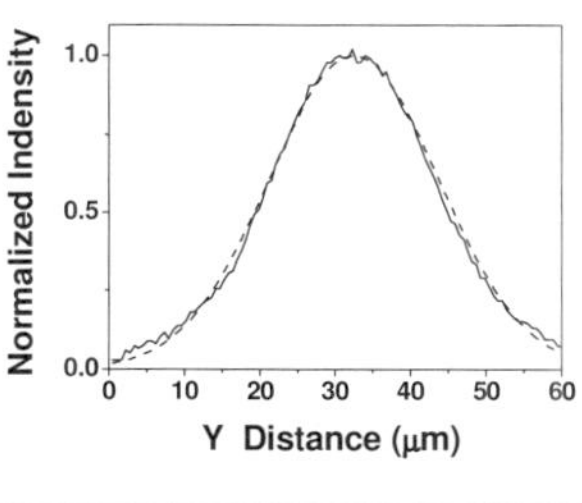	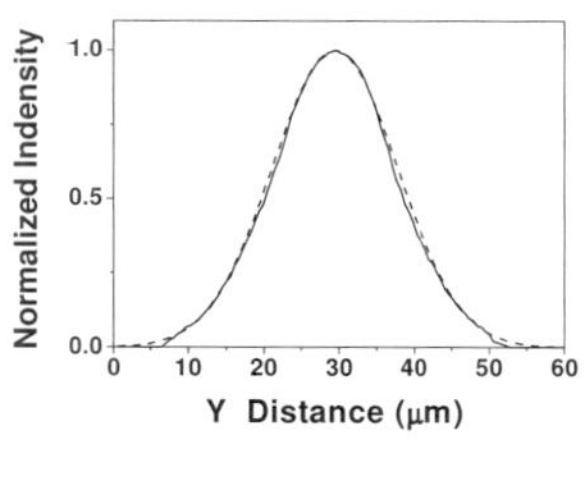	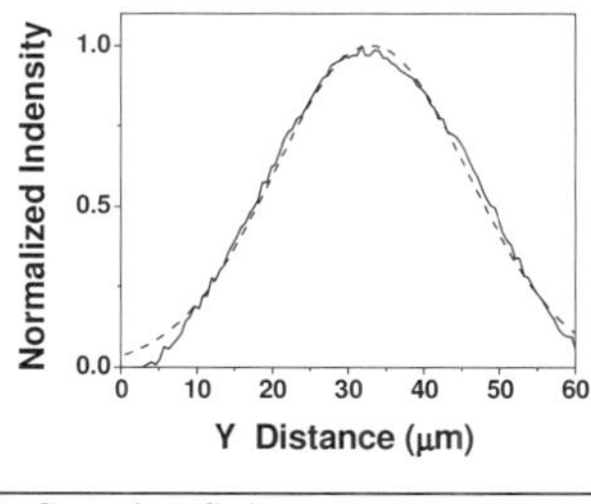

[a] Solid lines represent measured results and dashed lines represent their Gaussian-fitting.

generally match very well despite slight deviations on both tail ends of the distributions leading to discrepancies between the measured and Gaussian-fitted beam diameters as was shown in Figure 3. In addition, the circular shapes in the profile images as shown in Table 2 indicate high x and y symmetry of the beam profiles through all range of depth of field.

Fig. 4 (a) shows a fresh onion OCT image using a forward ball fiber lens (samples #13) probe with a swept-source OCT system. A schematic diagram of the swept-source OCT system is shown in Fig. 4 (b). The OCT system was described in detail elsewhere[25]. Briefly, the swept source (HSL2000, Santac) has a central wavelength of 1320 nm and a full scan wavelength range of 110 nm. The average output power and coherence length of the swept source is 7 mW and 10.8 mm, respectively. A repetition scan rate of 20 kHz was used in our system. Ninety percent of light source power was input into the sample arm while ten percent was fed into the reference arm by a 10/90 2x2 fiber coupler. A galvanometer (Blue Hill Optical technologies) scanner scanned the fiber probe light transversely on the sample over 4 mm at 40 Hz with 500 transverse pixels. The reference arm was arranged with a fiber collimator and a mirror. A variable attenuator was inserted between the collimator and mirror for adjusting the optical power on reference arm to achieve the shot noise limitation. Two fiber circulators were used in both the sample arm and reference arm to redirect the back-reflected light to a 3x3 fiber coupler with a 1/3 split ratio for a imaginary signal acquisition. Two polarization controllers (PC) in both reference and sample arms were used for adjustment to match the polarization state of the two arms. Two differential photo-detectors (2117-FC, NewFocus) were used with adjustable band-pass filter and gain. The two detector outputs were digitized using a data acquisition card (DAQ) (PCI 5122, National Instruments) with 14-bit resolution and acquired at a sampling speed of *100 MS/s*. A programmable wave number trigger (*k*-trigger) board (HSL-OPK-PKT, Santec) with *17.5 GHz k* interval was used for constant wave number *k* interval trigger. The k-trigger sampling number is 1024. The swept source generated a start trigger signal that was used to initiate the k-trigger and the function generator for the galvo scanner, in order to initiate the data acquisition process for each A-scan. An average background noise was removed, before the signal acquisition. A-scans with resolved complex conjugate artifact were obtained by inverse Fourier transformation (IFT).

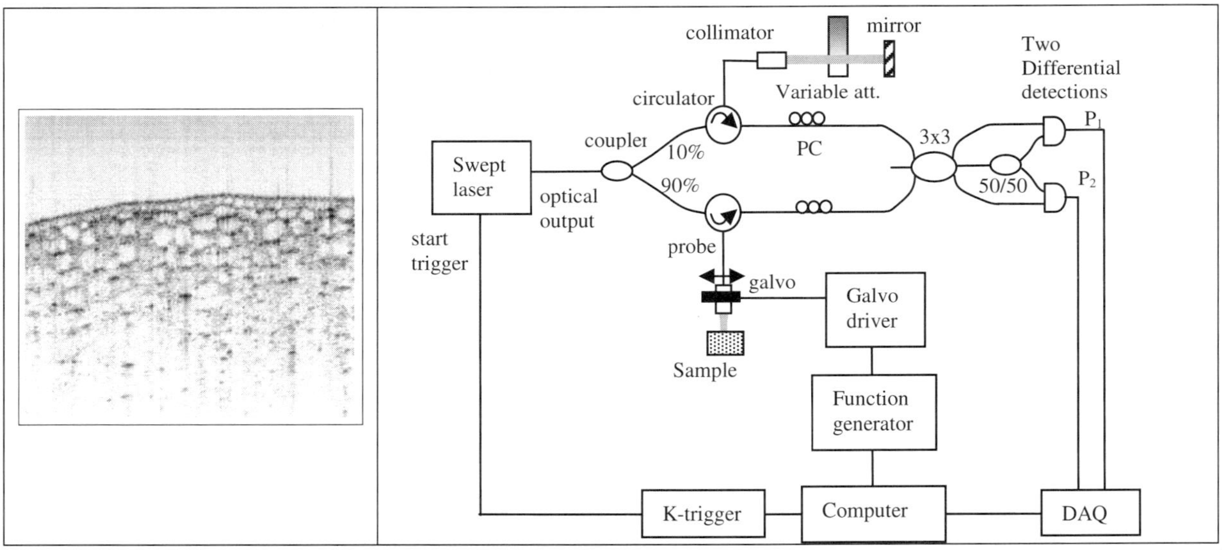

Figure 4: (a) A fresh onion OCT image (2x2 mm) using the ball fiber lens (sample #3) probe. (b) A schematic diagram of the swept-source OCT system used in this work.

4. Conclusion

We presented a design, construction and beam profile characterization of different variations of graded-index (GRIN) and ball fiber lenses, which were recently proposed for ultra-small OCT probes. Those fiber lens modules were made of single mode fibers and GRIN and ball fiber lenses with/without fiber spacers between them. We used fusion-splicing in between the fibers, lenses and spacers to ensure high quality light transmission. We found that beam-distance profiles (i.e. 0.4 - 1.2 mm of focus distance, 0.8 – 1.5 mm of depth of field, and 26 – 35 μm of spot size) can be obtained by precisely adjusting the lengths of the fiber spacer and the GRIN fiber lens or diameter of the ball lens for the different tissue imaging in human body. Using ZEMAX, optical design software, we modeled our optic probes which proved a precise approach. We obtained very high quality focused Gaussian beam profiles with high x and y symmetry using the conventional multi-mode GRIN fibers and home-made fiber ball lenses. Their high quality beam and ultra-small size make such fiber lens based probes very valuable for optical coherence tomography systems.

Reference

1. D. Huang, E. A. Swanson, C. P. Lin, J. S. Schuman, W. G. Stinson, W. Chang, M. R. Hee, T. Flotte, K. Gregory, C. A. Puliafito, and J. G. Fujimoto, "Optical coherence tomography," Science **254**, 1178-1181 (1991).
2. Z. P. Chen, T. E. Milner, D. Dave, and J. S. Nelson, "Optical Doppler tomographic image of fluid flow velocity in highly scattering media, Opt. Lett. **22**, 64-66 (1997).
3. J. A. Izatt, M. D. Kulkarni, S. Yazdanfar, J. K. Barton, and A. J. Welch, "*In vivo* bidirectional color Doppler flow imaging of picoliter blood volumes using optical cohorernce tomography", Opt. Lett. **22**, 1439-1441 (1997).
4. S. Yazdanfar, A. M. Rollins, and J. A. Izatt, "Ultrahigh-velocity resolution imaging of the microcirculation in-vivo using color Doppler optical coherence tomography" Proc. SPIE **4251**, 156-164 (2001).
5. V. X. D. Yang, M. L. Gordon, B. Qi, J. Pekar, S. Lo, E. Seng-Yue, A. Mok, B. C. Wilson, and I. A. Vitkin, "High speed, wide velocity dynamic range Doppler optical coherence tomography (Part I): system design, signal processing, and performance", Opt. Express **11**, 794-809 (2003).
6. B. R. White, M. C. Pierce, N. Nassif, B. Cense, B. H.Park, G. J. Tearney, B. E. Bouma, T. C. Chen, and J. F. de Boer, "In vivo dynamic human retinal blood flow imaging using ultra-high-speed spectral domain optical coherence tomography," Opt. Express **11**, 3490-3497 (2003).
7. S. Yazdanfar, A. M. Rollins, and J. A. Izatt, "Imaging and velocimetry of the human retinal circulation with color Doppler optical coherence tomography," Opt. Lett. **25**, 1448-1450 (2000).
8. Y. H. Zhao, Z. P. Chen, C. Saxer, S. H. Xiang, J. F. de Boer, and J. S. Nelson, "Phase-resolved optical coherence tomography and optical Doppler tomography for imaging blood flow in human skin with fast scanning speed and high velocity sensitivity," Opt. Lett. **25**, 114-116 (2000).
9. Y. Zhao, Z. P. Chen, C. Saxer, Q. Shen, S. Xiang, J. F. de Boer, and J. S. Nelson, "Doppler standard deviation imaging for clinical monitoring of *in vivo* human skin blood flow," Opt. Lett. **25**, 1358-1360 (2000).
10. V. X. D. Yang, S. Tang, M. L. Gordon, B. Qi, G. Gardiner, M. Cirocco, P. Kortan, G. Haber, G. Kandel, I. A. Vitkin, and B. C. Wilson, "Endoscopic Doppler optical coherence tomography in human gastrointestinal tract: initial experience," Gastrointest. Endosc. **61**, 879-890 (2005).
11. P. H. Tran, D. S. Mukai, M. Brenner, and Z. Chen, "*In vivo* endoscopic optical coherence tomography by use of a rotational microelectromechanical system probe" Opt. Lett. **29**, 1236-1238 (2004).
12. X. Li, C. Chudoba, T. Ko, C. Pitris, and J. G. Fujimoto, "Imaging needle for optical coherence tomography," Opt. Lett. **25**, 1520-1522 (2000).
13. V. X. D. Yang, Y. X. Mao, N. Munce, B. Standish, W. Kucharczyk, N. E. Marcon, B. C. Wilson and I. A. Vitkin, "Interstitial Doppler optical coherence tomography", Opt. Lett. **30**, 1791-1793 (2005).

14. J. G. Fujimoto, S. A. Bopart, G. J. Tearney, B. E. Bouma, C, Pitris and M. E. Brezinski, "High resolution in vivo intra-arterial imaging with optical coherence tomography", Heart **82**, 128-133 (1999).
15. L. J. Diaz-Sandoval, B. E. Bouma, G. J. Tearnay and I. Jang, "Optical Coherence Tomography as a tool for percutaneous coronary interventions", Catheter. And Cardio. Interventions **65**, 492-496 (2005).
16. E. Swanson, C. L. Petersen, E. McNamara, R. B. Lamport, D. L. Kelly, "Ultra-small optical probes, imaging optics, and methods for using same", U.S. Patent 6445939 (Sept. 3, 2002).
17. M. Shishkov, B. E. Bouma, and G. J. Tearney, "System and method for optical coherence tomography", U.S. Patent 20060067620A1 (Mar. 30, 2006).
18. W. A. Reed, M. F. Yan, and M. J. Schnitzer, "Gradient-index fiber-optic microprobes for minimally invasive *in vivo* low-coherence interferometry", Opt. Lett. **27**, 1794-1796 (2002).
19. M. S. Jafri, S. Farhang, R. S. Tang, N. Desai, P. S. Fishman, R. G. Rohwer, C. Tang, J. M. Schmitt, "Optical coherence tomography in the diagnosis and treatment of neurological disorders", J. Biomed. Opt. **10**(5) 051603 (2005).
20. H. Li, B. A. Standish, A Mariampillai, N. R. Munce, Y. Mao, S. Chiu, N. E. Marcon, B. C. Wilson, A. Vitkin, and V. X. D. Yang, "Feasibility of interstitial doppler optical coherence tomography for in vivo detection of microvascular changes during photodynamic therapy", Lasers in Surgery and Medicine **38,** 754–761 (2006).
21. A. W. Sainter, T. A. King, M. R. Dickinson, "Theoretical comparison of light sources for use in optical coherence tomography", Proc. SPIE **4619**, 289-299 (2002).
22. Y Mao, S, Chang, S, Sherif, and C. Flueraru, "Graded-index fiber lens proposed for ultrasmall probes used in biomedical imaging", App. Opt., 46, 5887-5894 (2007).
23. H. W. Kogelnik and T. Li, "Laser Beam and Resonators" Appl. Opt. **5**, 1550 (1966).
24. W. L. Emkey and C. A. Jack, "Analysis and evaluation of graded-index fiber-lenses", J. of Lightwave Tech., **LT-5**, 1156-1164 (1987).
25. Y. Mao, S. Sherif, C. Flueraru, and S. Chang, "3x3 Mach-Zehnder Interferometer with Unbalanced Differential Detection for Full Range Swept-Source Optical Coherence Tomography", Appl. Opt. submitted (Manuscript ID: 86374).

CROSS-SECTION IMAGING OF RICE TILLERS BY MCT SYSTEM*

ZHENG FANG, KUN BI, SHAOQUN ZENG, QINGMING LUO†, QIAN LIU

The Key Laboratory of Biomedical Photonics of Ministry of Education, Huazhong University of Science and Technology Wuhan, Hubei 430074,China

Tiller number is one important parameter of paddy morphology. The yield of rice depends remarkably on it. The usual roboticized way to get the plant morphology is taking photos by digital camera and performing digital image processing by software. But the tiller number is hard to obtain when it is rather largish. Because tillers are not isolated to each other in the pictures even if we change to any photo angle. On the other hand, computer tomography technology can solve this problem. Computer tomography generates the cross-section image from a serial of projections. And can easily count the tiller number on the cross-section image. This paper explains the hardware structure and the reconstruction algorithm, and then shows a perfect result of an experiment. Some details in experiment and the future research are discussed in the last part.

1. Introduction

Tiller number is an important agronomic trait of rice [1]. It is mainly controlled by the general genetic factor [2], and is also affected by the grow environment such as temperature and light. It plays a significant role in the formation of grain yield in rice, and is a weighty factor of fertilizer application timing [3]. Appropriate tiller number is the prerequisite for attaining desired yield in rice breeding programs. The traditional way to get this parameter is numerate the number of tillers by hand, which is a laborious work. There is no automatic method to get the parameter. Some researchers try to find the relation between tiller number and other morphological characteristic parameters of rice. For example, some papers describe the number and components of tillers and leaves and their relationships [4]. But these indirect ways are merely right in statistics, and the empirical formulas are got by large amount of experiment. We make an attempt at to get the cross-sectional image of rice tiller, which may lead to many further researches about tiller such as tiller number countering. The method bases on the technology of computerized tomography. A prototype of cone-beam CT is installed to show the slice image of rice tillers. The sample is Liangyoupeijiu, a super high-yield hybrid rice [5], which is widely planted in South China [6]. The experiment result exhibits a wonderful slice image of rice tillers, in which we can see even the anatomic structure of each tiller.

2. Materials and Methods

2.1. *Specimen*

The specimen is Liangyoupeijiu, provided by Huazhong Agricultural University, Wuhan, Hubei, China. It is an improved plant type with erect uppermost leaves. It has a high light transmission rate and a low light extinction coefficient from middle to late growth stage. With a high level of heterosis and plant ideotype, the hybrid showed high yield potential, good grain quality, and strong resistance to bacterial leaf blight and blast [7]. For yield potential and grain quality, Liangyoupeijiu is well known as the prototype of "super hybrid rice", which has been registered as the first two-line hybrid. It had spread in southern China and Southeast Asia, including Vietnam and the Philippines [8]. The specimen was planted in experimental field, and was dig up to plant in a culture pot with fertile soil when in the early procreation stage.

2.2. *Hardware*

To get the slice image of rice tillers, we build up a prototype of cone-beam CT. For a radiography imaging system, there are two key components: detector and ray source. The detector is PaxScan1313, a new product of amorphous silicon digital x-ray imagers manufactured by Varian Medical Systems, Inc. It is the latest development from Varian Medical Systems. This new product incorporates all of the latest high speed

* This work is supported by the National Natural Science Foundation of China (30727002) and the National High Technology Research and Development Program of China (2006AA020801).

† Work partially supported by Cheung Kong Scholars Programme.

electronics used in real-time imagers, but in a smaller format suitable for applications currently using 6 inch Image Intensifiers. Interfacing to this new panel is a simple task through the industry standard CameraLink port. The high resolution receptor combined with a software-based command processor capable of 20 frames per second makes this an ideal imager for microfocus X-ray applications. Computer tomography (CT) applications also benefit from the high dynamic range and uniform image quality across a wide X-ray dose range. An optional external power supply is available to simplify product integration. The detector area is 130mm×130mm, with 1024×1024 pixels. The pixel pitch is 127μm × 127μm, which provide the limiting resolution of 3.94lp/mm. PaxScan1313 can detect the x-ray with the energy range from 30 to 160 kVp. The A/D conversion is 14 bits. The most distinct character of this detector is that it can show the real-time image up to 30 fps. The micro-focus x-ray tube can work with the current of 0.25-0.55mA and with the voltage of 35-70kV. Its focus size is 0.05mm.

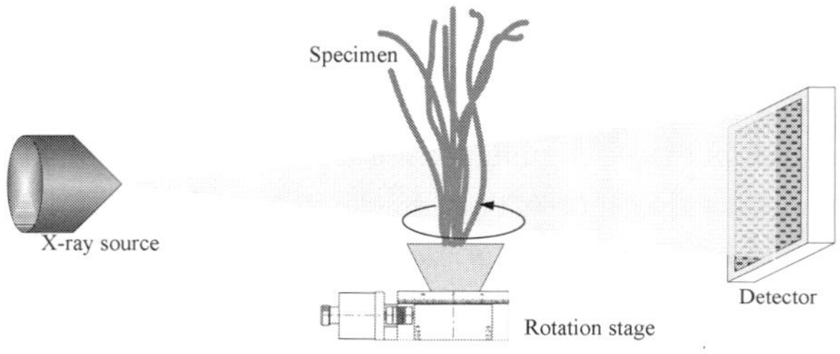

Figure 1. Sketching of the cone-beam system

Figure 1 is the sketching of the cone-beam CT system. The specimen is placed on the rotating stage. The line connecting the x-ray source focus and the center of the detector is perpendicular to the rotating axis. The plant containing the connecting line and perpendicular to the rotating axis is called mid-plane. We adjust the height of the specimen until the mid-plane can show the perfect image of all tillers. The rotation angle of each step is 1.8 degree. We take 200 shadow images of radiation, which cover the whole circularity scanning of the specimen. There is a computer not displayed in Figure 1, control the scanning procedure. The rotating stage rotates a step; the x-ray source begin to expose; then the detector save the raw date of shadow image into the hard disk of the computer; after above finished, the x-ray source is terminated. These procedures are repeated until the last shadow image is taken.

2.3. *Algorithm*

After acquiring slice images, what left to do is reconstructing slice image of the specimen. The key technology of CT system is algorithm, which can magically change projections into cross-sections. Although here is a cone-beam CT system, in this experiment we only need to reconstruct one cross-sectional image. We choose the mid-plane to reconstruct, and the algorithm can be simplified to fan-beam algorithm [9]. Iterative algorithm and backprojection algorithm are two genres of CT algorithm. Iterative algorithm came forth earlier, but backprojection algorithm prevails in applications soon after its naissance. We use backprojection algorithm to reconstruct the cross-sectional image.

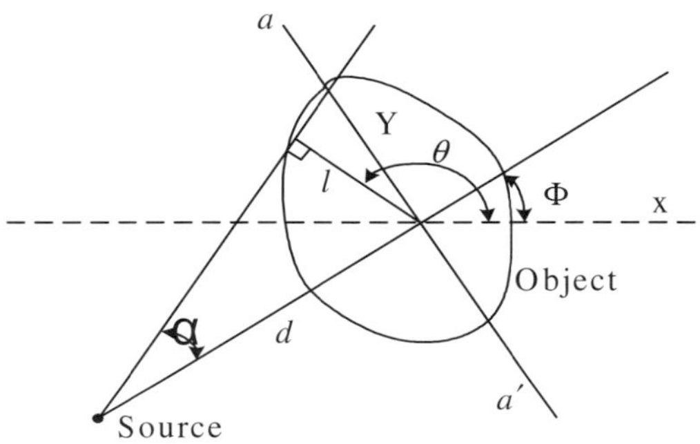

Figure 2. Sketching of the cone-beam system

The planar detector system may, for convenience, be represented by its projection on a plane parallel to it that contains the axis of rotation [10]. This axis is a distance d from the x-ray point source, as in Figure 2. Thereafter, we refer to the detector plane as if it were in the plane containing the rotation axis. The angle of rotation is Φ, and the coordinate along the detector that specifies the point of detection is Y. The perpendicular distance l from the origin (0, 0) to the ray that intersects the detector plane at Y is related to Y by

$$l = Yd \,/\, (d^2 + Y^2)^{1/2}. \quad (1)$$

The angle θ from the x axis to the perpendicular is given by

$$\theta = \Phi + \pi/2 + \alpha, \quad (2)$$

where

$$\alpha = \tan^{-1}(Y/d) = \tan^{-1}[l/(d^2-l^2)^{1/2}]. \quad (3)$$

From the projection data, the density can be reconstructed according to the formula [11]:

$$f(r,\phi) = \frac{1}{4\pi^2}\oint \frac{d^2}{[d+(r\cos(\phi-\Phi)]^2}\tilde{P}_\Phi[Y(r,\phi)]d\Phi, \quad (4)$$

where

$$Y(r,\phi) = dr\sin(\phi-\Phi)/[d+r\cos(\phi-\Phi)], \quad (5)$$

$$\tilde{P}_\Phi(Y) = \int_{-\infty}^{\infty} dY' \frac{d}{(d^2+Y'^2)^{1/2}} P_\Phi(Y')g(Y-Y'), \quad (6)$$

and

$$g(Y) = \mathrm{Re}\int_0^{\omega_{y0}} \exp(i\omega Y)\omega d\omega. \quad (7)$$

3. Result

Figure 3. Experiment result

(a) photo of specimen

(b) projection of rice tillers

(c) sinogram of mid-plane

(d) reconstruction of rice tillers in mid-plane

We take a pot of hybrid rice, Liangyoupeijiu, on the rotation stage. The computer control the rotation stage through a motor driver provide by the same company of the rotation stage. The distance from rotation axis to detector plane is 140mm, and the distance from rotation axis to x-ray focus is 1200mm. So the fan angle is 5.55 degree. One photo of the specimen is showed in Figure 1(a). The white foursquare box in Figure 1(a) illustrates the radiation area demonstrated in Figure 1(b). There are dark dots in the shadow images, which is caused by bolar granules adhered on the tillers and leaves. Because the bolar granules are not in the mid-plane, they will not affect the reconstruction result. After taking 200 projections, backprojection algorithm reconstructs the cross-sectional image of mid-plane. The first step is select one line of pixels in the mid-plane of each projection and save the alignment in the hard disk for the further processions. The alignment of these 200 mid-plane projections is called sinogram, which is displayed in Figure 3(c). The reconstructed cross-sectional image is showed in Figure 3(d). The result is fairly vivid. It provides a vantage for us to do further study on tillers.

Discussion and Conclusion

This paper illustrates the way to get the cross-sectional image of rice tiller by micro-CT system. The frame rate and the quality of the flat detector is a pair of contradictories: the quality of the projection will deteriorate when we increase the frame rate. In this experiment we chose the frame rate of 1fps, which leads to a perfect quality of the shadow images. If higher frame rate was selected, the noise in the projection will be intensified, which will destroy the backprojectional image. To speed up the rate of sampling and reconstruction, a short-scan fan-beam reconstruction of 180 degrees' rotation can be adopted [12, 13]. A high-pass filter is used in filtered backprojection. It amplifies noises significantly, resulting in obvious noises in the reconstructed image. If you want to get higher spacial resolution, a better x-ray source of smaller focus is prerequisite, which can provide a higher amplificatory ratio. Further work is under going, such as obtaining the tiller number from the slice image automatically and displaying the three-dimensional image of tillers.

Acknowledgments

I would like to thank Guoxing Chen (National Key Laboratory of Crop Genetic Improvement, Huazhong Agricultural University, Wuhan 430070), who provided the specimen.

References

1. Alexander J. Smart, Kenneth P. Vogel, Lowell E. Moser, Walter W. Stroup, "Divergent Selection for Seedling Tiller Number in Big Bluestem and Switchgrass", *CROP SCIENCE*, **43**, 1427-1433, (2003).
2. Wei-Ren Wu, Wei-Ming Li, Ding-Zhong Tang, Hao-Ran Lu, A. J. Worland, "Time-Related Mapping of Quantitative Trait Loci Underlying Tiller Number in Rice", *Genetics*, 151, 297-303, (1999).
3. LIU Gui-fu, XU Hai-ming, YANG Jian, ZHU Jun, "Genetic analysis on tiller number and plant height per plant in rice (Oryza sativa L.)", *Journal of Zhejiang University*, **32**(5), 527-534, (2006).
4. Alert J. Fischer, Hector V. Ramirez, Kevin D. Gibson, Beatriz Da Silveira Pinheiro, "Competitiveness of Semidwarf Upland Rice Cultivars against Palisadegrass (Brachiaria brizantha) and Signalgrass (B. decumbens)", *AGRONOMY JOURNAL*, **93**, 967-973, (2001).
5. REN Zhao-Yu, XU Xiao-Ming, WANG Shui-Cai, XIN Yue-Yong, HE Jun-Fang, KUANG Ting-Yun, "Properties of Picosecond Fluorescence of Super High-Yield Hybrid Rice", *Acta Botanica Sinica*, **45**(12), 1442-1446, (2003).
6. XIN Ye-yun, ZHENG Zhan, XIONG Yi-Ping, YUAN Long-ping, "Identification and Purity Test of Super Hybrid Rice with SSR Molecular Markers", *Rice Science*, **12**(1), 7-12, (2005).
7. Longxing Tao, Xi Wang Huijuan Tan, Junhui Shen, Shihua Cheng, "Effects of soil water saturation on rice growth and development", *Indian J. Crop Science*, **1**, 59-64, (2006).
8. XIE Wen-xia, YAN Li-jiao, WANG Guang-huo, "Simulation and Validation of Rice Potential Growth Process in Zhejiang Province of China by Utilizing WOFOST Model", *Rice Science*, **13**(2), 125-130, (2006).
9. BERTHOLD K. P. HORN, "Fan-Beam Reconstruction Methods", *PROCEEDINGS OF THE IEEE*, **67**(12), 1616-1623, (1979).
10. L. A. Feldkamp, L. C. Davis, J. W. Kress, "Practical cone-beam algorithm", *J Opt. Soc. Am*, **1**(6), 612-619, (1984).
11. PS Cho, RH Johnson, YW Griffin, "Cone-beam CT for radiotherapy applications", *Phys. Med. Biol.*, **40**, 1863-1883, (1995).
12. Shinichiro Mori, Masahiro Endo, Shuher Komatsu, Susumu Kandatsu, Tomoyasu Yashiro, Masayuki Baba, "A combination-weighted Feldkamp-based reconstruction algorithm for cone-beam CT", *Phys. Med. Biol.*, **51**, 1953-1965, (2006).
13. Kai Zeng, Zhiqiang Chen, Li Zhang, Ge Wang, "A half-scan error reduction based algorithm for cone-beam CT", *Journal of X-Ray Science and Techology*, **12,** 73-82, (2004).

DETECTION AND APPLICATION OF MICROFLUIDIC ISOTHERMAL AMPLIFICATION ON CHIP*

GUOLIANG HUANG [A, B *], CHAO HAN [A, B], XIAOYONG YANG [B], JIANG ZHU [A, B], SHUKUAN XU [A, B], CHENG DENG [A, B]

A. Medical Systems Biology Research Center, Tsinghua University School of Medicine, Beijing, 100084, China

B. National Engineering Research Center for Beijing Biochip Technology, Beijing, 102206, China

** Corresponding Author: Guoliang Huang, E-mail: tshgl@mail.tsinghua.edu.cn*

Loop-mediated isothermal amplification (LAMP) is a novel nucleic acid amplification method. Compared with the widely utilized polymerase chain reaction (PCR), LAMP has higher speed and efficiency as well as lower requirement for system temperature control, because the whole amplification process is isothermal and no efforts are needed to switch between different temperatures. In this paper, we designed and fabricated different kinds of PC microfluid chips, explored appropriate reaction condition for LAMP in microenvironment (1nL→10μL), and developed a microfluidic isothermal amplification and detection system. DNA optimal amplification temperature is obtained, the time of starting exponential amplification of DNA is put forward farther. An application of testing clinic MRSA was performed. The optimal condition of DNA amplification in microenvironment, a little reaction materials used and early starting exponential amplification time of DNA are very important for clinic DNA detection and the application of Lab on a Chip.

1. Introduction

Nucleic acid amplification and analysis has so far been employed in various fields. The first developed amplification protocol is the polymerase chain reaction (PCR) (Saiki et al., 1985), which is also the most widely used method nowadays. With this method, we can amplify a specific nucleic acid sequence from a complex sample in a cyclic process, and generate a large number of identical copies, which is available for further analysis.

Loop-mediated isothermal amplification (LAMP) is a novel nucleic acid amplification method, which is first described by Notomi et al. (2000). Since this first application, LAMP has been employed in various fields, including the detection of virus (Fukuta et al., 2003; Imai et al., 2006; Hagiwara et al., 2007) and bacteria (Aoi et al., 2006; Minami et al., 2006; Hara-Kudo et al., 2007), etc. Compared with conventional amplification methods such as PCR, LAMP has several advantages: First, the whole amplification reaction can be carried out under isothermal conditions, which makes the temperature control system relatively easy to design and reduces the cost. Second, the reaction is highly specific for the target. Furthermore, because no extra time is needed for LAMP reaction to wait for thermal changes, the amplification efficiency is greatly enhanced.

Microfluid chips have already been widely utilized in PCR, for their fast speed and small sample volume requirement, as well as their potential to be integrated in micro total analysis or 'Lab on a Chip' systems (Zhang and Xing, 2007). Temperature control systems for PCR microfluid chips has also been developed (Roper et al., 2007). However, the combination of microfluid chips and LAMP has not been reported so far.

In this paper, we designed and fabricated different kinds of PC microfluid chips, explored appropriate reaction condition for LAMP in microenvironment, and developed a microfluidic isothermal amplification and detection system. The system also passed performance testing and feasibility analysis. Using this system, we successfully carried out real-time fluorescence detection of LAMP. The optimal amplification temperature is obtained, the time of starting exponential amplification of DNA is put forward farther. An application of testing clinic sample was performed.

* This work is supported by the National Foundation of High Technology of China (2006AA020701 and 2006AA020803), National Program on Key Basic Research Projects 973 of China (2006CB705700), the Nature Science Foundation of Zhejiang Province (2006C21G3210005) and Tsinghua-Yuyuan Medicine Foundation (40000510B).

2. Materials and Methods

2.1. *Materials*

Thermo Buffer, $MgSO_4$, RPM_0691-Sau_femA_FIP, RPM_0691-Sau_femA_BIP, PM_0691-Sau_femA_F3, RPM_0691-Sau_femA_B3, dNTPs, dUTP, BSA, EvaGreen, Bst polymerase, UNG, Template (Methicillin Resistant Staphylococcus Aureus gene-TR558).

2.2. *Amplification Temperature Control Methods*

The process of DNA amplification temperature control in microfluid chips with a microenvironment (1nL→10μL) is as table 1.

Table 1 The control process of DNA amplification

Temperature(℃)	Time(min)	Remark
37	2	for ready
T	3	for preparation
T	60	for test in real time
80	5	for put out life activity

In Table 1, T is a temperature changed from 50℃ to 70℃ for different microenvironment (1nL→10μL) of microfluid chips. When the microenvironment of microfluid chips is set, for example as 1μL, we can change DNA amplification temperature T and explore the optimal condition of DNA amplification.

2.3. *Microfluid Chip Fabrication*

The microfluid chips are designed as Figure 1. At first, the microfluid chips with different structure and microenvironment 1nL→10μL was designed as masking diagram by machine design software *Solidworks* 2006. Secondly, these designed masking diagrams were fabricated in PC materials with thickness 0.5mm, 1.0mm and 1.5mm etc, and PC masking was obtained. Thirdly, two PC films with thickness 0.1mm were bonded to PC masking surface from up and down directions by using heat-compress technique with a heating temperature 150℃ and a compressive stress 100Kg about 30 minutes. Finally, microfluid chips was cut edge tidily for encapsulation, and respectively added two small holes at the top and bottom of microenvironment for injecting sample.

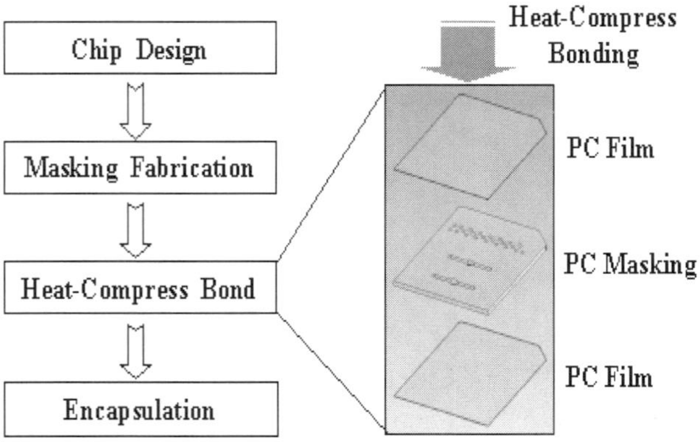

Figure 1. The fabrication of microfluid chips

2.4. *Microfluid Chip Isothermal Amplification Detection System*

In order to detect a real time fluorescent signal of DNA amplification with high sensitivity and high resolving power, a confocal optical system was designed and optimized by using the rays tracing calculation. The optical geometry structure of system was described in Figure 2, and whole optical design structure parameters were written in table 2(units: mm).

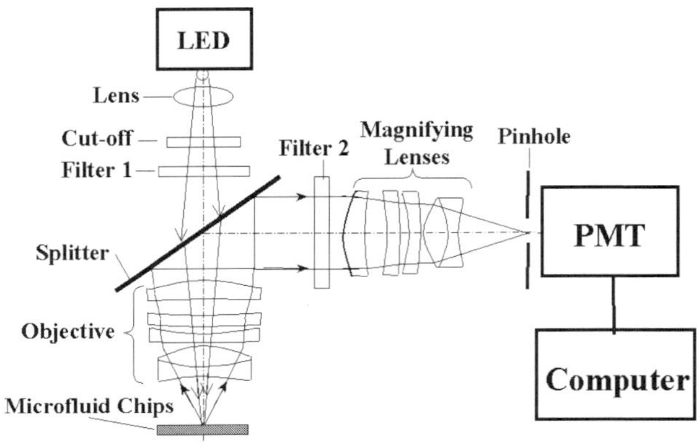

Figure 2. The optical geometry structure of system

Table 2 The whole optical design structure parameters (units:mm)

Surface	Radius	Thickness	Glass
1	-26.3	2.20	ZF2
2	20.1	6.10	ZK7
3	-12.7	0.21	
4	79.3	2.70	ZK7
5	32.5	2.00	
6	183.5	3.10	ZF2
7	73.2	4.20	
8	-354.5	5.50	ZK7
9	-25.5	100.00	
10	25.5	5.50	ZK7
11	354.5	4.20	
12	-73.2	3.10	ZF2
13	-183.5	2.00	
14	-32.5	2.70	ZK7
15	-79.3	0.21	
16	12.7	6.10	ZK7
17	-20.1	2.20	ZF2
18	26.3	16.50	

In Figure 2, the optical detection system is composed of a set of objective and magnifying lenses. The objective set is a combination of five lenses including one doublet and uses only two glass materials, ZF2 and ZK7 shown in the table2. The objective has a high numerical aperture of 0.4 for collecting the signal from the chips, a focal length of 32.0 mm, and a front focal length of 16.5 mm to provide a long working distance. There is a parallel ray path between the objective and the magnifying lenses, to which it is convenient to also add the filter and other optical elements. The magnifying lenses consist of five lenses with one set of doublets also. The light of LED was assembled by lens and filtered by the filter1 with a center wavelength of 532nm and width of 20nm, and an irradiated beam was obtained. The irradiated beam passed the splitter was focused on microfluid chips and the fluorescent signal was excited. A cut-off was used to control the exposure time and reduced the photo-bleaching from the irradiated beam. The fluorescence of DNA amplification on the microfluid chips was collected by the objective, reflected by splitter, filtered by the emission filter 2 with a center wavelength of 570nm and width of 50nm, and focused finally by the magnifying lens set on the pinhole with diameter 100μm. PMT collected the fluorescence and sent to the computer for digital image processing by a data acquisition unit as Figure 3.

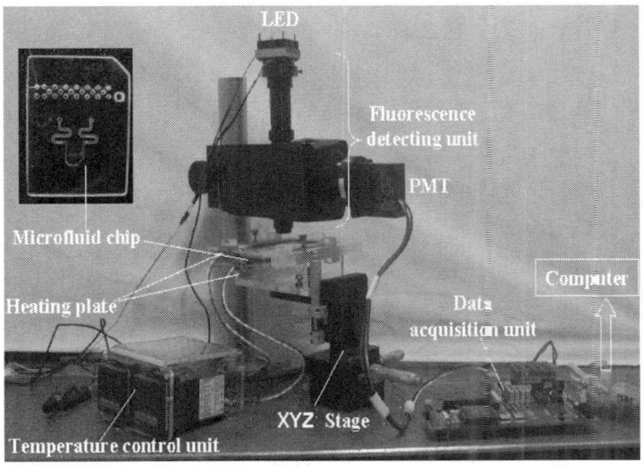

Figure 3. The microfluidic isothermal amplification and detection system

Based on above optical design and microfluid chips, a microfluidic isothermal amplification and detection system was set as Figure 3, which integrated XYZ stage positioning unit, temperature control unit, heating plate, microfluid chip, fluorescence detecting unit and data acquisition unit. Temperature control and real time fluorescent signal detection of isothermal amplification in microfluid chip were supervised by a computer.

As the process of DNA amplification temperature control in the Table 1, when the microenvironment of microfluid chips was set in Figure 3, T was changed with different temperature from 50℃ to 70℃ for DNA amplification, and the optimal condition of DNA amplification was explored.

3. Results and Discussion

With above developed microfluid chips and microfluidic isothermal amplification and detection system, the clinic sample of Methicillin Resistant Staphylococcus Aureus gene-TR558 was identified. The whole process of temperature control and real time fluorescent signal detection of isothermal amplification in microfluid chip were performed as Figure 4.

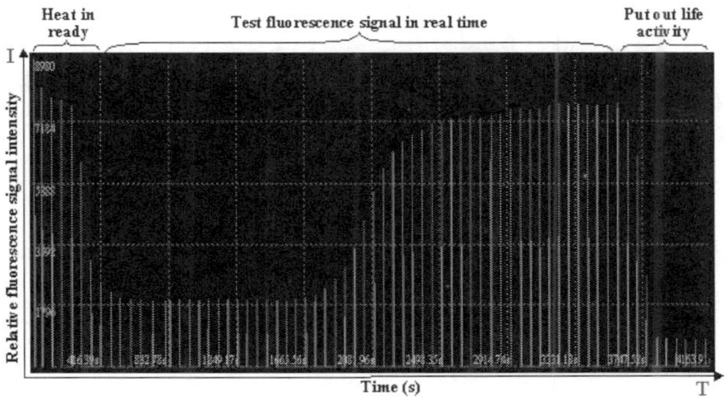

Figure 4. The process monitor of DNA isothermal amplification in microfluid chips

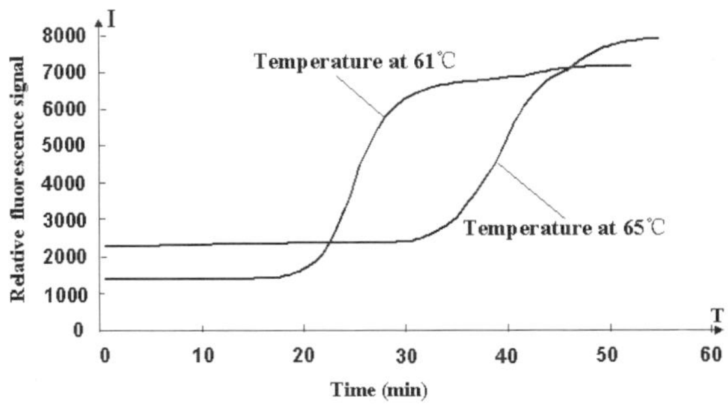

Figure 5. The real time fluorescence signal of DNA amplification at different temperature

When the microenvironment of microfluid chips with an efficient detection volume 1nL→10μL was set, DNA amplification with different temperature T also was done from amplification temperature T=50℃ to T=70℃ respectively, we explored the optimal condition of DNA amplification. For example, when a microfluid chip was set with the efficient detection volume 1μL, the real time fluorescence signal of DNA amplification for clinic sample of Methicillin Resistant Staphylococcus Aureus gene-TR558 at temperature T=65℃ and T=61℃ were obtained as Figure 5. In Figure 5, the time of starting exponential amplification of DNA at temperature 65℃ is at about 32'th minute and the ratio of the maximum relative fluorescence signal to the original relative fluorescence signal is about 4.2. When amplification temperature was changed to 61℃, the time of starting exponential amplification of DNA was put forward to about 18'th minute, the ratio of the maximum relative fluorescence signal to the original relative fluorescence signal was about 5.1. The shape of real time fluorescence signal of DNA amplification at 61 ℃ was good as the exponential function curve.

As a comparison, the commercial real time fluorescence PCR instrument (Capitalbio Corp., Beijing, China) was used for testing DNA isothermal amplification of the same clinic sample of Methicillin Resistant Staphylococcus Aureus gene-TR558 with a volume 20μL, the real time fluorescence signal of DNA amplification was obtained as Figure 6. In Figure 6, the time of starting exponential amplification of DNA was at about 34'th minute and the ratio of the maximum relative fluorescence signal to the original relative fluorescence signal was about 5.

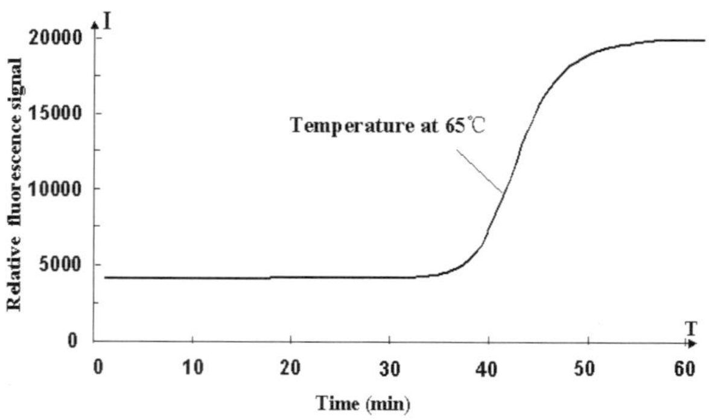

Figure 6. The real time fluorescence signal of DNA amplification by using RT-PCR instrument

From Figure 5 and Figure 6, when using the microfluid chips and optimal condition of DNA amplification, a little reaction materials used and early starting exponential amplification time of DNA can be seen obviously.

4. Conclusion

We have demonstrated a detection and application of microfluidic isothermal amplification on chip in this paper. The method of designing and fabricating PC microfluid chips was introduced, a new microfluidic isothermal amplification and detection system was developed, and DNA optimal amplification temperature for LAMP in microenvironment (1nL→10μL) was explored. The experiment showed that the time of starting exponential amplification of DNA was put forward when reaction materials were reduced. Under a 1μL efficient detection volume, the optimal temperature condition of DNA amplification was at 61℃. Using the microfluid chips and optimal condition of DNA amplification, the reaction materials and testing time of DNA amplification are very economical, which are very important for clinic DNA detection and the application of Lab on a Chip.

Acknowledgements

We thankfully acknowledge Wang Guoqing and Zhang Rong in biological experiment, Zhou Cheng in temperature control, Hu Yumin and Li Guanxing in chip fabrication. Keith Kmitchelson is thankfully acknowledged for checking English.

References

1. Aoi Y, Hosogai M and Tsuneda S. *J Biotechnol*, 2006, 125: 484-491.
2. Fukuta S, Kato S, Yoshida K et al. *J Virol Methods*, 2003, 112: 35-40.
3. Hagiwara M, Sasaki H, Matsuo K, et al. *J Med Viro*, 2007, 79(5): 605-615.
4. Hara-Kudo Y, Nemoto J, Ohtsuka K, et al. J Med Microbiol, 2007, 56: 398-406.
5. Imai M, Ninomiya A, Minekawa H, et al. *Vaccine*, 2006, 24: 6679-6682.
6. Minami M, Ohta M, Ohkura T, et al. J Clin Microbiol, 2006, 44(11): 4032-4037.
7. Notomi T, Okayama H, Masubuchi H, et al. *Nucleic Acids Res*, 2000, 28(12): e63.
8. Roper M, Easley C, Legendre L, et al. *Anal Chem*, 2007, 79: 1294-1300.
9. Saiki R, Scharf S, Faloona F, et al. *Science*, 1985, 230: 1350-1354.
10. Zhang Chunsun and Xing Da. *Nucleic Acids Res*, 2007, 35(13): 4223-4237.

HIGH-QUALITY INTERPOLATION AND PARALLEL IMPLEMENTATION ON FPGA

JINTAO WANG[1, 2†], YONGQUAN LU[1], PENGDONG GAO[1], RUI LU[1], WENHUA YU[1]
1. High Performance Computing Center,Communication University of China, Beijing 100024, China

GUANGYING GE[2]
2. College of Physics Science and Information Engineering, Liaocheng University, Liaocheng 252059, China

If single sensor (*CCD* or *CMOS*) is employed in image capturing system, the simple methods of interpolating results in boundary distortion, complicated ways are usually implemented by Host(*PC*), thus reducing the "real-timing" performance, on the other hand restricting application in embedded system. This article presents a parallel algorithm FPGA based, which implements *High-Quality Interpolation* for raw images that is divided into sub-blocks in real-time. Furthermore, a low-cost solution which parallel processing with FPGA is proposed, and evaluate the capability of other hardware system.

Key words: demosaicing; color interpolation; parallel algorithm; Pipeline; Operation Block

1. Introduction

CCD or CMOS sensor currently use a color filter array(CFA),called the Bayer pattern(Fig.2[1]., in Real-Time Image Acquisition System as shown in Fig.1. Convenient for completing algorithms, the raw Bayer image should be transformed to RGB or YUV format. The Bayer image data are firstly sent to PC via USB bus, 1394 bus or PCI bus, then achieved demosaicing and Color Space Converting. This general method has restricted application domain of embedded system. In order to implement this function in an embedded device, considering some situation which image identification required catching Full Size of image, demanding for CPU performance and incidental expenses of system might be increased. Moreover, some applications just need an ecumenic processor except FPGA chip in a practical system, such as TI's 2000 family DSP's, MCU and ARM7. As a result of decrease in cost and increase resource in current FPGA chip, however, to deal with image using FPGA is a smart choice.

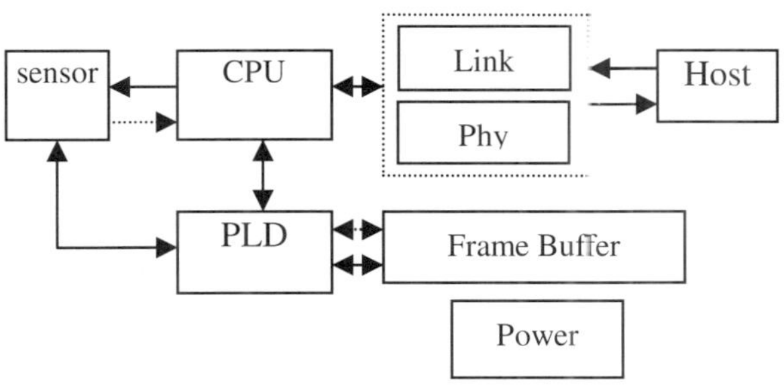

Fig.1 Capture Image System in real-time

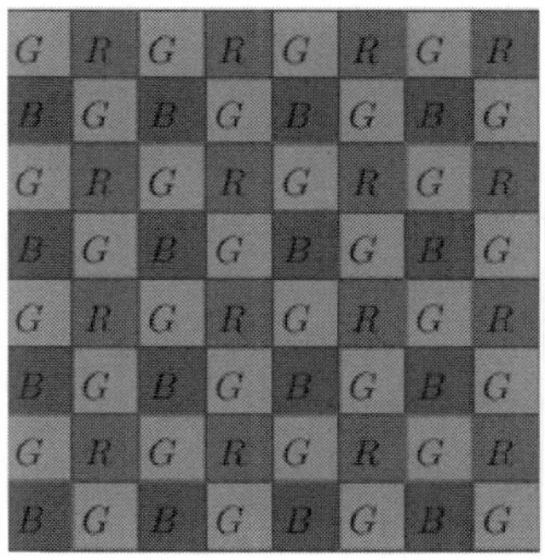

Fig.2 Bayer pattern.

The majority of existing methods solving problems described above which usually include ***Nearest neighbor, Bilinear, Bicubic, B2Sp line, Adaptive Algorithms***, and ***Mathematical Recovery Methods***[2-5]. Although these methods are simply, fast and easy-to-implement by early hardware, but its application may result in edge distortion in an image. Recent years, academicians have proposed rather complex methods in order to improve demosaiced image quality. Owing to

† JINTAO WANG is with *High Performance Computing Center, Communication University of China, Beijing 100024, China*
E-mail: wangjt@cuc.edu.cn

these methods which applied strong correlation between each color component with these ways, the acquired image quality has been enhanced more or less.

2. Algorithms and Methods

Optimized Demosaicing:

By studying and comparing, we may obtain perfect results that adopted *Optimal Recovery* approach which are proposed by *D. Darian Muresan* and *Thomas W. Parks*[4], and absolutely carried out on FPGA chip. Thus, the algorithm scheme looks as follows:

Looking at Fig.3, achieve the interpolation of greens possibly using fine scale training vectors.

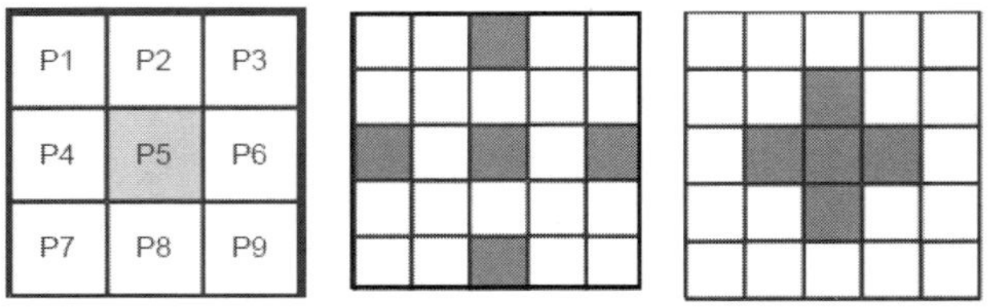

a. Plan to Interpolate Pixel *P5* b. Coarse scale training vector c. Fine scale training vector

Fig.3 Interpolating Pixel and scale vectors

The blues are processed in two steps:

Recovery the missing blues at the red place

$$B_5 = G_5 + \frac{\sum_{i=1,3,7,9} E_i(B_i - G_i)}{E_1 + E_3 + E_7 + E_9} \quad (1)$$

Recovery the missing blues at the green place

$$B_5 = G_5 + \frac{\sum_{i=2,4,6,8} E_i(B_i - G_i)}{E_2 + E_4 + E_6 + E_8} \quad (2)$$

Deal with red the same method as blue.

Add the correction step below and repeat three times:

Correct the Green to fit the green over blue ratio:

$$G_5^B = B_5 + \frac{\sum_{i=2,4,6,8} E_i(G_i - B_i)}{E_2 + E_4 + E_6 + E_8} \quad (3)$$

$$G_5^R = R_5 + \frac{\sum_{i=2,4,6,8} E_i(G_i - R_i)}{E_2 + E_4 + E_6 + E_8} \quad (4)$$

a) Average the interpolation results between the Blue and Red

$$G_5 = \frac{G_5^B + G_5^R}{2} \quad (5)$$

b) Correct the Blue and Red values by the ratio

$$B_5 = G_5 + \frac{\sum E_i(B_i - G_i)}{\sum E_i}, i \neq 5 \quad (6)$$

$$R_5 = G_5 + \frac{\sum E_i(R_i - G_i)}{\sum E_i}, i \neq 5 \quad (7)$$

c) End of loop.

The Application of Parallel Algorithm with FPGA:

Take full advantage of FPGA resources, we can store the gathered Bayer image data into inner four SRAMs, which contain five lines of image (1280 x 5=6400 Bytes). The arithmetic resources, MAC (Multiply-Accumulate Unit) and DSP(*Digital Signal Processor*) within FPGA which may be divided into four OBs(Operation Block) that appends some of logic circuits, parallelized processing four parts of image within RAM1-4. One line of data from processed image became 3 lines of RGB data. Therefore the RAM5-8 should be than 3840 Bytes (3 lines of data). The RAMs, OBs, FIFO and other Control Logic Block were connected each other via inner Bus Switch's in FPGA chip as shown in Fig.4.

For increasing operation frequency of logic circuit, we should shorten the data-path of longest delay in FPGA chip. The larger combinatorial logic would be split into smaller units which insert some triggers. It's called "pipeline" technology.

The address of each pixel data where it is at a picture is perspicuous. We needed neighborhood edge pixels while processing the edge pixel of sub-blocks in an image. The inner logic would synchronously access multi-memory, and transfer data to same OB. In other word, this system is not only able to divide a full-size image to four parts, parallelized operation, but also to access the edge data of sub-block in real-time. The proceed algorithm would then be: *If* the address of pixel prepared deal with is on threshold (for instance, 3 lines or 3 columns). *Then* save neighborhood edge pixels into secondary buffers. *Else* don't hold any pixel data in those buffers.

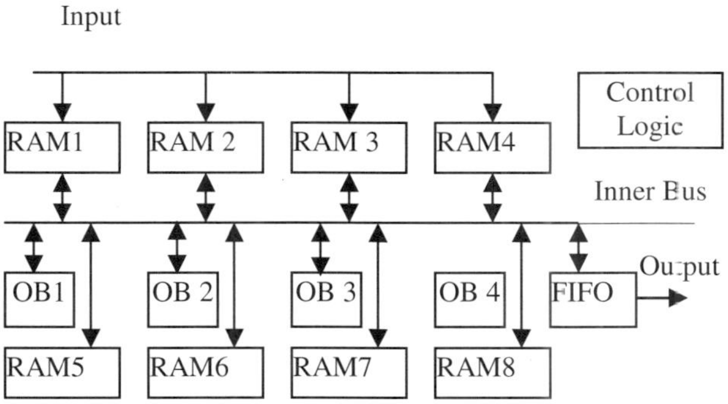

Fig.4 Operation Structure.

The end.

After treating a whole picture, the pixel sequences were orderly delivered to peripheral FIFO chips according as the location which is at the picture, and were sent to Host (*PC*) for display.

The sensor in the system was Sony ICX285AQ, number of active pixels: 1280 (H) × 1024 (V) approx. 1.40M pixels; all pixels' signals to be output independently within approximately 1/15 second. The FPGA chip was Lattice LFECPIIM50. Embedded Memory: 4147Kbits, Distributed Memory:101Kbits, 18x18 Multipliers: 88, sysDSP Blocks: 22. The CPU in this system was TI's TMS320LF2407A, which was used to control and management. During evaluation, for illustrating problems we have completed plenty of "testing images", and compared with results of another board that caught image using TMS320C6713B DSP chip. The data were tabulated in table 1.

Table 1. Operation time

	LF2407A	FPGA (non-parallel)	FPGA (parallel)	C6713B
System clk	30MHz	50MHz	50MHz	300MHz
Test1 (s)	2.131	0.014	0.0046	0.098
Test2 (s)	2.189	0.016	0.0042	0.090
Test3 (s)	2.156	0.013	0.0045	0.093

The comparison of FPGA's used resource were tabulated in table 2. We listed both situation of parallel and non-parallel algorithm construction within FPGA, which has PFU (Programmable Functional Unit) registers, PIO(pin of Input and Output), DSP Blocks, and Embedded Memory(not include Distributed Memory).

Table 2. Used Resource in FPGA

		PFU Registers	PIO	DSP Blocks	Embedded Memory
FPGA (non-parallel)	Used	1240	296	4	52KB
	Rate	3.36%	77.69%	18.2%	10.03%
FPGA (parallel)	Used	6582	296	16	216KB
	Rate	17.85%	77.69%	72.7%	41.67%

Figure 5 shows capturing images containing a "testing image" generated by itself of FPGA in (a), longer time of deal with causing losing frame or line synchronous signal in (b), exchanged period of line in (c), losing several line data in (d), losing synchronous signal at random in (e), a full-color image caught with developed industry high-definition camera in (f).

3. Conclusions

The results in Table 1 and table 2 indicate that parallel algorithm is able to improve real-timing of system, which avoid overtime of processing, losing frames, losing lines, losing synchronous signal .etc. as illustrated in Fig.5.

In practice, algorithm implement on FPGA using VHDL hardware language is not enough according with habits of designers, furthermore, effect efficiency of developing. We would try to use similarity C language application tools, Handel-C and System-C, write out primary codes and optimize it on special platform.

The sensor in this system ICX285AQ has number of available pixels: 1280x1024, frame rate: 15/s. The total data values are 1280x1024x15=18.75MB, which would be 18.75x3=56.25MB after Interpolation. The limited speed of Firewire bus (1394A) we used is 480Mb/s, 60MB/s, considering stability, frame rate is set up 7-10/s generally as shown in table 1. The methods in this paper would be adopted to rapidly perform pretreatment of image with hardware, and applied in multi-channel image catch, or used in *High-Rate Camera* area.

References

1. LIU Xiao-song. A Color Interpolation Method for Digital Image Sensors [J]. Journal of Image and Graphics. 2003, Vol. **8(A)**, No.5:517-518.

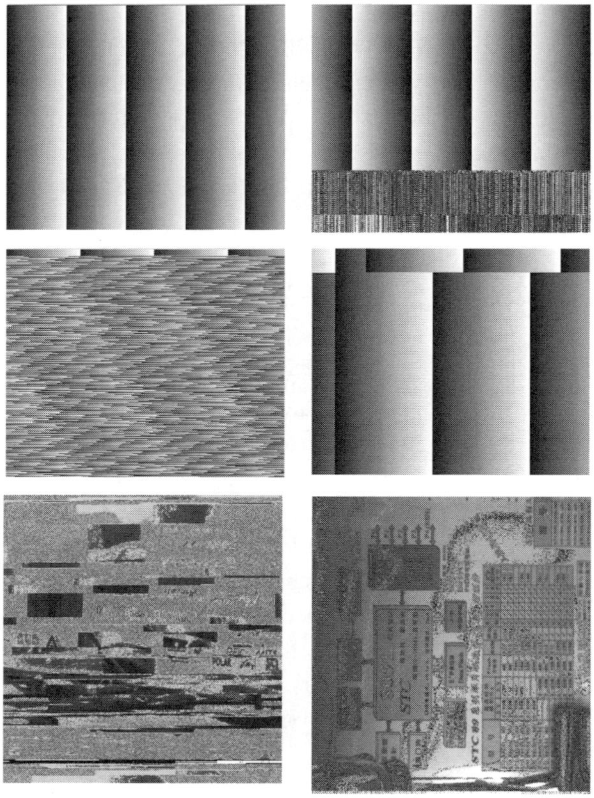

a	b
c	d
e	f

Fig.5 Real-timing Images

Fig.5 a. gray testing image b. lose frame sync. and Line sync. c. changed frame period d. lose lines e. aperiodic losing sync. f. real-time full color image

2. A. Lukin, High-Quality Algorithm for Bayer Pattern Interpolation [J]. *Programming and Computer Software, 2004,* **Vol.30**, *No.6: 347–358.*
3. Kimmel R., Demosaicing: image reconstruction from co lo r CCD samples [J]. IEEE Transactions on Image Processing, 1999, **8(9)**: 1221-1228.
4. D. D. Muresan, Optimal Recovery Approach to Image Interpolation [C]. ICIP2001, Greece, 2001.
5. A. Lukin, AN IMPROVED DEMOSAICING ALGORITHM [C], 14-th International Conference on Computer Graphics, GraphiCon' 2004 proceedings.

NEUTROPHIL MEDIATED ANTIMICROBIAL PHOTOBIOMODULATION

TIMON CHENG-YI LIU[*1,2], LING ZHU[1], KANGTAO LV[1, 2] AND SONG-HAO LIU[3]

1 Laboratory of Laser Sports Medicine, South China Normal University, Guangzhou, GD 510631, China
2 College of Life Science, South China Normal University, Guangzhou, GD 510631, China
3 School for Information and Optoelectronic Science and Engineering, South China Normal University, Guangzhou, GD 510631, China

Aiming to bacteria themselves, traditional treatment paradigms combating antimicrobial resistance have been largely unsuccessful. Aiming to polymorphonuclear neutrophils (PMNs), the key players in the initial response against invading pathogens, this paper reviewed antimicrobial photobiomodulation (PBM). There are tree kinds of monochromatic light or laser irradiation (LI), low intensity LI (LIL), ~10 mW/cm^2, moderate intensity LI (MIL), $10^{2\sim3}$ mW/cm^2, and high intensity LI, ~10 W/cm^2. PBM is a modulation of LI on biosystems. PMNs might be dysfunctional for the diseases of antimicrobial resistance. The PBM of LIL might kill bacteria by phototransduction mediated rehabilitation of PMNs. Intravascular low energy laser therapy (ILELT), a PBM of MIL, might kill bacteria by reactive oxygen species mediated rehabilitation of PMNs. There might be no antimicrobial resistance for cellular rehabilitation of PBM so that we have found photonic paradigms combating antimicrobial resistance, which might be of importance in clinic applications.

1. Introduction

Antibiotic resistance is an increasing global problem. Many diseases have been found to be of antibiotic resistance. Tuberculosis (TB) was one of the most well known ones. Eight million people develop TB each year and 5,000 of them will die from the diaease every day throughout the world. Although treatable and curable, globally about 3% of all newly-diagnosed patients have multi-drug-resistant TB (MDR TB) making their treatment complicated, expensive and uncertain [1]. The worldwide rise in antibiotic resistance necessitates the development of novel antimicrobial strategies. However, traditional treatment paradigms combating antimicrobial resistance have been largely unsuccessful. It might be due to its aiming to bacteria themselves. This paper would review a candidate way combating antimicrobial resistance.

Polymorphonuclear neutrophils (PMNs) are one of the main types of effector cell in the innate immune system and were first shown to effectively kill microorganisms by phagocytosis more than 100 years ago [2]. Recently, however, it has been found that stimulated neutrophils can also produce extracellular structures called neutrophil extracellular traps (NETs), which consist of chromatin and granular proteins that capture and kill microorganisms [3]. PMNs endowed with an impressive antimicrobial armamentarium [4] are one of the first lines of defense against invading microbes [2]. However, PMNs might be dysfunctional for the diseases of antimicrobial resistance [5, 6]. Interferon-gamma (IFN-γ) has been found to significantly augment antimicrobial activities against antimicrobial resistance [7].

Monochromatic light or laser irradiation (LI), apart from conventional methods, has been shown to have potential to eliminate bacteria. There are tree kinds of LI, low intensity LI (LIL), ~10mW/cm^2, moderate intensity LI (MIL), $10^{2\sim3}$ mW/cm^2, and high intensity LI (HIL), ~$10^{1\sim2}$ W/cm^2. Photobiomodulation (PBM) is a modulation of LI on biosystems, which stimulates or inhibits biological functions but does not result in irreducible damage [8]. The antimicrobial PBM of LIL and MIL would be discussed in this paper.

2. Polymorphonuclear Neutrophils

PMNs, the key players in the initial response against invading pathogens, kill microbes by several mechanisms. Well known for engulfing microorganisms in phagosomes, where they are subjected to high concentrations of ROS and antimicrobial peptides discharged from granules, activated PMNs also snare and kill microbes in NETs [3]. Fuchs *et al* [9] used live-cell imaging to simultaneously monitor the morphology and viability of individual PMNs and the appearance of NETs. Following activation with the phorbol ester phorbol 12-myristate 13-acetate (PMA), PMNs underwent a series of morphological changes; eventually cells ruptured (indicated by the simultaneous

* corresponding author. Email: liutcy@scnu.edu.cn, timontcm@sina.com

loss of the dye calcein blue and exposure of phosphatidylserine). NETs only appeared with cell death. Morphological analysis of PMNs fixed at various times after stimulation with PMA or *Staphyloccus aureus* indicated that the process of cell death that led to NET formation, which involved dissolution of the nuclear envelope and granular membranes so that chromatin could mix with granule contents, was distinct from apoptosis and necrosis. Moreover, cells activated to produce NETs did not undergo DNA fragmentation as detected by TUNEL (terminal deoxynucleotidyl transferase-mediated deoxyuridine triphosphate nick end labeling). NET production was enhanced by catalase inhibition and did not occur in the PMNs of individuals with chronic granulomatous disease (associated with mutations in NADPH oxidase) and was blocked by pharmacological inhibition of NADPH. In both cases, NET production was rescued by hydrogen peroxide. Thus, Fuchs *et al* [9] conclude that NET production involves a distinct form of active cell death that requires ROS generation and enables even dead PMNs to retain antimicrobial activity.

PMNs might be dysfunctional for the diseases of antimicrobial resistance. There was a multidrug-resistant Pseudomonas aeruginosa (r-Pa) in a lethal case of sepsis in a bone marrow transplant recipient, and Mencacci *et al* [5] found that the patient's PMN killing activity and nitric oxide production against r-Pa were profoundly impaired. Fiorenza *et al* [6] have found that tuberculosis (TB) inpatients had a lower respiratory burst (RB) response than healthy controls (HCo) for PMNs ($p<0.01$) regardless of heat-killed Mycobacterium tuberculosis (HK Mtb) stimulation; compared to HCo, PMNs from TB patients presented a reduced CD11b expression, with the two subject groups showing a decrease in this marker expression following HK. Mtb was added to both cell cultures; whereas fewer IL-8 and TNFα receptors were found when studying PMNs from TB patients, antigen stimulation significantly raised the expression for both cytokine receptors.

The emergence of multidrug-resistant enterococci presents a major therapeutic challenge since there is currently no clearly effective antimicrobial therapy for these infections. Onyeji *et al* [7] have studied the effects of interferon-γ(IFN-γ) on antienterococcal activity of PMNs, and found a significant IFN-γ-induced anti-enterococcal effect can be obtained with a concentration as low as 1 U/ml, after 24 h treatment; a maximal effect was achieved with a IFN-γ concentration of 100 U/ml, however, this appeared not to be significantly different from the effect of 1 U/ml. IFN-γ had no direct inhibitory effect on the organism at any of the concentrations used.

PBM on volunteer PMN RB has been studied in our laboratory [10]. The neutrophils were isolated from peripheral blood of 13 volunteers (10 ordinary persons, 3 athletes) and treated by red light emitting diode irradiation (640±15nm) (RLED) at 50, 100, 300, 500 and1000 J/m^2 and fixed 100s duration. Blood samples of athletes were extracted at different time in the 10 km non-interrupted long-distance running, before running, 1 hour after running beginning, just finishing the running, resting for 1 hour and 2 hours after running. It has been found that there were three types of modulation of RLED on the RB of three types of PMNs, respectively, promotion for the one of subactivated neutrophils, inhibition for the one of overactivated neutrophils and none for the neutrophils in homeostasis.

3. LIL-rehabilitated Bactericidal Action

LIL-rehabilitated PMN-mediated bactericidal action has been studied mainly in China and Russia.

Wang *et al* [11] have studied therapeutic effects of low intensity He-Ne laser irradiation (LHNL) on rabbit skin flaps infected with 1 ml *Staphylococcus aureus* (9×10^8 bacteria/ml), and found the skin flaps in the treatment groups were much better than those in the control groups, and PMN count of treatment groups were much lower than the one of the control groups. This kind of phenomena has also been found by Gao *et al* [12], Chen *et al* [13] and Guo [14], respectively. The animal model of suppurative mice inflammation was obtained with *Staphylococcus aureus* injection, and then improved with LHNL [12]. PMN count of treatment groups was significantly lower than the one of the control groups. The bacterail infection wound of 15 buffaloes, 4 scalpers, 5 dogs and 1 rhinopithecus was improved with LHNL, but their counts of white blood cells decreased [14]. Human dermatopathy was also improve with LHNL [13]. PMN count of bacterial dermatopathy after the treatment was significantly lower than the one before the treatment, but PMN count of virual dermatopathy after the treatment was significantly higher than the one before the treatment [13]. It indicated that it might be NET-mediated bactericidal action since NETs were produced from dying PMNs.

PMN of the cervical secretion has been found to be dysfunctional since the content of active intracellular forms of oxygen was decreased in females suffering from Chlamydia cervicitis [15]. Its PBM studies found that LIL stimulated not only phagocytosis but also intracellular generation of active oxygen forms so that it eliminates dysfunction of cervical secretion PMNs [15]. Positive clinical effect of the local use of the LIL for Chlamydia cervicitis treatment was accompanied by positive changes in immunological indices of cervical secret, normal concentration of cytokines in cervical secret, restoration of the number and functions of PMNs [16]. Local use of the LIL contributed to decreased number of opportunistic pathogenic microorganisms and their associations, and restored local flora [16].

4. Intravascular Low Energy Laser Irradiation Therapy

Intravascular low energy laser therapy (ILELT) was put forward for cardiocirculatory diseases by in USA in 1982 [17], was popular in Russia in 1980s, and then in China in 1990s. The order of the intensity at the tip of the optic fiber with diameter 0.2mm used in Chinese ILELT is about 10^3mW/cm^2 [18], but the blood velocity in vein is about 1-10cm/s [19] so that the order of the radiation time of each blood cell is about 10ms, and then the order of the dose of each blood cell is about 100 J/cm^2. As an intravascular PBM of MIL, ILELT can bi-directionally modulate reactive oxygen species (ROS) level. Ishina *et al* [20] have shown that ILELT activates function of oxygen-dependent bactericidal system of blood PMNs and inhibited activity of non-oxygen-dependent system for patients with bronchial asthma.

Karniushina *et al* [21] have found ILELT on 100 patients decreased the level of endogenous intoxication, stimulates the defensive forces of the organism, improves the condition of the patient, contributes to quicker recovery and may be actively used in treatment of patients with a surgical infection.

ILELT might be an effective treatment on acute pneumonias or TB. Korochkin *et al* [22] have used ILELT to treat patients with acute pneumonias, and found the therapeutic approach has had a favourable effect on the clinical course of acute pneumonias, accelerating the terms of pneumonia resolution and promoting and earlier and more complete restoration of the blood stream and normalization of the hemostasis, in contrast to routine therapy. Eshankhanov *et al* [23] have reported a bacteriostatic effect of ILELT on TB growth. ILELT used in 85 TB patients induced no side effects. Ten-twelve sessions relieved the symptoms of TB-related intoxication, reduced the infiltration and destruction, promoted abacillosis. Laser irradiation holds promise in management of torpid infection and hepatotoxicity induced by isoniazid and rifampicin. The highest effect occurred in infiltrative tuberculosis. Parmon *et al* [24] have studied ILELT on patients with urinary TB, and found that ILELT activates local system of T-helpers which after specific antigenic impact differentiate into T-helpers-1 which synthesized in loco INF-γ, TNF-α and β and IL-2 stimulating bactericidal mechanisms directed at destruction of M. TB and resolution of the infection focus.

Gushcha *et al* [25] have revealed a close relationship between the severity of pancreatogenous endotoxicosis and red cell membrane hyperlipoperoxidation and spontaneous red cell hemolysis, which allows three degrees of blood red cell lipoperoxidation to be distinguished in the period of acute endotoxemia: mild, moderate, and severe, and found inclusion of ILELT into the management of acute pancreatogenous toxemia in moderate and severe degrees of blood red cell hyperlipoperoxidation reduced authentically the peroxide damage of the red cell membranes and the marked character of hemic hypoxia.

Autotransfusion of blood treated by ultraviolet irradiation (ABUVI) is also a ROS-mediated therapeutic approach. Imbalance of cellular and humoral immunity factors was the main factor determining the severity of infants' acute purulent destructive pneumonia (APDP). It also influences the completeness of humoral immunity and reduction of T-lymphocyte deficiency in the acute phase of the disease. ABUVI has been found to be an effective measure for correcting the immune response of the child's organism to the bacterial aggression through adequate production of monocytic phagocytes and plasma cells of the blood [26]. ABUVI also raised the efficacy of complex treatment of toxicoseptic forms of APDR, reduced 1.7-fold the terms of treatment, and reduced considerably the mortality rate of this disease in young children [26].

5. Mechanism of Photonic PMN-mediated Bactericidal Action

There are two kinds of pathways mediating cellular PBM [27], one kind is specific, which is mediated by the resonant interaction of LI with singlet oxygen or endogenous photosensitizers such as cytochrome nitrosyl complexes of mitochondrial electron transfer chain, riboflavin, hemoglobin or endogenous porphyrines, the other kind is non-specific, which is mediated by the non-resonant interaction of LI with membrane proteins. Mechanism of photonic PMN-mediated bactericidal action would be discussed from the viewpoints of the two kinds of pathways.

The specific pathways were mediated by singlet oxygen or endogenous photosensitizers mediated ROS [28, 29]. An abundance of scientific literature exists demonstrating that oxidative stress influences the mitogen-activated protein kinase (MAPK) signaling pathways [30]. It has been shown that different ROS levels activate different MAPK pathways, that is, low level ROS activates extracellular signal-regulated kinase(ERK) mediated MAPK, a signal for proliferation, differentiation and survival, moderate level ROS activates the stress signals Jun N-terminal kinase (JNK) or p38 mediated MAPK, which leads to cell survival or apoptosis, and high level ROS leads to cellular apoptosis or necrosis [31, 32]. ROS can also induce formation of NETs from PMNs by NADPH oxidase-dependent way and NADPH oxidase-independent way, respectively [9]. Although the molecular basis for priming of the PMN NADPH oxidase is unknown, a number of signaling pathways have been found to be implicated in the priming process including the phosphatidylinositol 3-kinase pathway and several MAPK pathways [33]. Although the present studies were very primary, they have provided a clue to understanding bactericidal action of ILELT, that is, the specific pathways of PBM might mediate bactericidal action of ILELT. Of course, the further studies should be done.

The non-specific pathways were mediated by signal transduction [34]. Extraocular circadian phototransduction was put forward by Campbell *et al* [35] in their report that 3 h of bright light exposure to the area behind the knee caused phase shifts of the circadian rhythms of both body temperature and saliva melatonin in humans. However, Yamazaki *et al* [36] found no evidence for extraocular photoreceptors in the circadian system of the Syrian hamster, and Wright *et al* [37] found absence of circadian phase resetting in response to bright light behind the knees. Therefore, extraocular cellular phototransduction should be further studied. Since the original work by Duan *et al* [38], many extraocular cellular phototransduction phenomena [39, 40, 41, 42] have been observed. Among them, the original work by Duan *et al* [38] is very important in this paper. Duan *et al* [38] have probed signal transduction pathways of RB of bovine PMNs which were induced by He-Ne laser irradiation at a dose of 300J/m^2 by using the protein tyrosine kinases (PTKs) inhibitor, genistein, the phospholipase C (PLC) inhibitor, U-73122, and the protein kinase C (PKC) inhibitor, calphostin C, respectively, and found the inhibitor of PTKs can completely inhibit the He-Ne laser-induced respiratory burst of neutrophils, and PLC and PKC inhibitors can obviously reduce it, but not fully inhibit it. These results suggest that PTKs play a key role in the He-Ne laser-induced respiratory burst of neutrophils and [PTK-PLC-PKC-NADPH oxidase] signal transduction pathways may be involved in this process. As NADPH oxidase might mediate NET formation, the non-specific pathways might mediate LIL-rehabilitated PMN-mediated bactericidal action.

As discussed above, there might be two kinds of pathways mediating antimicrobial PBM, the one mediated by low level ROS in ILELT, the other mediated by phototransduction in PBM of LIL. Both of them, ILELT [29] and PBM of LIL [43], are homeostatic. In other words, they are a kind of cellular rehabilitation of PMN.

6. Discussion

Light, a nearly ubiquitous environmental signal, regulates myriad developmental and behavioral responses in plant, fungal, bacterial, and animal cells. Photosensitive proteins abound in the bacterial kingdom, but their cellular functions often remain a mystery. Swartz *et al* [44] identify a functional role for a new type of blue light sensor in bacteria--light, oxygen, or voltage (LOV) histidine kinase. The LOV domain is rendered sensitive to blue light through a noncovalently bound flavin cofactor. In the notorious pathogen Brucella abortus, blue light increases the enzymatic activity of this kinase, which, remarkably, increases virulence of the bacterium. Flavin-binding LOV domains are broadly distributed across the plant, fungal, and bacterial kingdoms, and thus, serve as evolutionarily conserved sensory modules. At this point, it should be pointed out that Swartz *et al* [44] should be paid attention to when the blue light is used in PMN-mediated bactericidal action.

Acknowledgments

Our study is supported by National Science Foundation of China (60478048, 30470495, 6017800 and 6027812); National 973 basic project (2005CB523502); Group project (20003061), Key Project (2001480), and General Project (20031526) of Guangdong Science Foundation of China; the Opening Project of MOE Key laboratory of Laser Life Science, South China Normal University, Guangzhou 510631, China; and Guangdong Qianbaishi Fellow Foundation (Q02087).

References

1. M. C. Raviglione and I. M. Smith, *N Engl J Med.* **356**:656(2007).
2. C. Nathan, *Nat. Rev. Immunol.* **6**:173–182(2006).
3. V. Brinkmann and A. Zychlinsky, *Nat Rev Microbiol.* **5**:577(2007).
4. B. E. Steinberg and S. Grinstein, *Sci STKE.* pe11(2007).
5. A. Mencacci, E. Cenci, A. Repetto, R. Mazzolla, F. Bistoni, F. Aversa, T. Aloisi and A. Vecchiarelli, *J Infect.* **53**:e259(2006).
6. G. Fiorenza, M. A. Farroni, C. Bogue, D. Selenscig, D. M. Lamas and D. Dlugovitzky, *Arch Med Res.* **38**:526(2007).
7. C. O. Onyeji, D. P. Nicolau, C.H. Nightingale and L. Bow, *Int J Antimicrob Agents.* **11**:31(1999).
8. H. Friedmann and R. Lubart, *Proceedings of SPIE - The International Society for Optical Engineering*, **v 2630**, 60-64(1996).
9. T. A. Fuchs, U. Abed, C. Goosmann, R .Hurwitz, I. Schulze, V. Wahn, Y. Weinrauch, V. Brinkmann and A. Zychlinsky, *J Cell Biol.* **176**:231(2007).
10. M. Wu, L. Zhu, B. Hu, T. C. Y. Liu and D. L. Rong, *Lasers Surg Med.* **39**: 82(2007).
11. L. Wang, Y. Wang, D. Cheng, Y. He, M. An and L. Luo, *Chinese J Reparative and Reconstructive Surgery.* **12**: 55(1998) (in Chinese).
12. M. Gao, H. Wang, S. Qiu and W. Li, *Chinese J Laser.***23**: 519 (1996) (in Chinese).
13. G. Chen and T. Li, *Chinese J of Rehabilitation.* **11**: 163 (1996) (in Chinese).
14. D. Guo, *Chinese J Traditional Chinese Veterinarian.* **4**: 1 (1999) (in Chinese).
15. O. A. Gizinger and I. I. Dolgushin, *Vopr Kurortol Fizioter Lech Fiz Kult.* **5**:20(2006).
16. I .I. Dolgushin, O. A. Gizinger and L. F. Telesheva, *Zh Mikrobiol Epidemiol Immunobiol.* **4**:105(2006).
17. I. Kertesz, M. Fenyo, E. Mester and G. Bathory, *Opt Laser Technol.* **14**: 31(1982).
18. M. Tong, Y. F. Liu, X. N. Zhao, C. Z. Yan, Z. R. Hu and Z. X. Zhang, *Laser Med Sci* .**15**: 201(2000).
19. D. J. Dowsett, J. T. Ennis, C. T. Collum and R. B. De Jong, *J Nucl Med.* **26**:859(1985).
20. T. I. Ishina, I. M. Kakhnovskii, O. V. Makarova, A. S. Solomatin and M. E. Alekseeva, *Ter Arkh.* **73**:15(2001).
21. N. L. Karniushina, L. K. Veniaminov, O. V. Inozemtseva, L. N. Voro'beva amd E.G. Grigo''rev, *Vestn Khir Im I I Grek.* **143**:48(1989).
22. I. M. Korochkin, T. K. Platonova, G. M. Kapustina, A. M. Belov and O. G. Alekseeva, *Sov Med.* 7:22(1989).
23. M. E. Eshankhanov, M. E. Khodzhaeva, M. I. Kozarez and Sh. T. Aliev, *Probl Tuberk.* **2**:27(1993).
24. E. M. Parmon, V. V. Borshchevskii and L.G. Bortkevich, *Urologiia.* **6**:13(2001).
25. A. L. Gushcha and S. V. Tarasenko, *Khirurgiia (Mosk).* **1**:43(1993).
26. V. N. Kalinkin, G. D. Mezentsev, E. A. Kashuba, L. A. Konovalova and L. N. Shatilovich, *Khirurgiia (Mosk).* **8**:14(1991).
27. T. C. Y. Liu, J. L. Jiao, X. Y. Xu, X. G. Liu, S. X. Deng and S. H. Liu, *Progress in Biomedical Optics and Imagine.* **5**: 185(2005).
28. R. Lubart, M. Eichler, R. Lavi, H. Friedman and A. Shainberg, *Photomed Laser Surg.* **23**:3(2005).
29. T. C. Y. Liu, Y. Q. Wang, X. Y. Xu, X. F. Zhao and S. J. Hu, *Proceedings of 2007 IEEE/ICME international Conference on Complex Medical Engineering.* IEEE Catalog No. **07EX** 1719: 1135(2007).
30. J. A. McCubrey, M. M. Lahair and R. A. Franklin, *Antioxid Redox Signal.* **8**:1775(2006).
31. C. Bladier, E. J. Wolvetang, P. Hutchinson, J. B. de Haan and I. Kola, *Cell Growth Differ.* **8**:589(1997).
32. E. D. Owuor and A. N. Kong, *Biochem Pharmacol.* **64**:765(2002).
33. G. E. Brown, M. Q. Stewart, S. A. Bissonnette, A. E. Elia, E. Wilker and M. B. Yaffe, *J Biol Chem.* **279**:27059(2004).
34. T. C. Y. Liu, J. L. Jiao, R. Duan, Y. Li, Y. Y. Yeung and S. H. Liu, *Lasers in Medicine, Surgery and Dentistry.* 83(2003).

35. S. S. Campbell and P. J. Murphy, *Science*. **279**: 396(1998).
36. S. Yamazaki, M. Goto and M. Menaker, *J Biol Rhythms*. **14**:197(1999).
37. K. P. Wright and C. A. Czeisler, *Science*. **297**: 571 (2002).
38. R. Duan, T. C. -Y. Liu, Y. Li, H. Guo and L. -B. Yao, *Lasers Surg. Med.* **29**: 174(2001).
39. Y. O. Zhang, S. P. Song, C. C. Fong, C. H. Tsang, Z. Yang and M. S. Yang, *J Invest Dermato*. **120**: 849(2003).
40. G. Y. Luo, C. Y. Huang, T. C. Y. Liu, Y. P. Zhao and S. H. Liu, *Lasers Surg Med*. **17S**: 60(2005).
41. G. Y. Luo, T. C. Y. Liu, Y. P. Zhao and S. H. Liu, *Lasers Surg Med*. **18S**: 51(2006).
42. M. Wu and T. C.-Y. Liu T, *Photomed Laser Surg*. **25**:129(2007).
43. T. C. Y. Liu, J. Q. Yuan, Y. F. Wang, X. Y. Xu and S. H. Liu, *Proc SPIE*. 6534: 653403-1~-6 (2007).
44. T. E. Swartz, T. S. Tseng, M. A. Frederickson, G. Paris, D. J. Comerci, G. Rajashekara, J. G. Kim, M. B. Mudgett, G. A. Splitter, R. A. Ugalde, F. A. Goldbaum, W. R. Briggs and R. A. Bogomolni, *Science*. **317**:1090(2007).

INFECTIVE PHOTOBIOMUDULATION IN LASER SURGERY AND PHOTODYNAMIC THERAPY

KANGTAO LV[1, 2], TIMON CHENG-YI LIU[*1,2], CHANG-XIONG RUAN[1] AND SONG-HAO LIU[3]

1 Laboratory of Laser Sports Medicine, South China Normal University, Guangzhou, GD 510631, China
2 College of Life Science, South China Normal University, Guangzhou, GD 510631, China
3 School for Information and Optoelectronic Science and Engineering, South China Normal University, Guangzhou, GD 510631, China

Anti-infection prophylaxis is necessary when postoperative infection of laser surgery (PILS) or photodynamic therapy (PIPT) is possible. However, PILS with herpes simplex virus (HSV) was observed only in 7.4% regardless of prior HSV history in an evaluation of CO_2 laser resurfacing of 500 patients. PDT do not usually require anti-infection prophylaxis. PILS and PIPT were studied from laser beam intensity distribution in this paper. The central part of high intensity laser irradiation (HIL) beam in laser surgery or moderate intensity laser irradiation (MIL) beam in PDT is so high that it could damage the target cells, microbes or viruses. However, the marginal part of the HIL or MIL beam is so low that it could induce photobiomodulation on the undamaged tissue cells, microbes or viruses, which was called marginal photobiomodulation (MPBM) for convention. MPBM modulated microbes or viruses might survive so that the undamaged microbes or viruses should be killed by immune cells or anti-infection treatment. MPBM were dependent on HIL or MIL beam intensity distribution which can be designed by changing HIL or MIL power. MPBM of HIL or MIL might rehabilitate immune cells and then inhibit PILS or PIPT, and no drugs of anti-infection might be needed. MPBM of HIL might also induce dysfunction of immune cells so that the infection might be almost left free, and drugs of anti-infection might be needed.

1. Introduction

Anti-infection prophylaxis is necessary for laser surgery when the risk for postoperative infection is significant or when the risk of infection is moderate but the consequences of infection are significant [1, 2]. However, postoperative infection with herpes simplex virus (HSV) was only observed in 7.4% regardless of prior HSV history in an evaluation of carbon dioxide laser resurfacing of 500 patients [3]. Moreover, photodynamic therapy (PDT) do not usually require anti-infection prophylaxis including antibiotic prophylaxis [2]. Postoperative infection of laser surgery or PDT were studied from laser beam intensity distribution in this paper.

2. Laser Beam Intensity Distribution

The central part of the laser irradiation beam of high intensity laser irradiation (HIL) in laser surgery is so intense that it could damage the target cells, microbes or viruses. More recently, it has been suggested that HIL might be useful for destroying microbes and viruses. Lee *et al* [4] have used a CO_2 laser to treat bacteria-infected cutaneous wounds, and found the HIL sterilization in the surgical treatment of sternal wound infections decreased morbidity, mortality, and cost. Jawhara *et al* [5] have monitored bactericidal action of HIL (810nm, 6.5W/cm^2, 20, 30 and 40s) by in vivo imaging of bioluminescent E. coli in a cutaneous wound infection of rats, and found loss of viability of E. coli depends on laser fluence, bioluminescent bacteria at 48 h later after HIL were not detected (100% loss of viability) in the wound irradiated at 260 J/cm^2. Smirnov *et al* [6] have investigated the inactivation of VEE virus with laser UV impulses of nano- and picosecond duration, and found there is a decrease of the inactivation cross-section with the rise of irradiation intensity in both cases.

The central part of the laser irradiation beam of moderate intensity laser irradiation (MIL) in PDT is so intense that its photodynamic effects could damage the target cells, microbes or viruses. Photodynamic effect (PDE) uses the combination of non-toxic photosensitizers and harmless visible light to produce reactive oxygen species (ROS) that can kill mammalian and microbial cells [7, 8] and viruses [9]. Hamblin *et al* [10] have reported on the first use of a photochemical approach to destroy bacteria infecting a wound in an animal model. Following topical application, a targeted polyationic photosensitizer conjugate between poly-L-lysine and chlorine (e6) penetrated the gram (-) outer bacterial membrane, and subsequent activation with 660

* Corresponding author. Email: liutcy@scnu.edu.cn, timontcm@sina.com

nm laser light rapidly killed Escherichia coli infecting excisional wounds in mice. To facilitate real-time monitoring of infection, they used bacteria that expressed the lux operon from Photorhabdus luminescens; these cells emitted a bioluminescent signal that allowed the infection to be rapidly quantified, using a low-light imaging system. There was a light-dose dependent loss of luminescence in the wound treated with conjugate and light, not seen in untreated wounds. Treated wounds healed as well as control wounds, showing that the photodynamic treatment did not damage the host tissue. Their study pointed to the possible use of this methodology in the rapid control of wounds and other localized infections. Demidova *et al* [7] have used genetically modified bioluminescent bacteria to follow the PDE in infected wounds, burns, and soft tissue infections in mice, and found that not only were bacteria infecting wounds, burns, and abscesses killed, but mice were saved from death due to sepsis and wound healing was improved.

As discussed above, the central part of the laser irradiation beam of HIL in laser surgery or MIL in PDT is so intense that it could damage the target cells, microbes or viruses. However, the marginal part is so low that it could induce photobiomodulation (PBM) on the undamaged tissue cells, microbes or viruses, which was called marginal PBM (MPBM) for convention. MPBM in laser surgery has been supported by our cellular model [11, 12] and clinical review [13, 14]. Huang *et al* [11, 12] have used monolayer human skin fibroblasts (HSFs) as models to the long-term effects of laser surgery, and found high pulse energy irradiation results in small survival rate and large damage radius, HSF proliferation was significantly promoted 24 hours/48 hours after 1,064 nm laser treatment at 100 and 150 mJ pulse energy, but significantly inhibited 24 hours/48 hours after 532 nm laser treatment at 100 mJ pulse energy. O'Donnell *et al* [13] have studied the effect of Er:YAG laser holes on osteoinduction in demineralized rat calvarial allografts, and found that laser ablation may allow maintenance of structural integrity while altering the surface geometry in such a way as to promote ingrowth of new bone.

MPBM in PDT has been also studied. Light-activated hematoporphyrin derivative (Hpd) was 99.9% bactericidal against Staphylococcus aureus, Streptococcus faecalis, Bacteroides fragilis, Streptococcus M-G intermedius, Streptococcus mutans, Peptostreptococcus anaerobius, Peptococcus magnus, and Clostridium perfringens, but rendered 98% of PMNs inviable [15]. However, combinations of light-activated Hpd and PMNs had a 4-log (99.99%) bactericidal effect on both intra- and extracellular S. aureus so that the ability of Hpd to localize in inflammatory tissues may have therapeutic applications in the treatment of abscesses [15]. Modulating effects of photoactivated chlorin on interaction of blood cells with bacteria have been studied so that the strongest bactericidal effect was found when the cells and photoactivated chlorin act together, and selective bactericidal regime of photosensitizer application which does not affect cell viability was demonstrated, which suggested that this photosensitizer may enhance cell natural defense capabilities [16].

3. Marginal photobiomodulation

PBM is a modulation of monochromatic light or laser irradiation on biosystems, which stimulates or inhibits biological functions but does not result in irreducible damage [17, 18, 19]. PBM has been shown to act on immune system cells in a number of ways [20, 17, 21], and the cell-specific PBM depends the intensity of monochromatic light or laser irradiation [17, 18, 19]. MPBM also depends on the average beam intensity of HIL in laser surgery or MIL in PDT beside the damage radius [12] of the laser beam. Obviously, the more intensive the laser irradiation, the larger the damage radius, and the weaker the average beam intensity [12]. In other words, the average beam intensity beside the damage radius can be designed by changing the power of HIL in laser surgery [12] or MIL in PDT so that the needed MPBM can be chosen.

Although there is MPBM on the undamaged microbes [17] and viruses [22], MPBM is a cellular rehabilitation [19] so that the undamaged microbes and viruses might survive and should be killed by immune cells or anti-infection treatment. Taravella *et al* [23] have studied whether live virus can withstand excimer laser ablation and pose a possible health hazard to medical personnel. In their experiment, fibroblasts infected with oral polio vaccine virus were ablated with an excimer laser. The plume was collected using a smoke evacuator and bubbled through viral culture media. They found live virus in the material trapped from the laser plume. This result indicated that oral polio vaccine virus can survive excimer laser ablation.

MPBM might induce dysfunction of immune cells so that the infection might be left free after laser surgery and drug of anti-infection was needed. Excimer laser keratectomy appears to be an efficient trigger for the reactivation of latent HSV-1 in the New Zealand rabbit ocular model [24] and clinical excimer laser photokeratectomy reactivates latent herpes simplex virus [25] because MPBM of excimer laser always induce dysfunction of immune cells. For herpes simplex keratitis following excimer laser application, antiviral treatment was administered [26, 27]. Treatment of cervical intraepithelial neoplasia in the presence of human papilloma virus infection results in higher failure rates and this is significantly increased when the treatment is by CO_2 laser ablation rather than by loop excision of the transformation zone, which might be due to MPBM inducd dysfunction of immune cells.

Although simultaneous exposure of neutrophils to MC540 and light inhibited phagocytosis, random migration, chemotaxis, hydrogen peroxide production, and oxygen consumption, but the ability of neutrophils to kill engulfed bacteria and to produce superoxide radical was not compromised [28]. Although the magnitude of damage inflicted on the tumour by aminolevulinic acid-based PDT (ALA-PDT) does not depend on the presence of neutrophils in the tumour or circulation, but the number of neutrophils within the illuminated tumour and in the circulation increased significantly following therapy [29]. PDT do not usually require anti-infection prophylaxis including antibiotic prophylaxis [2], which might be because MPBM of PDT is reactive oxygen species (ROS) mediated and is always beneficial to immune cells.

There was no evidence of herpes simplex virus inactivation by laser irradiation [30]. However, Schindl *et al* [31] have demonstrated that a total of 10 irradiations with low-intensity laser therapy significantly lowers the incidence of local recurrence of herpes simplex infection on the base of a randomized double-blind placebo-controlled study. Baggish [32] have treated viral venereal infections with the CO_2 laser at the intensity ranged from 311 to 660W/cm^2, and found patients with primary herpes were "cured" in 83% of cases and condyloma acuminatum could be eliminated in virtually every woman. These two studies and PDT studies suggested that MPBM might rehabilitate immune cells so that the infection was inhibited after laser surgery and no drug of anti-infection was needed. One of the main advantages of laser surgery is its bactericidal effect which reduces the risk of postoperative infections. Several study designs have been set to investigate this effect. It is rare the postoperative infection of phototherapeutic keratectomy, the most feared complication of the surgery [33]. The carbon dioxide laser stripping technique of hidradenitis suppurativa, a cicatrising and persistent disease of apocrine gland-bearing areas in adults, caused no additional spread of bacteria into the bloodstream.

4. Discussion

ROS production was always mediated by PDE. Lubart *et al* [34] found a new way that the Er-YAG laser irradiation dissociates water and generates OH radicals. The dissociation of water needed to generate OH radicals is possibly due to intermolecular vibrational (V-V) energy transfer in water, competing with vibrational relaxation, thus leading to water dissociation [34]. High amounts of oxygen radicals (e.g., hydroxyl groups) have a sterilization effect, whereas low concentrations of ROS stimulate fibroblasts, causing collagen and extracellular matrix formation, and PMNs, promoting antimicrobial or antivirus effects, so that ROS formation may explain the rapid wound repair following Er:YAG treatment and its bactericidal effect [34], no infections or hypertrophic scars in Er:YAG laser skin resurfacing [35], and no adverse effects such as scarring, bacterial infection, or contact dermatitis in resurfacing of pitted facial acne scars with a long-pulsed Er:YAG laser [36]. At this point, the MPBM of Er:YAG laser surgery might be always anti-infection as PDT MPBM, and Er:YAG laser surgery might be better than PDT since no photosensitizers are needed.

Acknowledgments

Our study is supported by National Science Foundation of China (60478048, 30470495, 6017800 and 6027812); National 973 basic project (2005CB523502); Group project (20003061), Key Project (20011480), and General Project (20031526) of Guangdong Science Foundation of China; the Opening Project of MOE Key laboratory of Laser Life Science, South China Normal University, Guangzhou 510631, China; and Guangdong Qianbaishi Fellow Foundation (Q02087).

References

1. S. Gilbert and E. McBurney, *Dermatol Surg*. **26**:50(2000).
2. M. S. Nestor, *J Drugs Dermatol*. **4**:s20(2005).
3. C. A. Nanni and T. S. Alster, *Dermatol Surg*. **24**:315(1998).
4. J. S. Lee, S. K. Tarpley and A. S. Miller, *South Med J*. **92**:380(1999).
5. S. Jawhara and S. Mordon, *Lasers Med Sci*. **21**:153(2006).
6. IuA. Smirnov, S. P. Kapitulets, IaIa. Tsilinskii, R. O. Esenaliev, P. G. Morev, A. A. Oraevskii and D.N. Nikogosian, *Biofizika*. **34**:570(1989).
7. T. N. Demidova and M. R. Hamblin. *Int J Immunopathol Pharmacol*. **17**:245(2004).
8. T. N. Demidova, F. Gad, T. Zahra, K. P. Francis and M. R. Hamblin, *J Photochem Photobiol B*. **81**:15(2005).
9. M. Wainwright, *Photochem Photobiol Sci*. 3(5):406(2004).
10. M. R. Hamblin, D. A. O'Donnell, N. Murthy, C.H. Contag and T. Hasan, *Photochem Photobiol*. **75**:51(2002).
11. P. Huang, T. C. Y. Liu, J. L. Jiao, Z. Li, R. Duan, G. H. Xie, K. Y. Li, S. X. Deng, W. Hu, Q. Guo and S. H. Liu, *Lasers Surg Med*. **S16**: 40(2004).
12. P. Huang, C. Y. Liu, X. Y. Xu, J. L. Jiao, L. Zhu, Z. Li and S. H. Liu, *Zhong guo Ji guang/Chinese Journal of Lasers*, **v 32**, 723(2005).
13. R. J. O'Donnell, T. F. Deutsch, R. J. Flotte, C. A. Lorente, W.W. Tomford, H. J. Mankin and K.T. Schomacker, *J Orthop Res*. **14**:108(1996).
14. T. C. Y. Liu, D. L. Rong, J. Huang, X. Y. Deng and S. H. Liu, *Proc SPIE,* 6026: 60260C (2006).
15. F. R. Venezio, C. DiVincenzo, R. Sherman, M. Reichman, T. C. Origitano, K. Thompson and O. H. Reichman, *J Infect Dis*. **151**:166(1985).
16. D. Lapotko, T. Romanovskaya, G. Kutchinsky and V. Zharov, *Cytometry*. **37**:320(1999).
17. T. Karu, *Amsterdam: Gordon and Breach Science Publishers*. (1998).
18. T. C. Y. Liu, J. L. Jiao, R. Duan, Y. Li, Y. Y. Yeung and S. H. Liu, *Lasers in Medicine, Surgery and Dentistry*. 83(2003).
19. T. C. Y. Liu, J. Q. Yuan, Y. F. Wang, X. Y. Xu and S. H. Liu, *Proc SPIE,* 6534: 653403-1~-6 (2007).
20. T. Tadakuma, *Keio J Med*. **42**:180(1993).
21. R. Duan, T. C. -Y. Liu, Y. Li, H. Guo and L. -B. Yao, *Lasers Surg. Med*. **29**: 174(2001)
22. T. C. Y. Liu, C. C. Zeng, J. L. Jiao and S. H. Liu, *Proc SPIE*. **5254**: 150(2003).
23. M. J. Taravella, A. Weinberg, M. May and P. Stepp, *Ophthalmology*. **106**:1498(1999).
24. D. K. Dhaliwal, DA. Jr. Barnhorst, E. Romanowski, P. G. Rehkopf and Y. J. Gordon, *Am J Ophthalmol*. **125**:488(1998).
25. T, Deai, M. Fukuda, Y. Tomoda, S. Higaki, K. Hayashi and Y. Shimomura, *Jpn J Ophthalmol*. **48**:570(2004).
26. J. Levy, R. Lapid-Gortzak, I. Klemperer, T. Lifshitz, *J Refract Surg*. **21**:400(2005).
27. C. K. Lu, K. H. Chen, S. M. Lee, W. H. Hsu, J. Y. Lai, Y. S. Li, *J Refract Surg*. 2006 May;22(5):509.
28. O. M. Smith, D. L. Traul and F. Sieber, *Exp Hematol*. **20**:1278(1992).
29. H. S. de Bruijn, W. Sluiter, A. van der Ploeg-van den Heuvel, H. J. Sterenborg and D. J. Robinson, *Photodermatol Photoimmunol Photomed*. **22**:238(2006).
30. R. Körner, F. Bahmer and R. Wigand, *Hautarzt*. **40**:350(1989).
31. A. Schindl and R. Neumann, *J Invest Dermatol*. **113**:221(1999).
32. M. S. Baggish, *J Reprod Med*. 27(12):737(1982).
33. P, Fagerholm, *Acta Ophthalmol Scand*. **81**:19(2003).
34. R. Lubart, G. Kesler, R. Lavie and H. Friedmann, *Photomed Laser Surg*. **23**:369(2005).
35. L. S. Bass, *Ann Plast Surg*. **40**:328-34(1998).
36. J. T. Jeong, Y. C. Kye, *Dermatol Surg*. **27**:107(2001).

LAYERED-RESOLVED AUTOFLUORESCENCE IMAGING OF PHOTORECEPTORS USING TWO-PHOTON EXCITATION

LINGLING ZHAO, JUNLE QU, DANNI CHEN, HANBEN NIU

Key Laboratory of Optoelectronic Devices and Systems of Ministry of Education, Institute of Optoelectronics, Shenzhen University, Shenzhen 518060, China

In this paper, we present our investigation on the morphological and autofluorescence characteristics of the cones and rods using two-photon excitation with a femtosecond Ti: sapphire laser. The results show that the microstructures of the photoreceptor layers can be visualized at submicron level without any staining or slicing. The morphology and spatial distribution of the cones and rods can be resolved by autofluorescence imaging. The autofluorescence in the photoreceptor outer segments is much stronger than other layers, but susceptible to light-induced damage.

1. Introduction

The dysfunction of photoreceptors is one of most important factors of ocular fundus diseases, such as retinitis pigmentosa (RP), age-related macular degeneration (AMD), glaucoma, and so forth.[1,2] Retinitis pigmentosa is the leading cause of inherited blindness, which is characterized by progressive loss of visual function related to death of rods then cones, and leads to the breakdown of the photoreceptor outer segment disc membranes. Patients with retinitis pigmentosa (RP) show various symptoms. The onset is often gradual and insidious, and many patients fail to recognize the manifestations of this condition until it has progressed significantly. When patients do report symptoms, they commonly include difficulty with night vision (nyctalopia) as well as loss of peripheral vision. In addition, Age-related macular degeneration (AMD) is a progressive disease that is related to the abnormal accumulation of lipofuscin, which is located in the retinal pigment epithelium (RPE) cells and is a byproduct of incomplete digestion of photoreceptor outer segment disc membranes. Generally, the people who suffer from AMD have an early abnormal condition and experience minor visual loss. For many of these people, macular degeneration will not progress to a more serious condition. But for the others, macular degeneration will lead to complete visual loss.[3]

Fortunately, with the advent of near infrared femtosecond laser, multiphoton excitation fluorescence microscopy, which is based on simultaneous absorption of two or more photons, has become a novel and powerful tool in biomedical imaging.[4] It enables the possibilities to investigate the autofluoresecence and spatial distribution of photoreceptors at early stage of ocular diseases, since it can increase the penetrability through the thick biological tissue and provide high-resolution imaging of endogenous fluorophores in biological tissue.[5, 6] And also, Multiphoton microscopy can facilitate the simultaneous excitation of different endogenous fluorophores in biological specimens at submicron level, which benefits to the identification and evaluation of the morphological and autofluorescence characteristics of the photoreceptors. Several endogenous fluorophores exist in the photoreceptors, e.g., all-trans-retinol, NAD(P)H, A2-PE, FAD etc, which are all related to the metabolism and visual cycle.[7,8,9]. Therefore, multiphoton microscopy can be used to investigate the autofluorescence characteristics of photoreceptors.

In this study, layered-resolved autofluorescence imaging using two-photon excitation has been performed, which can provide not only structural, but also functional information about the different layers of photoreceptors, particularly photoreceptor inner and outer segments which are vital functional layers in the photoreceptors and significant for the early diagnosis of the retinal diseases.

2. Materials and Methods

2.1. *Materials*

Fresh porcine eyes are supplied by the local slaughterhouse and extracted immediately after slaughtering within one hour. The eyeballs are transported to the lab in PBS (PH 7.16)/normal saline solution and kept in a freezing box (0-4℃). The eyes are kept in the PBS (PH 7.16)/normal saline solution for less than 30 minutes before experiment. A 6mm-diameter retina-RPE-choroid-sclera complex near the macula (both the rods and cones are in existence) is trephined and the neurosensory retina is peeled off gingerly. The side that connects with the RPE cells is kept down to contact the slip of a specially designed chamber. All the specimens are freshly prepared without fixing, slicing and labeling.

2.2. *Methods*

The experiment is performed on an inverted confocal laser scanning microscope (Leica TCS SP2). A mode-locked femtosecond Ti: sapphire laser (Coherent Mira 900F) is coupled to the microscope for two-photon excitation of the specimen. The Ti: Sapphire laser produces ultrashort laser pulses with tunable wavelength from 700nm to 980nm, a pulse width of about 120 fs and a repetition rate of 76 MHz. The excitation wavelength in this experiment is 800nm. A 63×/ NA1.32 oil-immersion objective is used to focus the excitation light onto the specimen and collect the autofluorescence from the photoreceptors.

3. Results and Discussions

There are two basic types of photoreceptors, rods and cones, which are of different shapes and involved in different visual functions. Rods are highly sensitive to the weak and faint light, which are used for vision under dark-dim conditions at night. Cones are the basis of our color perceptions since they have different wavelength sensitivity and the consequent pathways of connectivity to the brain.

Both the rod and cone photoreceptors are consisted of four parts, i.e., the outer segment, the inner segment, the nuclear region and the synaptic terminal. The outer segment is filled with stacks of membranes containing the visual pigment molecules such as rhodopsins. The inner segment contains mitochondria, ribosomes and membranes where opsin molecules are assembled and passed to the outer segment discs. The nuclear region contains the nucleus of the photoreceptor cell and the synaptic terminal is responsible for the neurotransmission to second order neurons.

Usually rod photoreceptors are slim rod-shaped structures with their inner and outer segments filling the area between the larger cones in the subretinal space and stretching to the pigment epithelium cells. Rod cell bodies make up the remainder of the outer nuclear layer below the cone cell bodies. Cone photoreceptors, on the other hand, are robust conical-shaped structures that have their cell bodies situated in a single row right below the outer limiting membrane and their inner and outer segments protruding into the subretinal space towards the pigment epithelium. Apical processes from the pigment epithelium envelope the outer segments of both rods and cones.

In our study, we obtain the image of rod and cone photoreceptors nicely aligned along vertical and horizontal directions using two-photon excitation fluorescence microscopy.

Fig.1 shows the cross-sectional autofluorescence image of photoreceptors, in which individual cones and rods are clearly resolved. The XZ image of the different photoreceptor layer structures are distinguished accurately with submicron resolution. The autofluorescence in the photoreceptor outer segments is much stronger than other layers, since the outer segment is abundant in many fluorophores, such as all-trans-retinol, A2-PE, FAD, etc.

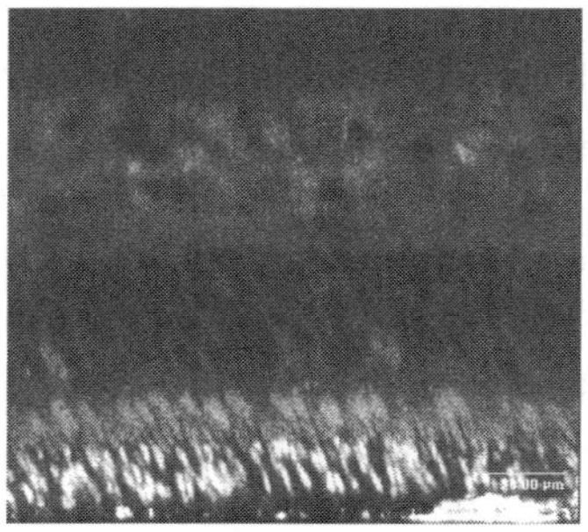

Fig.1. Cross-sectional autofluorescence image of photoreceptors.

The XY image of the photoreceptor outer segment is shown in Fig. 2. Because the photoreceptor outer segment is connected with the RPE cells, when the photoreceptor outer segments is separated from the RPE cells, the outer segment at top may be disarranged, but the pillar-like shape can be observed (Fig. 2 (a)). When focusing the imaging plane slightly into the specimen, we can see the rod is aligned annularly, round in shape and 2 μm in diameter (Fig.2 (b)). The center is dark. But when focusing the objective further into the specimen, the autofluorescence of cone outer segment, which locates in the center, can be detected, and the size of the cones is becoming larger and larger (Fig.2 (c) and Fig. 2 (d)), whereas the autofluorescence of rod outer segment weakens.

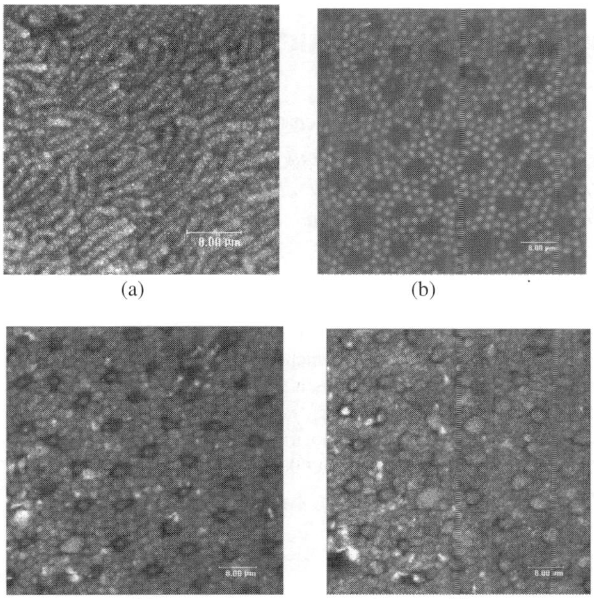

(a) (b)

(c) (d)

Fig.2. Autofluorescence image of photoreceptor cut segment.

Fig.3 shows the autofluorescence image of photoreceptor inner segment. The bigger cell is corresponding to the cone inner segment whose diameter is around 6 μm. Rod inner segment is around the cone and the diameter is about 2μm. Fig. 4 is the autofluorescence image of photoreceptor nuclear region, which contains the nucleus of the photoreceptor cell. The cone nucleus layer is above that of the rods.

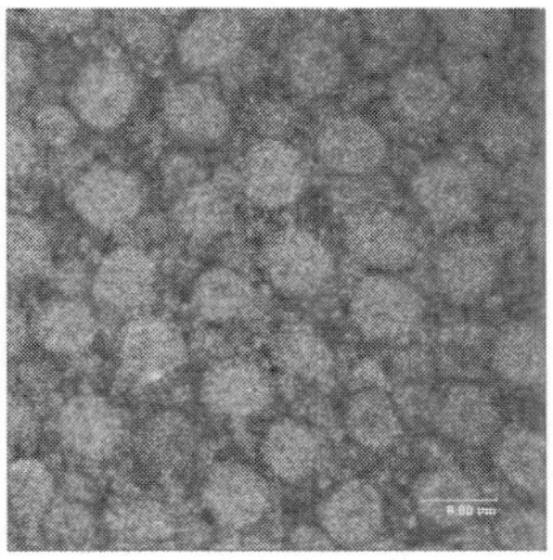

Fig.3. Autofluorescence image of photoreceptor inner segment

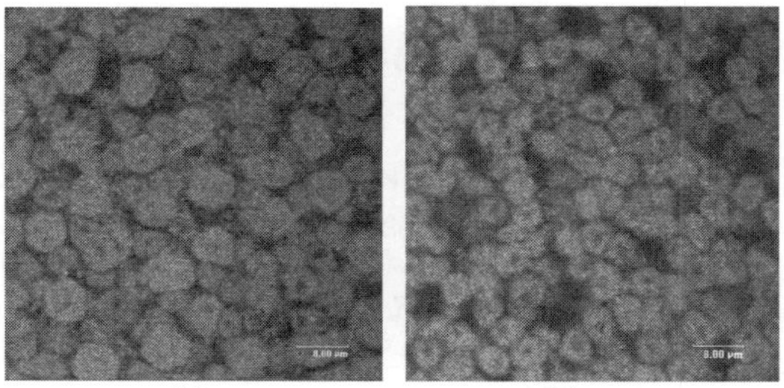

(a) (b)

Fig.4. Autofluorescence image of cone (a) and rod (b) nucleus layer.

4. Conclusions

Our results demonstrate, for the first time, that different layers of photoreceptors can be identified by autofluorescence imaging using two-photon excitation. Although much research remains to be done, it appears that this technique has great potential to bridge the gap between clinical examination and invasive biopsy and thus facilitate the early detection and diagnosis of ocular diseases. The distribution and intensity of autofluorescence may provide an insight into the sequence of events that leads to retinal damage and may help elucidate pathological mechanisms.

Acknowledgments

This work is supported by the National Natural Science Foundation of China (NSFC) under contract No. 60627003 and No. 60408011, and is also supported in part by Guangdong Natural Science Foundation grant No. 5010500.
Junle Qu's email address is jlqu@szu.edu.cn.

References

1. G. Jiangang and C. Kyeongmi, PNAS, **99 (8)**, 5698 (2002).
2. P. Kayatz, G. Thumann, T. T. Luther, et al., Invest. Ophthalmol. Vis. Sci., **42**, 241 (2001).
3. J. Hopkins, A. Walsh and U. Chakravarthy, Invest. Ophthalmol. Vis. Sci., **47**, 2269 (2006).
4. GJ. Brakenhoff, J. Squier, T. Norris and AC. Blito, J Microsc., **181**, 253 (1996).
5. W. Denk, JH. Strickler, Science, 248, 73 (1990).
6. M. Han, G. Giese, Journal of Biomedical Optics, **12(2)**, 024012 (2007).
7. C. Chunhe, Biophysical Journal, **88**, 2278 (2005).
8. S. R. Kim and K. Nakanishi, Experimental Eye Research, **82**, 828 (2006).
9. L. Laura E., S. John D., Photochemistry and Photobiology, **79(2)**, 127 (2004).

MODELING OF PASSIVE Q-SWITCHED Tm: Ho LASER*

CHUYUN HUANG [1,2†], CHENGXIAN DENG [3], YUCHENG YAO [2], ZHENGJIA LI [2]

1. School of Science, Hubei University of Technology, Wuhan, China, 430068

2. School of Optoelectronics Science and Engineering, Huazhong University of Science and Technology, Wuhan, China, 430074

3. Physics School, South China University of Technology, GuangZhou, China, 510641

2μm solid laser running at room temperature can be widely applied in many fields like optical communication, medical cure, air pollution monitoring, and is also an ideal pumping source for mid-infrared optical parameter oscillator, further more, it's safe for human eyes. To achieve above objectives, higher peak power and shorter pulse duration laser is desirable. Q switched pulse laser is an effective approach to meet the above requirements. For the development and optimization of laser system, it is a practical choice to use laser modeling to predict the influence of various parameters for laser performance. For given parameters, rate equation for Q switched Tm:Ho laser pumped by diode was obtained and used to predict the change of population densities in various manifolds. When pump pulse energy is 14J and pulse width is 40μs, output signal pulse energy and width are 1.19mJ and about 600ns, respectively. The model can be used to find the potential laser systems that meet application requirements.

Keywords: Tm:Ho laser, Q-Switch, rate equation.

1. Introduction

The Tm:Ho laser that wavelength is about 2μm has attracted considerable attention in the recent years for its great usefulness in many fields like medical application, optical communication, air pollution monitoring, and industrial processes. Furthermore, the 2μm laser can pump mid-infrared optical parameter oscillation (OPO) material. For above applications and further potential explore, higher peak power and shorter pulse width are very desirable. Q-switched laser is a most practical way to satisfy those goals. Actively Q-switched Tm and Ho lasers were developed during the early 1990s[1-4]. The advantages of saturable absorber Q switches (SAQSs) compared with active Q switches are well known and consist primarily of simplicity, reliability, and economy. At 2μm, it has been difficult to obtain SAQSs. Ho^{3+} doped crystals were used to Q switch Tm^{3+}:YAG lasers[5,6]. Good repeatable Q-switching results (10mJ, 60ns) could be obtained by Ho:YLF/YVO_4/CaF_2 SAQSs for Tm^{3+}:YAG lasers. However, intracavity focusing was required for these Q-switched lasers. To obtain better laser performance, it is necessary to look for the crystal dispensing with intracavity focusing as saturable absorber. Although there were reports about these[7,8], only few crystals were used as saturable absorber for Q-switched laser. The Cr^{2+}:$Cd_{0.85}Mn_{0.15}Te$ used in this paper demonstrates the laser performance for Tm:Ho lasers in theory.

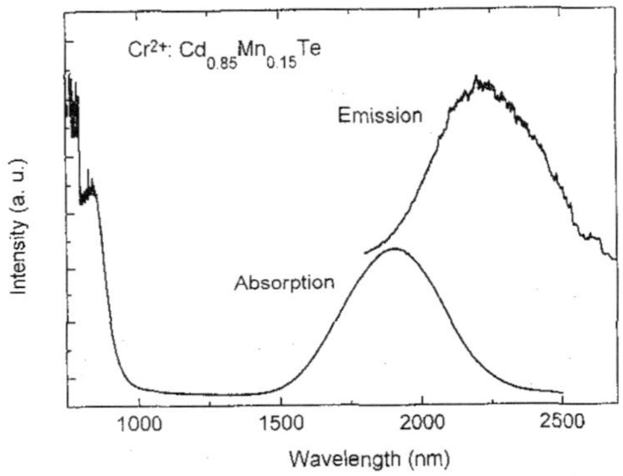

Fig. 1. Room temperature absorption and emission spectra of Cr^{2+}:$Cd_{0.85}Mn_{0.15}Te$[9]

The adjacencies between the absorption peak and laser wavelength should be taken into account in choice of saturable absorber. Absorption and emission spectra of Cr^{2+}:$Cd_{0.85}Mn_{0.15}Te$ at room temperature are shown in figure **1**[9]. The absorption spectrum exhibits a broad band centered at 1900nm, which presents a high absorption for 2μm wavelength. The absorption cross section σ_a of Cr^{2+}:$Cd_{0.85}Mn_{0.15}Te$ is ~10 times the emission cross section σ_e of Tm:Tm:YLF at 2060nm. Compared with the previous Ho^{3+} SAQSs,

† Postal addresses of authors:
Chuyun HUANG: tel: +86 27-63133530 e-mail: chuyunh@163.com

Cr^{2+}:$Cd_{0.85}Mn_{0.15}Te$ has a much lower saturation energy fluence ($E_s=h\upsilon/\sigma_a$) resulting in a reduced vulnerability to damage during the Q-switching operation. Furthermore, because of the short decay lifetime (~1.4μs), Cr^{2+}:$Cd_{0.85}Mn_{0.15}Te$ can provide a much faster repetition rate for Q-switched pulses than Ho^{3+} materials (τ_a >10 ms).

To fully understand the laser performance, it is necessary to know the lasing processes. Many researchers have done a lot of work in the Ho:Tm lasing processes. On the base of researches in theory and experiments, Norman P B[10]and Chang J L[11]et al established theoretical model for calculating energy transfer parameters in Ho:Tm laser and rate equation, and calculated the parameters which influence the energy transfer process, and population densities in various manifolds which influence the laser process, and proved them in experiments. Their theses are ones of the comprehensive articles in the field of holmium laser. In consideration of the four manifolds in both Tm and Ho for a total of 8 manifolds which can influence the laser process, Brain M W[12] et al have calculated and experimented on Tm:Ho:YLF and Tm:Ho:LuLF, and concluded that Tm:Ho:LuLF is superior to Tm:Ho:YLF.

In the Q-switching holmium laser, Sousa J M [13] et al studied the active Q-switch Tm,Ho:YAG and Tm:YAG, and established the rate equation. Zheng Jia[14] et al also studied rate equation model for A-O Q-switch Cr:Tm:Ho:YAG. But the model only simply processed laser upper and lower manifolds, without involving complicated energy transfer process.

This thesis established rate equation model for passive Q-switched Tm:Ho:YLF laser by using saturation absorber and give numerical calculation.

2. Rate Equations Model

To understand the laser performance, the usual approach is to construct a set of rate equations governing the populations of all the manifolds that have an impact on the laser performance. The rate equation method is also used in this article. The energy-transfer processes involved in the 2-μm laser operation of the codoped Tm–Ho materials have been thoroughly described by several authors [15-18]. In the case of Tm and Ho co-doped systems, the quantities of manifolds to consider can number 10 or more. This requires knowledge of a great many energy transfer and decay parameters. Inclusion of all these parameters in laser models becomes problematic. The main difficulty is in accurately measuring many of these parameters. At the other hand, the energy transfer processes among the high manifolds have only a negligible effect on lasing[19]. The approach is then to make some assumptions regarding the dynamics of interaction of the Tm and Ho ions. As far as the manifolds lying above the Tm $^3\boldsymbol{H}_4$ and Ho $^5\boldsymbol{I}_5$ are concerned, they are not included in the model since it is well established[20,16] that the upconversion processes that excite the Tm^{3+} and Ho^{3+} ions to these states are relatively unimportant as compared to the upconversion process involving the Tm $^3\boldsymbol{F}_4$ and Ho $^5\boldsymbol{I}_7$ multiplets.

Therefore, We have considered only the 4 lowest manifolds in both Tm and Ho for a total of 8 manifolds which can interact in the Tm:Ho laser system. Here we use the main lines of these analyses with reference to the energy-level diagram of Tm and Ho presented in Fig. **2**. To simplify the notation, the manifolds of Tm $^3\boldsymbol{H}_6$, $^3\boldsymbol{F}_4$, $^3\boldsymbol{H}_5$, $^3\boldsymbol{H}_4$ are marked as 1, 2, 3, 4, and Ho $^5\boldsymbol{I}_5$, $^5\boldsymbol{I}_6$, $^5\boldsymbol{I}_7$, $^5\boldsymbol{I}_8$ as 5, 6, 7, 8, specifically[9]. In most applications, the Tm $^3\boldsymbol{H}_4$ manifold is used as the pump manifold, while lasing occurs between levels in the Ho $^5\boldsymbol{I}_7$, and $^5\boldsymbol{I}_8$ manifolds.

Although energy transfer can occur between the eight manifolds, according the principle of parsimony, only the processes shown in fig. **2** are used in the model. The processes are clarified briefly as below: (i) self-quenching of the pump Tm $^3\boldsymbol{H}_4$ manifold producing two ions, both in the Tm $^3\boldsymbol{F}_4$ manifold, and the reverse process; (ii) energy transfer and back-transfer between the Tm $^3\boldsymbol{F}_4$ and the Ho $^5\boldsymbol{I}_7$ manifolds, while the Ho ion in $^5\boldsymbol{I}_7$ is promoted to the $^5\boldsymbol{I}_5$ manifold.; (iii) radiative and nonradiative decay of every manifold of one ion to all the manifolds below belong to the same ion; (iv) energy transfer through cross relax between the Tm $^3\boldsymbol{F}_4$ and the Ho $^5\boldsymbol{I}_7$; and also between Tm $^3\boldsymbol{H}_5$ and Ho $^5\boldsymbol{I}_6$.

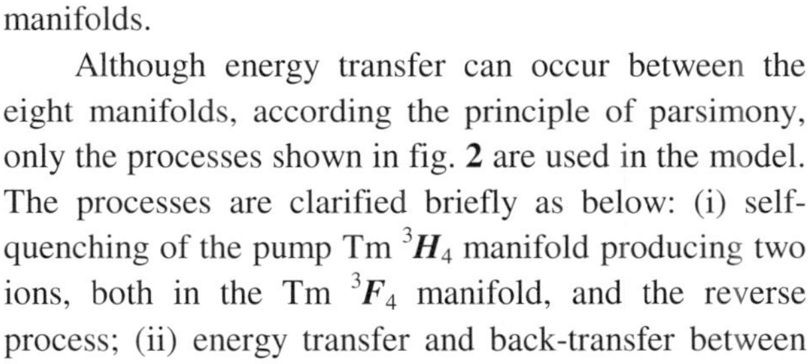

Fig. 2. Energy transfer and decay processes in Tm:Ho co-doped materials

For passive Q-switching, the change of absorber population densities needs to be put into consideration.

Combining all the processes, rate equations for the relevant manifolds can be developed. With the notation introduced above, the rate equations can be written as below:

$$\frac{dn_1}{dt}=-Rp[1-\exp(-\sigma_a\ell_p n_1)]+\frac{n_2}{\tau_2}+\frac{\beta_{41}}{\tau_4}n_4+n_2n_8P_{28}-n_7n_1P_{71}-n_4n_1P_{41}+n_2^2P_{22}$$
$$+n_2n_7P_{27}-n_5n_1P_{51}-n_6n_1P_{61}+n_3n_8P_{38} \tag{1}$$

$$\frac{dn_2}{dt}=-\frac{n_2}{\tau_2}+\frac{\beta_{32}}{\tau_3}n_3-n_2n_8P_{28}+n_7n_1P_{71}+2n_4n_1P_{41}-2n_2^2P_{22}-n_2n_7P_{27}+n_5n_1P_{51} \tag{2}$$

$$\frac{dn_3}{dt}=-\frac{n_3}{\tau_3}+\frac{\beta_{43}}{\tau_4}n_4+n_6n_1P_{61}-n_3n_8P_{38} \tag{3}$$

$$\frac{dn_4}{dt}=Rp[1-\exp(-\sigma_a l_p n_1)]-\frac{n_4}{\tau_4}-n_4n_1P_{41}+n_2^2P_{22} \tag{4}$$

$$\frac{dn_5}{dt}=-\frac{n_5}{\tau_5}+n_2n_7P_{27}-n_5n_1P_{51} \tag{5}$$

$$\frac{dn_6}{dt}=-\frac{n_6}{\tau_6}+\frac{\beta_{56}}{\tau_5}n_5-n_6n_1P_{61}+n_3n_8P_{38} \tag{6}$$

$$\frac{dn_7}{dt}=-\frac{n_7}{\tau_7}+\frac{\beta_{67}}{\tau_6}n_6+n_2n_8P_{28}-n_7n_1P_{71}-n_2n_7P_{27}+n_5n_1P_{51}$$
$$-c\sigma_{se}(f_7n_7-f_8n_8)\phi \tag{7}$$

$$\frac{dn_8}{dt}=\frac{n_7}{\tau_7}-n_2n_8P_{28}+n_7n_1P_{71}+c\sigma_{se}(f_7n_7-f_bn_8)\phi \tag{8}$$

$$\frac{dn_{gs}}{dt}=-\frac{A}{Asa}c\sigma_{gs}\phi n_{gs}+\frac{n_{cr}-n_{gs}}{\tau_{cr}} \tag{9}$$

$$\frac{d\phi}{dt}=-\frac{\phi}{\tau_c}+c\frac{l}{L_{opt}}\sigma_{se}(f_7n_7-f_8n_8)\phi+c\frac{l}{L_{opt}}\frac{n_7}{\tau_7}B-c\frac{\sigma_{gs}n_{gs}l_a}{L_{opt}}\phi+c\phi\sigma_{es}\frac{n_{Cr}-n_{gs}}{\tau_{Cr}} \tag{10}$$

Where, τ_c is photon lifetime in the cavity; $\frac{1}{\tau_c}=\frac{l}{\varepsilon L_{opt}}[\ln(\frac{1}{R_mR_L})+2\varepsilon^2(1-T_s)]$ and $R_p=\frac{\eta_p\lambda_pE_p}{hc\pi w_r^2 lt_p}$ represents the inverse of the cavity lifetime and the pump rate, respectively. The Q-switch laser oscillation energy, E_{LO}, is calculated from the photon density, Φ, according to

$$E_{LO}=\frac{1}{2}T_m\frac{hc}{\lambda_e}\pi w_l^2L_{opt}\phi \tag{11}$$

The terminology in the above equations is the following:

$\boldsymbol{N_i}$ is the population density in the various manifolds, where $\boldsymbol{i}$ represent the manifold number assigned in fig. 2; and $\boldsymbol{\Phi}$ is the laser photon density; $\boldsymbol{n_{gs}}$ is the ground state population density of absorber, and $\boldsymbol{n_{cr}}$ is the concentration of Cr^{2+} in the absorber.

$\boldsymbol{P_{ij}}$ is the energy transfer parameter among various manifolds. The correspondence embodied as follow:

P_{28}:Tm($^3F_4\rightarrow{}^3H_6$); Ho($^5I_8\rightarrow{}^5I_7$);

P_{71}:Ho($^5I_7\rightarrow{}^5I_8$); Tm($^3H_6\rightarrow{}^3F_4$);

P_{41}:Tm($^3H_4 \rightarrow ^3F_4$); Tm($^3H_6 \rightarrow ^3F_4$);
P_{22}:Tm($^3F_4 \rightarrow ^3H_6$); Tm($^3F_4 \rightarrow ^3H_4$);
P_{61}:Ho($^5I_6 \rightarrow ^5I_8$); Tm($^3H_6 \rightarrow ^3H_5$).
P_{38}:Tm($^3H_5 \rightarrow ^3H_6$); Ho($^5I_8 \rightarrow ^5I_6$).
P_{51}:Ho($^5I_5 \rightarrow ^5I_7$); Tm($^3H_6 \rightarrow ^3F_4$):
P_{27}:Tm($^3H_4 \rightarrow ^3H_6$); Ho($^5I_7 \rightarrow ^5I_5$);

τ_i is the lifetime of manifold i (i=2,3,4,5,6,7), τ_{Cr} is the lifetime in emission manifold of Cr^{2+} in the absorber; β_{ij} is probability that an ion in manifold i transitions to j (ij = 43, 56, 67); f_7 is the Boltzmann fraction of the Ho $^5I_7$ upper laser level, and f_8 is Ho $^5I_8$ lower laser level correspondingly; σ_a is the absorption cross section of the laser rod at the pump wavelength; σ_e is the spectroscopic emission cross section of the laser rod at the laser wavelength; σ_{se} is the stimulated emission cross section of the laser rod at the laser wavelength ($\sigma_{se}=\sigma_e/f_7$); σ_{gs} is the absorption cross section of the absorber at the laser wavelength; η_p is the laser pump distribution efficiency λ_p is the laser pump wavelength and λ_e is the laser emission wavelength; Ep is the laser pump energy; τ_c is photon lifetime in the cavity; ε is 2 for a linear cavity and 1 for a ring cavity; t_p is the current pump pulse length; w_r is the laser rod radius, w_l is the laser mode radius; l is the laser rod length and l_p is the pumped length; L_c is the total laser cavity length; L_{opt} is the laser cavity optical length, $(n-1)l + L_c + (n_A-1)l_A$; l_A is the absorber length; T_s is the laser rod surface transmission; R_m is the reflectivity of the output coupler, and R_L represents the passive losses; B is the spontaneous emission factor; A is the beam area of laser rod and A_{sa} is of absorber correspondingly.

The parameters that used in the model were listed in table 1.

Table 1 Input parameter for Q-switched Tm:Ho YLF laser model

symbol	value	symbol	value
P_{28}	$1.68\times10^{-22}cm^3/\mu s$	P_{71}	$1.28\times10^{-23}cm^3/\mu s$
P_{41}	$3.13\times10^{-24}cm^3/\mu s$	P_{22}	$3.48\times10^{-25}cm^3/\mu s$
P_{38}	$7.64\times10^{-23}cm^3/\mu s$	P_{61}	4.09×10^{-22} $cm^3/\mu s$
P_{27}	1.47×10^{-22} $cm^3/\mu s$	P_{51}	$1.16\times10^{-21}cm^3/\mu s$
f_7	0.0851	f_8	0.0258
η_p	1.3	σ_a	$3.4\times10^{-19}cm^2$
σ_{esa}	$0.4\times10^{-19}cm^2$	σ_{se}	$1.4\times10^{-19}cm^2$
β_{ij}	1	τ_{Cr}	1.4μs
τ_2	15000μs	τ_3	1μs
τ_4	2000μs	τ_5	20μs
τ_6	2200μs	τ_7	16000μs
R_m	0.95	R_l	1
W_l	1.2mm	W_r	0.2cm

Although the rate equation in this paper is for the purpose of passive Q-switch Tm:To YLF laser system, its fundamental principle can also be used in other systems. In this model, the Tm:Ho laser system uses linear cavity, ε=2. The cavity length is 20.0cm. The concerned parameters are as the followings: output mirror reflectivity as 0.95; laser rod length as 4.0cm; diameter as 0.4 cm; the Tm concentration as 6.0% and Ho concentration as 0.4%. Side diode pumping is applied in this system, with pumping wavelength as 0.79μm. The saturation absorber's length, l_A, is 0.4cm and diameter is 2.0cm.

We solved numerically the differential equations described above using an adaptive step fourth-order Runge-Kutta algorithm.

The laser performances predicted by the model were shown in figure 3. For rectangle pulse pump which duration is decuple microsecond, laser pulse appears through delay of a period of time. The more low pump energy, the more long delay time. For pump energy of 14J and pump width of 40μs, the delay time is about 150μs.

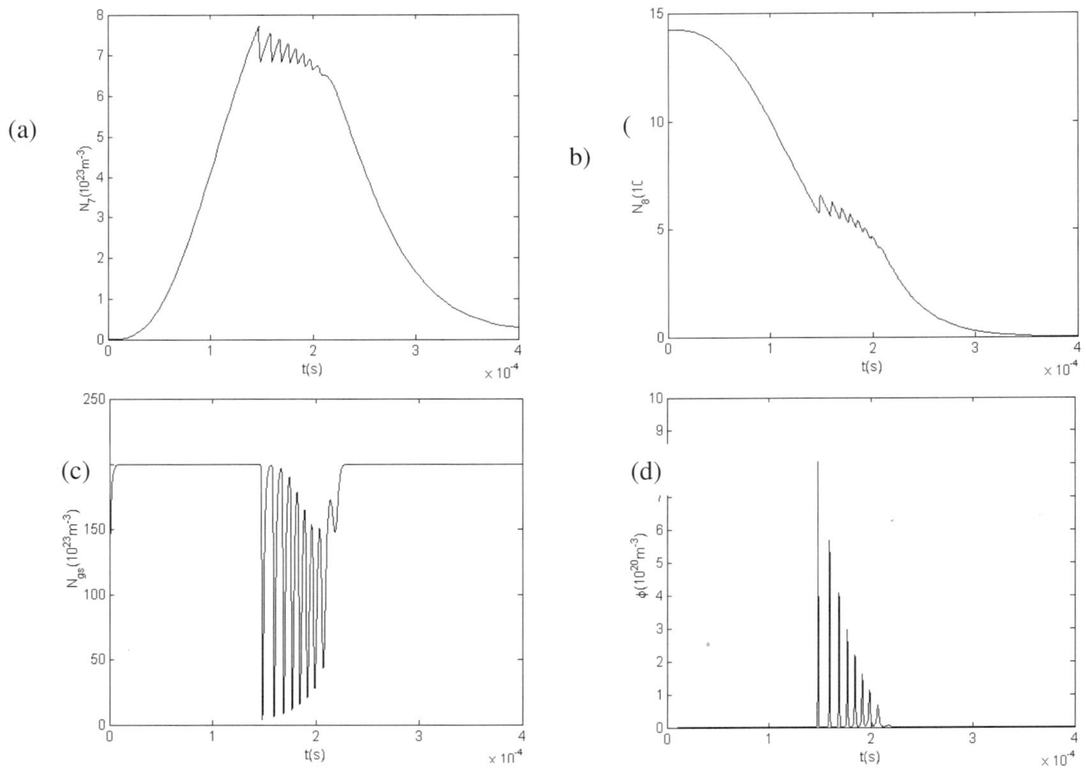

Fig. 3. Performance of passive Q-switched Tm:Ho:YLF laser. (a) The upper manifold of Ho population, (b) The lower manifold of Ho population, (c) The ground state of absorber Cr population, (d) The photon density

Shown in figure **4** are pulse energy dependence on pumping energy and pumping pulse width. From fig. **4**, it can be seen that the optimal pump energy and width are 14J and 0.4×10^{-4}s, respectively. According to parameter above, signal pulse output energy is 1.19mJ, and the half width is 600ns, and the pulse space is about 11μs.

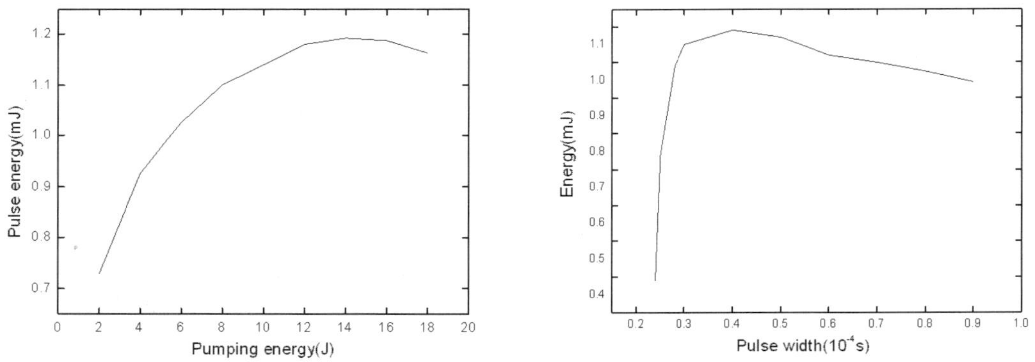

Fig. 4. Output pulse energy. (a) Energy dependence on pump energy, (b) Energy dependence on pump pulse width

As we known so far, there are no reports about Cr^{2+}:$Cd_{0.85}Mn_{0.15}Te$ using as Q-switcher for 2μm laser, the predictions by the model can't compared with experiment. The next we intend to is experimenting passive Q-switch Tm:Ho: laser with the absorber above. There were reports about Cr^{2+}:$Cd_{0.55}Mn_{0.45}Te$ using as passive Q-switcher for 2μm laser[7], the absorber used in our model has a higher absorption for 2μm compared with Cr^{2+}:$Cd_{0.55}Mn_{0.45}Te$, the reasonable result that Cr^{2+}:$Cd_{0.85}Mn_{0.15}Te$ has a better performance in saturable absorber can be predicted.

3. Summary

A laser model has been presented and used to model the performance of passive Q-switched Tm:Ho:YLF laser operating at 2.06μm. The results indicate that the Cr^{2+}:$Cd_{0.85}Mn_{0.15}Te$ crystal can be used as saturable absorber to Q-switch Tm:Ho laser. The optimal pump energy and pump pulse width are 14J and 40μs for selected parameter in our model. With above pump condition, Output signal pulse energy and width are 1.19mJ and about 600ns, respectively. The model can be used in other passive Q-switched laser system.

References

1. B. T. McGuckin, R. T. Menzies, and H. Hemmati. Efficient energy extraction from a diode-pumped Q-switched Tm, Ho:$YLiF_4$ laser. Appl. Phys. Lett., 1991; **(59)**:2926-8.
2. S. R. Bowman, M. J. Winings, S. K. Searles, and B. J. Feldman. Short-pulsed 2.1μm laser performance of Cr, Tm, Ho:YAG. IEEE J. Quantum Electron. 1991; **27**:1129-31.
3. T. S. Kubo and T. J. Kane. Diode-pumped lasers at five eye safe wavelengths. IEEE J. Quantum Electron. 1992; **28**:1033-40.
4. S. R. Bowman, J. G. Lynn, S. K. Searles, B. J. Feldman, J. McMahon, W. Whitney, C. Marquardt, D. Epp, G. J. Quarles, and K. J. Riley. High-average-power operation of a Q-switched diode-pumped holmium laser. Opt. Lett. 1993; **18**:1724-6.
5. Y.-K. Kuo, M. Birnbaum, and W. Chen. Ho:$YLiF_4$ saturable absorber Q-switch for the 2-μm Tm, Cr:$Y_3Al_5O_{12}$ laser. Appl. Phys. Lett. 1994; **65**:3060-2.
6. Y.-K. Kuo and M. Birnbaum. Ho:YVO_4 solid-state saturable absorber Q switch for a 2-μm Tm, Cr: $Y_3Al_5O_{12}$ laser," Appl. Opt. 1996; **35**:881-4.
7. A. V. Podlipensky, N. V. Kuleshov. V.I. Levechenko, et al. Cr^{2+}:$Cd_{0.55}Mn_{0.45}Te$ passive Q-switcher for 2μm lasers. Conference on Lasers and Electro-Optics Europe - Technical Digest 2000, CTuK40, p106.
8. Tzong-Yow Tsai and Milton Birnbaum. Q-switched 2-μm lasers by use of a Cr^{2+}:ZnSe saturable absorber. Appl. Opt. 2001; **36(40)**:6633-7.
9. U. Hömmerich, X Wu, V. R. Davis, et al. Demonstration of room-temperature laser action at 2.5μm from Cr^{2+}:$Cd_{0.85}Mn_{0.15}Te$, Optics Lett. 1997; **22(15)**:1180-2.
10. Norman P B, Elizabeth D F, Clyde A M, et al. Ho:Tm laser I: Theoretical. IEEE J. of Quan. Elec. 1996; 32(1):103.
11. Chang J L, Gooywan H, and Norman P B. Ho:Tm laser II: experiments. IEEE J. of Quan. Elec. 1996; **32(1)**:104-11.
12. Brian M. Walsh, Norman P. Barnes, Mulugeta Petros, et al. Modeling of pulsed lasers for remote sensing. Proc. of SPIE **Vol. 5653**:24-32.
13. J. M. Sousa, J. R Salcedo, V. V. Kuzmin. Simulation of laser dynamics and active Q-switching in Tm, Ho:YAG and Tm:YAG lasers. Appl. Phys. B 1997; **64**:25-36.
14. Zheng Jia, Liu Jingjiao, Tang yi, et al. A-O Q-Switch of 2.1μm Laser. Proc. of SPIE **Vol. 5627**: 147-53.
15. T. Y. Fan, G. Huber, R. L. Byer, P. Mitzcherlich. Spectroscopy and diode laser-pumped operation of Tm, Ho:YAG. IEEE J. Quan. Elec. 1988; **24**:924-33.
16. G. Rustad and K. Stenerson. Modeling of laser-pumped Tm and Ho lasers accounting for upconversion and ground-state depletion. IEEE J. Quan. Elec. 1996; **32**:1645-56.
17. R. R. Petrin, M. G. Jani, R. C. Powell, and M. Kokta. Spectral dynamics of laser-pumped $Y_3Al_5O_{12}$:Tm, Ho lasers. Opt. Mater. 1992; **1**:111-24.
18. N. P. Barnes, W. J. Rodriguez, and B. M. Walsh. Ho:Tm:YLF laser amplifiers. J. Opt. Soc. Am. B 1996; **13**:2872-82.
19. M. Falconieri, A. Lanzi, G. Salvetti, A. Toncelli. Fluorescence dynamics in Tm, Ho:YLF following 800 nm pulsed laser excitation. Appl. Phys. B 1998; **66**:153-62.
20. M. Falconieri, A. Lanzi, G. Salvetti. Spectroscopic investigation of the visible and mid-infrared emission in Tm- and Ho-doped YAG and YLF crystals. Appl. Phys. B 1996; **62**:537-46.

RESEARCH OF MICRO-EXPLOSION MODEL IN LASER LITHOTRIPSY

HUANG CHUYUN[1,2]

1. *School of Science, Hubei University of Technology, Lijiadeng, Hongshan Wuhan, Hubei 430068, China*
2. *School of Optoelectronics Science and Engineering, Huazhong University of Science and Technology Wuhan, Hubei 430074, China*

YAN XUDONG

School of Science, Hubei University of Technology, Lijiadeng, Hongshan Wuhan, Hubei 430068, China

YANG TAO

School of Science, Hubei University of Technology, Lijiadeng, Hongshan Wuhan, Hubei 430068, China

YAO YUCHENG

School of Optoelectronics Science and Engineering, Huazhong University of Science and Technology Wuhan, Hubei 430074, China

Laser was been applied widely in calculi surgical therapy. The ablative mechanism varies with laser wavelength change. The laser induced thermodynamic micro-explosion of the tissue surface water has the important effect to the tissue ablation in laser lithotripsy. With the multi-material Euler algorithm, the numerical model was developed. The high pressure shock and thermal deposition on the tissue surface during the micro-explosion process were calculated quantitatively, and the laser eradiation intensity and water film thickness effect to the shock pressure have been analyzed. The results indicate that the high energy and short pulse laser can make high pressure shock on the tissue surface, which is well accord with the correlative experiments. The research can provided guidance and reference for lithotripsy and other laser medical application.

1. Introduction

Laser lithotripsy is a perfect tiny wound surgery in urinology. The factors, as pulse laser energy, action time and the distance between fiber end and tissue, and so on, affect the lithotripsy effect. Malcom Sperrin[1] et al pioneered in research on mechanism of Ho:YAG lithotripsy. They thought that laser energy brings on instantaneous vacuole in water, and the calculi would be damaged by the shock wave induced by vacuole collapse. Different opinion was delivered by Steren A. Schafer[2] in depth research for lithotripsy mechanism. The vacuole under water was photographed by high speed camera, and the evolvement of vacuole from emergence to break could be photographed by changing delay time. The water around calculi absorbs laser energy and induces vacuole form and collapse. The laser energy through vacuole is absorbed by calculi and induces spattering of calculi. The spattering materials solidify rapidly and are swept away by water bubble. The excited spectrum of Ho:YAG action is similar to black radiation without fine configuration, and there are not plasma.

The above researches indicated that thermodynamic micro explosion on surface between laser and hard tissue takes an important effect on process. Thermodynamic micro-explosion even if can't crash calculi directly, the impulse of micro explosion pressure to calculi fragment group and laser energy deposition after micro explosion vacuole forming effect weightily calculi ablation. Therefore, research on micro explosion would help understanding to ratio of laser lithotripsy, definition, and so on.

The researches for micro explosion at present concentrate mainly on experiment process, and merely on qualitative analysis. Quantitative analysis would be taken only with experiment parameter. Numerical simulation can overcome the localization of experiment method and obtain delicate and clear analysis results for action process. By simulation, the influence of parameter, as impact pressure, duration, quantity of heat and incidence, on micro explosion could be understood, and the simulative results could used to afford guidance to parameter select and experiment.

There are many theoretic and experimental researches about micro explosion mechanism of water film or liquid drop on tissue surface under laser irradiation[3,4]. With improvement of computer, numerical models and methods were developed largely[5], and micro explosion process enables to be simulated.

Although Euler method is limited to material interface, the advantages that Lagrange method haven't are[6]: (1) allowing nature surface forming and (2) allowing overfall and great distortion random. On thermodynamic micro-explosion of middle infrared laser ablating hard tissue under water, the calculating region involves water and air, and involves interaction of water, tissue and air if more analysis. All kinds of material would deform greatly in explosion, the "multi-material" Euler method is suitable to simulate thermodynamic micro-explosion of laser ablation on hard tissue.

2. Model of Micro-Explosion

Model scheme of laser lithotripsy was shown in fig.1. The action process was being surrounded by water. There is a tiny distance between fiber end and calculi surface, and laser from fiber end first irradiates water and induces micro explosion. Although calculi surface shows goose egg shape, as the fiber end is thin and the micro explosion region is tiny, the calculi surface can be thought as rigid plane.

The assumptions about working substance in explosion, water and air, are as follows:

- The material consist of a large numbers of tiny agglomerates which size are unimportant comparing with flow field, but are bigger than molecule freedom distance and structure size of solid crystal;
- Each agglomeration is in thermodynamic equilibrium, and all conclusions of equilibrium thermodynamics can be used for it, as there is thermodynamics statistic average estimate for it.
- Each agglomeration consists of same substance that the physics and chemical property is isotropy.

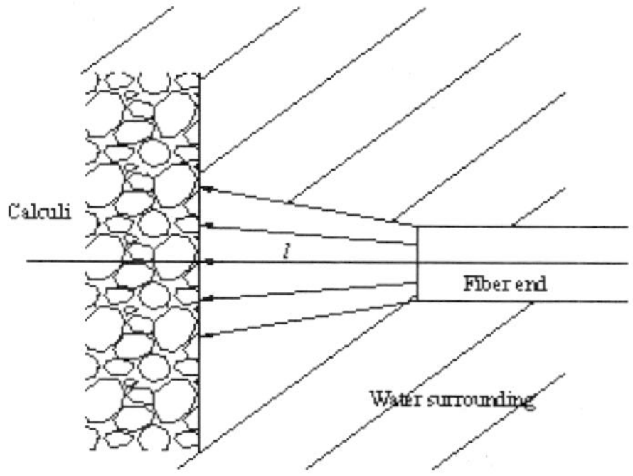

Figure 1 Scheme of laser irradiation

2.1. *Control Equation*

2.1.1. *Conservation equation*

Two dimensional nonconstant flows depend on except time variable, t, also on space variable, r and z. basing on the assumption above and ignoring heat conduction, the dynamics control equations about explosion were gained as follow:

$$\frac{\partial\rho}{\partial t}+u\frac{\partial\rho}{\partial z}+v\frac{\partial\rho}{\partial r}+\rho[\frac{\partial u}{\partial z}+\frac{1}{r}\frac{\partial}{\partial r}(rv)]=0 \quad (1)$$

$$\rho(\frac{\partial u}{\partial t}+u\frac{\partial u}{\partial z}+v\frac{\partial u}{\partial r})=-\frac{\partial p}{\partial z} \quad (2)$$

$$\rho(\frac{\partial v}{\partial t}+u\frac{\partial v}{\partial z}+v\frac{\partial v}{\partial r})=-\frac{\partial p}{\partial r} \quad (3)$$

$$\rho(\frac{\partial e}{\partial t}+u\frac{\partial e}{\partial z}+v\frac{\partial e}{\partial r})=-p[\frac{\partial u}{\partial z}+\frac{1}{r}\frac{\partial}{\partial r}(rv)]+S \quad (4)$$

In above equations, $\boldsymbol{u}$, $\boldsymbol{v}$ denote velocity of $\boldsymbol{z}$ and $\boldsymbol{r}$ direction individually, and $\boldsymbol{E}$ is total ratio energy. $\boldsymbol{p}$, $\boldsymbol{\rho}$, $\boldsymbol{s}$ denote pressure, density, and laser energy absorbed by tissue in unit time.

2.1.2. *State equation*

1. State equations of water and steam tiny agglomerates

   The water is at multi-state of liquid, steam and concomitant liquid-steam. The Bender Eq.[7,8] was used to describe the thermo physic state, and the state equation is as follow:

$$p=\rho T[R+B\rho+C\rho^2+D\rho^3+E\rho^4+F\rho^5+(G+H\rho^2)\rho^2\exp(-a_{20}\rho^2)] \quad (5)$$

In eq.5, T is water temperature and R is state constant. B, C, D, E, F, G, H depends on temperature, T and density, ρ, and the expressions refer to literature [7,8].

The relationship of temperature T and ratio inner energy was shown as follow:

$$e = e_{\rho 0} + c_v(T - T_0) \tag{6}$$

c_v is ratio thermal capacity and T_0, $e_{\rho 0}$ are basic temperature and corresponding ratio inner energy, individually.

2. State equations of steam

The state equation of air within calculation region was described approximately by it of ideal gas as follow:

$$p = (\gamma - 1)\rho e \tag{7}$$

γ is adiabatic index and equals 1.4 for air.

2.1.3 *Heat quantity equation for laser*

The beam of laser was described as Gauss fundamental module. Ignoring influence of laser absorption and dispersion induced by water temperature and phase change, according to Lambert-Beer law, the underwater transfer equation of laser was:

$$I(r,z) = I_0 \exp(-k_0 \rho z)\exp(-2r^2/w^2) \tag{8}$$

In eq.9, z is irradiation deepness of laser in water; I_0 is luminous intensity of laser in centre on water surface; k_0 is absorption coefficient of water for laser and the value is 25cm^{-1} when laser wavelength is 2100nm. w is light spot radius and r is the distance from laser midpoint to centre. The heat quantity absorbed by unit volume water in unit time can be denoted as follow:

$$S(r,z) = -\frac{\partial I(r,z)}{\partial z} = k_0 \rho I(r,z) \tag{9}$$

2.2. *Numerical Calculation*

2.2.1. *Gridding disperse for calculating region*

The calculating region was fixed on rectangle according to model showed as fig.1. To improve calculating precision and minish calculating quantity, unequal space step length was introduced to calculating grid.

The time step length, $\Delta t^n = t^{n+1} - t^n$, was ensured by the result of n step and CFL(Courant-Friedrichs-Lewy) stability condition.

$$\Delta t^n \le a \cdot \min\{\min(\frac{\Delta z_i}{u_{ij}^n + C}), \min(\frac{\Delta r_j}{v_{ij}^n + C})\} \tag{11}$$

a is the adjusting coefficient ($0<a<1$) and the lesser value could improve the stability.

2.2.2 *Different disperse*

All physics units, ρ、u、v、e、p, were defined in grid centre to capture artificial glutinosity, qz, of surge wave, and q_z was defined in grid top and midpoint of right boundary. From partial differential eq.1-6, forward difference was adopted to time and centre difference to space.

2.2.3 *Mixed grid interphase of two kinds of substances and calculation of transportation*

Using Youngs' interface disposal thought and basing on volume proportion of substance in eight adjoining grid, the interface of mixed grid was ensured.

3. Results and Analysis for Micro-Explosion

Refer to fig. 1, the symmetric centre, surface of fiber and tissue were fixed as original boundary of calculating region which was dispersed as rectangle 200×200 sub-region. The minimum space step length, $\Delta z_1 = \Delta r_1 = l/30$, i.e. 1/30 of the distance between fiber end and calculi surface. Fiber diameter is 0.5 mm.

The water pressure fields induced by micro-explosion versus time were shown in fig. 2. The laser intensity I_0 equals 10^5w/cm^2 and the distance between fiber end and tissue surface is 0.5mm. The pressure increase at the start because the water in fiber end surface absorbs more energy and micro-explosion came into being subsequently in the region between fiber end and calculi and gas agglomerates diffuses all around. The pressure of fiber centre to calculi surface achieves peak after 1.5μs and negative pressure vacuole emergences lastly.

The pressure at calculi surface centre, i.e. point A, versus time was shown in fig. 3. The laser intensity I_0 equals 10^5w/cm^2 and the distance between fiber end and tissue surface is 0.5mm. It can be seen that the different laser parameter also induces a specific pressure impact. Duo to micro explosion impact, the pressure in tissue surface increases rapidly, and the pressure at point A achieved the highest about 67MPa with above parameter.

(a) t=0.1μs

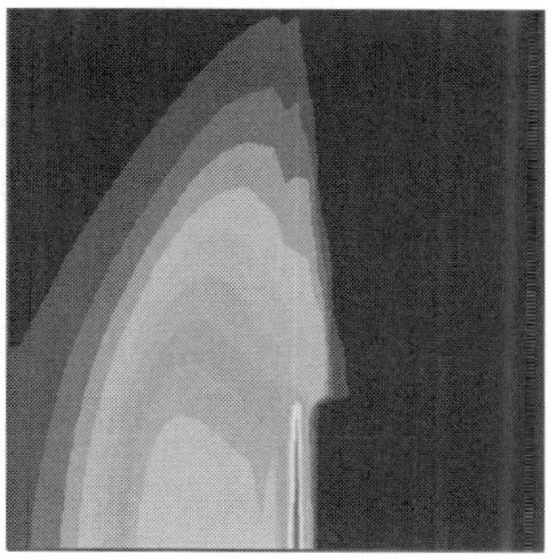

(b) t=0.5μs

(c) t=1.0μs

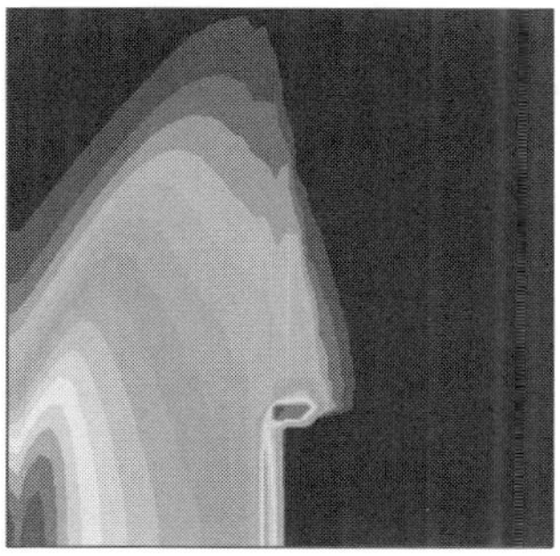

(d) t=1.5μs

Figure 2 The pressure fields of micro-explosion about calculi versus time irradiated by Ho:YAG laser

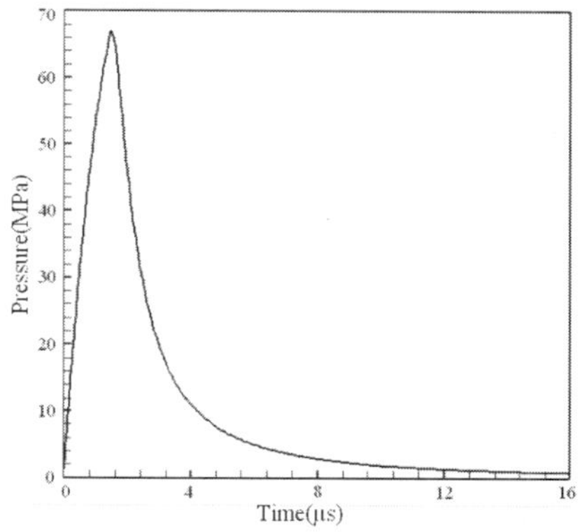

Figure 3 Variety of pressure at centre point A

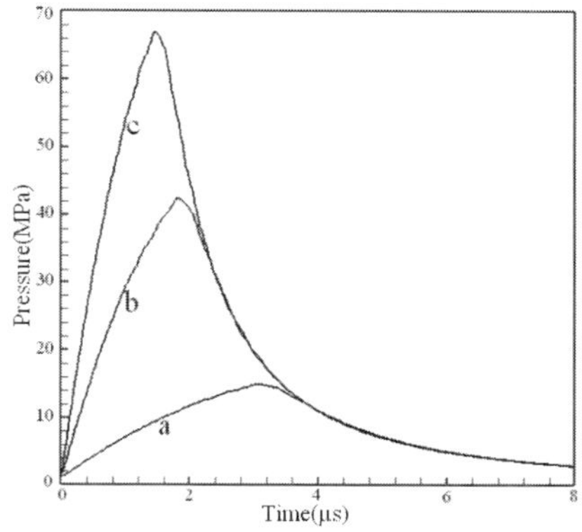

Figure 4 Variety of pressure at A with different laser intensity

Fig. 5 indicated the variety of pressure at point A versus time with different distance between fiber and tissue surface. The power density I_0 is 10^5w/cm^2. The distance, l, is 0.2mm in curve ***a***, 0.5mm in curve ***b*** and 1.5mm in curve ***c***. It can be seen that the duration of micro explosion impact pressure prolongs along with the increase of distance between fiber end and tissue surface under certain range and the stronger impact pressure appears with suitable distance. The duration time of impact is 2μs when the distance is 0.2mm and prolong to 6μs when the distance is 1.5mm. The higher impact pressure achieves 67MPa while l=0.5mm.

Steven A.S. observed the action of vacuole induced by micro explosion on calculi while fiber parallels tissue surface thereby laser swept the tissue[2]. It is found that the vacuole can't crush calculi and the conclusion is that the micro explosion pressure can't crush calculi with Ho:YAG laser. But simulation indicated that the pressure on tissue surface is very different with laser obverse irradiation and side irradiation. The impact pressure at point A with different irradiation mode was shown in fig. 6. The curve ***a*** indicated pressure at point A with obverse laser irradiation and ***b*** with side irradiation. It can be seen that the pressure with obverse irradiation is much larger than it with another irradiation mode. Therefore the micro explosion could crush calculi in laser lithotripsy.

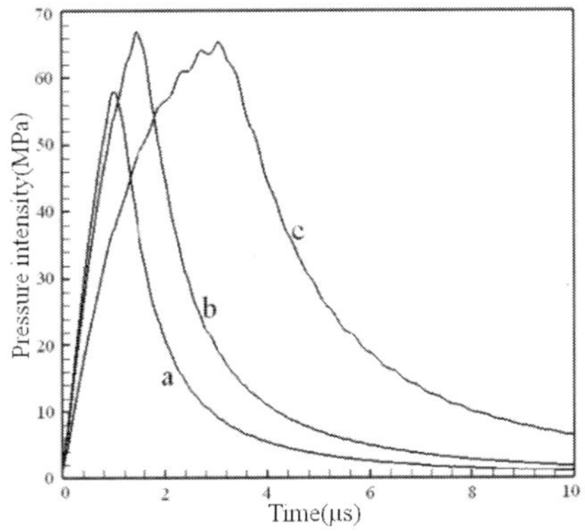

Figure 5 pressure at point A with different distance

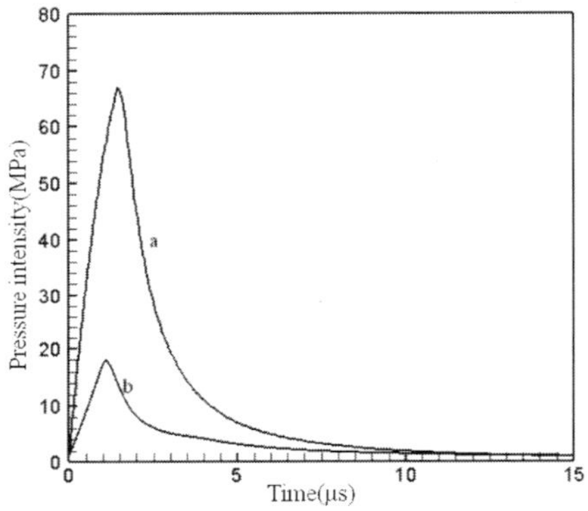

Figure 6 pressure of different irradiation mode

4. Conclusion

The dominative equation for micro-explosion were derived from conversation of mass, momentum and energy and the numerical model of micro explosion of water under laser irradiation was presented by using multi-ingredient Euler method, and the micro explosion in Ho:YAG laser lithotripsy was simulated and analyzed. The results indicate that the micro explosion of water induced by laser irradiation brings quite high pressure in tissue surface. High laser power density, suitable distance between fiber end and tissue surface can bring very high impact pressure. The pressure impact is far higher in obverse irradiation than it in side irradiation, and the high pressure can crush calculi.

References

1. Malcom S and Keith R., *IEE Colloquium*, **10**, 1-5 (1999).
2. Steven A.S., Frederic M.D., Bahaeddin J. et al., *IEEE Trans. Biomed. Eng.*, **41**, 276 (1994).
3. Raimund Hibst and Ulrich Keller, *Proc. SPIE*, **1880**, 156 (1993).
4. Junya Kohno, Fumitaka Mafune and Tamotsu Kondow, *J. Phys.Chem.* **A108**, 971 (2004).
5. Meyers M., *App. Mech. Rev.*, **48**, 27 (1995).
6. Benson and David J., Computer *Methods in App. Mechanics and Eng.*, **99**, 235 (1992).
7. Polt A. and Maurer G. *Fluid Phase Equilibria*, **73**, 27 (1992).
8. Kuskova N.I., Tkachenkoy S.I. and Koval S.V., *Journal of Physics: Condens. Matter*, **9**, 6175 (1997).
9. Lilei H., Yanqi W. and Jiqin C., et al., *Acta Optica Sinica*, **14**, 931 (1994).

RESEARCH ON NONLINEARITY MODIFIABLE ARITHMETIC OF 2D-PSD

HONGMENG YAN TIANZE LI
School of Electric and Electronic Engineering, Shandong University of Technology, Zibo 255049, China

KUN LIU
Department of Electronic Engineering, Dezhou Vocational Technical School, Dezhou 253034, China

Expound the work principle of semiconductor position sensitive detector (PSD) and caused reasons of nonlinearity of PSD. Analyze the bilinear interpolation and neural network, and utilize the two means to modify the nonlinearity error of PSD and compensate it. Under the premise of not increasing cost and equipments complexity, the linearity and degree of confidence of area B are advanced, the measure range and precision are expanded. At the end, the merits and flaws of the two means are given out, some conclusions are also elicited.

1. Introduction

Position sensitive detector (short form PSD) having been applied in the field of photoelectric detection widely, because it has some excellences, for example, the output signal only relate to the center position of optical point energy, wide frequency response, fast response speed, simple processing circuit. But due to PSD's shortcomings, for example uneven surface resistance, large nonlinearity error of PSD;s the output will be produced. Although in the recent years, the study of PSD is deepening in and out nation, and the principle and production process have improved, but the nonlinearity of PSD does not avoid. For advancing the credibility of data and measurement range of PSD, demarcating the nonlinearity error of PSD and designing the corresponding algorithm is necessary.

2. Principle of PSD

Improved tetra-lateral 2-D PSD adopts the curved electrode and tetra-lateral construction, its signals are elicited from the diagonal. It is not merely that the non-linearity error of output position can be minished, the excellence, for example lower dark current, loading reverse bias voltage easily, high speed response and small edge deformity etc, can be preserved. It shows in Fig1.

When the incidence point has different distances from the four electrodes, the current controlled by the four electrodes are also different. Each current controlled by the electrodes is consistent with the optical point position of photosensitive surface. Taking photosensitive surface of the geometric center as the origin of the two-dimensional x-y coordinates, the output current of PSD's electrodes are $I_{x1}, I_{x2}, I_{x3}, I_{x4}$. After processing the output current of electrode, the output signal is the single-valued function of position coordinates. Its equation is:

$$P_x = \frac{(X_2 + Y_1) - (X_1 + Y_2)}{X_1 + X_2 + Y_1 + Y_2} = \frac{x}{L} \quad (1)$$

$$P_y = \frac{(X_2 + Y_2) - (X_1 + Y_1)}{X_1 + X_2 + Y_1 + Y_2} = \frac{y}{L} \quad (2)$$

Where X_1, X_2, Y_1, Y_2 are output signals (optical current) of each electrode, x, y are the position coordinates of optical points, P_x, P_y are the output signals of the position of x and y direction. It is corresponding with the coordinates.

In the formulas (1) (2), it is supposed that the size and shape of the PSD's electrodes are ideal components, which are contacted through point-to-point and geometric dimensions are symmetrical. But actually the size and shape of output electrodes of improved 2D-PSD have its inherence, and geometric dimensions are dissymmetrical, thus the non-linearity error will not be avoided Many factors lead non-linearity of PSD. In

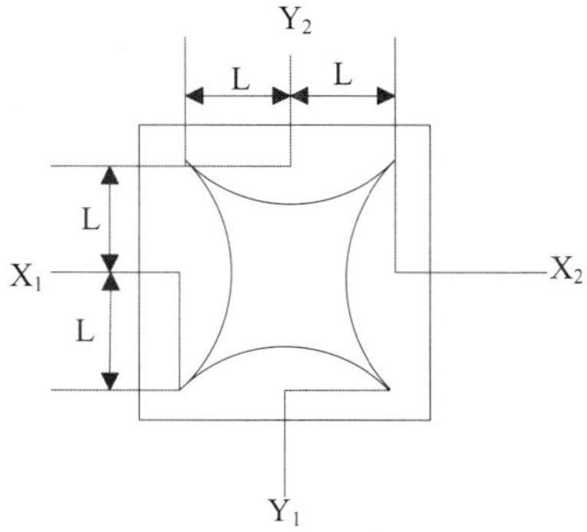

Fig.1 The structure of improved 2D-PSD

practice, the optical sensitive surface of PSD would be divided into area A and B. In fig.2, it is shown that the position accuracy of area A is more excellent than B.

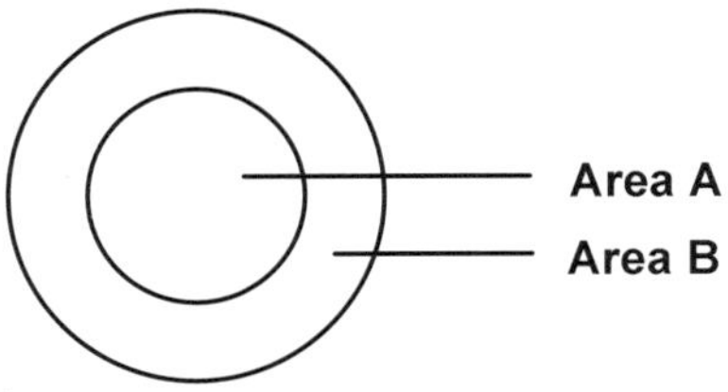

Fig.2 Area A and area B of PSD

3. Nonlinear Amendment of PSD

Because the nonlinearity error of PSD is larger than optical system and circuit, for advancing the output performance and large-scale measurement accuracy of PSD, the position output error of PSD must be amendment. The non-uniformity of silicon materials used to produce PSD in process emerges a slow gradient change, so the position error function $E(x, y)$ of whole PSD is a small gradient curve surface. If this function is discrete, series error function value of grids lattice are get, error function value of non-grids lattice can be approximated by means of using the interpolation method and neural network, so the tested approximate value could be obtained.

3.1. *Discretization of Error Function*

Because PSD's nonlinearity error in x, y direction is relatively independent, each point on optical sensitive surface of PSD has corresponding error $E_x(x, y)$ in x direction and $E_y(x, y)$ in y direction. The derivation process of $E_x(x, y)$ is same to $E_x(x, y)$, so we only take $E_x(x, y)$ as the example to derivate. Firstly, put PSD on a high precision numerical control displacement platform, then fix a laser on it so as to irradiate on the optical sensitive surface of PSD. According to a certain step S, control the platform to move along x,y direction. In this way, series error function value of $E_x(x_i, y_j)$ and $E_y(x_i, y_j)$ on the grids lattice:

$$E_x(x_i, y_j) = x_i' - x_i'' \tag{3}$$

$$E_y(x_i, y_j) = y_i' - y_i'' \tag{4}$$

Where (x_i', y_j') is the measurement output value of PSD on grid lattice; (x_i'', y_j'') is exact location reading of CNC (computerized numerical control) machine.

3.2. *The Method of Duo-Linear Interpolation*

On the optical sensitive surface of PSD, a series of discrete points form a rectangular grid, take random rectangular region $A_1A_2A_3A_4$, in fig.3 it is shown. Supposed the error function value of vertex A_1、A_2、A_3、A_4 are $E_x(A_1)$, $E_x(A_2)$, $E_x(A_3)$, $E_x(A_4)$, then calculate its interpolation function $E_x(x,y)$, thereinto (x,y) is the point of rectangular area $A_1A_2A_3A_4$.

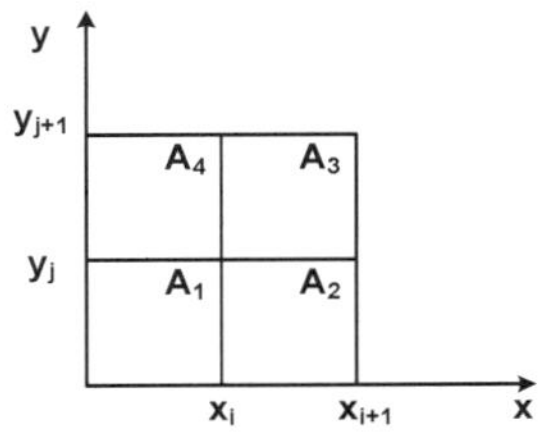

Fig.3 Duo-linear interpolation area

Due to the four points information of interpolation condition, takes the interpolation function G as:

$$a + bx + cy + dxy \tag{5}$$

The eqs.(5) duo-linear equation. According to this, the duo-linear expression of random point of rectangular area is

$$E_x(x, y) = \sum_{i=1}^{4} E_x(A_i)\varphi_i(x, y) \tag{6}$$

Utilize the obtained error function value and measurement output value of PSD, actual position value of PSD can be calculated:

$$\begin{cases} X_R = X_{out} - E_x(x, y) \\ Y_R = Y_{out} - E_y(x, y) \end{cases} \tag{7}$$

Where X_R, Y_R actual position value; X_{out}, Y_{out} is the measurement output value of PSD.

$$x_i = \langle x \rangle, y_i = \langle y \rangle \tag{8}$$

$$x_{i+1} = x_i + sign(x)\Delta x,\ y_{j+1} = y_j + sign(y)\Delta y \tag{9}$$

Where Δx、Δy is the step on x,y direction, and here $\Delta x = \Delta y = S$.

3.3. *The Method of Neural Network*

The non-linear compensation principle of neural network is shown in fig.4. The signal from sensor by plastic, filtering, amplification, computing, calibration sent into neural network to compensate. Due to the output x,y of eqs. (1),(2) is non-linear, it can not be shown by using the expression, but it can realize by using feed-forward neural network which has the characteristic of approximate the non-linear function in arbitrary precision. In this way, so as put the output x,y from sensor on the input of feed-forward neural network, in the output of neural network the anticipant linear output value can be obtained.

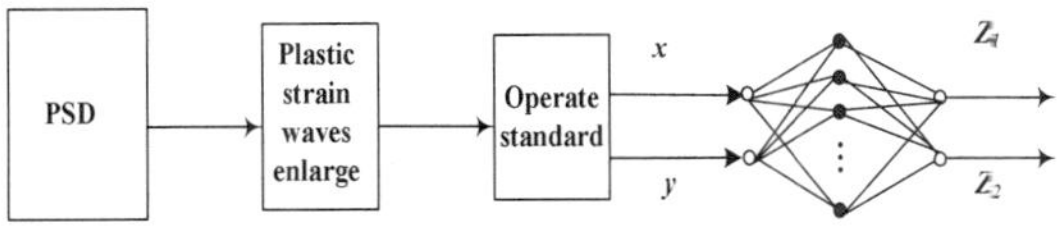

Fig.4 PSD nonlinear compensation principle

For the neural network construction, it is as simple as possible to realize the effective compensation, it need the confirm the number of network hidden layers and neurons of all network layers. Taking the double-input-output system as a example, because the neural network need to receive the two signals x and y from the sensors, and then export the two signals, the number of output layers and input layers' neurons is 2. It is proved that the feed-forward neural network, including one hidden layer, can realize the arbitrary nonlinear mapping, so in order to obtain the better compensation effect, we adopt a three layers neural network including one hidden layer ,which includes 14 neurons.

For training the sample data of neural network, we put PSD on the 2D high precision CNC, then use the stepper motor to sample the output voltage value and utilize the coordinates to calibrate them. In this way, the sample data can be get. The training sample selection should cover the entire measurement space, and the distance among the sample data could not be too large, or else it would effect the extend of network Take x, y in these samples as the input of neural network and T as the expectation output of neural network to train the neural network. While training, compare the actual output Z of neural network with expectation output constantly, and according to the difference between Z and T, adjust the network weight value by means of training arithmetic of neural network to approximate T constantly, until accord with the error square sum or achieve the expectation cycle number of training. When the training finish, on the weight value of neural network the linear information and nonlinear compensation of sensor have been stored.

3.4. *Analysis of Nonlinear Amendment*

3.4.1 Duo-linear interpolation amendment

The output value x,y of PSD measured by homemade W202 is shown in fig.5, the actual optical points position are on the grids. Form the fig.5, we can see that the linearity of PSD's area A (the range of geometric centre $2mm \times 2mm$) is better than B, there is obvious non-linear error output value in area B, the error is very serious on the edge especially. In the fig.6, the amendment output result of PSD is shown by the duo-linear interpolation method. Form fig.6, we can see that on the amendment optical sensitive surface of PSD $9mm \times 9mm$ the error is reduced.

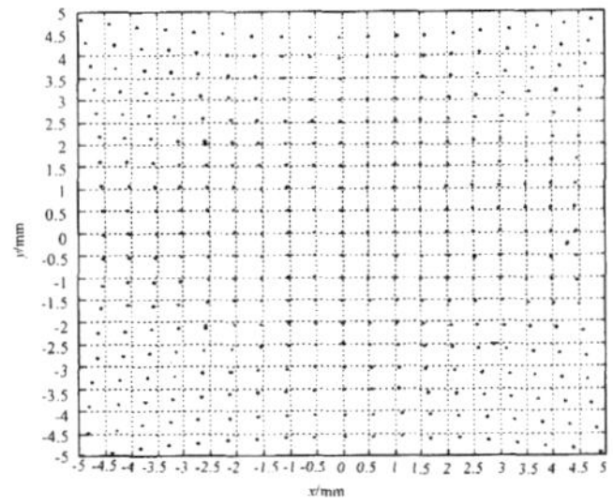

Fig.5 Output of PSD before amendment

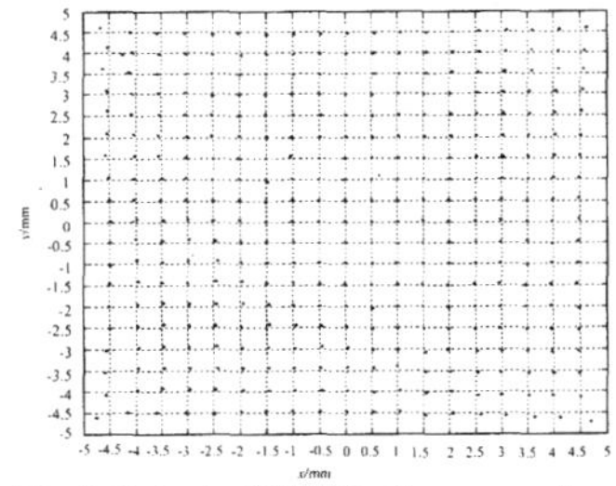

Fig.6 Output of PSD before amendment

3.4.2 Neural network amendment

Take the quantitative scope data of PSD as an example to adjust by using the neural network. This network's structure should be duo-input and duo-output, the hidden layer is consist of 14 neural network structure, using S type transfer function, in the output layer use the linear function, the precision can be achieved after training. It is shown in fig.7:

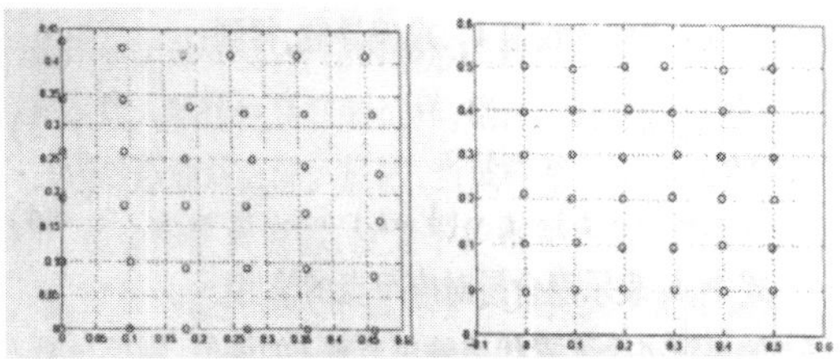

Fig.7 Output of PSD before and after asjusting (Unit： mm)

4. Arithmetic Analysis and Compare

4.1. *Interpolation Method*

The characteristic of interpolation method are: (1)simple operate;(2)easy program and fast operation speed;(3) use the data process ability of computer fully, and test measurement results by using software and adjust real-time. When analysis the interpolation arithmetic error, reducing the discrete function step can minish the interpolation arithmetic error, but if the step is minished, calibration workload is increased, and the device precision require also advance, the program is more complex and spending of system is increased too.

4.2. *Neural Network Method*

The method of neural network have provided some new effective means, so under reducing the require of sensor's material and performance, the detection performance of PSD can be advanced by means of neural network. We also can use the neural network toolbox of Matlab to simulate the improved arithmetic of BP network, in this way the results can be satisfied, weight value training workload can be reduced, the work efficiency is advanced. BP network(Error Back Propagation) is one of most popular neural network. But there are also some shortcomings:(1) convergence speed of study arithmetic is slow;(2)the local minimum value problem;(3)the weight value amendment problem.

5. Conclusion and Prospect

5.1. *Conclusion*

(1)When we adjust and compensate the nonlinear output, because there isn't interpolation area of edge output value on the edge part, the amendment effective of edge measurement value is worse. When compensating the nonlinear output by means of neural network, the process is more complex. But we can use the neural network module of Matlab to obtain the higher precision.

(2)Utilize the interpolation arithmetic and neural network arithmetic to adjust the nonlinearity of PSD, enlarge the use area of PSD, advance the output precision, improve the work efficiency effectively.

(3)According to the difference precision require on location, we can choose the difference amendment compensation method.

5.2. *Prospect of PSD*

In the recent years, PSD is used widely in every industry. Except the single crystal silicon PSD is used in commercial area in great force, large area, flexible film, organic plastic and 1D,2D array's appearance satisfy the position detection require on the special location. Otherwise, as the semiconductor process application apply in the optical part and integrate circuit develops, PSD based on COMS has appeared. So, we can see that smaller unit area and more compositive COMS-PSD would appear, and its detection performance would be advanced. The control circuit will be integrated with the sensor. For satisfying the require of high speed detection and signal collection, high speed and fast response PSD would be a kind of research direction, and larger, more sensitive, nice linear PSD would be the emphasis to study for the future.

References

1. Zhou Baoyu, Chen Peifeng, Zhu Mingzhu. Study on nonlinear error amendment experiment of position sensitive detector [J]. Optical instruments, 2005, 27(5):**11-15**.
2. Mo Changtao, Chen Changzheng, Zhang lili. Study on the neural network arithmetic amendment of 2D-PSD [J]. Photoelectron ▪ Laser, 2003, 14(4):**342- 344**.
3. Shi Longzhao, Huang Meizhen, Yang Xiaoling. 2D-PSD and signal processor [J]. Sensor Technology, 2002, 21(8):**20-22**.
4. Noriega JR, Wang H. IEEE Transactions Neural Networks, 1998, 9(1):**27-34**.
5. Hou XL, Chen CZ, Yu HJ. et al. Optimum method a bout weights and thresholds of nerve network [J]. Journal of Nor the as tern University (Natural Science), 1999, 20(4):**447-450**. (in Chinese)
6. Wang Xu, Wang Hong, Wang Wenhui. Artificial neural network theory and application [M]. ShenYang: Northeastern University Press, 2000, **51-54**.
7. Wang XD, Ye MY RBF neural network application in the surface rough fiber optic sensor [J]. Journal of Optoelectronics ▪ Laser, 2001, 12(10):**936-940**. (in Chinese)

HIGH RESOLUTION CIRCULAR SCANNING ENDOSCOPIC OPTICAL COHERENCE TOMOGRAPHY

JIANAN LI, PING XUE*

Department of Physics and Laboratory of Atomic and Molecular Nano Sciences, MOE

Tsinghua University, Beijing, 100084, P.R. China

Optical coherence tomography (OCT) is an emerging technology for micron-scale cross-sectional imaging of biological tissue and materials. OCT is analogous to ultrasound imaging except that it uses low coherence light instead of sound waves. In this paper a combination of OCT and endoscope technique, namely endoscopic OCT, is demonstrated for high resolution circular scanning OCT imaging, showing high potential in the application of clinical diagnosis from GI tract, respiratory system to urinary and reproductive tract, as well as vascular and neural system. We demonstrate the design and fabrication of a miniature circular scanning probe for endoscopic OCT. A hollow ultrasonic motor is for the first time employed for driving the rotation of optics and also allowing the light delivery fiber, objective lens and motor along the same side, shortening the distal end of the probe and also enhancing the lateral imaging resolution. As the diameter of the motor is only 1mm, a catheter of 1.8mm in diameter is constructed. We demonstrate preliminary result of OCT imaging with this probe. Imaging studies using this system and further improvement are in progress.

1. Introduction

Optical Coherence Tomography (OCT) is a new optical technique which permits cross-sectional tomographic imaging in highly turbid media. [1] OCT is analogous to ultrasound imaging except that it uses low coherence light instead of sound waves. Compared with conventional X-ray computed tomography, magnetic resonance imaging and ultrasound imaging, OCT is non-invasive, safe, portable and of high resolution.

Since the combination of OCT and endoscope technique has dramatically extended its application field, the endoscopic OCT (EOCT) has been greatly developed in the past few years. [2] Fig.1 shows the basic scheme of a typical time domain EOCT system.

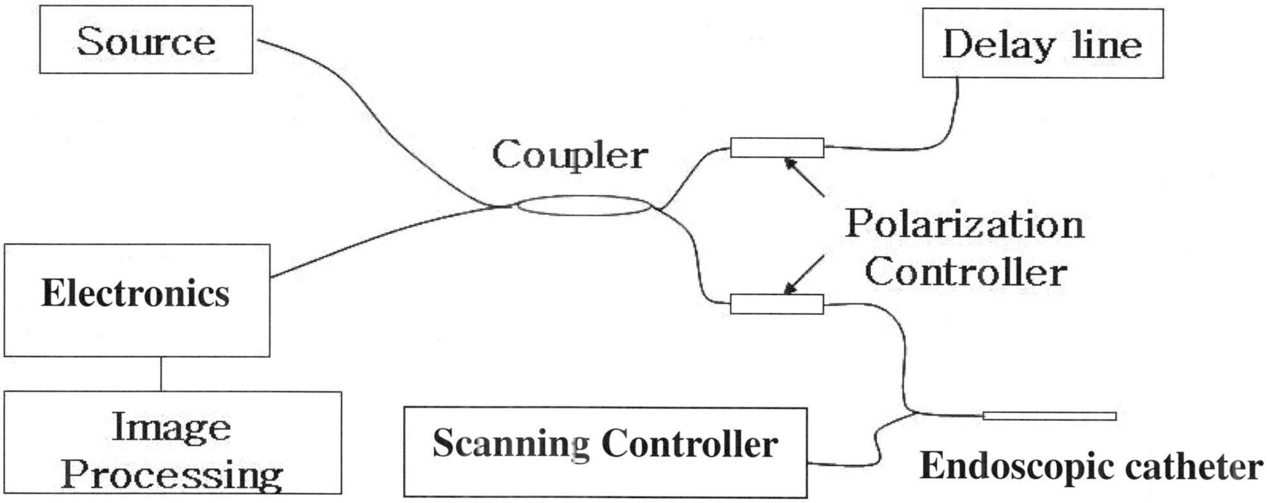

Fig.1

2. Endoscopic Probe for OCT

2.1 *Motivation*

A primary task of building an EOCT system is the fabrication of the endoscopic probe. According to the process of acquiring cross-sectional images, conventional circular scanning endoscopic probes for OCT usually have a structure as Fig.2 shows. [3] This structure have two inevitable drawbacks for clinical utilization: A. The working distance of the lens limits the N.A. of the optical system and consequently the lateral resolution is hard to be enhanced; B. The rigid tip of the probe is relatively long, which could bring problems in clinical applications. With the consideration of avoiding these disadvantages, a probe with a different layout is constructed in our work.

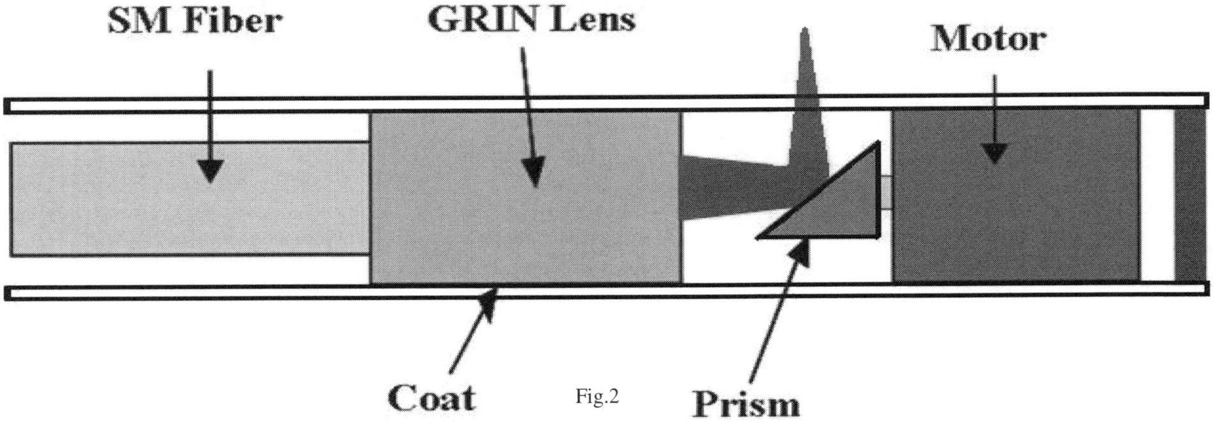

Fig.2

2.2 *Design & Fabrication*

A tiny ultrasonic motor (USM) is built by using a hollow piezoelectric ceramic cylindrical rod as a vibration generator of the stator (Fig.3). The fabricated tiny motor is 1mm in diameter, 8mm in length and only 36mg in weight. The resonant frequency is 32 kHz with one fixed boundary or 83 kHz with two free boundaries. By using the bending vibration mode of the stator, the motor shows a performance with a speed of 1800rpm and a stalled torque of 4μNm. The motor is then used for driving an optical prism for circular scanning in endoscopic OCT system.

The layout of the probe is shown in Fig.4. A single mode fiber (SMF) is placed in the axial of the hollow motor. Infra-red light with a 1310nm central wavelength coming out from this SMF is collimated by a self-focusing lens, then reflected and focused by a right-angled prism with an optimized aspheric surface - we calls it “the lens-prism” - to the sample. The back-scattered light from of the sample is collected in a confocal mode, which means that the signal light comes back along the same path as the light of incidence. The lens-prism, which is driven by the 1 mm cylindrical rod USM, rotates at a speed of about 6r/s. When packaged

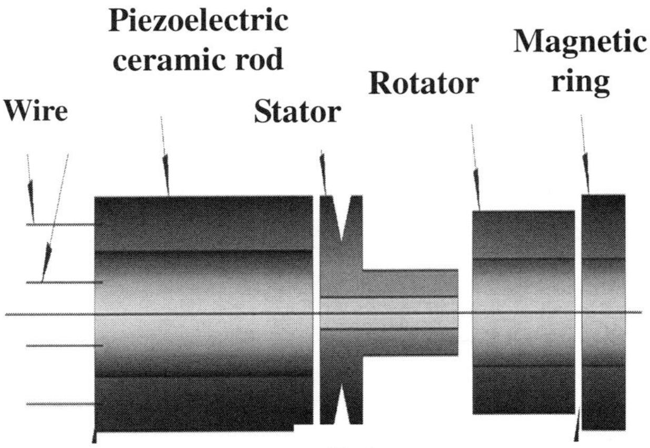

Fig.3

in a quartz tube, the outer diameter of the probe is 1.8mm, which is good for endoscopic OCT imaging. Compared to MEMS, Galvo system and other driven sources, the ultrasonic motor is cheap, robust and easy to control as well as of high torque, very suitable for OCT endoscopic probe.

2.3 *Result & Discussion*

The length of the GRIN lens as well as the space between the SMF and the lens are chosen appropriately to make sure the out coming light is parallel. The

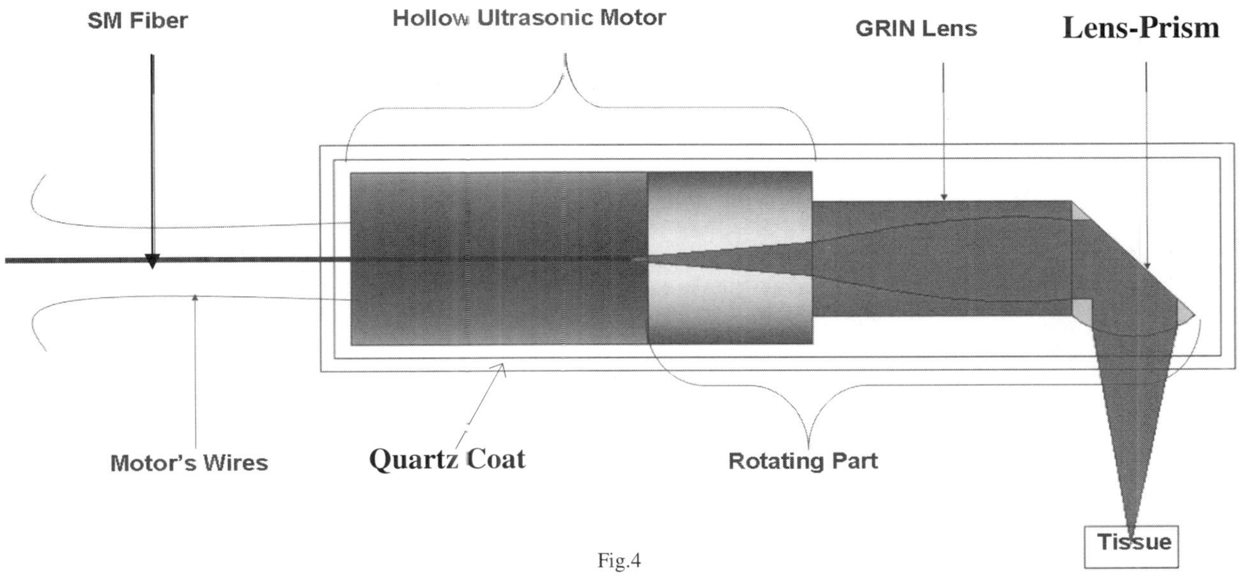

Fig.4

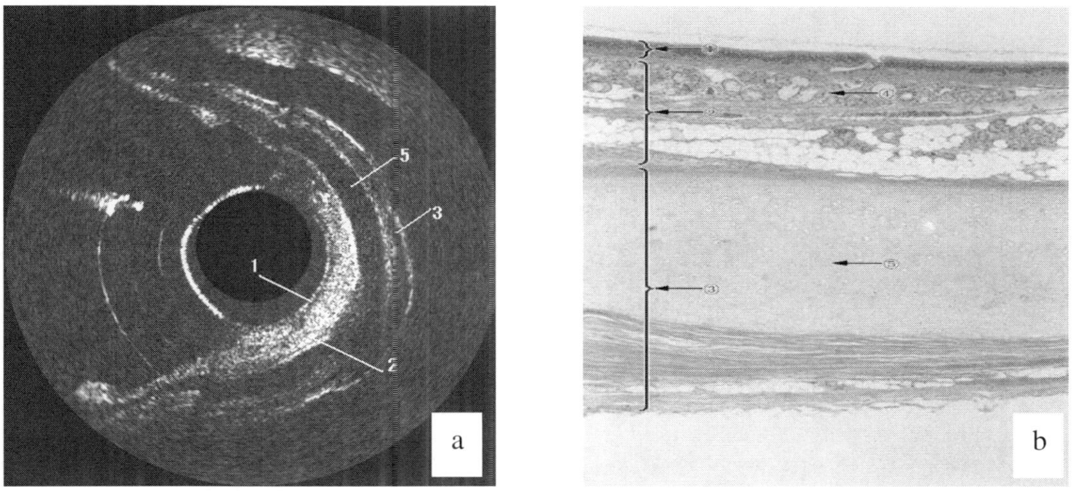

Fig.5 The micrometer-scale tomographic scanning images by using the endoscopic OCT. (a) The image of chicken trachea; (b) The image of its biopsy: 1, mucous membrane. 2, underlayer of mucous membrane. 3, ectoblast. 4, trachea grand. 5, cartilage.

precision of the lens-prism is crucial, as it effects significantly to the final performance of the entire optical system. We manufacture it from a prototype of *Lightpath Ltd.* aspheric lens product. When making a 0.8mm edges, 45°prism, now we could control the dimension tolerance to 0.02mm and the angel tolerance to 2 arc minutes. The designed lateral resolution of this probe is about 5μm, while the inaccuracy of fabrication will certainly reduce this estimation. Imaging studies using this probe and further improvement are in progress. As a reference, Fig.5 is the image achieved by our group with a similar USM probe just 30% bigger in diameter.

Fig.6, 7 and 8 are the photographs of the ultrasonic motor, the aspheric lens-prism and the packaged probe.

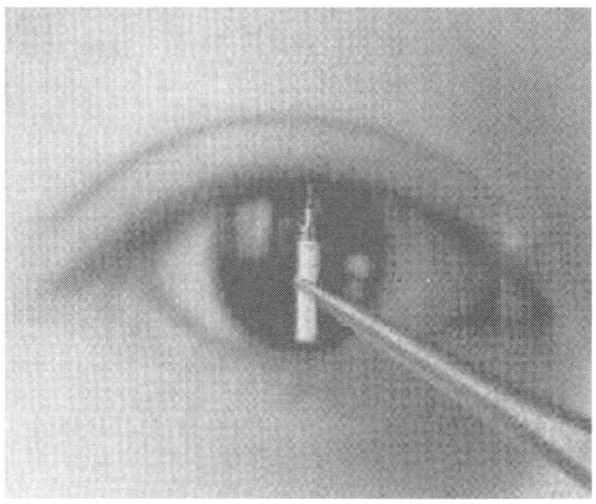

Fig.6

Fig.7

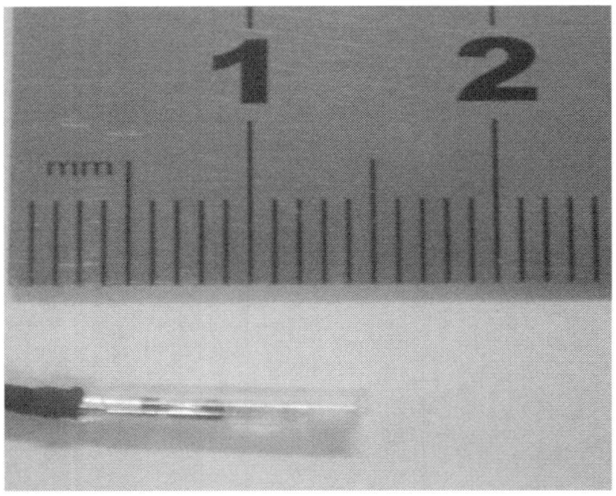

Fig.8

3. Conclusion

In this paper, a tiny ultrasonic motor is used to implement the circular scan in endoscopic OCT system. A lens-prism is also utilized to shorten the working distance. We expect the new design could enhance the N.A. of the OCT probe in order to gain a better lateral resolution than the formers. Since the diameter of the ultrasonic motor is only 1mm, a catheter relatively small in diameter (1.8mm) is achieved. Imaging studies using this system and further improvements are in progress.

Acknowledgements

This work was supported in part by Ministry of Science and Technology of China under contract No. 001CB510307 and 2006AA02Z472, National Science Foundation of China under grant No. 90508001 and 10574081, and Ministry of Education of China under grant No. 306020. The authors thank Professor Tieying Zhou for the help in making the ultrasonic motor.

References

1. D. Huang et al. "Optical coherence tomography". *Science*, **254**, 1178-1181 (1991).
2. Z. Yaqoob et al. "Methods and application areas of endoscopic optical coherence tomography". *J. Biomedical Optics*, **11**, 063001 (2006).
3. Q. Miao, "OCT and its applications in medicine". *M.S. Thesis* (2006).

FORECAST OF BLOOD FAT CONCENTRATION BASED ON SERUM UV ABSORPTION SPECTRA AND NEURAL NETWORK*

WEIHUA ZHU[1,2], ZHIMIN ZHAO**[1], XIN GUO[1], LEXIN WANG[1], HUI CHEN[1]

[1]*College of Science, Nanjing University of Aeronautics and Astronautics, Nanjing 210016, China*

[2]*College of Science, Hohai University, Nanjing 210098, China*

Blood plays an important role in the clinical diagnosis and treatment, the analysis of blood will be of very important practical significance. Study shows that the UV absorption spectrum is of complex shape, there is more absorption peak in 200 to 300nm, it shows that there is a complex absorption phenomenon in blood group macromolecules; there is a degree of displacement of absorption peak to different samples; there is no significant correlation between absorbance at some a wavelength and blood fat concent, but random. Based on the evident correlation between serum UV absorption spectrum and blood fat concentration in the wave band of 265 to 282nm, a neural network model was built to forecast the blood fat concentration, and a relatively good prediction was obtained. It provides a new spectral test method of blood fat concentration.

Keyword: Blood fat concentration; UV absorption spectrum; Neural Network (NN); Serum

1. Introduction

Organism (especially the human organism) is complex life systems, which carries important spectral information (fluorescence spectroscopy, IR and UV spectra). Research shows that as biological tissue abnormality, the spectral features will be changed. According to the report, the Swiss scientists found that nerve cell degeneration will lead Alzheimer's disease. Through the analysis of fluorescence information of degeneration process of tissue, early diagnosis of the disease can be made [1]. Thus, it is a new challenging research topics to make effective analysis and diagnosis of disease by using spectrum information of biological tissue particularly blood analysis technology. It has attracted the attention of scholars, and there have been some meaningful research [2-8].

Although there are a lot of reports for blood spectra at home and abroad, a deep and general results are still lack because there is little blood samples, and the samples are not of representative. [3,5]. The actual blood components are very complex, there are a lot of group macromolecules; the absorption spectrum is consist of multiple spectral components, therefore the spectrum contained blood component information.

The analysis of spectrum is to obtain the information of the components from the distribution and spectral characteristics. The aim will eventually be able to detect the actual blood components, which required a maths model that can be expected to obtain information from complex spectra.

2. Neural Network Structure and Methods

Neural network can simulate human learning to handle highly nonlinear problems. The advantages of the technology make it can be widely applied in complex systems forecast and achieve the effect of nonlinear mappings that traditional algorithms are difficult to achieve. In this paper, a neural network model based on the absorption spectrum of serum was built to forecast the blood fat concentration, and good results was obtained.

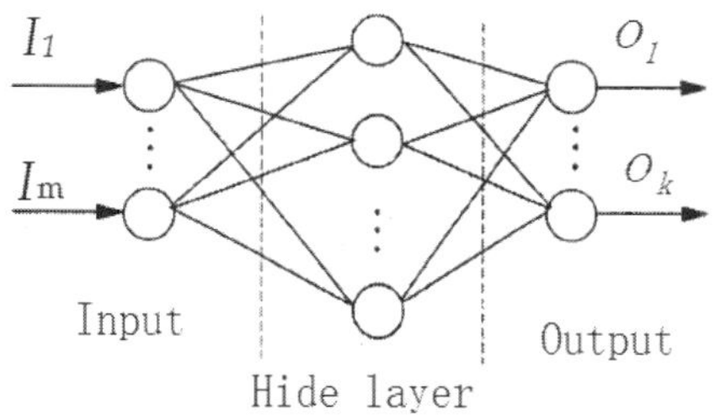

Fig. 1 Diagram of a BP neural network

Figure 1 shows a BP neural network structure widely applied. It is a feed-forward neural network, where I is for importation and O is for output. The neurons of this network structure normally only receives one output before the neurons. There is no connection between neurons in the same layer. Study shows that three layer feed-forward neural network will be able to approach any continuous function [9].

* The work was supported by the Innovation Foundation of Nanjing University of Aeronautics and Astronautics.

1st Author Introduction: weihua_zhu@126.com, Tel: 025-83725972

** Corresponding author: Zhimin Zhao; Prof.: E-mail: zhaozhimin@nuaa.edu.cn; Tel: 025 - 84892011

In this paper, BP neural network was used to determine the blood fat content in the blood. Its structure contains three layers, corresponding to the system input, hide layer and output layer respectively. There is a node in the output layer, which represents blood fat content (expressed by concentration).

The absorbance of 265 to 282 nm wave band is for Network input. Hide layer node was determined by the correlation coefficient between the the target and the output. The two transfer function is different. S-type function was used for input layer and hide layer, and specific function here was hyperbolic form. The linear function was used for hide layer and output layer.

In order to improve the efficiency of neural network training, the absorbance spectrum were pretreatment:

(1) Normalization will make the treated input and target data appear normal distribution;
(2) The treated sample data was used for PCA to eliminate redundant data of sample composition and to reduce the number of data dimension.

In order to increase network capacity of generalization and recognition, "stay ahead" approach was used in training .The sample was divided training sample data collection, validation sample collection and testing sample collection.

3. The Laboratory Equipment and Testing Conditions

(1) The Japanese company SHIMADZU UV3600 spectrophotometer was used in the investigations.
(2) Wavelength scanning range: 220 ~ 900nm, the scanning speed: fast, internal 0.5nm.
(3) Experimental blood samples were chosen from the Hospital of Nanjing University of Aeronautics and Astronautics.

4. The Experimental Results and Analysis

4.1 *Absorption Spectra*

The serum UV absorption spectra of different human with blood fat content are shown in Figure 2. Abscissa is wavelength, and longitudinal coordinates is absorbance.

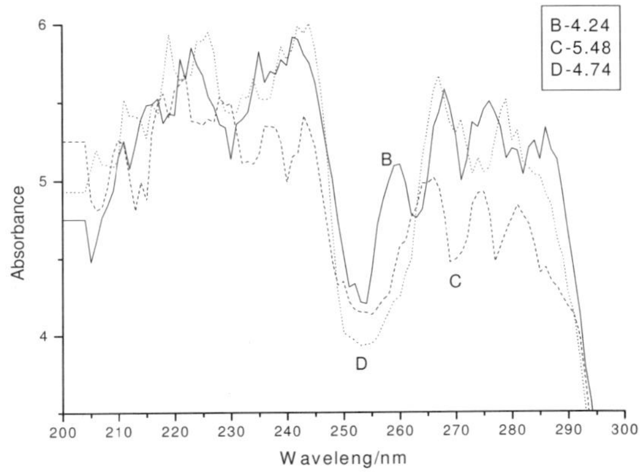

Fig.2 Absorption spectra of different human

Figure 2 shows that absorption spectrum is of the following characteristics in the wave band of 200 ~ 300 nm: (1) The absorption spectrum is of complex shape, there is more absorption peak in UV-band, it shows that there is a complex absorption phenomenon in blood group macromolecules; (2) the curve shapes of absorption spectra are similar to different samples ,the reason is that the spectrum is synthesis of some group macromolecules absorption spectrum, therefore, the spectral distribution contains information such as the blood fat content.; (3) there is a degree of displacement of absorption peak to different samples.

4.2 *Neural Network Prediction Cholesterol Content*

As noted above, the serum UV absorption spectrum is the synthesis result of many components. At each wavelength, there is information of many elements. Therefore, the spectral distribution contains information such as the blood surgar content. In this paper, the neural network model based on a certain UV absorption spectra is used to obtain information of the blood fat contents. By calculating, when the spectrum data of 265 nm to 282 nm is used for network iuput, we can find a good network forecast model.

Figure 3 shows the network training error curves, with the abscissa being the training number and ordinate corresponding to MSE. As seen from the Fig.3, the validation error, error testing trend is basic agreement. It indicates that the samples division is basic reasonable. Network training early stop at step 4, which is because that the testing error becoming larger. From the training curve error, training errors is relatively satisfying.

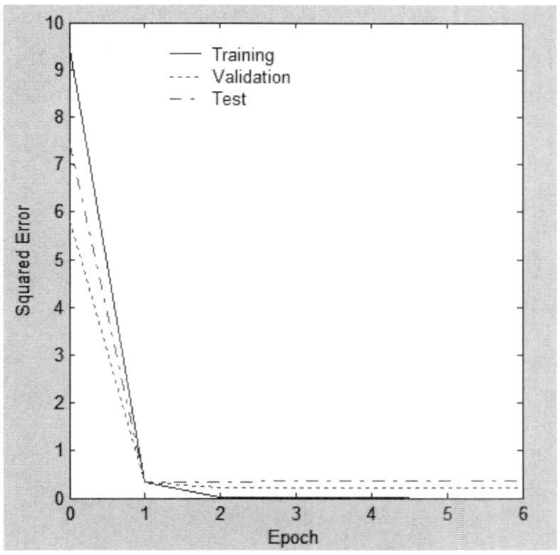

Fig.3 Error training curves

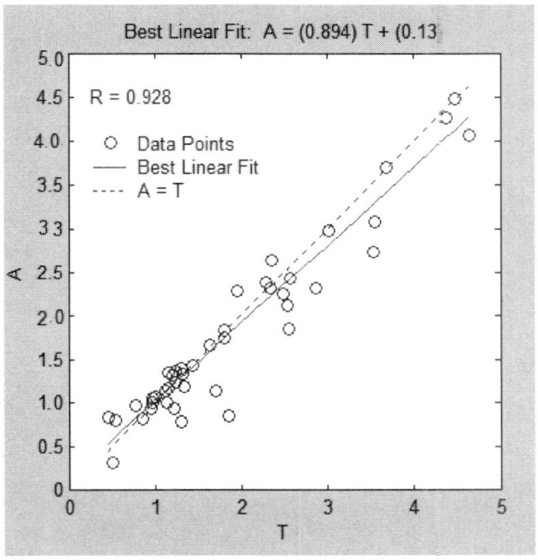

Fig.4 Regression analysis result of network output (A) blood fat content T

Figure 4 is regression analysis result of network output blood fat concentration. The correlation coefficient between predicted blood fat content output A and objectives T reaches 0.928, which can be regard as better prediction. It also shows that the blood fat content information is implicit in absorption spectrum. Neural networks provide an effective means to access spectral information.

5. Conclusion

(1) The UV absorption spectrum is of complex shape, there is more absorption peak in 200 to 300nm, it shows that there is a complex UV absorption phenomenon in blood group macromolecules.

(2) The UV absorption spectrum of serum is the synthesis result blood surgar and other spectral components, and the information is contained at each wavelength. Therefore blood surgar content and other information are contained in the spectral distribution, which is the basis of blood test based on spectra analysis.

(3) There is a degree of displacement of absorption peak to different samples.

(4) Based on the evident correlation between serum UV absorption spectrum and blood fat concentration in the wave band of 265 to 282 nm, a neural network model was built to forecast the blood fat concentration, and a relatively good prediction was obtained.

References

1. SHUN Zhou. Nanjing Daly, 2004-02-20.
2. WANG Le-xin, ZHAO Zhi-min, YAO Hong-bing, et al. Infrard Absorption Spectrum Analysis and Its Application to Blood [J]. Spectroscopy and Spectral Analysis, 2002, **22(6)**: 980~982.
3. ZHAO Zhi-min, CHEN Yu-ming, YU Xiao-lei. Characteristic Analysis and Application of Blood Fluorescence Spectra [J]. Spectroscopy and Spectral Analysis, 2003, **23(5)**: 922~925.
4. ZHAO Zhi-min, GUO Lin-feng, YU Xiao-lei, et al. Analysis and Application of Excitation Fluorescence Intensity of Blood [J]. Spectroscopy and Spectral Analysis, 2005, **25(12)**: 2056~2060.
5. LAN Xiu-feng, LIU Jian-gang, LIU Ying, et al. Spectroscopy Research on Cholesterol in Hypercholesterolemia Serum [J]. Spectroscopy and Spectral Analysis, 2006, **26(13)**: 467~470.
6. YU Chang-qing, LIU Fang-guang, GU Ying, et al. Laser Induced Fluorescence Spectra of Human Skin and Blood [J]. Chin J Laser Med Surg., 2004, 13: 41~444.
7. LI Jian-dong, DING Jian-hua, WU Min, et al. Diagnosis of Hepatocirrhosis by Fluorescence Spectrum of Serum [J]. APPLID LASER, 2006, **24(2)**: 112~114.
8. MEN Ji-wu, ZHENG Rong-er, ZHANG Yun-dong, et al. Cancer Detection by Serum Fluorescence Analysis: A pectroscopic Study [J]. Chin J Laser Med Surg., 2003, **1(1)**: 43~47.
9. MENG JOO ER. Face Recognition with Radial Function (RBF) Neural Networks [J].IEEE Transaction on Neural Network. 2002, **13(3)**: 15~17.

PHOTO BLANCHING OF HYPOCRELLIN B AND PHOTODAMAGE OF GELATIN IN AQUEOUS SOLUTION *

JUCHENG ZHANG

College of Science, Honghe University, Mengzi, Yunnan Province

Key Laboratory of Natural Pharmaceutical & Chemical Biology of Yunnan Province, Honghe University, Mengzi, Yunnan Province, Mengzi 661100, P. R. China

PEI ZHANG

College of Technology, Honghe University, Mengzi, Yunnan Province

College of Science, Honghe University, Mengzi, Yunnan Province, Mengzi 66110, P. R. China

WEI LIU†, CHANBA NG ZHANG, YING LI, ZHAOLONG HUANG

College of Science, Honghe University, Mengzi, Yunnan Province

Key Laboratory of Natural Pharmaceutical & Chemical Biology of Yunnan Province, Honghe University, Mengzi, Yunnan Province, Mengzi 661100, P. R. China

Gelatin is common biological materials used for drug preparations, such as hypocrellin B-gelatin nanoparticles. Interaction of hypocrellin B (HB) with gelatin and the photo-blanching of HB and photodamage of gelatin in aqueous solution was studied in this work. It was found the absorption peak of HB was red-shifted from 470nm to 514nm and appearance an isobestic point at around 480nm interacted at dark at pH 7.4. The spectroscopic responses of the photo-blanching of HB and photodamage of gelatin in aqueous solution were studied by irradiation, the wavelengths longer than 470 nm, with air saturated, oxygen-saturated, argon-saturated respectively. The absorbance of HB decreased, and the fluorescence of HB was quenched in aqueous solution at pH=7.4 by illumination, the content of HB in aqueous solution is 8.0 μM. The gelatin was photodamaged via photosensitization of HB in Phosphate buffered saline（PBS）at pH=7.4. Furthermore, photo-bleaching of HB and photodamage of gelatin are both oxygen-dependent.

1. Introduction

Hypocrellins, including hypocrellin A (HA) and hypocrellin B (HB) (Fig. 1), which are the photo-sensitizers naturally present in a fungus *Hypocrella bambusae* grown mainly in Yunnan province, China., were proved to be promising for photodynamic treatment of some vascular diseases [1]. They are well known for the high photodynamic activity to tumors and other biological targets but very low dark-toxicity [2-9]. The chemical and physical properties of HA and HB have been extensively studied and reviewed [10, 11]. The other research shown the hypocrellins and its derivatives can bind to some biology molecule, such as human serum albumin (HSA), hyaluronan etc [12-14], or can recognition of various biomolecules by the environment- sensitive spectral responses of HB [15, 16].

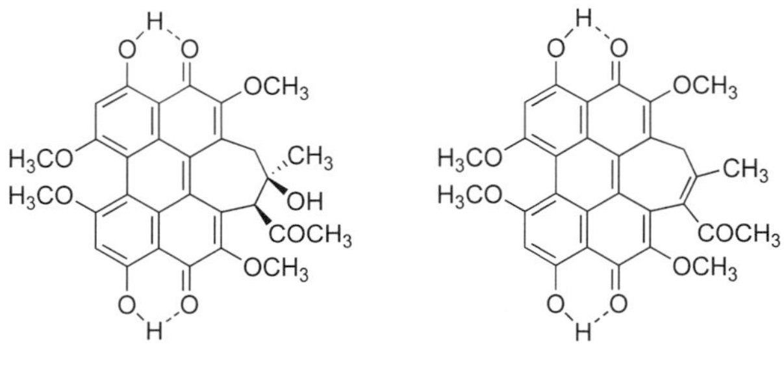

Fig. 1. The molecular structures of HA and HB.

Gelatins are denatured proteins derived from collagens, use as drug carriers, specially, HB-gelatin nanoparticles were prepared to improve the solubility of hypocrellins in aqueous solution [17]. In the current paper, the interaction HB with gelatin and also the photo-blanching of HB and photo damage of gelatin were studied. The photo-blanching of HB and photo

* This work is supported by the Natural Science Foundation of Yunan province. (No. 2005C0078M) and the Key Discipline Foundation of Yunnan province.

† Corresponding author: E-mail:hhusci@163.com, Tel: +86-873-3694923, Fax: +86-873-3694923.

damage of gelatin in aqueous solution (pH 7.4) were studied by absorption spectra and fluorescence spectra. The sample solution was irradiated with the wavelengths longer than 470 nm and the fluence rate was about 100 mW/cm^2, with air-saturated, oxygen-saturated, argon-saturated respectively. The absorbance of HB decreased, and the fluorescence of HB was quenched in aqueous solution at pH=7.4 by illumination, the content of HB in aqueous solution is 8.0 μM. The gelatin was photo damaged via photo-sensitization of HB in Phosphate buffered saline (PBS) at pH=7.4. Furthermore, photo-bleaching of HB and photo damage of gelatin are both oxygen-dependent.

2. Materials and Methods

2.1. *Materials*

HB was quantitatively produced by the dehydration of HA (HA was extracted from the fungus of Hypocrella bambusae and crystallized twice from chloroform–petroleum ether before use.). The purities of HB were determined by TLC (thin layer chromatography). Gelatin was purchased from Sigma Chemical and was used as the received. Phosphate buffered saline (PBS) pH=7.4 was prepared by the use of 50 mM KH_2PO_4 and 50 mM K_2HPO_4. The working solutions were prepared immediately before use.

2.2. *Preparation of HB–gelatin Solution*

The gelatin solution was prepared in 50 mM phosphate buffer (pH=7.4). A concentrated solution of HB in dimethylsulfoxide (DMSO) (2 mg/mL) was added to the gelatin solution in microliter quantities and mixed quickly by shaking 10 min.

2.3. *Spectroscopy Measurements*

Steady-state absorption and fluorescence spectra were recorded with a UV-1601 UV–Vis spectrophotometer (Hitachi, Japan) and F-4500 spectrofluorimeter (Hitachi, Japan), respectively. For the fluorescence measurement, HB was selectively excited at 470 nm, and gelatin was excited at 300 nm. Samples were dark-adapted at room temperature for 10 min prior to fluorescence measurement.

2.4. *Irradiation*

The HB–gelatin solution was air-saturated, oxygen saturated or bubbled with argon according to the experimental requirements, and then irradiated in a 1×1 cm quartz cuvette. A 450 W medium-pressure sodium lamp equipped with a filter cutting off the wavelengths shorter than 470 nm was used for irradiation. The fluence rate was about 100 mW/cm^2 measured by a UV-Radiometer.

3. Results and Discussion

3.1. *Photo Blanching of HB in PBS Gelatin Solution*

HB is a hydrophobic compound and tends to aggregate in an aqueous solution, however, its solubility in PBS can be promoted by adding bio-molecule.

To prepare HB–gelatin PBS solution, a concentrated solution of HB in DMSO was added slowly into the gelatin PBS solution by using a micro syringe with stirring in dark. The content of HB in this solution is 8.0μM. The solution was air-saturated, oxygen saturated or bubbled with argon according to the experimental requirements, then the solution was irradiated with different time. After in the dark for 10 min to room temperature, the solution was detected by UV-Vis spectrophotometer and fluorescence spectra. The UV-Vis spectra and the fluorescence emission spectra for HB (8.0μM) in the PBS (pH 7.4) containing gelatin are shown in Fig. 2 and Fig. 3 respectively. Compared Fig. 2 a to c, it can be seen that the photo-bleaching is less for the air-saturated sample than that for the oxygen-saturated one. Plots d shown that the

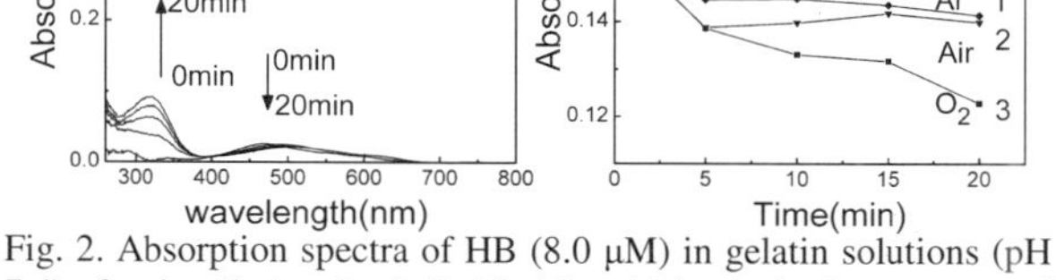

Fig. 2. Absorption spectra of HB (8.0 μM) in gelatin solutions (pH 7.4) after irradiation for 0, 5, 10, 15 and 20 min in the presence of oxygen (a), argon (b), and air (c). (d). Plots of absorbance at 470 nm against irradiation time. Line 1: argon saturated; Line 2: air saturated; Line 3: oxygen saturated.

absorbance has descend slowly in the absence of oxygen (argon saturated, Fig.2 d), suggesting that the photo-bleaching of HB is oxygen-dependent. From

Fig.3 the curves shown the HB was photo blanching with prolonging the irradiation time, the fluorescence emission spectra shown the same results as absorbance spectra.

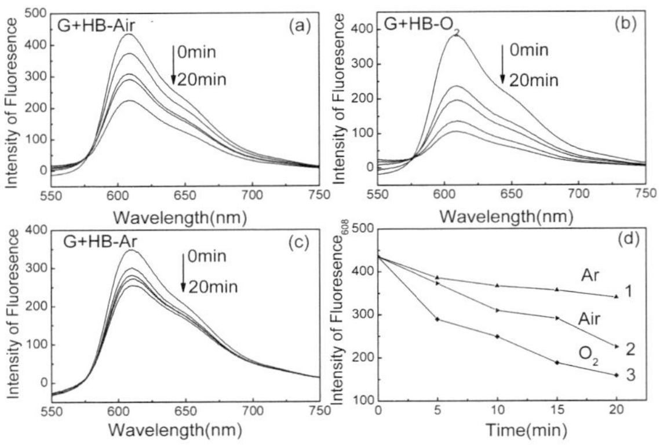

Fig. 3. Fluorescence spectra of HB (8.0 μM) in gelatin solutions (pH 7.4) after irradiation for 0, 5, 10, 15 and 20 min in the presence of air (a), oxygen (b), and argon (c). (d). Plots of intensity of fluorescence at 608 nm against irradiation time. Line 1: argon saturated; Line 2: air saturated; Line 3: oxygen saturated.

3.2. *Photo Damage of Gelatin by HB in Aqueous Solution*

Gelatin use as drug carriers, it is a denatured proteins derived from collagens. The gelatin was used as model proteins to study the photo damage by HB which as photosensitizer in aqueous solution. The UV-Vis spectra can't know distinctly the damage of gelatin, so the fluorescence emission spectra for gelatin with HB (8.0 μM) in the PBS (pH 7.4) was used to detected the damage. The gelatin is denatured proteins, so the

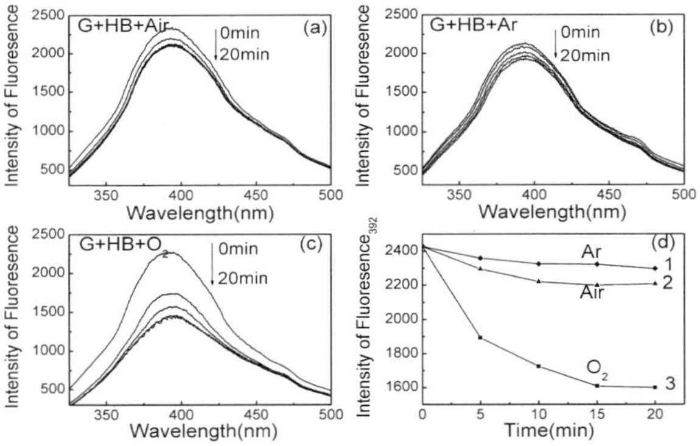

Fig. 4. Fluorescence spectra of the gelatin (pH 7.4) when HB in solution after irradiation for 0, 5, 10, 15 and 20 min in the presence of air (a), argon (b), and oxygen (c). (d). Plots of intensity of fluorescence at 392 nm against irradiation time. Line 1: argon saturated; Line 2: air saturated; Line 3: oxygen saturated.

excited wavelength at 300nm. The fluorescence of gelatin by irradiation is shown in Fig. 4. From Fig. 4 (a–c), it can be seen that a prolonged irradiation reduces the fluorescence for all the samples. The intensity of fluorescence was apparently descent for the oxygen-saturated in all samples, which suggesting photo-damage of gelatin by HB in aqueous solution.

4. Conclusions

In the current work, both the absorption spectra and fluorescence spectra suggest there was photo-reaction under irradiation, the irradiation wavelength higher than 470nm. The absorption spectra shown the HB (8.0 μM) was photo-blanching in gelatin PBS (pH 7.4) solution, and the fluorescence emission spectra suggest the gelatin was photo damage by HB (8.0 μM) under irradiation in PBS (pH 7.4). The results shown the photo blanching and photo damage are oxygen-dependent, which suggest that photo generated reactive oxygen species ($^1O_2$, $O_2^{\cdot-}$, and $^{\cdot}OH$) play the most important role.

Acknowledgments

This research was supported by the Natural Science Foundation of Yunan province. (No. 2005C0078M), and the Key Discipline Foundation of Yunnan province.

References

1. L. Zhang, Y. Gu, F.G. Liu, Y.Y. Liu, X.M. Zhao, J. Zeng, L.N. Han, W.D. Dai, *Chin. J. Laser Med. Surg.*, **12**, 69 (2003).
2. DiWu Z.J. *Photochem. Photobiol.*, **61**, 529 (1995).
3. G.A. Kraus, W.J. Zhang, M.J. Febr, J.W. Petrich, Y. Wannemuehler, S. Carpenter, *Chem. Rev.*, **96**, 523 (1996).
4. J.B. Hulson, J. Zhou, J. Chen, L.Harris, L. Yip and G.H.N. Towers, *Photochem. Photobiol.,* **60,** 253 (1994).
5. M.J. Fehr, S.L. Carpenter, Y. Wannemuehler, J.W. Petrich, *Biochem.*, **34**, 15845 (1995).
6. Jianghua Ma, Jingquan Zhao, Lijin Jiang, *Photochem. Photobiol.*, **74**, 143 (2001).
7. Y.Z. Hu, J.Y. An, L. Qin, L.J. Jiang, *J. Photochem. Photobiol., A Chem.*, **78**, 247 (1994).
8. S.J. Xu, S. Chen, M.H. Zhang, T. Shen, *J. Org. Chem.,* **68**, 2048 (2003).
9. J. Moan, K. Berk, *Photochem. Photobiol.*, **55**, 931 (1992).
10. L.J. Jiang, *Kexue Tongbao*, **21**, 1608 (1990).

11. L.J. Jiang, *Kexue Tongbao*, **22**, 1681 (1990).
12. K. Das, A.V. Smirnov, J. Wen, P. Miskovsky, J.W. Petrich, *Photochem. Photobiol.*, **69**, 633 (1999).
13. Baozhong Zhao, Jie Xie and Jingquan Zhao, *Bioorganic and Medicinal Chemistry*, **14**, 2428 (2006).
14. Liming Song, Baozhong Zhao, Jie Xie, Jingquan Zhao, *Biochim. Biophys. Acta–Genercl.* **1760**, 333 (2006).
15. Jin Xuanye, Zhao Yuewei, Xie Jie, Zhao Jingquan, *Science in China* B, **47**, 335 (2004).
16. Liming Song, Jie Xie, Chunxi Zhang, Cong Li, Jingquan Zhao, *Photochem. Photobiol Sci.*, **6**, 683 (2007).
17. Baozhong Zhao, Jie Xie and Jingquan Zhao, *Biochim. Biophys. Acta*, **1670**, 113 (2004).

THE DIFFERENCE OF PHOTOSYNTHETIC CHARACTERISTICS AND CHLOROPHYLL FLUORESCENCE PARAMETERS IN TWO CUCUMBER VARIETIES WITH HEAT TOLERANCE UNDER HIGH TEMPERATURE STRESS

JIANJIAN LI

Colleg of Life Science and Technology, Honghe University, Yunnan Mengzi, 661100, China;

YAJUN CHANG

College of Life Science, Lanzhou University, Lanzhou, 730000, China;

JIHUA YU

Department of Horticulture, Gansu Agricultural University, Lanzhou, 730070, China;

CANBANG ZHANG*

College of Science, Honghe University, Yunnan Mengzi, 661100, China

Two different heat-tolerance varieties of cucumber (*Jinchun No.4*, heat tolerance variety; *Zhongnong No.12*, heat sensitive variety) were used to study the difference of photosythetic characteristics and chlorophyll fluorescence induction kinetics under high temperature stress (40/25°C). The results showed that the net photosynthetic rate (Pn), Chlorophyll content, the chlorophyll fluorescence Fv/Fm and qp of the two varieties decreased under high temperature stress, but the decrease in heat tolerance variety (*Jinchun No.4*) were less than in heat sensitive variety (*Zhongnong No.12*); at the same time, Fo increased, but the increasing extents of *Jinchun No.4* were less than *Zhongnong No.12*. Except that all indexes in heat tolerance variety (*Jinchun No.4*) could resume to some extent after the plants had been resumed for two days, still that could not resume getting the level of untrearment; And all indexes in heat sensitive variety (*Zhongnong No.12*) almost could not resumed.

1. Introduction

Crop plants experience various stresses during their life cycle, such as high or low temperature, drought and salinity.High temperature is a common stress for plants, restricting their growth and productivity in many regions[1]. Models of global climate change predict a further 1.5–5.5°C warming for this century as a result of the increased atmospheric concentrations of CO_2 and other trace gases[2]. Extreme high temperature events are also anticipated to increase greatly in frequency[3], thus plants are increasingly likely to be limited by high temperature stress.So an understanding of the mechanisms involved will be critical for the breeding of more tolerant genotypes[4].

Photosynthetic functions are well known to be particularly sensitive to heat stress. High temperature directly influences the formation and functions of photosynthetic apparatus in plants. Under high temperature stress, the constituents of the thylakoid membrane in a photosynthetic apparatus are easily harmed and the photosynthetic rate and photosynthetic quantum efficiency in plant leaves drop, which results in photoinhibition[5].

Chlorophyll fluorescence measurements as an indication of PSII photochemistry and electron transport activity at molecular level. Presently, measurements of Chl a fluorescence by Pulse Amplitude Modulated (PAM) fluorometer and Plant Efficiency Analyser (PEA) are used as non-destructive, rapid and efficient methods for in vivo study of photosynthetic activity when plant has been exposed to different environmental stress [6,7,8].

The objective of this study was to define the mechanisms of two varieties of cucumber in response to long-term high temperature stress. Chlorophyll fluorescence and gas exchange measurements were used to investigate and compare the photosynthetic responses to high temperature stress in varieties in order to detect the protective or hurt mechanisms in differentially adapted cultivars.

* Corresponding author: Can-bang Zhang, email: cbzhang@vip.km169.net

2. Materials and Methods

2.1. *Plant Materials and Experimental Treatments*

Seeds of *Cucumis sativus* L *Jincun No.4* (heat tolerance variety) and *Zhongnong No.12* (heat sensitive variety) were obtained from cucumber research institute in Tianjin of China and vegetable and flowers research institute of Chinese Academy of Agricultural Sciences. The whole experiments were done in Gansu Agricultural University Laboratory of Physiology and Biotechnology. Seeds surface were sterilized by a 30s immersion in 80% (v/v) ethanol, followed by being washed three times with sterilized distilled water. The seeds were imbibed for 36h at 18• in darkness to synchronize germination, and then were sown on sterilized moist vermiculite.

Plant seedlings were grown routinely in the greenhouse of department of horticulture (Gansu Agricultural University) at 23–25°C with a 12h photoperiod and a relative humidity of 75%. The seedlings grew till the third leaves mature and then transferred them from the greenhouse to artificially lighted growth cabinets for high temperature treatment. The temperature in cabinet I was concrolled at 40°C in day time and 25°C at night (40/25°C, d/n) with 12h photoperiod (PFD 350μmol·m^{-2}·s^{-1}) and a relative humidity of 80%. The seedlings moved to another cabinet II [with temperature 25/13°C (d/n), photoperiod and relative humidity was identical with cabinets I] were taken as the control. The temperature in cabinet I was shifed to 25/13°C (d/n) after 5days stress, and photoperiod and relative humidity maintained unchanged.

3.3. *Photosynthesis Characteristic Parameters Measurements*

Using a portable open-flow gas analyzer in conjunction with a portable leaf microclimate control system temperature chamber (CIRAS-2, PPsystem Company, UK). Photosynthesis characteristic parameters measurements (photosynthesis) were taken on the second leaf at a CO_2 concentration of 350uL/L, a relative humidity of 80%, and at a photon flux density of 600μmol·m^{-2}·s^{-1} (PAR) before treatment (0d) and after 1,3,5 days and 2 days recovered. Five times measurements were made per plant treatment and took five values each time. Then averaged.

3.3. *Chlorophyll Content and Chlorophyll Fluorescence Measurements*

Chlorophyll fluorescence measurements at room temperature were carried out with a handy plant pulse modulated fluorometer (Hansatech FMS2, UK). Prior to fluorescence measurement, treated plant leaves (the 2nd) were adapted 30 min to darkness for obtaining an oxidized steady state of PSII[9]. The initial fluorescence (Fo) was determined with a weak modulated light (0.04μmol·m^{-2}·s^{-1}). The maximal fluorescence (Fm) was monitored with a 2s pulse of saturating white light (6000μmol·m^{-2}·s^{-1}) passed through an infrared heat filter. The primary photochemical efficiency of photosysten II (PSII) (Fv/Fm), the photochemical quenching coefficient (qP) were calculated according to Gray et al.[10] and Schreiber et al.[11].

At the same time, chlorophyll contents were measured from five times measurements according to J. R. McWilliam and A. W. Naylor[12].

3.3. *Statistical Analyses*

The experiment was conducted in a completely randomized design. Photosynthetic characteric and chlorophyll fluorescence values are the mean of seven independent replicates of each plant treatment.

3. Results

3.3. *Effects of High Temperature Stress on Photosynthesis Characteristic Parameters in Leaves of Cucumber Seedlings*

The net photosynthetic rate (Pn) of the plants varied considerably with the length of high temperatures stress. Two varieties showed similar descending trend of Pn to the control time (Fig. 1A), but a remarkably smaller decrease was found in heat tolerance variety (*Jinchun No.4*) than in heat sensitive variety (*Zhongnong No.12*). After 5 days stress, *Jinchun No.4* had a reduction of about 59.84% in net photosynthetic rate, while a reduction of about 74.36% on Zhongnong No.12 was observed.

Stomatal conductance (Gs) of the plants was decreased under high temperature stress. The change situation of Gs was similar to Pn basically (Fig. 1D).

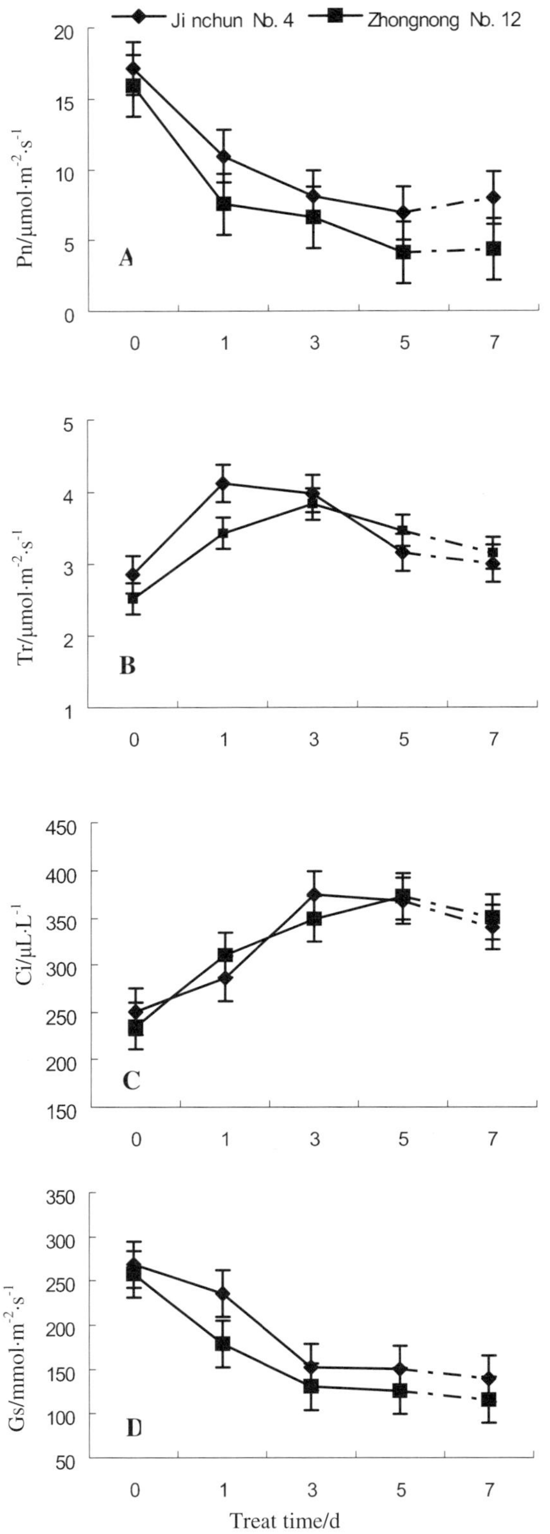

Fig. 1. Changes of Pn, Tr, Ci and Gs of two varieties of cucumber seedlings under high temperature stress.
Dotted line in figs represents 2 days resume of treated seedlings at general growth temperature.

On the contrary, both transpiratory rate (Tr) and Intercellular CO_2 concentration (Ci) of two varieties increased under high temperature stress (Fig. 1B and C), but Ci started decreasing after 3 days stress. What's more, the difference of change in two varieties was not notable.

Judging from the condition recovering(doted line in figs), The net photosynthetic rate had a light ascending trend to heat tolerance variety (*Jinchun No.4*), but that maintained at level before recovering to heat sensitive variety (*Zhongnong No.12*). Except Pn, other photosynthetic parameters Tr, Ci and Gs hadn't resume at all or resuming extents weren't notable with both varieties.

3.3. *Effects of High Temperature Stress on Chlorophyll Content in Leaves of Cucumber Seedlings*

Table 1.Changes of chlorophyll content in two varieties of cucumber seedlings under high temperature. (unit: $mg \cdot g^{-1}FW$)

Treat time(day)	Jinchun-No.4	Zhongnong-No.12
0d	2.0644A	1.9836A
1d	1.9289B	1.8907B
3d	1.5331C	1.4969C
5d	1.3209D	1.2108D
7d	1.3680D	1.2311D

Note: The different capital letters are significant difference in 1% level.

Table 1 shows the changes of chlorophyll content of two varieties of cucumber under the stress. The chlorophyll content decreased notably at both varieties. The chlorophyll content of *Jinchun No.4* decreased 36.02% after 5days treat, and lower than that in *Zhongnong No.12* (38.96%). Futher more, the chlorophyll content of heat tolerance variety (*Jinchun No.4*) was higher than that of heat sensitive variety (*Zhongnong No.12*) under same condition. Both varieties had little rising degree on chlorophyll content after 2 days recoving.

3.3. *Effects of high temperature stress on chlorophyll fluorescence parameters in leaves of cucumber seedlings*

The chlorophyll fluorescence technique is an effective way to investigate the efficiency and kinetics of photosynthetic apparatus in plants under various conditions. Figure 3A shows the changes in the primary

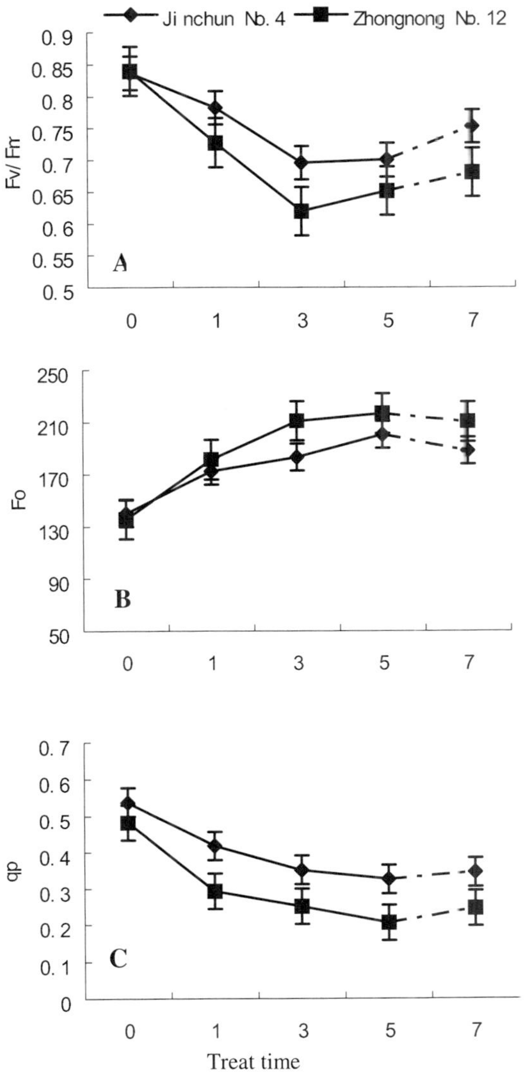

Fig. 3. Changes of chlorophyll fluorescence Fv/Fm,Fo and qp of two varieties of cucumber seedlings under high temperature stress. Dotted line in figs represents 2 days resume of treated seedlings at general growth temperature.

photochemical efficiency of PSII (Fv/Fm) and qp in leaves of cucumber under relative long-term treatment with 40/25°C (d/n). Fv/Fm in both heat tolerance variety (*Jinchun No.4*) and heat sensitive variety (*Zhongnong No.12*) declined with time increase of the treat, but a remarkably smaller decrease was found in *Jinchun No.4* than *in Zhongnong No.12*. After 5 days treatment, Fv/Fm in *Jinchun No.4* was 83.76% of that of untreatment, higher than that in *Zhongnong No.12* (77.57%), which demonstrated that PSII in heat tolerance variety was more tolerant to high temperatures than that in heat sensitive variety.

The base fluorescence Fo has been widely used as an indicator of heat damage to plants[5]. It reflects the diversion efficiency of light energy in the presence of pigments and the capture efficiency of excitation energy. Figure 3B shows the Fo-treat time curves in two varieties of cucumber. With the continuous increase in treat time, Fo-treatment curves had a rising trend in two varieties. but a notably higher increase as found in *Jinchun No.4* than in *Zhongnong No.12* after 1days stress. Which suggests that PS II reactive center was destroyed under the treat.

High temperature stress induced the changes of chlorophyll fluorescence quenching coefficients. qP indicates that the fluorescence level originated from the photochemical reactions of photosynthesis[13]. Figure 3C shows that qP in *Jinchun No.4* and *Zhongnong No.12* dropped with the prolong of the treatment time. After 5 days stress, qP in *Jinchun No.4* and *Zhongnong No.12* decreased 39.52% and 57.40% respectively. The results indicated that high temperature of 40/25°C might lead to the deactivation of the PSII reaction center and a rapid decrease in the relative level of the PQ sink.

Besides that, chlorophyll fluorescence parameters Fv/Fm, Fo and qp were all resume under 2days recovering and had a obvious recovery process for heat tolerance variety (*Jinchun No.4*), while that didn't recover for heat sensitive variety (Zhongnong No.12) except Fv/Fm.

4. Discussion

High temperatures affect photosynthesis by altering the excitation energy distribution by changing the structure of thylakoids[14,15] and by changing the activity of the Calvin cycle and other metabolic processes such as photorespiration and product synthesis. It has been reported previously that temperature limits CO_2 availability or alters the functioning of Rubisco by varying the substrate specificity of the enzyme[16,17] and its activity[18,19].

This study with cucumber seedlings under high temperature stress indicates that Phytosynthesis parameter Pn and Gs, chlorophyll content, chlorophyll fluorescence parameter Fv/Fm and qp were all reduced during treat time, while Ci and Tr, Fo increased. Net photosynthetic rate (Pn) of two varieties was affected by high temperature stress, although Intercellular CO_2 concentration and transpiration were higher under high temperature stress and, on contrary, stomatal conductance lower.

The primary photochemical efficiency of PSII (Fv/Fm) decreased largely under high temperature stress, which demonstrated that cucumber leaves occur to photoinhibition likely. Moreover, Fo increased indicated that photosystme II reaction centre was destroyed under the high temperature stress. At the same time, chlorophyll fluorescence quenching coefficients decreased showed the quantities of QA oxidize were reduced, which could lead to decrease transfer rate of electron on the entire electron transfer chain, and affect the ATP and NADPH formation further.

Differences between two varieties were observed in the photosynthetic characteristic and Chlorophyll fluorescence parameters under the stress. And these suggested heat tolerance variety have stronger resistivity or Adaptability to high stress than heat sensitive variety under identical condition. Besides, Judging from the recovering condition, it indicated same conclusion.

The chlorophyll content of cucumber seedlings under high temperature decreased remarkably (table 1), the main reason likely is that the stress aroused it decomposition or the rate of photodestruction of chlorophyll and associated pigments occurs more rapidlly than synthesis. Another possibility is that high temperature stress resulted in inhibition of Chl biosynthesis. Meanwhile, the chlorophyll content of heat tolerance variety (*Jinchun No.4*) had a relative higher chlorophyll content than heat sensitive variety (*Zhongnong No.12*) under identical condition, demonstrating that *Jinchun No.4* had stronger Pn so that could resisting high temperature ability and delayed the chlorophyll decomposition to a certain extent.

Acknowledgments

This study was funded by the Key project program of Gansu Provincial Science and Technology Department (2GS052-A41-007-05) of PR China.

References

1. Boyer JS, Plant productivity and environment. *Science*. **218**, 443–448 (1982).
2. Houghton JT, Ding Y, Griggs DJ, Climate change 2001: *the scientific basis*. Cambridge: Cambridge University Press (2001).
3. Wagner D, Scenarios of extreme temperature events.*Climatic Change*. **33**, 385–407 (1996).
4. Oleg Stasik1,2 and Hamlyn G. Jones, Response of photosynthetic apparatus to moderate high temperature in contrasting wheat cultivars at different oxygen concentrations. *Journal of Experimental Botany*. **58**, 2133–2143 (2007).
5. Ling Shao, Zhan Shu, Shu-Lan Sun,Chang-Lian Peng, Xiao-Jing Wang and Zhi-Fang Lin, Antioxidation of Anthocyanins in Photosynthesis Under High Temperature Stress.*Journal of Integrative Plant Biology*. **49(9)**: 1341–1351 (2007).
6. Popovic, R., Dewez, D., Juneau, P., Application of chlorophyll a fluorescence parameters in ecotoxicological studies of pollutants: heavy metals, herbicides and air pollutants. *In: Toivonen, P., DeEll, J. (Eds.), Practical Applications of Chlorophyll Fluorescence in Plant Biology.* Kluwer Academic Publisher, Dordrecht, pp. 152-179 (2003).
7. Schreiber, U., Pulse-amplitude (PAM) fluorometry and saturation pulse method. *In: Papageorgiou, G.C., Govindjee (Eds.), Chlorophyll a Fluorescence: A Signature of Photosynthesis.* Springer, Dordrecht, pp. 279~319 (2004).
8. Strasser, R.J., Tsimilli-Michael, M., Srivastava, A., Analysis of the chlorophyll a fluorescence transient. *In: Papageorgiou, G.C., Govindjee (Eds.), Chlorophyll a Fluorescence: A Signature of Photosynthesis.* Springer, Dordrecht, pp. 321~362 (2004).
9. Horton, P., Hague, A., Studies on the induction of chlorophyll fluorescence in isolated barley protoplasts. IV. Resolution of non-photochemical quenching. *Biochimica et Biophysica Acta*. 932, 107-115 (1988).
10. Gray GR, Chauvin LP, Sarhan F, Huner N. Cold acclimation and freezing tolerance: A complex interaction of light and temperature. *Plant Physiol.* **114**, 464–474 (1997).
11. Schreiber, U., Schliwa, U., Bilger, W., Continuous recording of photochemical and non-photochemical chlorophyll fluorescence quenching with a new type of modulation fluorometer. *Photosynthesis Research*. **10**, 51-62 (1986).

12. J. R. McWilliam and A. W. Naylor, Temperature and Plant Adaptation. I. Interaction of Temperature and Light in the Synthesis of Chlorophyll in Corn. *Plant Physiol.* **42**, 1711-1715 (1967).
13. Bilger W, Fisahn J, Brummet W, Kossmann J, Willmitzer L. Violaxanthin cycle pigment contents in potato and tobacco plants with genetically reduced photosynthetic capacity. *Plant Physiol.* **108**, 1479–1486 (1995).
14. Berry J, BjÖrkman O, Photosynthetic response and adaptation to temperature in higher plants. Annu Rev *Plant Physio.* **l31**: 491-543 (1980)
15. Weis E, Berry JA, Plants and high temperature stress. In SP Long, FI Woodward, eds, Symposia of the Society for Experimental Biology Number XLII. *The Company of Biologist,* Cambridge,UK, pp. 329-346 (1988).
16. Jordan DB, Ögren WL, The CO_2/O_2 specificity of ribulose 1,5-bisphosphate carboxylase/oxygenase. *Planta.* **161**: 308-313(1984).
17. Brooks A, Farquhar GD, Effect of temperature on the CO_2/O_2 specificity of ribulose-1,5-bisphosphate carboxylase/oxygenase and the rate of respiration in the light. *Planta.* **165**: 397-406 (1985).
18. Weis E, Reversible heat-inactivation of the Calvin cycle: a possible mechanism of temperature regulation of photosynthesis. *Planta.* **151**: 33-39 (1981).
19. Santarius KA, Exner M, Thebud-Lassak R, Effects of high temperature on the photosynthetic apparatus in isolated mesophyll protoplasts of Valerianella locustu (L.) Betcke. *Photosynthetica.* **25**: 17-26 (1991).

MORPHOLOGICAL ANALYSIS OF THE NUCLEUS PULPOSUS CELLS FROM YOUNG AND OLD RABBIT INTERVERTEBRAL DISC*

LIU XIAOYUN, YANG SHUHUA[ξ], LI JIN, ZHANG YUKUN, FU DEHAO, WU QIANG

Department of Orthopedics, Union Hospital, Tongji Medical College, Huazhong University of Science and Technology, Wuhan, 430022, China

Objective: To compare the cellular and extracellular matrix (ECM) characteristics of the nucleus pulposus (NP) cells of young and old rabbits. Methods: The structure and arrangement of the NP cells and ECM in young and old rabbits were investigated both the tissue and ultrastructure scale by using confocal laser scanning microscopy and transmission electron microscopy. Results The young rabbit discs were Thompson's grade I. The NP cells, which had a large size containing numerous large inclusions, often formed as clusters with a diameter of 266 ± 167 μm. The collagen fibril was homogeneous. The old rabbit discs were Thompson'sgrade III with less cells and cell clusters. The average diameter of the cell clusters was 94±42 μm. Compared with the NP cells from the young rabbit discs. There are a fewer inclusions in the NP cells from the old rabbit. Collagen fibril often cross-linked and arranged irregular. Conclusions: There were striking differences in the cell architecture of young versus old nuclei pulposi, these features of the aging and degenerating disc contribute to decreased biomechanical function.

Key words: Intervertebral disc; Rabbit; Nucleus pulposus cell; Morphology

1. Introduction

Intervertebral disc degeneration is the most common cause of lower back pain. The intervertebral disc is a closed structure without blood vessels. Along with the decline in nutrition of nucleus pulposus cells, a variety of factors such as chronic stress and heredity of intervertebral disc, as the comprehensive factor, made intervertebral disc easy to be degenerative [1]. Morphology and composition changes of nucleus pulposus cell are the initiating agents for degeneration [2]. Previous studies mainly targeted at the matrix changes of aging and degenerative intervertebral disc and lacked of a systematic comparative observation and analysis of the morphological changes in intervertebral disc cells [3,4].A study of the nucleus pulposus cells and matrix structure of intervertebral disc will contribute to a better understanding of the biological function of intervertebral disc cells and the reason of intervertebral disc degeneration and find a potential method for repairing damage of intervertebral disc, delaying and reversing the disc degeneration [5]. Accordingly, our study used confocal microscopy and transmission electron microscopy to observe and analyze the nucleus pulposus cells and matrix of juvenile and old rabbits.

2. Materials and Methods

10 New Zealand white rabbits were used in our study. 5 young rabbits of 5 to 7 weeks old were in the first group and 5 old rabbits of 3 to 3.5 years old were in the second group (provided by Hubei Academy of Medical Sciences animal experiments Center). The grades of disc were categorized according to Thompson grading system [6]. Animals were executed and immediately taken out the whole lumbar spine. The soft tissue around the lumbar spine was removed, then the lumbar spine was treated as followed.

The grain-sized nucleus pulposus from L45 and L56 of each rabbit were prepared as electron microscopy specimens. The remaining organization was immediately frozen and embedded at -20°C, and then cut into 10 μ m thick slices which were fixed by acetone for 10 min, then observed by confocal microscopy. 8 high power fields (× 40 times) of each slice were random selected and used to observe the cell morphology and distribution. Zeiss LSM5 image browser software was used to measure the size of cell cluster.

The electron microscope specimens were fixed immediately with 2.5% glutaraldehyde for 2 hours, then flushed by 0.1 mol / L phosphate buffer, and fixed by

* This work was supported by the National Science Foundation of China (NO: 30672130), China Postdoctoral Foundation (20060390249).

ξ E-mail: xylxyhzzh@126.com

1% osmium acid for 2 h, gradient dehydrated by ethanol, embedded by epoxy resin. Ultrathin sections was obtained, and then double stained by lead and uranium. Sections were placed in copper networks and observed via Hitachi H-7000FA transmission electron microscope at 75 kV.

Statistical analysis: Results are presented as means ± SD of 5 animals in each group. The statistical significance of the data was determined by t-test. ($\alpha = 0.05$).

3. Results

3.1. *The Anatomization of Disc*

The disc nucleus pulposus of first group was tremeloid. The boundaries between nucleus pulposus and annulus fibrosus were clear. The structure of annulus fibrosus was clear and lined up in order. Thompson grade of the disc nucleus pulposus was grade I (Figure 1A). Calcified spots scattered around the nucleus pulposus of the intervertebral disc in second group which surround by gray and white fibrous tissue. The boundaries between nucleus pulposus and annulus fibrosus were not clear. The structure of annulus fibrosus was punch-drunk. Thompson grade of the disc nucleus pulposus was grade III (Figure 1B).

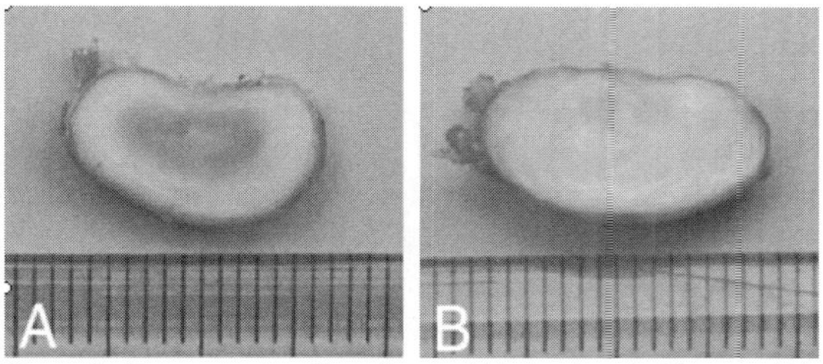

Figure 1. Representative discs from rabbit lumber A: Young rabbit disc, Thompson's grade I; B: Older rabbit disc, Thompson's grade III.

3.2. *Morphological Structure of Nucleus Pulposus Cells and Matrix*

Cells of Group 1 were distributed as a cluster. The cell clusters was capsuled by lamellar matrix (Figure 2A). The average diameter of cell clusters was 266 ± 167 μ m. The cell clusters of Group 2 were smaller or the cells distributed sporadically. The density of cell clusters was significantly lower than the first group (Figure 2B). Matrix contained large amounts of fibrous tissue. The average diameter of cell clusters was 94 ± 42 μ m (excluding single cells) which was significantly less than the group 2($P < 0.05$).

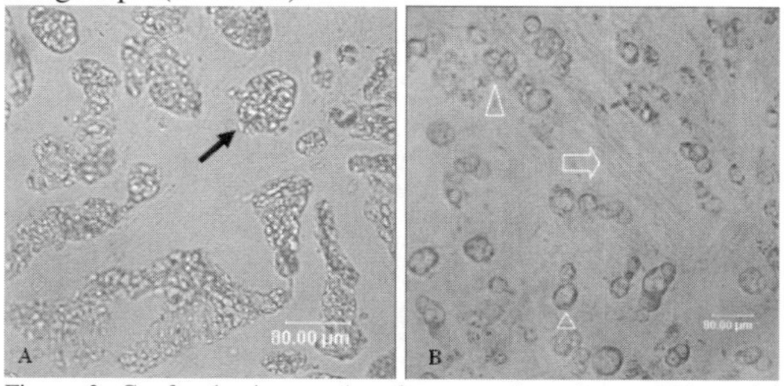

Figure 2. Confocal micrographs of cyro-sections from rabbit nuclei pulposi. A: NP from young rabbit contained much cells in large clusters surrounded by a thin capsule of matrix material (black arrow). B: NP from older rabbit contained relatively few cells in small clusters or alone (triangle) and contained a large amount of fibril (white arrow).

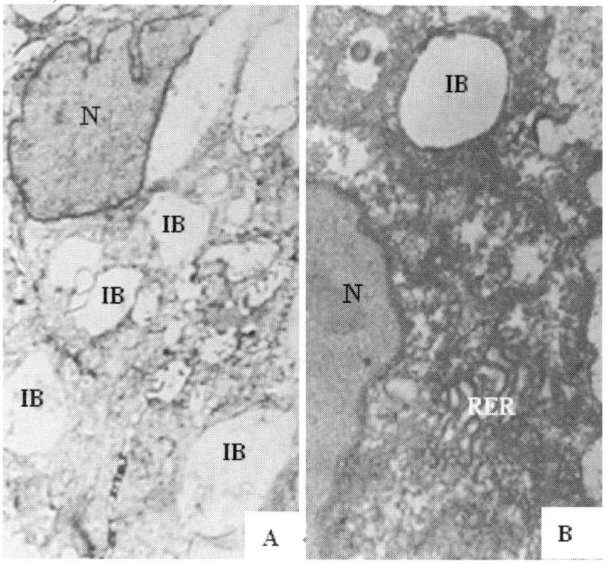

Figure 3. Transmission electron micrographs of rabbit nuclei pulposi cells. IB: Inclusion body; N: Nucleus; RER: Endoplasmic Reticulum; A: Yong rabbit (×5000); B: Older rabbit (×7000).

3.3. *The Ultrastructure of Nucleus Pulposus Cells*

Transmission electron microscopy revealed that a lot of vacuole-like inclusion bodies were resided in the nucleus pulposus cells of group 1. The inclusion body was not besieged by membrane which was round or circular on the section. The endoplasmic reticulum was rarely observed in most of the cells (Figure 3A). Collagen fiber lined up in order, the thickness of which was uniform. Cyclical stripes of collagen fiber were observed (Figure 4A). Gap junction between the nucleus pulposus cells was ubiquitous (Figure 5A). The number and size of inclusion bodies in the nucleus pulposus cell of group 2 was smaller. RER was clearly observed (Figure 3B). The diameter of collagen fiber was various. A large number of cross-linked were resided in the nucleus pulposus cells and lined up without order (Figure 4B). Gap junction between the nucleus pulposus cells was not observed.

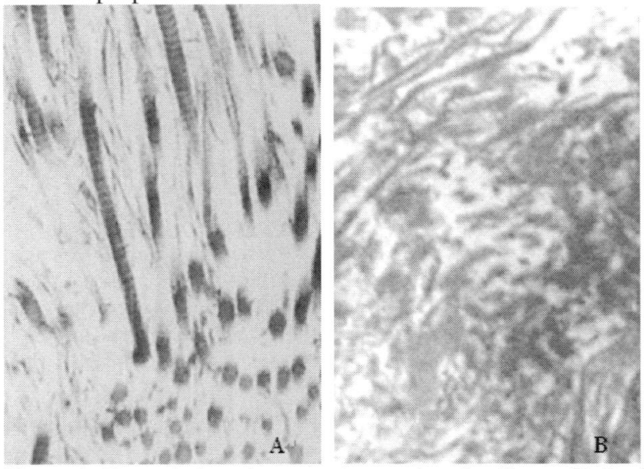

Figure 4. Transmission electron micrographs of collagen fibril of rabbit extracellular matrix. A: Yong rabbit (×20000); B: Older rabbit (×10000).

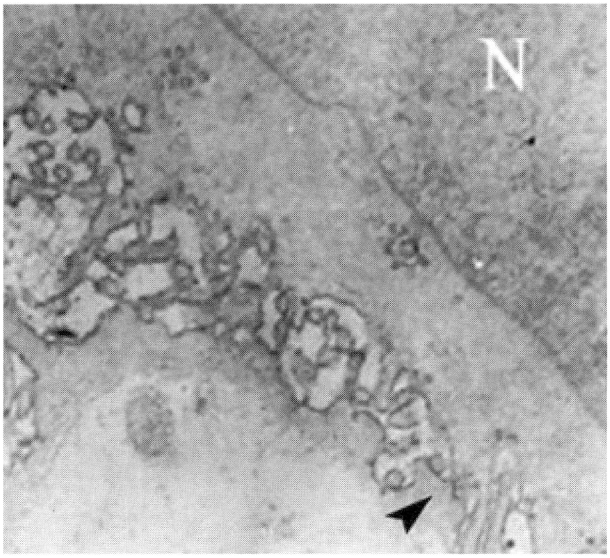

Figure 5. Transmission electron micrographs of gap junctions from rabbit nuclei pulposi cells (×12000). N: Nucleus (black arrow): gap junctions.

4. Discussions

Nucleus pulposus tissue of disc is important to maintain the normal function of intervertebral disc because nucleus pulposus cells of disc can produce extracellular matrix and constantly engage in self-repairing, updating and remodeling which are the biological basis of load endurance, stress scattering, of oscillation buffer [7, 8]. Along with the growth of age, the intervertebral disc began to be degenerative. The earliest change is the amount and structure of nucleus pulposus cells, cell types, as well as extracellular matrix of changes, which resulted in a series of changes, such as biomechanical function, of disc, then the secondary degeneration was produced [8]. Therefore, the study of morphological changes in the nucleus pulposus of the intervertebral disc is important for the awareness and understanding of disc degeneration.

Studies indicated that nucleus pulposus tissue of young rabbit nucleus was a notochord nucleus pulposus which was transformed into fibrocartilage nucleus pulposus at 2-year-old, which was similar to human beings [9]. Therefore, we used the young rabbits of 5 to 7 weeks old and old rabbits of 3 to 3.5 years old for observed study, which could be good to respond the characteristic of morphological changes of aging and degenerative disc.

Research indicated that nucleus pulposus tissue of young rabbit was a notochord nucleus pulposus. The contents of the proteoglycan and type II collagen in matrix are very rich, which made disc resist various stress and self-repair and remodel [2]. So the Thompson grade of the disc in young rabbits was grade I. With the aging, the contents of the proteoglycan and type II collagen in matrix decreased significantly, and the contents of type I collagen increased. This led to the decline of hydratable ability which resulted in flexibility reducing. Due to the weak capacity of in anti-tensile and anti-compression, disc was prone to impaired fibrosis [10]. This study using confocal microscopy confirmed fibrosis in the degenerative nucleus pulposus matrix, So,old rabbit disc was Thompson grade III.

This study showed that the cluster distribution of nucleus pulposus cells in young rabbit was dense. The size and the number of cell clusters were large. The cell density of nucleus pulposus inold rabbit was low. The cell clusters was in small size or single distribution. The reasons for this change occurred possible was associated with the apoptosis [11] and the change of cell population [9]. Along with the growth of age, under the bad biology conditions of long-term nutritional deprivation, low oxygen, low pH environment, nucleus pulposus cells of disc long-term born mechanical load lead to cell apoptosis [11], which reduced the number of cells and size of cell clusters. At the same time, as age increased, the population of the nucleus pulposus would be changed. Chondrocyte-like gradually replaced notochord cells. Notochord cells and Chondrocyte-like were two categories of the cells with different biological characteristics [9]. Therefore, the nucleus pulposus cells of juvenile and old rabbit had different morphology and distribution characteristics.

Ultrastructural observations of this study found that there were a lot of cell gap junctions in the first group, but not in the second group, which indicated that the gap junctions might be the basis structure of cluster distribution of the cells in the notochord nucleus pulposus. The gap junctions of cells allowed the cell membrane potential and molecules less than 1 KDa to pass through [12]. Much information could be transmitted directly and rapidly in the entire cell netwoke, thus cells could respond to all kinds of stimulus in the entire netwoke, which may be one of the reasons that notochord nucleus pulposus possessed the ability of stress resistance that could not occurred degeneration easily [13]. The collagen fibers in nucleus pulposus tissue in young rabbits lined up in order. The thickness of collagen fibers was uniformed. Cyclical stripes could be observed, which indicated that the molding capability of collagen was powerful. The collagen fibers in nucleus pulposus tissue in young rabbits lined up without order. A large number of cross-linked were resided in the nucleus pulposus cells. The diameter of collagen fiber was various. The results indicated that the molding capability of collagen was weak and the collagen fibers were easily degraded, which led to the fibrosis of nucleus pulposus at the time of disc degeneration [14]. A lot of vacuole-like inclusion bodies were resided in the nucleus pulposus cells of juvenile rabbit, and the number and size of inclusion bodies in the nucleus pulposus cell of old rabbit was smaller which consistent with the literature [15]. Since the disc was a closed structure without vascular, the tissue repair and remodeling need energy, which indicated that inclusion bodies may be the energy of reserve [8]. The nucleus pulposus of young rabbit mainly contained notochord nucleus pulposus cell which contained large amount of energy for tissue repair, updating and remodeling. So notochord nucleus pulposus did not prone to degeneration. The nucleus pulposus of old rabbit had changed into fibrocartilage nucleus pulposus. Cells lacked of energy reserves, the weak ability to self-repair and update, which resulted in naturally disc degeneration.

This study conducted compared observation of the juvenile and old rabbit nucleus pulposus cells and matrix in the morphological aspects and initially revealed the morphological changes characteristic of disc degeneration and aging. Disc degeneration is a very complicated physiological and pathological process. Further exploring the mechanism of population changes in nucleus pulposus cells of disc to, maintaining the long-term existence of notochord cells in the nucleus pulposus may be a potential new method to prevent and retard disc degeneration.

Acknowledgments

The authors would like to thank Yang Jie and Luo Weihua for technical support in operation of confocal microscopy.

References

1. S. R. Garfin, H. N. Herkowitz, In: Wiesel SW, Weinstein JN, eds. The Lumbar Spine.M Philadelphia: *WB Saunders*, 369 (1990).
2. R. H. Pearce and B. J. Gimmer, *J Orthop Res.* **5,** 198 (1987).
3. R. J. Errington, K. Puustjarvi and I. R. White, *J. Anat.*; **192**, 369 (1998).
4. A. J. Freemont, A. Watkins and C. Le Maitre, *J Pathol.* **196**, 374 (2002).
5. H. E. Gruber and E. N. Hanley, *Spine* **28**, 186 (2003).
6. J. P. Thompson, R. H. Pearce and M. T. Schechter, *Spine* **15**, 411(1990).
7. H. E. Gruber and E. N. Hanley, *Spine* **27**, 798 (2002).
8. C. J. Hunter, J. R. Matyas and N. A. Duncan, *J Anat.* **202**, 279 (2003).
9. J. J. Trout, J. A. Buckwalter and K. C. Moore, *Anat Rec.* **204**, 307 (1982).
10. K. W. Kim, T. H. Lim and J. G. Kim, *Spine* **28**, 982 (2003).
11. H. E. Gruber, H. J. Norton and E. N. Hanley, *Spine* **25**, 2153 (2000).
12. C. J. M. Hunter, J. R. Duncan and A. Neil, *Spine* **29**, 1099 (2004).
13. H. J. Donahue, *Bone* **26,** 417 (2000).
14. H. L. Pokhama and F. M. Phillips, *Spine* **23**, 1645 (1998).
15. S. Roberts, J. Menage and V. Duance, *Spine* **16**, 1030 (1991).

A WIRELESS ECG MONITORING SYSTEM WITH NEW WAVEFORM RECOGNITION METHOD

LINQING LIANG
Huazhong University of Science and Technology
Wuhan, Hubei, 430074, China
Email: llq19840313@126.com

YUXING YANG, JIE LUO, LINMING ZHOU, WUCHENG HONG
Huazhong University of Science and Technology
Wuhan, Hubei, 430074, China
Email: yyx999@sina.com

The high price of traditional electrocardiogram (ECG) monitoring systems has limited their usage area to large hospitals. The low-cost and high-performance of our portable wireless remote ECG monitor will allow for small and medium hospitals to have access to ECG technology without breaking their budgets. Our ECG utilizes Derivative of Zero-Domain (DZD), a new recognizing method that greatly improves the data recognition accuracy.
Keywords: PTR5000 wireless module; DZD; ECG Monitoring

1. Introduction

Heart disease continues to be the leading cause of mortality in the world, ECG monitoring systems are the main tool for doctors to diagnose heart disease. However, traditional ECG monitoring systems are always expensive and bulky. We propose a portable low-cost and high-performance wireless remote ECG monitoring system.

The newly proposed algorithm DZD can accurately recognize the feature point of not only the QRS complex, but also P wave and Q wave. The traditional algorithm can only recognize the QRS complex, which limits their real-time data processing capacity while our design does not.

The application of USB interface with PC has greatly increased the data transmission rate. Compared with other similar devices, its structure is easier, its cost is cheaper, and its function is complete.

2. Hardware Design of the System

The ECG monitor consists of a circuit of four beds, and a transmitting module, a receiving module and a PC interface.

The ECG detecting circuit includes the Preliminary stage, Amplify-Filtering stage and Voltage-Adjusting stage. The Amplifier MAX194 in the Preliminary stage is powered by 3V batteries. The frequency in the Amplify-Filtering stage ranges from 0.05HZ to 100HZ with a gain of 500 times. The Voltage-Adjusting stage can adjust the signal within the range of A/D transmission. The structure of the circuit is simple and requires little power to run.

The block diagram of ECG wireless [1] circuit can be seen in Fig1. The wireless circuit consists of transmission module PTR5000 and receiving module EEPROM 25AA320. PTR5000 required little power and has a high transmission rate of 1Mbps. The wireless operating frequency 2.4GHz is available globally with the GFSK modulation mode.

The USB interface circuit consists of FT245BM and 93C46, FT245BM is to transform the parallel data in PTR5000 module to parallel date of USB interface.

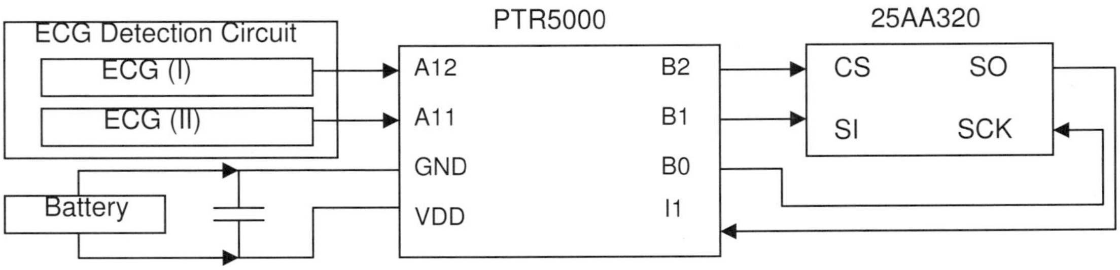

Fig.1. Block Diagram of ECG Wireless Circuit

The FIFO inside the chip is to highly transmit data inter-computer. The EEPROM (93C46) is to store the information of USB device, including VID, PID, serial number and power parameters, etc.

The transmission module is powered by twc 7th batteries, which can supply power for 72 hours continuously. And the receiving module is powered with the style of USB BUS.

3. Software Design of the System

Software system functions are composed of a man-machine interface, displays of the ECG Waveform, recognition of the ECG waveform feature point, calculation of ECG parameters and analysis of arrhythmia, data storing and playback, and other functions. This is a multi-purpose multithreading real-time software application that is based on the VC++6.0 platform.

3.1. *Man-Machine Interface & Displaying of ECG Waveform*

The software has two interfaces, the first for displaying the four beds waveform, the second for a single bed's waveform. Each bed is controlled by a series of commands. Commands include setting patient information, start monitoring [2], stop monitoring, display single-bed data, storing data, Filtering ECG data, etc. The sampling [3] frequency is 500Hz, and the 2-lead ECG are sampled with 12 bits

3.2. *Theory of Derivative of Zero-Domain (DZD)*

The new algorithm Derivative of Zero-Domain (DZD) is based on the method of recognizing Zero-pass-point (ZPP) and Zero-near-point (ZNP), which is simple and effective, and it can accurately recognize the feature points. Fig.2(a) shows a unimodal model, which usually appears in P-wave, T-wave and single R-complex, while Fig.2(b) shows a bimodal model, which always appears in QRS-complex, and sometimes in P-wave and T-wave. The derivative graphs of Fig.2(a) and Fig.2(b) are shown in Fig2(a) and Fig.2(b) blow their respective model. The narrow area near the Zero-line is called the Zero-domain, "A" stands for "ZPP", and "B" stands for "ZNP".

There are one ZPP and two ZNPs in the derivative graph of Fig.2(c). One ZPP of the maximum positive

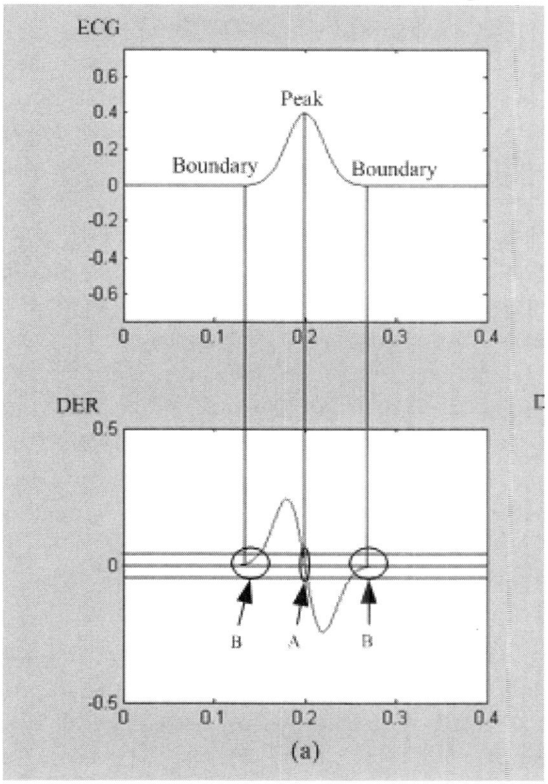

Fig.2. (a) ZPP 与 ZNP Unimodal model

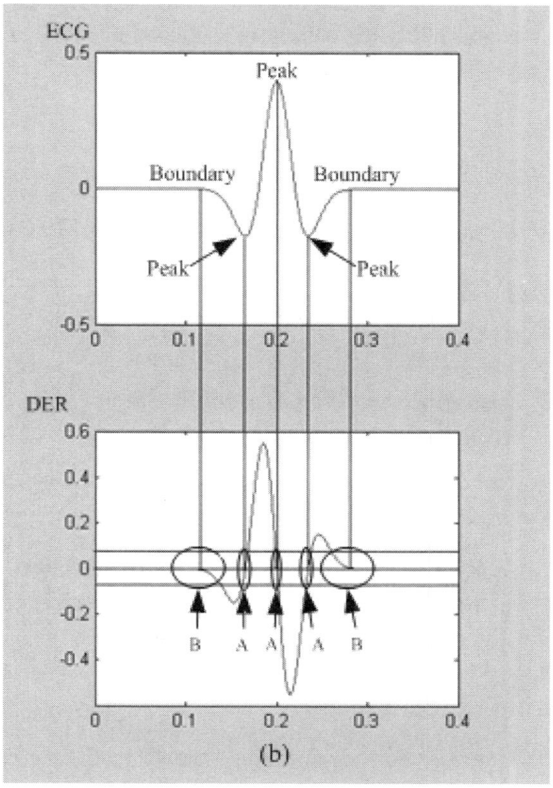

Fig.2. (b) ZPP 与 ZNP Bimodal model

wave corresponds to peak of the wave (like R-wave, P-wave and T-wave), and two ZNPs to the beginning point of P-wave and ending point of T wave.

There are three ZPPs and two ZNPs in Fig.2(b). One ZPP of the maximum positive wave corresponds to R-peak, another two ZPPs to Q-peak and S-peak respectively, two ZNPs to the beginning point and ending point of model (like QRS complex in Fig.2(b)) respectively.

This method is suitable for any QRS-complex, P-wave and T-wave that conform to the model in Fig.2. Those unconformable waveforms should be omitted.

There are many types of QRS-complexes in actual ECGs, some of these types are shown in Fig.3, where (a) is for QRS, (b) for QR, (c) for RS and (d) for classical negative wave. The first three waves can be accurately recognized using the above method, while (d) is always be omitted. We can apply ZPP and ZNP of the maximum negative wave, instead of the maximum positive wave, to recognize the (d) wave in Fig.3.

The algorithm of DZD can directly recognize the peak and boundary of QRS-complex without estimation, After analyzing the magnitude and direction of ZPPs between the two boundaries, we can locate P-waves and T-waves. ZNP helps greatly in recognizing low and flat P-waves and T-waves.

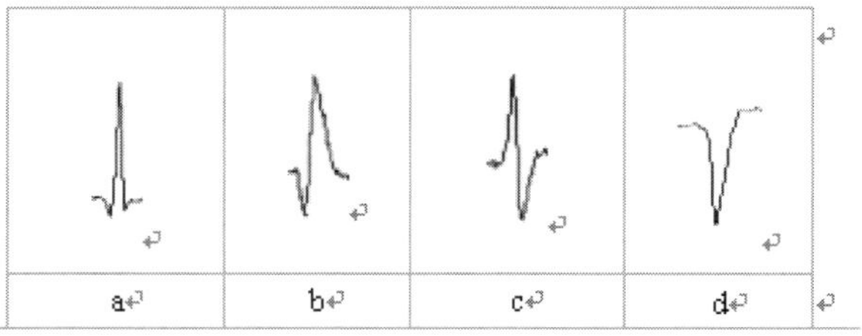

Fig.3. Several QRS-complexes with different morphologies

3.3. *Steps of Recognition Method Based on DZD Algorithm*

The following nine steps present the method of recognizing feature points based on DZD algorithm:

- Applying derivative and threshold to locate R-wave.

Firstly applying derivative to ECG waveform, and then filtering the derivative waveform with low-pass filter. Finally locate the R-wave according to the maximum derivative.

- Methods for determining the Range of Zero-Domain (RZD) and Time Interval (TI).

Determining RZD is for judging ZPP and ZNP, and its formula has been deduced by experience.

Here, MD stands for maximum derivative, ZD stands for Zero derivative. RZD_{QRS} stands for the magnitude of the QRS-complex,

For QRS-complex:

$$RZD_{QRS} = \frac{MD_R - ZD}{15 \sim 20} \quad (1)$$

$$TI_{QRS} = 10 \sim 20(\text{ms}) \quad (2)$$

For P-wave:

$$RZD_P = \frac{MD_P - ZD}{10 \sim 15} \quad (3)$$

$$TI_P = 15 \sim 30(\text{ms}) \quad (4)$$

For T-wave:

$$RZD_T = \frac{MD_T - ZD}{10 \sim 15} \quad (5)$$

$$TI_T = 15 \sim 30(\text{ms}) \quad (6)$$

- Methods for determining ZPP and ZNP.

Locating ZPP and ZNP depends on whether the derivative waveform is one-off or not. The character for ZPP is one-off, that is while passing zero the derivative of waveform is monotonically increasing, from positive to zero, and then from zero to negative. While the derivative of the waveform is not one-off, it is not monotonically increasing, and it is always very small and near the zero line, which is defined as ZNP. ZPP means crossing zero, while ZNP means nearing zero.

RZD for ZPP and ZNP has been depicted in Fig.4. There is a unimodal model in Fig.4(a) and a bimodal model in Fig.4(b), which stand for classic P-wave, T-wave and the QRS-complex. The above is the waveform of the model and the below is the derivative of the modal wave. In Fig.4(b) there are three short dotted lines(in red) that correspond to three ZPPs, that are one-off and monotonically increasing while crossing the zero domain. Another two long dotted lines (in black) correspond to the two ZNPs, whose values are rather small and are not one off while crossing the zero domain.

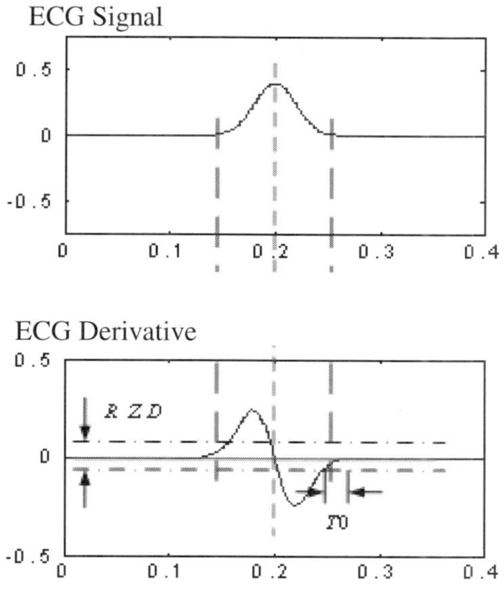

(a)

Fig.4. Range of Zero Domain and ZPPs and ZNPs (a) unimodal modal

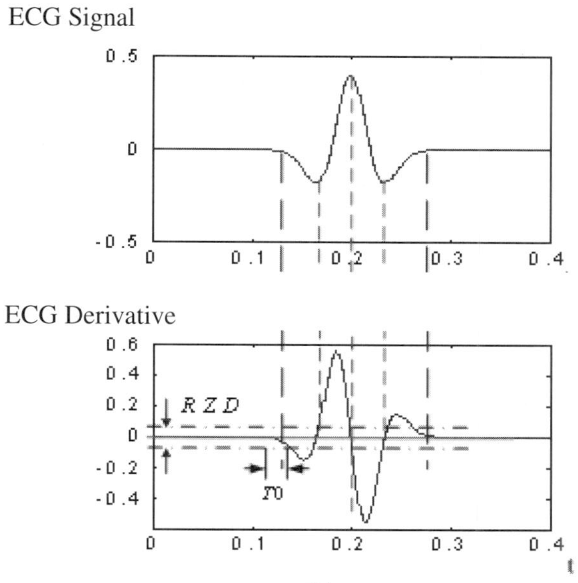

(b)

Fig.4. Range of Zero Domain and ZPPs and ZNPs (b) bimodal modal

The first step is to locate the R-wave, which is similar with traditional method. The second step is to find the forward and backward ZNPs. The last step we can judge whether there are ZPPs within the domain of ZNPs. If there ZPPs exist, they are multi-peak waves with the number of peaks depending on the number of ZPPs. If ZPPs does not exist, the wave is unimodal. This method is preferable for recognizing unimodal, bimodal and multimodal waves with reliability.

- Searching the ZPP from the right side of R-wave domain to determine the R-wave peak.

If ZPP exists in the zero domain, it is the located at the wave peak; if ZPPs does not exist, there are no positive waves in the QRS complex, and in this case, this lead will be omitted and the program will go to another lead.

- Searching the ZNPs to determine the boundaries of QRS-complex.

Search both the near right and near left of R-wave peak, which has been determined in the fourth step. According to the definition of ZNPs, there are two ZNPs at the boundary of QRS-complex, that are the beginning point and the ending point of the QRS-complex.

- Searching other ZPPS to determine the Q-wave peak and S-wave peak.

All feature points of QRS-complex can be determined by searching within the two ZNPs, the located ZPPs are the peaks of Q-wave and S-wave. Then all feature points of QRS-complex has been determined.

- Finding the P-wave and T-wave.

Finding the P-wave and T-wave can be accomplished by searching from the left of QRS-complex and right of QRS-complex respectively by utilizing searching windows.

The size of searching window for P-wave is:

$$[QRS_{onset}\text{-}250ms,\ QRS_{onset}\text{-}20ms] \quad (7)$$

RR stands for R-R inter-phase.

The ZPP with the maximum derivative in the P-wave and T-wave searching window is the P-wave peak and T-wave peak respectively. Its exact method is the same as detecting the R-wave peak, that is searching the ZPP to locate the wave peak. If the maximum peak is negative, then this lead will be omitted to detect another lead. Sometimes we can take the absolute value of the P-wave and the T-wave instead of omitting them, and then analysis according to the above theory to determine the P-wave peak and the T-wave peak.

- Searching the ZNPs to determine the boundaries of T-wave.

  Search both the near right and near left of T-wave peak, which has been determined in the 7th step. According to the definition of ZNPs, there are two ZNPs at the boundary of T-wave, which are the beginning point of the T-wave.

- Searching the ZNPs to determine the boundaries of P-wave.

  Search both the near right and near left of P-wave peak, which has been determined in the 7th step. According to the definition of ZNPs, there are two ZNPs at the boundary of P-wave, which are the beginning point of the P-wave.

The DZD Algorithm fully makes use of derivative's strengths and makes up for its shortcomings. It has greatly increased the accuracy recognition rate and the computation time has not risen obviously.

3.4. *Arrhythmia Analysis, Data Storage and Playback*

Traditional ECG monitor generally deals with heart rate analysis, calculation of ST section excursion and simple Arrhythmias analysis. This software is based on the new algorithm ZND to recognize the beginning point of P wave and QRS complex and the ending point of QRS complex and T wave, as well as calculating PR inter-phase and QT inter-phase and other ECG parameters. It can analyze more than ten types of familiar arrhythmia.

Data storage and playback function offers doctors a more convenient and effective way to analyze patients' problems.

4. Conclusion

Experiments show that the new algorithm can accurately recognize feature points and precisely calculate multi-lead ECG parameters. The device can monitor patients within 30 meters in a room, and batteries can supply as long as 72 hours.

Samples of wireless remote ECG monitor have been developed, and a mass of clinical experiments has been made in the main hospital of Huazhong University of Science and Technology and some other big hospitals. The waveform is displayed with clarity, the ECG parameter is calculated accurately, and the man-machine interface is convenient. The result of heart rate, ST excursion, ECG parameters and arrhythmia analysis can be shown in real-time, which is of importance for doctors to make judge immediately.

Acknowledgments

I'd like to thank PIBM conference organizers for this great opportunity. I'd like to thank my tutor Professor Yuxing Yang and my Colleagues Jie Luo and Wucheng Hong for their help in the process of preparing for this paper.

References

1. James Welch, Farzin Guilak, Steven D.Baker. A Wireless ECG Smarter Sensor for Broad Application in Lift threatening Event Detection. International conferences of the IEEE EMBS. September 1-5, 2004.
2. I.Korhonen, J.Parkka, M.van Gils, "Health Monitoring in the home of the future", IEEE Engineering in Medicine and Biology Magazine, Vol. 22, Number 3, pp. 66-73, 2003.
3. Wear .Well, IEEE Engineering in Medicine and Biology Magazine, Vol. 22, Number 3, 2003.
4. S.M.Szilagyi, L.Szilagyi and Z.Benyo, "Recognition of Various Events from 3-D Heart Model", 16th IFAC World Congress, 2005, Prague, paper #4695, pp. 1-6.
5. M.A.Watanabe and G.Schmidt, Heart rate turbulence: a 5-year review, Heart Rhythm, Vol. 1, pp. 732-738, 2004.
6. JOSE C.PRINCPE, etc., Developments in ECG Acquisition, Preprocessing, Parameter Measurement, and Rcording, IEEE on biomedical engineering, VOL. BME-33, No. 6, 1986.

TEST OF DRUG RESISTANCE ON BACTERIA WITH IMPEDANCE MICROORGANISM SENSOR*

HONGZHI WANG[1, 2], XIAOFENG PANG[1], YIMING WANG[1†]

College of Life Science and Technology University of Electronic Science and Technology of China, Chengdu China

AIHUA WANG[2]

College of Life Science and Technology Southwest University for Nationalities, Chengdu China

In this paper, test of drug resistance on bacteria was performed to medical clinic with impedance microorganism sensor. The impedance value of different bacterial cultures was measured using the home-made microorganism sensor and measurable setup of impedance, then all impedance values attained in the period of bacteria culture were recorded and the curve between impedance value and cultural time was drawn to judge the drug resistance of different bacteria. It was shown that the curve was obviously difference between liquid cultures that antibacterial drugs was added and cultures of normal bacteria. This method was significant for rapid detection of drug resistance, automatization and intelligentization of test of drug sensitivity in medical clinic.

Keywords: Drug resistance; Impedance; Microorganism sensor; Test.

1. Introduction

The phenomena of drug resistance were extensively existed in some nosogenetic bacteria because people have large amount antibiotics and some therapy was not scientific. Large dose medicine were used to treat the diseases, however it's effect was not ideal. These medicines were few or no curative effect in medical clinic. Therefore the antibiotics employed to clinical therapy were very important and it needed to be resolved instantly to the medicinal field [1-2].

The test of drug resistance was traditionally detected by culturing bacteria and observing the colonial morphology and size, but some intrinsic drawbacks were not avoid, for example, long-time bacterial culture, observation of bacteria with naked eyes and the limited number of samples which maybe delay the diagnosis of disease and influence the effect of treatment, even endanger the life [2]. In this paper, test of drug resistance on bacteria was performed to medical clinic with impedance microorganism sensor. During the bacterial culture, the impedance value will lower when the bacteria grows rapidly in the cultural solution. The impedance value of different bacterial cultures was measured using the home-made Microorganism sensor and measurable setup of impedance to indentify the drug resistance of bacteria. It finds a new path for the test of drug resistance in medical clinic.

2. Experiment

2.1. *Materials and Methods*

2.1.1. *Materials*

Staphylococcus aureus, E.Coli, and Salmonellas (College of life science and technology, Southwest University for nationalities, China), Benzylpenicillin Potassium (0.48g/flask) and Gentamycin (0.08g/flask) (Sichuan pharmaceutical factory, China). Cefazolin Sodium (0.5g/flask) (Hayao pharmaceutical factory, China). Home-made Mcrioorganism sensor.

2.1.2 *Treatmeat of bacterial species*

Pure *Staphylococcus aureus, E.Coli, and Salmonellas* were cultured in special medium, such as SM1H, SM2H and SM3H (see table 1). Afterward these bacteria were inoculated in Petri dish and continuously cultured at 37℃.

* This work is supported by etc, etc.

† Work partially supported by grant 2-4570.5 of the Swiss National Science Foundation.

TABLE I the inoculation of bacteria in SMIH medium

bacteria / medium / antibiotics	*E.Coli*	*Salmonellas*	Staphylococcus aureus	Control
Benzylpenicillin Potassium	SM1HDCB	SM1HSMB	SMIHPTB	-
Cefazolin Sodium	SM1HDCC	SM1HSMC	SMIHPTC	-
Gentamycin	SM1HDCG	SM1HSMG	SMIHPTG	-
Inoculation control	SM1HDCI	SM1HSMI	SM1HPTI	-
control	-	-	-	SM1HB

Notes: Only SM1H medium are listed. I=1, 2, 3

2.1.3 *Samples of antibiotics added*

Different antibiotics were added to the cultural solution after 4 h and bacteria were cultured till 20 h. Samples were collected from above cultures every 10 minutes to measure the impedance value (Ω cm) of samples with microorganism sensor.

2.1.4 *Impedance of inoculation control groups via control*

The impedance value of different inoculation control groups (SM1HDCI, SM1HSMI and SM1HPTI) was changed during the culture. Ω cm obviously increases in all control groups via control (SM1HB), and the curve was different, they are related with the growth of bacteria. The curve of SM1HB was basically straight line after culture 6 h, it showed that the bacteria didn't grow.

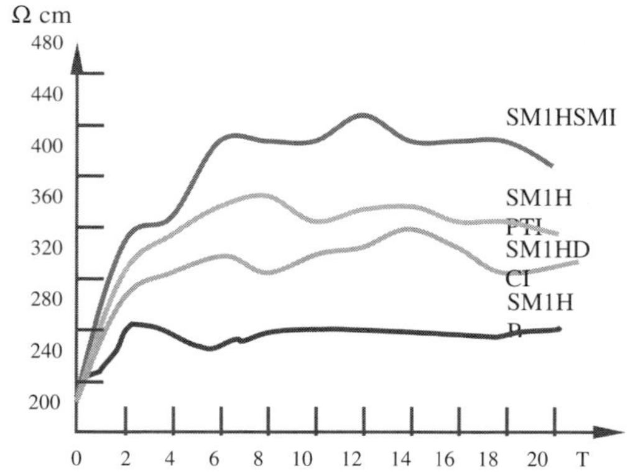

Fig.1 the impedance value of inoculation control groups

The impedance value of different inoculation control groups (SM1HDCI, SM1HSMI and SM1HPTI) was changed during the culture. Ω cm obviously increases in all control groups via control (SM1HB), and the curve was different, they are related with the growth of bacteria. The curve of SM1HB was basically straight line after culture 6 h, it showed that the bacteria didn't grows.

2.1.5 *Impedance of E. coli, Salmonellas and Staphylococcus aureus after adding the different antibiotics*

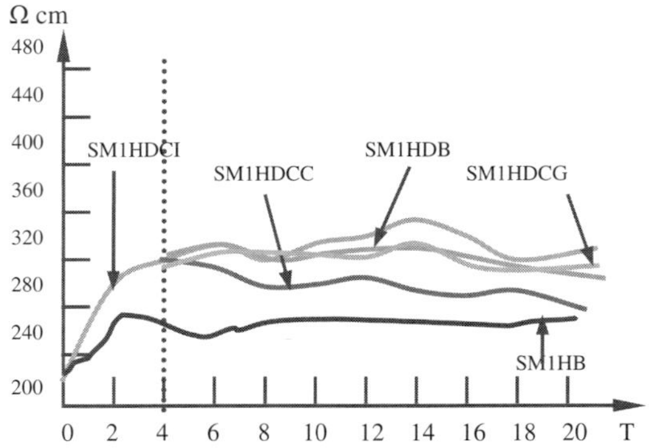

Fig. 2 Test of drug resistance on *E. coli* adding three different antibiotics

Fig. 2 showed that Ω cm had changed when three different antibiotics was added in cultural medium, they all were lower than SM1HDCI's. Three antibiotics can restrict the growth of E. coli, the effect is not same. Among them, Benzylpenicillin Potassium and Gentamycin were weak to E. coli. Their relationship is Cefazolin Sodium > Benzylpenicillin Potassium ≥ Gentamycin.

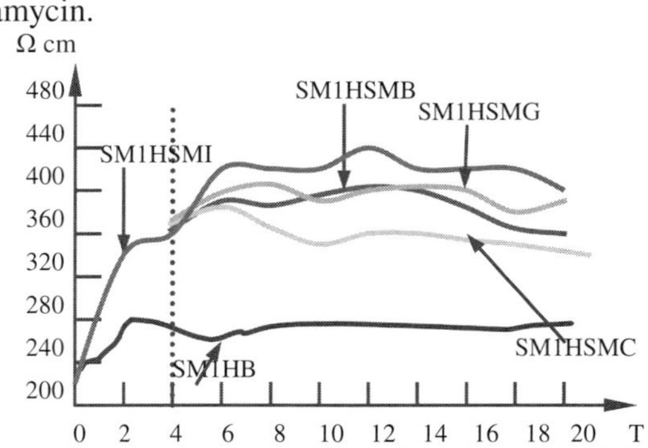

Fig. 3 Test of drug resistance on *Salmonellas* adding three different antibiotics

Fig. 3 showed that Ω cm had changed when three different antibiotics was added in cultural medium. Three antibiotics can restrict the growth of *E. coli*, the effect of Benzylpenicillin Potassium and Gentamycin were partially same to *Salmonellas*. Their relationship is Cefazolin Sodium > Gentamycin ≥ Benzylpenicillin Potassium.

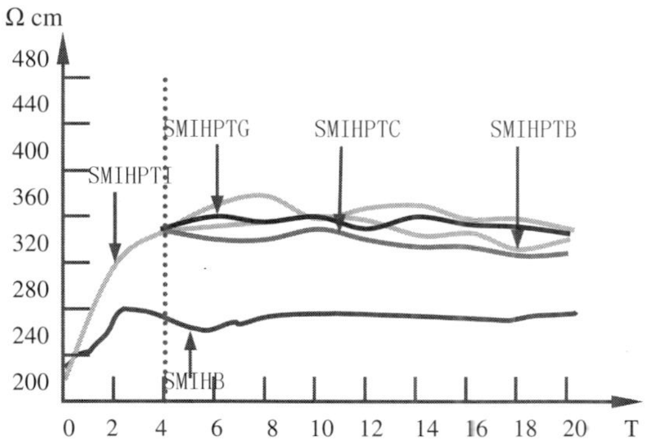

Fig. 4 Test of drug resistance on *Staphylococcus aureus* adding three different antibiotics

Fig. 4 showed that Ω cm had changed when three different antibiotics was added in cultural medium. Three antibiotics can restrict the growth of *Staphylococcus aureus*, the effect of Benzylpenicillin Potassium and Gentamycin were similar to *Staphylococcus aureus*. Their relationship is Cefazolin Sodium > Benzylpenicillin Potassium ≥ Gentamycin.

According to the test, Cefazolin Sodium have good drug resistance to three different bacteria. It shows the method is feasible and reliable for medical clinic, and some diseases can be rapidly diagnosed, it is significant for rapid detection of drug resistance, automatization and intelligentization of test of drug sensitivity in medical clinic. It is a new path for the test of drug resistance for future medical measurement

References

1. Leanne B. Gasink MD, MSCE, Theoklis E. Zaoutis MD, MSCE, and Warren B. Bilker PhD *et al.* The categorization of prior antibiotic use: Impact on the identification of risk factors for drug resistance in case control studies. American Journal of Infection Control, **35(10)**, pp638-642, 2007.
2. Aun Turkmani, Anna Psaroulaki, Athanasia Christidou, George Samoilis et al. Uptake of ciprofloxacin and ofloxacin by Brucella strains and their fluoroquinolone-resistant variants under different conditions. An in vitro study. Diagnostic Microbiology and Infectious Disease, **59**, pp447-451, 2007.
3. Turkmani A, Ioannidis A, Christidou A, Psaroulaki A, Loukaides F, Tselentis Y. In vitro susceptibilities of Brucella melitensis isolates to eleven antibiotics. Ann Clin Microbiol Antimicrob, **5(1)**, pp24-27, 2006.
4. Hongzhi Wang and Aihua Wang. Testing Microorganism with Resistance Measurement. Journal of Southwest Nationalities College (Natural Science Edition). **23(4)**, pp416-417, 1997.

EXPERIMENTAL CORRELATION OF OCT SPECKLE CONTRAST RATIO WITH THE DETECTABLE DEPTH AND SCATTERING COEFFICIENT OF SAMPLE

ZHIFANG LI, HUI LI *, YOUWU HE, SHOUDONG CAI

Key Laboratory of Optoelectronic Science and Technology for Medicine, Ministry of Education

Fujian Provincial Key Laboratory of Photonic Technology

School of Physics and OptoElectronics Technology, Fujian Normal University

Fuzhou, Fujian 350007, China

**Corresponding author: hli@fjnu.edu.cn*

We applied a blind deconvolution technique to restore OCT speckle image under the condition of optical coherence tomography (OCT) speckle contrast ratios at the different depth achieving the maximum value. We experimentally found that the speckle contrast ratio CR is of linear function for the detectable depth z of biotissue, that is $CR = \alpha z + \beta$. And the proportionality coefficient α depends linearly on scattering coefficient, which suggests that the proportionality coefficient α is a potentially feasible means of discriminating the different optical properties among biotissues. So it was firstly found that the OCT speckle contrast ratio CR is proportional to the detectable depth z and scattering coefficient μ_s of biotissue.

1. Introduction

In the turbid media, OCT [1] signals are formed by the coherent addition of a light backscattered from tissue and a references. The measurement technique, based on interferometry, not only brings out a high spatial resolution, about 10um in coaxial direction, but also gives rise to the speckle. The types of speckle can be classified into the signal-carrying speckle and the signal-degrading speckle [2]. Considerable attention was devoted to methods for the speckle reduction, improving the signal-to-noise ratio [2-6]. Speckle-reduction techniques fall into four main categories: polarization diversity, spatial compounding, frequency compounding, and digital signal processing. A less common perspective in OCT has been the exploitation of speckle as a source of useful information, although it is well known to be so in other areas of optical metrology. To our knowledge, there are two methods involved to attempt to correlate speckle with the optical properties of samples. One, advocated by T.R. Hilman[7], was based on fully developed speckle pattern, which showed that OCT speckle contrast ratio is dependent on the local density of scattering particles in a sample. According to Mie theory, the approach could establish the relationship between speckle contrast ratio and scattering coefficient. The other was based on texture analysis on the OCT images, which found that texture analysis features may be capable of differentiating tissue phantoms containing various sizes and the distributions of scatters [8-9].

However, both methods have some of deficiency that they could not explain that speckle contrast ratio, the ratio of OCT intensity's standard deviation to its mean at the same depth, increases with the detectable depth, as Fig.2. The main objective of this paper is to investigate the phenomenon and its application. The next section applied blind deconvolution technique to restore the OCT speckle image. Then, we experimentally establish a correlation between OCT speckle contrast ratio and the sample's detectable depth and scattering coefficient μ_s , using Intralipid solution of various concentrations.

2. OCT Image Preprocessing

Material dispersion [10], the lens in the system [11] and the scanning widen the point-spread-function (PSF) of optical coherence tomography system. And OCT image will appear blurred. But we cannot obtain prior knowledge of the blurring point-spread-function. However, the blind deconvolution techniques can be used effectively when no accurate information about the blurring PSF is known. The blind deconvolution techniques are the constrained iterative restoration algorithm, and could restore the OCT speckle image and the blurring PSF simultaneously. Constraint we choose is those OCT speckle contrast ratios achieve the maximum value. In the real application, we need to rerun blind deconvolution technique, experimenting with PSFs of different sizes, until we obtain the satisfactory result.

3. Experimental Setup and Material

3.1. *Experimental Setup*

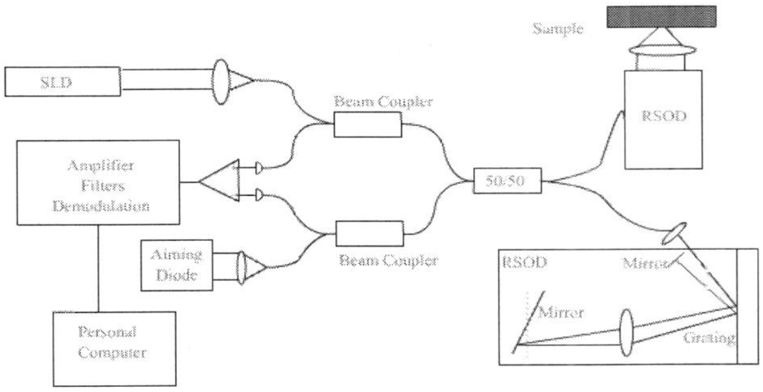

Figure 1 Schematic diagram of OCT system

Figure 1 depicts the schematic diagram of the experiment setup. The OCT system used for this study employs a broadband source, superluminescent SLD-HP-371 diode with a centre wavelength of 842.5nm and the spectrum width of 49nm, and the axial resolution in the air is $\sim 10um$, correspondingly. In both arms of Michelson interferometer, sample and reference, light is guided through single-mode fibers. In the reference arm, fast scanning is achieved with use of a Fourier domain rapid scanning optical delay line. An aiming beam of a CW light-emitting diode with the wavelength of 632.8nm is used to track the scanning position of sample arm. The operation of OCT scanning is automated and controlled using PC. Light scattered from the sample and light reflected from the reference arm mirror form an interferogram, which is detected by a photodiode. The signal from photodiode is preamplified, electronically filtered around the Doppler frequency of the OCT (42 KHz). Also, light from sample arms is focused by a 90mm focal distance lens, the diameter of the focal waist of light beam is roughly 17um, and the lateral resolution is $\sim 17um$, correspondingly. The OCT system was used to capture the interferometric information of depth with $2.6mm(\text{axial}) \times 2.0mm(lateral)$ resulting in images with 256 axial pixels and 160 lateral pixels. In all images, the light was focused on the surface of the scattering medium.

3.2. *Sample*

The Intralipid solution is a homogeneous, turbid suspension, and thus it can also be used as a tissue-simulating phantom [12]. In the research, 2%-10% Intralipid solutions were used to study the depth-induced and scattering-induced changes in the speckle contrast ratio. The samples were diluted from the 10% Intralipid solution (Sino-Swed Pharmaceutical Corp. LTD).

4. Results

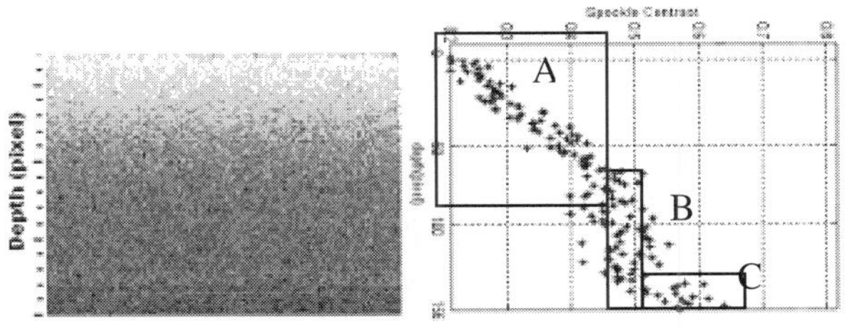

(a) (b)

Figure 2 (a) OCT image of Intralipid-10% solution
(b) Contrast ratio of speckle vs. depth

Figure 2 (a) shows a deblurred OCT image of a homogeneous Intralipid-10% solution, and the OCT image is of a nonhomogeneous speckle restored by using the blind deconvolution technique from the OCT image. Figure 2 (b) shows a plot of the speckle contrast ratio versus the depth, correspondingly. Figure 2 (b) can be divided into three parts. In the part A, the contrast ratio increases with the increase in the probing depths. Then the contrast ratio approaches to 0.5 in part B, independent of the detectable depth. Contrast ratio at part C location resulted from the system noise because of no scattered light coherent with the reference beam. Characterizations of contrast ratio in the part A maybe contain the optical properties and the detectable depth of sample.

Figure 3 shows a plot of the speckle contrast ratio versus the axial position in the OCT image of Intralipid-10% solution. The part A is region of interest (ROI, black frame inset in Figure 3). Figure 3 also shows that a good agreement between the fitting curve based on equation (1) and the experimental data points.

$$CR = \alpha z + \beta \qquad (1)$$

where CR denotes contrast ratio, and z is detectable depth, α represents proportionality coefficient, and β is constant.

Figure 4 shows a plot of proportionality coefficient α versus the concentration C of Intralipid solution. The fitting curve by $\alpha \propto C$ is consistent with the experimental data. The scattering coefficient μ_s can be described in terms of the concentration C and the

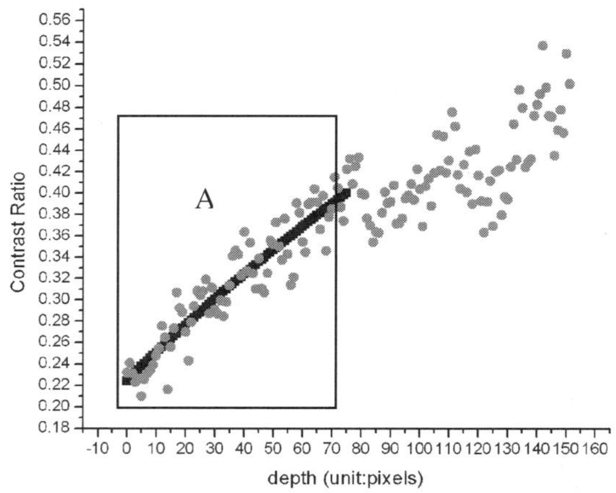

Figure 3 Contrast ratio of OCT speckle versus detectable depth in the Intralipid-10% solution

scattering cross section σ_s such that $\mu_s = C\sigma_s$. So we can obtain $\alpha \propto \mu_s$

According to Mie theory, the scattering coefficient μ_s of 2%-10% concentrations of Intralipid solution is between $40cm^{-1}$ and $205cm^{-1}$ [12]. This range includes many typical scattering coefficients of tissue types. Since a biotissue is a strong scattering media, $\mu_s \gg \mu_a$, so $\mu_t \approx \mu_s$. Therefore, the proportionality coefficient α is a potentially feasible means of discriminating among different samples.

It is not difficult to obtain a conclusion that the OCT speckle contrast ratio CR is proportional to the detectable depth z and scattering coefficient μ_s.

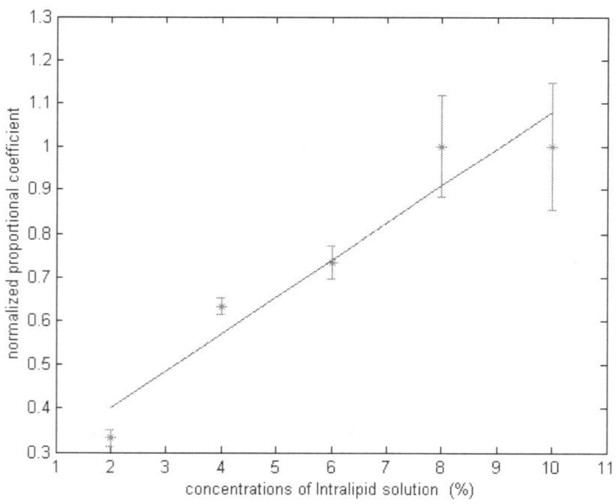

Figure 4 The proportionality coefficient α versus concentrations of Intralipid solution

5. Conclusion

We experimentally found that the speckle contrast ratio CR is of linear function for the detectable depth z and the scattering coefficient μ_s of biotissue, that is $CR \propto \mu_s z + \beta$. The empirical formulation of the contrast ratio CR can be expressed as $CR = \alpha z + \beta$, and the proportionality coefficient α depends linearly on the scattering coefficient. Therefore, the proportionality coefficient α is a potentially feasible means of discriminating the different optical properties among biotissues.

Acknowledgments

This project was sponsored by in part by the National Natural Science Foundation of China under Grant No. 60078024/60578056, and in part by the Program for New Century Excellent Talents in University under Grant No.NCET-04-0615 by the Ministry of Education of China.

References

1. D.Huang, E.A. Swanson, C.P. Lin, J.S. Schuman, W.G. Stinson, W.Chang, M.R. Hee, T. Flotte, K. Gregory, C.A. Puliafito, and J.G. Fujimoto, *Science* 254, 1178 (1991).
2. J.M. Schmitt, S.H. Xiang, and K.M. Yung, *J. Biomed. Opt.* 4, 95 (1999).
3. J.M. Schmitt, *Phys. Med. Biol.* 42, 1427 (1997).
4. N.Iftimia, B.E. Bouma, G.J. Tearney, *J. Biomed. Opt.* 8, 260 (2003).
5. D.C. Adler, T.H. Ko, and J.G. Fujimoto, *Opt. Lett.* 29, 2878 (2004).
6. J. Kim, D. T.Miller, E.Kim, S.Oh, J. Oh, T.E. Milner, *J. Biomed. Opt.* 10, 064034-1 (2005).
7. T.R. Hillman, S.G. Adie, V.Seemann, J.J. Armstrong, S.L. Jacques, D.D. Sampson, *Opt. Lett.* 31, 190 (2006).
8. K.W. Gossage, T.S. Tkaczyk, J.J. Rodriguez, J.K. Barton, *J. Biomed. Opt.* 8, 570 (2003).
9. K.W. Gpssage. C.M. Smith, E.M. Kanter, L.P. Hariri, A.L. Stone, J.J. Rodriguez, S.K. Williams, J.K. Barton, *J. Biomed. Opt.* 51, 1563 (2006).
10. D.L. Marks, A.L. Oldenburg, J.J. Reynolds, and S.A. Boppart, *Appl. Opt.* 42, 204 (2003).
11. T.S. Ralston, D.L. Marks, F.Kamalabadi, S.A. Boppart, *IEEE Trans. on Image Processing* 14, 1254 (2005).
12. Hugo J. van Staveren, Christian J. M. Moes, Jan van Marle, Scott A. Prahl, Martin J. C. van Gemert, *Appl. Opt.* 30, 4507 (1991).

EXPERIMENT AND CLINICAL APPLICATION OF KTO GREEN LASER VAPORIZATION OF BIOLOGIC TISSUE UNDER WATER

HE YANYAN, LI ZHENGJIA, HUANG CHUYUN, YAO YUNCHENG

School of Optoelectronics Science and Engineering, Huazhong University of Science and Technology Laboratory Wuhan 430074, China

Potassium-titanyl-phosphate (KTP) green (frequency double Nd:YAG) laser vaporizationwas a safe and effective procedure in urology. This study demonstrated the potential and feasibility of the domestically manufactured green laser to treat patients with Benign Prostate Hyperplasia (BPH). The porcine muscles were vaporized with different power (7.5-57W) under water by the domestically manufactured KTP laser system. The fresh canine prostate tissues were also vaporized with 43W green laser. All biology tissues were weighed, measured, serially sectioned, and whole mounted for histologic analysis. The laser vaporization in human prostate tissue was performed clinically. The porcine muscles were vaporized under water for ten seconds. Increasing laser power appeared to increase depth and diameter of vaporization while maintaining or decreasing depth of necrosis. With domestically manufactured 60W KTP laser system, the laser prostatectomy was performed in 15 BPHs. All patients received spinal anesthesia. Isotonic saline at room temperature was used as irrigation solution. Postoperative insertion of an indwelling catheter was mandatory. The initial study of the domestically manufactured KTP laser vaporization in animal tissue and human prostate tissue confirms its technical feasibility in most cases and good intraoperative and early postoperative effect.

1. Introduction

Benign Prostate Hyperplasia (BPH) is a condition wherein continued growth of the prostate restricts the passage of urine through the lower portion of the bladder and the urethra [1]. With the improvement of our living standards and the rapid increase of aging population, BPH is seen with an increasing frequency in elderly men [2, 3]. During the last decade, there has been an eruption of research and development in an attempt to discover less-invasive alternative to standard transurethral resection of the prostate (TURP) for treatment of obstructive BPH. Despite its prevalence and its gold standard appeal, TURP is costly and associated with some morbidity [4]. Over the years, various laser therapies have been investigated as alternatives to TURP. The best procedure for removal of excess prostate tissue is transurethral KTP laser photoselective vaporization of the prostate [5]. Its benefit is being able to successfully discharge patients on the day of surgery without a catheter and blood transfusion [6]. Initial evaluation of pure KTP laser vaporization was using a 38W system on the canine prostate model [7] and then reported using the 60 W high-power system [8]. The short-term and long-term results of 80 W KTP laser prostatectomy have also been reported recently [9-10]. In above reports, all the KTP laser systems are from Laserscope Co.

This study sought to evaluate for functional, pathologic, and histologic effects of the domestically manufactured KTP laser vaporization of the porcine muscle and canine prostate. The effects of laser power upon vaporization size and the depth of necrosis were studied, the bottleneck phenomenon of porcine muscles underwent green laser vaporization under water were discussed. In addition, the domestically manufactured green laser has been used to treat BHP in clinic, and the clinical results were reported.

2. Material and Methods

All experiments were performed using the domestically manufactured KTP green laser system, which operated 532nm in a quasi-continuous mode at power levels up to 60W from the tip of optics fiber. The laser was pumped by xenon flash lamp and produce a constant train of Q-switched pulses at 40kHz. During the experiment, anterograde laser vaporization was performed in fresh porcine muscles and canine prostates. Vaporization animal's tissues were carried out with a 600-μm end fiber. The laser fiber was contact with the tissue surface under water. The porcine muscles were vaporized with different power (7.5-57W) under water. The canine prostate was vaporized with fixed 43W power from the outer surface. In the clinical procedure, a 600-μm end fiber was used to vaporize prostate tissue and the laser fiber was contact with the tissue. Normal saline was the irrigant.

At first, the experiments were carried out on 12 pieces of the fresh porcine muscles, weighing about 20g.

The porcine muscles were vaporized with different power under water. Immediately, after vaporization the tissues were removed, transversely bisected, measured, photographed, and fixed in formalin. In order to determine vaporization diameter, slide calipers was used to measure the most, the least and 2 middle diameters of vaporization hole, then the average value was taken as the vaporization diameter. Vaporization depth was the distance from the surface of the tissue to the lowest of the vaporization hole, which was measured with slide calipers on the transverse of the hole. The depth of necrosis was measured from the surface of the vaporization hole to the interface of the different color tissue. In addition, the tissue pathologic was observed through the microscope. The fresh canine prostate tissues were also vaporized with 43W green laser.

3. Results

The various vaporization sizes for different laser power upon fresh porcine muscles are plotted in Figure 1. With the laser power increasing, the vaporization depth and vaporization width increased significantly, but the depth of necrosis declined slightly.

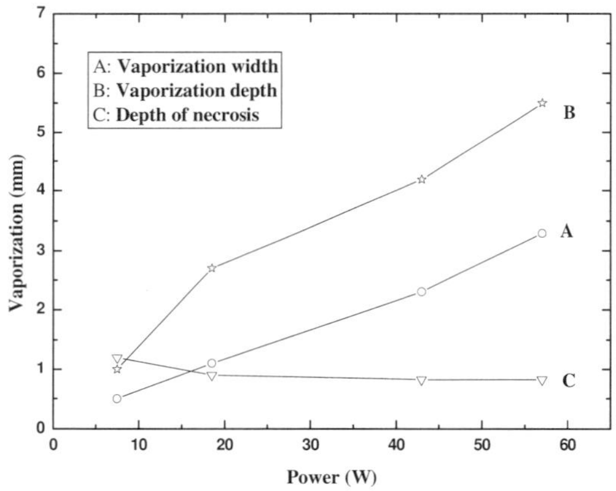

Figure 1. The vaporization sizes under different laser power. All fresh porcine muscles were vaporized under water and the laser parameter of 50Hz pulse frequency, 5ms pulse width, and 10s irradiation time was immovable in these vaporization experiment.

After ten seconds 43W green laser vaporization under water, laser produced a mean transverse cavity diameter of 4.18mm with a about 0.82mm thick rim of coagulation necrosis (Fig. 2a) and the hole depth of 2.30mm on porcine muscles (Fig. 2b). Microscopically, it is obvious that both the vaporization diameter and depth are far bigger than the necrosis thickness.

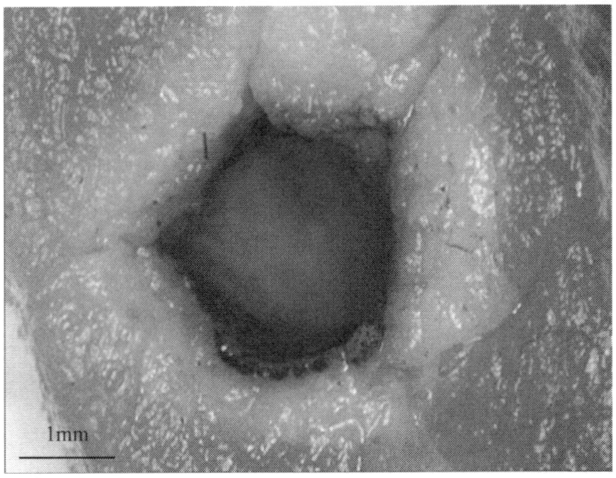

(a) Surface of fresh porcine muscle after green laser vaporization.

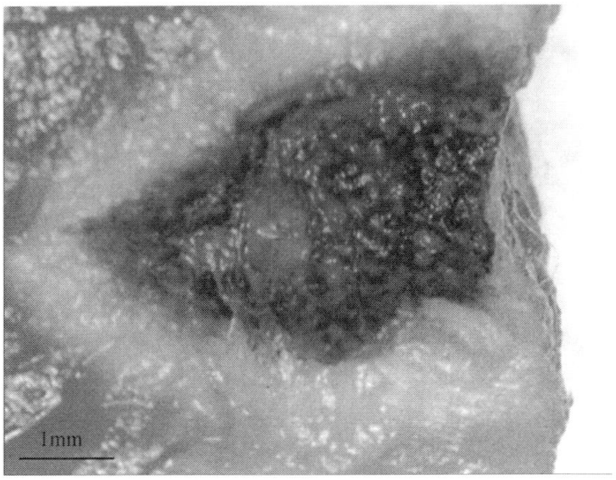

(b) Transverse section of fresh porcine muscle after green laser vaporization.

Figure 2. Photograph of fresh porcine muscle after 10s potassium-titanyl-phosphate laser vaporization with the laser parameters of 43W power, 50Hz pulse frequency and 5ms pulse width.

43W green laser vaporization after ten seconds under water produced a mean transverse cavity diameter of 2.05cm with a about 1.04mm thick rim of coagulation necrosis on canine prostate, as shown in Fig. 3a. The vaporization diameter and depth are also much larger than the necrosis thickness. This was similar to the results on porcine muscle. The typical morphology of KTP green laser vaporization tissue under water was available, as shown in Fig. 3b.

With domestically manufactured 60W KTP laser system, the laser prostatectomy was performed in 15 BHPs. All patients received spinal anesthesia. Isotonic saline at room temperature was used as irrigation solution. Postoperative insertion of an indwelling catheter was mandatory. Detailed patient characteristics are as follows.

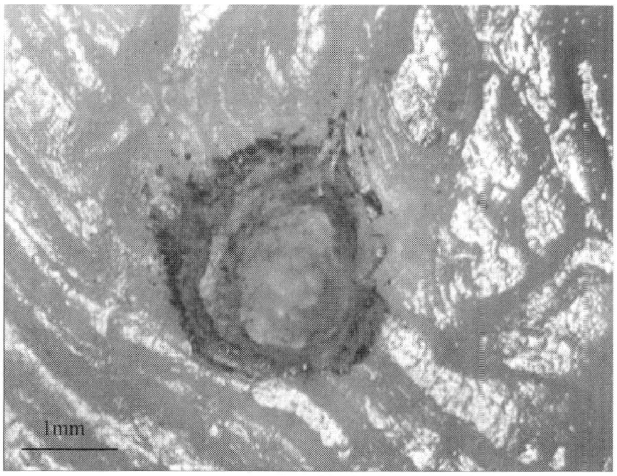

(a) Photograph of fresh canine prostate after green laser vaporization

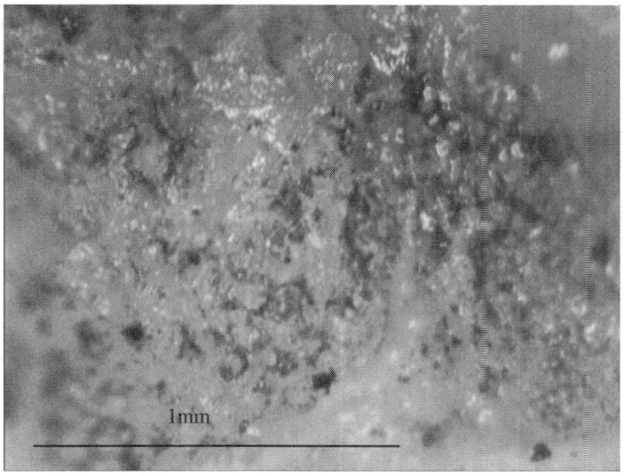

(b) Local magnification of A zone

Figure 3. Photographs of fresh canine prostates after 10s 43W potassium-titanyl-phosphate laser vaporization.

The mean age of patients who were suffering from BHP was 71.6±7.6 years (62-92), average prostate volume was 51.2±27.2 ml (29-150). Mean operating time was 56.9±29.7 min (30-145). No transfusion was required. The mean duration of indwelling catheterization was 2.4±0.9 day (2-4) and that of hospitalization was 3.2±1.2 days (3-7). Q_{max} (Maximum flow Rate) values in preoperative period and postoperative days 15, 30 were 6.8±3.2ml/sec(2-11), 13.6±2.6ml/sec(9-19), 18.8±2.2ml/sec(12-25).

4. Discussion

The vaporization of fresh porcine muscle showed that the higher the green laser power was, the better the effect of vaporization and the less the depth of necrosis the 532nm laser produced under water (Figure 1). The higher power was more efficient in inducing "char" formation. Although char formation is not essential to efficient rapid ablation, it was helpful because the darkened tissue was better at absorbing light. As a result, more efficient rapid ablation was brought out. When the power increasing, more and more laser energy was consumed by vaporization of tissue, the amount of laser energy leading to residual tissue coagulation gets smaller. While the power was too low, it cost too long time to absorb enough to induce char and it still absorb enough energy to vaporize. In this course, the energy absorbed by tissue penetrated around at large, so the large area necrosis was produced [9]. Increasing the volumetric power density absorbed in the tissue to be ablated has the result of decreasing the extent of residual injury of the surrounding tissue. So the use of higher power green laser systems, with greater levels of irradiance at the treatment area on the tissue can achieve the lower levels of adverse side effects and a quicker operation times [11].

The 43W 532nm Laser vaporization porcine muscles also produced minimal tissue necrosis and mainly tissue vaporized. A 4.18mm diameter and 2.30 mm depth of vaporization hole, with a 0.82mm thick zone of necrosis confined to the vaporized hole was observed microscopically after 43W green laser irradiation (Figure 2). The transverse section form of vaporization hole looked like a bottleneck (Figure 2(b)): the surface diameter of the hole was smaller than the diameter adjacent to hole's surface. Because the surface tissue was cooled by water, the efficiency of the surface tissue vaporization reduced [12]. The speed of the surface tissue vaporization was slower than that of the interior tissue and the diameter of the surface was smaller than that of the interior.

The canine prostate more strongly resemble human's than others, so the green laser vaporization canine prostates were studied under water. These vaporization results was the same as the porcine muscle except that the necrosis of canine prostate was little bigger than that of the porcine muscle (Figure 3). When the green laser vaporized canine prostate, the laser energy was absorbed first by a membrane covered at the surface of prostate. The membrane was white and its green light absorption coefficient was smaller than that of the red tissue, so the speed of vaporization membrane became slower, resulting in the bigger necrosis. The human prostate was also covered with a membrane, so the higher power green laser was selected to operate in order to increase speed of vaporization prostatectomy and reduce necrosis.

At present, the most commonly employed procedure for removal of excess prostate tissue has been transurethral resection of the prostate, also known as TURP. While effective, the TURP procedure was known to cause numerous side effects, including incontinence, impotence, retrograde ejaculation, prolonged bleeding [13] and TURP syndrome.

Laser energy has become increasingly popular as a urological tool over 15 years for both disease and soft-tissue applications [14]. Various lasers have been introduced as alternative to TURP, mainly the neodymium: yttrium aluminum garnet (Nd:YAG), the holmium(Ho):YAG, and the frequency-doubled Nd:YAG (potassium-titanyl-phosphate) KTP laser [15, 16].

The KTP lasers at a wavelength of 532nm is in the visible range of the electromagnetic spectrum, green in color, and produced by frequency-doubling of the 1064nm Nd: YAG laser. Green light is strongly absorbed by red tissue, as well as by oxy-hemoglobin, but almost not at all by the water. Thus, any blood-rich tissue or organ, such as prostate gland, will be an excellent target under water. Because of minimal tissue penetration (0.8mm), there was little heat dissipation from the treated site. These results in that laser energy are focused and more efficient for vaporization in prostate tissue.

Earlier KTP laser devices had a relatively low power output, it exist some disadvantages, such as long time and slow speed during the procedure. Therefore, laser power increased continuously from 38W, 60W, 80W and the recent 120W. More and more clinical results were reported [5-10].

The initial study of the domestically manufactured KTP laser vaporization BPH also confirms its technical feasibility and good intraoperative and early postoperative effect. In our clinic experiments, the end fiber, not side fiber was used, and the fiber tip was contacted with the tissue. The end fiber is cheap and easy available, although its learn curve is longer than that in side fiber.

5. Conclusions

In this paper, green laser vaporization the animal tissues and its clinic applications were studied. The animal tissue vaporization experiments showed that with the laser power increasing, the vaporization depth and vaporization width increased significantly, but the depth of necrosis declined slightly. Our clinical results indicated that the domestically manufactured green laser vaporization is a treatment method that can be adopted to treat patients with BPH, strait urethra, bladder tumor etc urology diseases, and be safe and effective, especially can be used in patients at high risk of anesthesia. Furthermore, the differences in stromal content between human and canine prostate has little impact on KTP laser vaporization. The initial study of the domestically manufactured KTP laser vaporization in animal tissue and human urologic tissue confirms its technical feasibility in most cases and good intraoperative and early postoperative effect.

Acknowledgments

This study was supported by Scientific Research Foundation of Huazhong University of Science and Technology under grant number HUST-SRF No. J2006M029B.

References

1. R. S. Kuntzmann, R. S. Malek and D. M. Barrett, *Mayo Clin. Proc.* **73**, 798 (1998).
2. W. M. Garraway, N. Collins and R. J. Lee, *Lancet.* **338**, 469 (1991).
3. R. J. Glynn, E. W. Campion, G. R. Bouchard and J. E. Silbert, *Am J. Epidemiol.* **121**, 78 (1985).
4. R. Tooher, P. Sutherland, A. Costello, P. Gilling, G. Rees and G. Maddern, *J. Urol.* **171**, 1773 (2004).
5. R. S. Malek, *In: Baba S, Ono Y, editors. Recent Advances in Endourology 8: Interventional Management of Urological Diseases. Tokyo: Springer-Verlag.* **103** (2006).
6. H. T. Andrew and J. G. Peter, *Curr. Opin. Urol.* **15**, 55 (2005).
7. R. S. Kuntzmann, R. S. Malek, D. M. Barrett and D. G. Bostwick, *Urology.* **48**, 575 (1996).
8. R. S. Kuntzmann, R. S. Malek, D. M. Barrett and D. G. Bostwick, *Urology.* **49**, 703 (1997).

9. A. E. Te, T. R. Malloy, S B. S. tein, J.C. Ulchaker, U.O. Nseyo, and M. A. Hai, *BJU. Intl.* **97**, 1229 (2006).
10. R. S. Malek, R. S. Kuntzman and D. M. Barrett, *J. Urol.* **174**, 1344 (2005).
11. S. A. Davenport, S. C. Murray, T. D. Coleman, H. Garlich, K. Arnold and K. Nahen, *US2005.* 0027286 A1 (2005).
12. H. J. Li, J. Liu and X. X. Zhang, *Space Med. Eng.* **15**, 103 (2002).
13. AUA Practice Guidelines Committee, *J. Urol.* **170**, 530 (2003).
14. L. C. Wilson and P. J. Gilling, *Nat. Clin. Pract. Urol.* **2**, 443 (2005).
15. R. M. Kuntz, *Eur. Urol.* 40, 961 (2006).
16. Y. S. Bao, *Int. Fascicule Biomed. Eng.* **17**, 328 (1994).

STUDY ON NONINVASIVE MONITORING OF TRAUMATIC BRAIN EDEMA *IN VIVO* BY NEAR-INFRARED SPECTROSCOPY

HE LIANG†, YANG TIAN MING

Department of Neurosurgery, Zhongda Hospital, Southeast University
No.87 Dingjia Bridge Street, Nanjing 210009, P.R.China

QIAN ZHIYU, GUO LINA, MAO WENLAN

Department of Biomedical Engineering, Nanjing University of Aeronautics and Astronautics
No.29 Yudao Street, Nanjing 210016, P.R.China

To explore a new method of noninvasive monitoring the development of traumatic brain edema in vivo, acute regional brain trauma was applied according to Feeney's apparatus. Changes of the water content in brain cortex were monitored by near-infrared spectroscopy (NIRS) technology and measured by the dry/wet weight method. Distinct edema in injured areas was observed at 6 hours after brain trauma, which reached a summit of severity at 72 hours later, then gradually decreased. Changes in the contralateral side were the same as those in injured areas, but to a lesser degree. Both the optical parameter of brain tissues (reduced scattering coefficient, μs'), monitored by NIRS, and the brain water content could reflect the changes in brain cortex after trauma. The μs' was a good indicator for monitoring traumatic brain edema. NIRS could monitor the development of traumatic brain edema in vivo.

Keyword: Near-infrared spectrometer (NIRS), traumatic brain edema, Noninvasive monitoring

1. Introduction

Traumatic brain edema, following the brain injury, is defined as the increasing of liquid content in brain substance. As the main change after trauma, brain edema causes brain volume expanding, which will lead to a series of complications, such as cerebral hernia, function failure of central respiratory and even death, if unable to be diagnosed and treated timely. In animal experiments, traditional means of evaluating brain edema are the dry/wet weight method and the special proportion calculation method [1-2], however, both can not be used for real-time monitoring the development of traumatic brain edema in vivo. In clinic, current methods of monitoring cerebral edema are divided into two categories, noninvasive and invasive. The former, CT and MRI included, costs a lot that aggravate the burden of patients and is unable to monitor the dynamic changes of brain edema continuously. While the latter is mainly about the intracranial pressure (ICP) monitoring technology, for invasive and indirect, has a risk of triggering cerebral bleeding, intracranial infections and other serious complications. We used the near-infrared spectroscopy (NIRS) to monitor the changes of optical parameter (reduced scattering coefficient, μs') of brain tissues before and after brain trauma, which may support a new method of monitoring the development of traumatic brain edema in vivo, real-time and noninvasive.

2. Materials and Methods

2.1. *Experimental Equipment and Materials*

In recent studies, we have developed a needle-like, near-infrared (NIR) probe with a small source-detector separation to identify cortex matter structures in the living animal [3-4]. The NIR probe, driven by an automatic step system, had an outer diameter of 0.6mm and contained two 0.1mm fibers to deliver light to and collect light from the brain tissue. The Jiangwan stereotactic instrument (type I) was used to fix the rats during the experiments. A computer with corresponding software was used to read the data.

† Work supported by grant 30671997 of National Nature Science Foundation of China.
Corresponding author. Tianming Yang Email: TMyangcn@yahoo.com

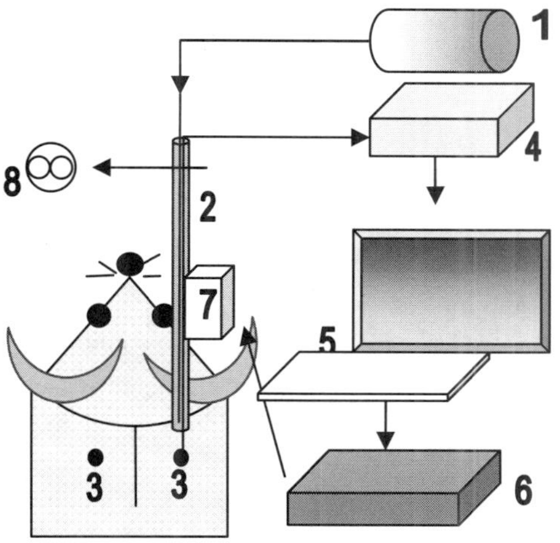

Fig.1. 1. a halogen light source; 2.the NIR probe; 3. bone windows in rats' head; 4. a USB optical spectrometer; 5. the computer and software; 6. a MID-7604 Stepper motor drive; 7. a PCI-7344 controller and stepper motor; 8. the section of the NIR probe

Fig 1 The structure of the NIR system

2.2. *Experimental Animals and Packet*

Sixty SD rats, weighting 250 ± 20g, half male and half female (supported by the Medical Animal Center of Southeast University, Nanjing, P.R.China), were divided into a control group and a traumatic group randomly. Each group was divided into five subgroups for study at different observation points after trauma: 1h, 6h, 24h, 72h and 120h with six in each.

2.3. *Animal Model Induction*

All rats were injected with 1% sodium pentobarbital (30mg/kg), then fixed in the stereotactic instrument, sheared the head skin and incised the median sagittal scalp to expose the skull. A hole, a diameter of 2mm, was drilled on each side of the skull (2mm besides the midline and 3mm after the coronary suture) with dental burr, maintaining the dura mater integrity. Rats in traumatic group were hit in preset windows on the top of left epidural by Feeney's free-falling body against devices [5] (30g in weight and 20cm in height, producing an impact of 600g • cm), causing a focal cerebral contusion in the left parietal lobe. It is apnea for a few seconds that indicate the success of inducing a brain injury model. Then the bone windows were closed with biological cement and rats were returned back to the cages when they were out of anesthesia. Rats in the control group were only drilled a bone window at the same position, but not injured. Rats in each subgroup were monitored by NIRS both in the injured side and the contralateral side at different time points after trauma: 1h, 6h, 24h, 72h and 120h for every 10min (control group were in the ipsilateral side and the contralateral side). At the end of each monitoring, all rats were killed immediately.

2.4. *Near-infrared Spectroscopy (NIRS) Monitoring*

After anesthesia gradually, all rats were fixed in Jiangwan stereotaxic instrument (type I) and exposed their skulls. NIRS built-in dual-fiber micro-probe, which had calibrated in 20% original emulsion calibration before, was installed on stereotaxic instrument [6]. Driven by an automatic step system, the probe was inserted from the surface of the skull downward by 0.7mm that had just arrived to the surface of dura mater. The optical parameter (μs') of the cortex was monitored by NIRS (2Hz in frequency, 10min) and read out in the compute.

2.5. *Measurement of Brain Water Content*

At the end of each monitoring, all rats' brains were taken out immediately. Cortexes tissues, about 1 mm^3 in volume, around the injury focus and the corresponding position in the contralateral side, were cut off and put on the electronic analytical balance to measure the wet weight (WW), then were put in oven at 100°C for 72h until the weight was constant (a difference of less than 0.2mg in twice weighting) to measure the dry weight (DW). Calculate the brain water content according to the formula: $[(WW - DW) / WW] \times 100\%$.

2.6. *Statistical Analysis*

All data were presented as mean ± S.D. of sixty rats. Significance of difference between data from two different groups was analyzed by ANOVA. Different indicators of monitoring traumatic brain edema, reduced scattering coefficient (μs') and brain water content, were analyzed by linear regression. A value of $P < 0.05$ was considered statistically significant. All statistical analysis was performed using SPSS statistical software package (SPSS 13.0 for Windows).

3. Results

3.1. *The Brain Water Content*

Table 1. shows the changes of brain water content in different groups. Compared with the control group, brain water content in traumatic group around the focus had increased at 1h after brain injury, but no significant differences were observed statistically. 6h later, brain water content was significantly higher than the control group (79.25±0.56%) versus (77.49±0.41%) (p <0.01). Then it increased rapidly and reached the peak of edema (82.43±0.81%) at 72h after brain injury (p <0.01). Afterwards brain water content in the injured side decreased slightly and arrived to a level (79.42±0.51%), still higher than normal, at 120h.

In the contralateral side, brain water content increased more smoothly and slowly than that in the injured side. There were also significant differences between the traumatic group and the control group at 24 ~ 72h after brain injury (p <0.01), however, the peak (79.20±0.66%) was far lower than that in the injured side (82.43±0.81%). 120h later, brain water content in the contralateral side returned back to the normal level basically(77.92±0.57%).

Significant differences were also observed between the injured side and the contralateral side in the traumatic group at 6h after brain injury (p <0.05). Then the margin was getting bigger and bigger with time and reached the maximum (3.62%) at 24h.

Table 1 The brain water content in different groups (mean ± S.D., n = 6)

		Brain water content(%)				
		1h	6h	24h	72h	120h
Control	Ipsilateral side	77.5±0.37	77.49±0.41	77.56±0.47	77.39±0.53	77.46±0.40
	Contralateral side	77.60±0.32	77.62±0.39	77.43±0.36	77.47±0.41	77.52±0.36
Traumatic	Injured side	78.08±0.53	79.25±0.56^{**f}	82.10±0.59^{**ff}	82.43±0.81^{**ff}	79.42±0.51^{**ff}
	Contralateral side	77.71±0.49	78.10±0.58	78.48±0.43**	79.20±0.66**	77.92±0.57

$^{*}p$ <0.05, $^{**}p$ <0.01 *vs.* control group.

^{f}p <0.05, ^{ff}p <0.01 *vs.* contralateral side.

3.2. *The Reduced Scattering Coefficient (μs')*

As shown in Table 2, compared with the control group, μs' of the injured side in traumatic group had risen to 10.30±1.01cm^{-1} at 1h after brain injury when significant differences were observed (p <0.05). From then on, the μs' increased continuously with time and reached the summit (18.12±1.92cm^{-1}) at 72h after brain injury (p <0.01). Then the us' decreased gradually, however, significant differences, compared with the control group, were still observed statistically at 120h (p <0.01).

By contrast, μs' of the contralateral side in traumatic group had ascended more smoothly and slowly than that of the injured side at the same time. However, significant differences were also observed between the traumatic group and the control group from 6 ~ 72h after brain injury (p <0.01). The peak of μs' appeared at 72h (11.82±1.44cm^{-1}) when the biggest margin (8.92cm^{-1}) was given as well. After that the μs' started to descend and returned back to the normal level (10.42±1.16) at 120h.

Significant differences were also observed between μs' of the injured side and the contralateral side in traumatic group from 6h ~ 120h after brain injury (p <0.01). The maximal value of margin (6.30cm^{-1}) appeared at 72h.

Table 2 The μs' in different groups (mean ± S.D., n = 6)

		μs'(cm^{-1})				
		1h	6h	24h	72h	120h
Control	Ipsilateral side	9.23±0.42	9.26±0.38	9.25±0.51	9.20±0.39	9.22±0.41
	Contra-lateral side	9.21±0.36	9.19±0.46	9.23±0.37	9.25±0.46	9.24±0.37
Traumatic	Injured side	10.30±1.01*	13.21±1.20^{**ff}	15.10±1.72^{**ff}	18.12±1.92^{**ff}	13.30±1.28^{**ff}
	Contra-lateral side	9.65±0.85	10.49±1.12*	10.78±1.25*	11.82±1.44**	10.42±1.16

$^{*}p$ <0.05, $^{**}p$ <0.01 *vs.* control group.

^{f}p <0.05, ^{ff}p <0.01 *vs.* contralateral side.

3.3 *Linear Regression Analysis*

As shown in Fig.2, two parameters, the average reduced scattering coefficient (μs') of rats' local cerebral cortex during the 10min and the brain water content of this region, were analyzed by linear regression. We received the fundamental equations of functions: μs' = 1.6145 × (brain water content × 100) - 115.75 R = 0.964, F = 326.21, p <0.001, marked linear relationship was observed between the two parameters.

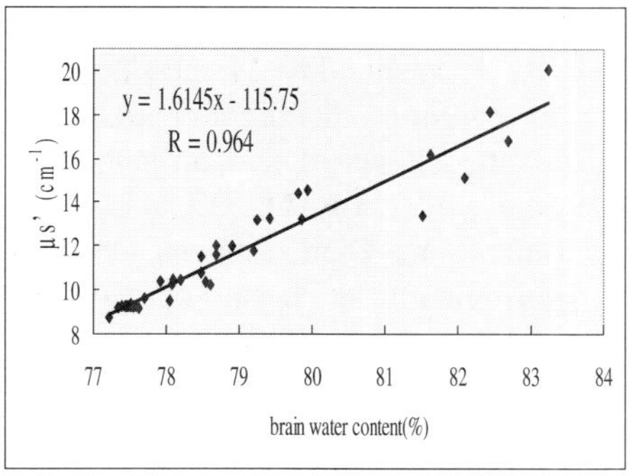

Fig.2. The relationship between reduced scattering coefficient (us') and the brain water content

4. Discussion

Traumatic brain edema, following brain injury usually, could lead to a series of complications, such as cerebral hernia, function failure of central respiratory and even death, which increase the mortality of brain injury remarkably, if diagnosis and treatment could not be given in time. In previous studies on brain trauma, typical means of evaluating the edema following the brain injury are the dry/wet weight method and the special proportion calculation method [1-2]. Both have many disadvantages, such as sacrifice of animals, error in weight measurement and poor time and space resolution, therefore they can not be used for real-time monitoring the development of traumatic brain edema in vivo. There are also two kinds of methods of monitoring cerebral edema in clinic at present, noninvasive and invasive. The former, including CT, MRI and other imaging tools, aggravates the burden of patients for its high price and is not applicable to the bedfast patients. The latter, mainly about the intracranial pressure (ICP) monitoring technology, has a risk of causing a series of complications, such as triggering cerebral bleeding, intracranial infections and so on. In this study, we used a special needle-like, near infrared (NIR) probe to measure the reduced scattering coefficient (μs') of rats' local cerebral cortex for 10min in each side, and observed the changes of values at different time points after brain trauma to prove whether μs' could reflect the development of brain edema.

Near-infrared spectroscopy (NIRS) is a new technique of detecting the structure and dynamic functions of tissues, based on a theory that biological tissues is relatively transparency to the near-infrared light (780 ~ 2526nm). In principle, different biological tissues have different optical properties [7]. Numerous studies have documented that the absorption and scattering properties of brain cortex differ significantly in the near infrared region [8-9]. Candidates for light scattering centers in tissue are the cell membrane and organelle membrane, the nuclei, and other intracellular organelles, including mitochondria [10]. Thus, the reduced scattering coefficient (μs') can be a possible marker or indicator for changes of tissue constituent content. Because of recent advances in NIR technology, uses of NIRS to monitor and understand neuronal functions and physiological changes in vivo have become a more recognized and frequently utilized research approach in the area of neuroscience [11]. The results reported here represent the first attempt to use a NIR technique to monitor the development of traumatic brain edema following the injury in vivo.

In this study, the results showed that traumatic brain edema had occurred in the injured side just at 1 hour after brain injury, then aggravated rapidly and reached the summit at 72 hours. After that, the amount of brain water began to decline gradually, but was still higher than normal level 120 hours later. This process of brain edema changes, in general, was the same as what reported in a previous study [12]. As shown in Table 1 and 2, traumatic brain edema had also happened in the contralateral side, although it could be observed only at 24 ~ 72 hours during the whole experiment and returned back to the normal level smoothly 120 hours later. Compared with the injured side, the degree and range of brain edema in the contralateral side were much smaller, however, the development law of edema was similar. It is possible that edema in the contralateral cortex was caused by fluid leakage, means fluid effusing from the hydropic region around the traumatic focus in the injured cerebral. Because the blood-brain barrier of the contralateral side was relatively integrity and the cerebrospinal fluid refluxed smoothly, this kind of edema was absorbed earlier and faster than the primary focus. In previous studies, Petty, et al, had also found the marked increasing of brain water content in the contralateral side [13], however, Haipeng Liu's group drew an opposite conclusion [14]. In consideration of different impacts given in these studies (600g•cm, 253.3~283.7kPa and 211kPa), it may be the different degrees of brain injury that determine the occurrence of edema in the contralateral side whether or not. Further work must be done to explore the mechanism of this contralateral edema. Compared with the dry/wet weight method, the golden standard of detecting the brain edema, we found that NIRS reflect the dynamic changes of brain edema more sensitive and earlier (96.57% *vs.* 6.51%, 1h *vs.* 6h). The simultaneous brain water content and μs' measurements, as shown in table 1 and 2 and presented in Fig. 2, clearly showed very close agreement between the two parameters. We also received a fundamental equation of functions: $\mu s' = 1.6145 \times (\text{brain water content} \times 100) - 115.75$ with a correlation value of $R = 0.964$ ($p < 0.001$) between the brain water content and μs' measured. We believe that this strong relationship between the two parameters provides validation that μs' was a good indicator for monitoring the dynamics changes of traumatic brain edema. For noninvasive, we are able to measure the μs' of brain tissues as long as we need, as a result, the water content changes within the brain can be monitored by NIRS in vivo.

5. Conclusion

This study demonstrates that NIRS has the ability to monitor the development of traumatic brain edema following injury in vivo. The results suggest that the NIR technique may have a potential for neural application to identify the brain edema before causing a series of severe complications. We believe that such a noninvasive technique with simple, low-cost, and portable aspects could be utilized for multiple applications in the brain studies, such as dynamic function of brain and drug delivery monitoring. However, since this study only examined the relationship between NIRS and the dry/wet weight method in monitoring the development of traumatic brain edema without any drug, further work must be done to understand the relationship between the two methods when treatments including drugs are given.

Acknowledgments

We appreciate Qi Wu, Yeguo Cheng and Kuanzheng Li for their excellent work in installation of NIR system and animal experiments. This paper was supported by National Nature Science Foundation of China (Grant No. 30671997). All the work has been done at the Functional Neurosurgery Laboratory of SEU and the Biophotonics Laboratory of NUAA.

References:

1. K Okiyama, DH Smith, MJ Thomas, et al. Evaluation of a novel calcium channel blocker, (S)-emopamil on regional cerebral edema and neurobehavioral function after experimental brain injury. J Neurosurg 1992; **77**: 607-612.
2. F Bareyre, F Wahl, TK Mcintosh, et al. Time course of cerebral edema after traumatic brain injury in rats: effects of riluzole and mannitol. J Neurotrauma, 1997; **14(11)**: 839-849.
3. HanLi Liu, ZhiYu Qian, Renwen Chen, et al. In vivo determination of tissue optical properties: reduced scattering coefficient (μs'). J Nanjing University of Aeronautics & Astronautics, 2004; **36(3)**: 369-372.
4. ZhiYu Qian, Yueqing Gu, HanLi Liu, et al. In vivo and real time measurement of rat brain reduced scattering coefficient (μs'). Chinese Journal of Medical Physics, 2005; **22(2)**: 463-465.
5. DM Feeney, MH Boyeson, RT Linn, et al. Response to cortical injury: Methodology and local effects of contusions in the rat. Brain Res, 1981; **211**: 67-77.
6. Rong Li, ZhiYu Qian, LiJuan Dai. Automatic Measurement System of NIR Spectrum of Biological Tissue Based on Virtual Instruments [J]. Computer Measurement & Control, 2007; **15(2)**: 154-156.
7. FC Wai, AP Scott, JW Ashley. A review of the optical properties of biological tissues. IEEE J of Quant. Elec. 1990; **26(12)**: 2166-2185.
8. CA Giller, M Johns, Hanli Liu. Use of an intracranial near-infrared probe for localization during stereotactic surgery for movement disorders. J Neurosurg. 2000; **93**: 498–505.
9. CA Giller, Hanli Liu, P Gurnani, et al. Validation of a near-infrared probe for detection of thin intracranial white matter structures. J Neurosurg. 2003; **98**: 1299–1306.
10. AK Senapatia, H Radhakrishnanb, Hanli Liu, et al. Detection of degeneration in rat sciatic nerve by in vivo near infrared spectroscopy. Brain Res, 2005; **14**: 119–125.

11. M Haller, SL Mironov, DW Richter. Intrinsic optical signals in respiratory brain stem regions of mice: neurotransmitters, neuromodulators, and metabolic stress. J Neurophysiol, 2001; **86**: 412.
12. AF Samdani, RJ Tamarago, DM Long. Brain edema: a review of current ideas. Neurosurg Quarterly, 1999; **9(2)**: 120-137.
13. MA Petty, P Poulet, A Haas, et al. Reduction of traumatic brain injury induced cerebral oedema by free radical scavenge. Eur J Pharmacol; 1996, **307**: 149-156.
14. Haipeng Liu, Zhengguo Wang, Kecheng Zhang, et al. The changes of brain edema following brain injury of rats by gas percussion. Chinese Journal of Neurosurgery, 2001; **17(6)**: 375-377.

THE DESIGN OF A PDT-LED LIGHTING SOURCE CONTROLLED BY SCM

HELING ZHOU, XIAOYUAN LI*, SHUWEI DU, YAN LIU, JIA LIU, LIN LÜ

Dept of Biomedical Engineering, Zhongshan School of Medicine, Sun Yat-Sen University, Guangzhou, CHINA, 510080

Photodynamic therapy (PDT) is a promising new method for cancer, which damages healthy tissue significantly less than other methods. The most popularly accepted light source is laser, while the rapid development of LED provides a much better choice for superficial lesions. Our group has been working to develop a new PDT light source of LED array, with the regulable parameters: irradiation power, time for irradiation and intermission. The irradiation power of LED is regulated with PWM using the relatively high frequency output by SCM. The relatively low frequency outputs use the time of up levels to regular the time of irradiation while the time of low levels to regular that of intermission. The light sources that we designed are tested with a light-intensity distributing apparatus to learn about the effectiveness of PWM. The consequences prove a pro rata relationship between photocurrent and duty-ratio. We also take experiments of PDT with human hepatocellular carcinoma cell line HepG2, and justify the validity of our design by the inhibition ratio shown in MTT.

1. Background & Objectives

Photodynamic therapy (PDT) is a relatively new therapeutic modality for both neoplastic and non-neoplastic diseases, involving the administration of photosensitizing compound, which selectively accumulates in the hyperproliferative target cells followed by local irradiation with visible light of lesion[1]. The clear attraction of this method is the possibility of some targeting of the disease by drug and by the area irradiated[2]. which, therefore renders the significantly less damage of healthy tissue near the lesions. Photosensitizers are compounds that absorb energy from light of specific wavelengths and are capable of using that energy to induce reactions in other non-absorbing molecules. So, all three components (photosensitizer, light and oxygen) have to be available for the treatment to occur [3]. Specifically, PDT is a photochemical process that shows complex interactions between the fluence, fluence rate, photosensitiser concentration and oxygen concentration[4].

Lasers have become the standard light sources for PDT applications, due to their monochromatic character, high power output, and ease of coupling to fiber optics for endoscopic light delivery within a body cavity or for interstitial implants[3]. However, the high costs associated with the purchase, maintenance, and operation of lasers may become one of the biggest obstacles to the clinical acceptance of this new treatment. In addition, the laser system might not be necessary for superficial and wide-spread oral dysplasial lesions[5]. LEDs are compact, lightweight and require significantly less energy to produce the desired wavelengths of light[6]. In addition, they are also much less expensive than laser light sources[7]. In recent years, improvements in semiconductor technology have substantially inproved the light output of LED chips. LED has provided an effective alternative to lasers for PDT[8].

In order to gain the clinical and laboratorial evidences, our group has been working to develop a new PDT light source of LED array, with the regulable parameters: irradiation power, time for irradiation and intermission.

2. Materials & Methods

The concern of our group is the light source, and for light source, any kind can be used for PDT as long as it has the required spectral characteristics. The light source must emit light of an appropriate wavelength, corresponding to the maximum absorption of the sensitiser, and must have sufficient energy to activate it. Meanwhile, it has been shown previously, that it is possible to increase the amount of necrosis produced by ALA induced PDT without increasing the administered dose of ALA or the applied light dose using light dose fractionation. It has been hypothesized that this results in a greater PDT effect as the break in illumination allows the treatment site to be reoxygenated[9]. Therefore, whether the light is fractionated or continuous also affects on the result.

* Correspondence author: Email: xyli@163.com

In short, the factors affecting the results of PDT and associated with lighting sources are listed as follows: wave length, fluence rate, fluence, continuity and discontinuity.

The ultra-high irradiation LED array is in series-parallel and is supplied by a voltage regulator. The array of power LED is supplied and controlled by the current regulator IC LT1618.

Both regulate the irradiation power of LED with PWM using the relatively high frequency outputs (around 52 KHz) by SCM (AT89S51). The relatively low frequency outputs (ranging from 1 KHz to 0.01Hz or even lower) use the time of up levels to regulate the time of irradiation while the time of low levels to regulate that of intermission.

The simplified control flow chart is shown in figure 1, users input orders by keyboard, according to which, SCM system regulates the irradiation intensity by the duty-ratio of the pulses it outputs, and at the same time regulates the time of irradiation as well as intermission by the time of rectangle waves and true low levers.

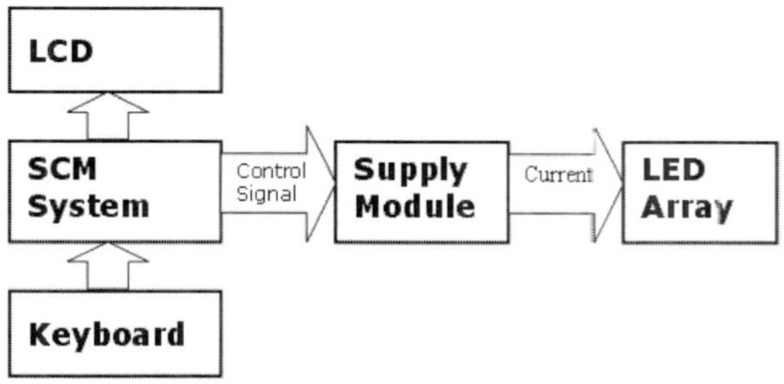

Figure 1. Sketch map of the lighting system

To meet different experimental and clinical needs, three different lighting sources have been made. Two of them, which are made of ultra-high irradiation LED array, are designed for wide-spread artificial lesions. And the one made of power LED is for the deeper lesions, where a relevantly high power rate should be maintained after the light is transmitted a certain depth.

3. Results & Discussions

The LED light sources are tested in electrical, optical and biological experiments. The optical results are from light-intensity distributing apparatus (Fig 2 to Fig 4). The biological results are from experiments of PDT with the human hepatocellular carcinoma cell line HepG2 in vitro (Fig 5 and Fig 6).

3.1. *Optical Experiments*

To measure the distribution, we put the detector 50mm away from LED arrays, move 0.5mm each time in the track paralleling the arrays, and record the photocurrent at each position.

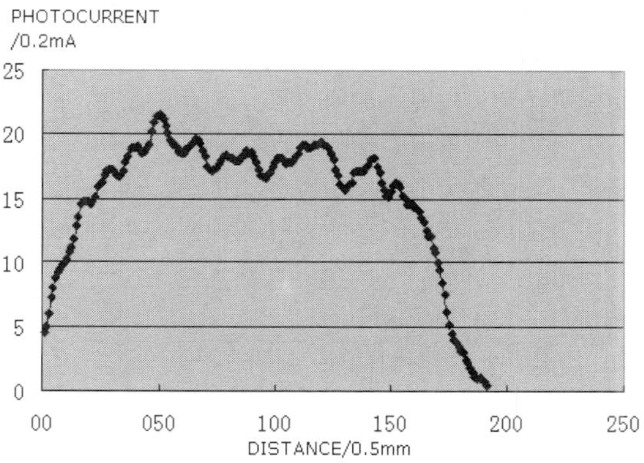

Figure 2. Fluence distribution of ultra-high irradiation LED array

Fig 2 shows the smooth irradiation in large area of the middle section of the ultra-high irradiation LED array, and a sharp decrease near the contour line of it. These characteristics make the light source convenient for wide-spread superficial lesions. And to focus the light with lens concentrator will be easier as well.

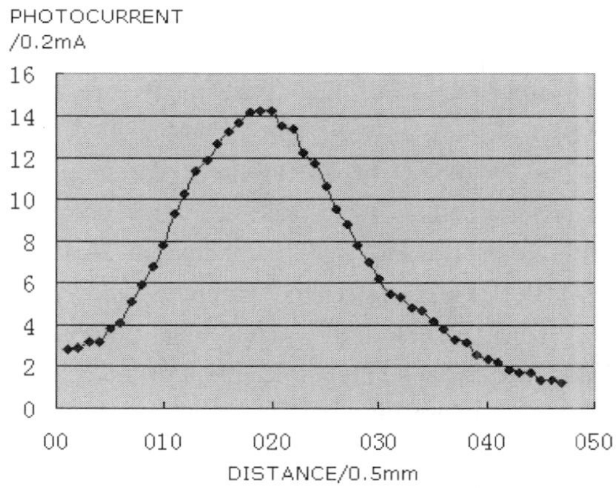

Figure 3. Fluence distribution of power LED array

Fig 3 shows focusing of power LED array is fairly successful, -- 50mm away from the lighting source, half the irradiation power distributes in a radius of 10mm. The focus is about 60mm away from the lens.

In order to test the regulation of irradiation, we fixed the detector at a certain location, change the duty-ratio of PWM only, and record the photocurrent at each ratio.

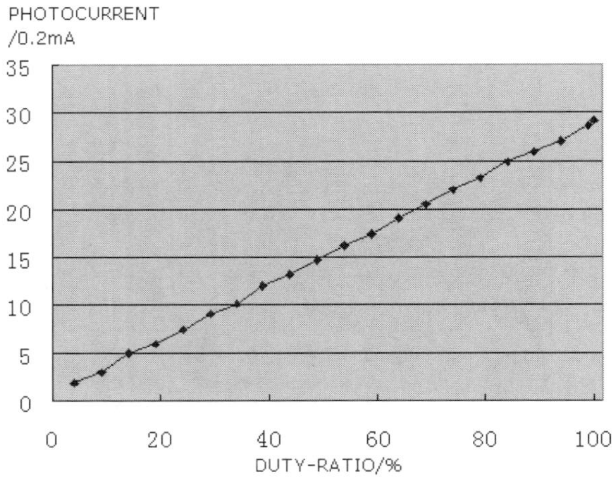

Figure 4. Fluence controlled by PWM

In Fig 4, the consequences prove a pro rata relationship between photocurrent and duty-ratio.

3.2. *In Vitro Studies*

3.2.1. The design of the experiments

In further tests, we use the LED-PDT lighting sources with the human hepatocellular carcinoma cell line HepG2 in vitro, and choose cancer cell inhibition rate as the parameter to estimate the effect of PDT and consequently the validity of our design. Inhibition rate is measured by MTT. The experiments are taken as follows.

The human hepatocellular carcinoma cell line(HepG2) cells were digested and collected in logarithmic phase and were incubated in 24 multiwell plates (1000μl per well), and incubated in serum-free culture 1640 for 24 hours. And then photosensitizer HMME was added into the medium and with the drugged cells were cultured for another 2 hours. Then the cells were washed in PBS three times before fresh medium was added, and irradiated by the LED-PDT lighting source. The concentration of photosensitizer was 10μg/ml or 20μg/ml, the irradiation time was 0s, 60s, 120s or 240s, and the duty-radio of PWM is 12.5%, 25%, 50%, or 100%. After the irradiation, HepG2 cells were incubated in dark for 24 hours and evaluated by a trypan blue dye exclusion test and determined by an MTT colorimetry assay. The inhibition rate is calculated by the following formula:

Inhibition rate = (OD of control group – OD of experimental group) / OD of control group*100%

3.2.2. The results of the experiments

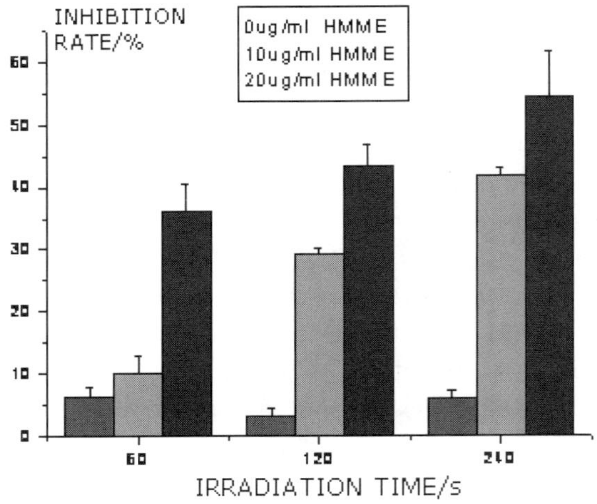

Figure 5. Inhibition rate under different irradiation time

Fig 5 shows that, with a various HMME concentration (10 μ g/ml, 20 μ g/ml), inhibition rate increases following the increase of the irradiation time.

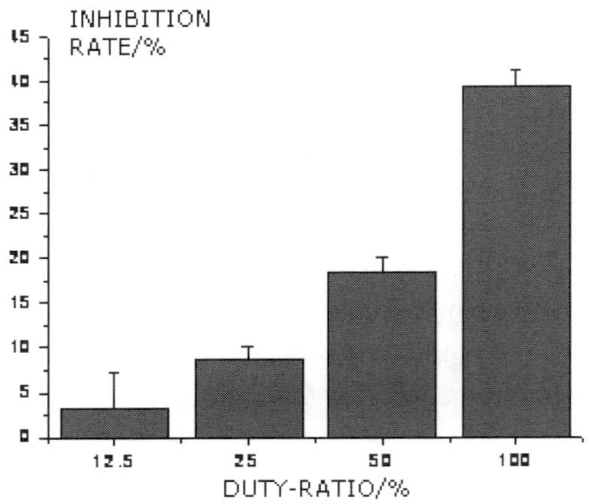

Figure 6. Inhibition rate under different pulse width

The results shown in Fig 6 prove the validity of the PWM control. With the photosensitizer HMME (10 μ g/ml) existing, and the irradiation for 120s each, inhibition rate increases following the increase of the pulse duty-ratio.

4. Conclusion

The results from electricity tests, lighting-intensity distributing apparatus, and experiments in vitro prove the validity of the light sources developed by our group. It is easy and accurate to control the regulable parameters (irradiation power, time for irradiation and intermission).

References

1. Živilė Lukšienė, *Medicina.* 3912, 1137 (2003).
2. Hugh Barr, Catherine Kendall, Janelle Reyesgoddard, et al, *Science Progress.* **8502**, 131 (2002).
3. C.H. Sibata, V.C. Colussi, N.L. Oleinick and T.J. Kinsella, *Braz J Med Biol Res.* **3308**, 869 (2000).
4. Keith Langmack, Ro Mehta, Paul Twyman, and Paul Norris, *J Photochem Photobiol.* **B60**, 37 (2001)
5. Jui-Chang Tsai, Chun-Pin Chiang, Hsin-Ming Chen, et al., *Lasers Surg Med*, **34**, 18 (2004).
6. Thomas S Mang, *Photodiagnosis Photodynamic Therapy.* **1**, 43 (2004).
7. Martijn Triesscheijn, Paul Baas, Jan H. M. Schellens and Fiona A. Stewart, *Oncologist.* **11**, 1034 (2006).
8. Harry T. Whelan, Ellen V. Buchmann, Noel T. Whelan, *Space Technology and Applications International Forum*, **CP522**, 35 (2001).
9. A. Curnow, J.C. Haller, S.G. Bown, *J Photochem Photobiol*, **B58**, 149 (2000).

EFFECTS OF REACTIVE OXYGEN SCAVENGER ON SONODYNAMIC KILLING OF RAT C6 GLIOMA CELLS

JIAN-HUA LI, DA-YONG SONG, WU YUE*
Department of Neurosurgery
The Fourth College Hospital of Haerbin Medical University
Harbin, P. R. China

ZHENG HUANG
Department of Radiation Oncology
Colorado University of Health Sciences Center
Denver, Colorado, USA

Correspondence to: yuewu@vip.163.com

Objective: To study the effects of reactive oxygen species (ROS) and their scavengers on rat C6 glioma cells in sonodynamic treatment (SDT). ***Materials and Methods:*** A photosensitizer hematoporphrin monomethyl ether (HMME, 10 μg/ml) was used as a sonosensitizer. Its sonodynamic effect on the glioma cells were examined in the absence or presence of various ROS scavengers (e.g. NaN_3, CAT, SOD and $C_6H_{14}O_6$). Sonication was delivered by a planer ultrasound transducer at 1 MHz and 1 W/cm^2. Cell viability was analyzed by MTT assay. ***Results***: HMME-mediated SDT could cause significant cell killing. The presence of NaN_3 showed a strong protective effect (nearly 50% reduction). Both CAT and SOD showed minor protective effect (~13% reduction). $C_6H_{14}O_6$ caused a slight reduction of 6% but no statistic significance. ***Conclusion***: Singlet oxygen ($^1O_2$) plays an important role in HMME-mediated SDT.

1. Introduction

Several major modalities have been used clinically for the treatment of glioma. However, they are still far from ideal in terms of tumor-inhibition and side effects. Ultrasound has an appropriate tissue attenuation coefficient for penetrating and reaching deep-seated tissues while maintaining the ability to focus energy onto a small volume. This is a unique advantage in the application of noninvasive treatment of deep-seated tumors when compares to electromagnetic modalities, such as laser beams. Sonodynamic therapy (SDT) is a promising new modality for cancer treatment. SDT is based on the ultrasonic activation of a sonosensitizer.[1, 2] Interestingly, it has been found that some photodynamic therapy (PDT) photosensitizers can be activated by ultrasound, leading to significant antitumor cytotoxic effects, both in vitro and in vivo.[3-6] Recently, we found that photochemically active porphyrins such as hematoporphrin monomethyl ether (HMME) can induce apoptosis of glioma cells when activated by ultrasound.[7] Reactive oxygen species (ROS) might be a significant mediator of SDT.[8-11] In order to delineate the mechanisms of SDT action of HMME, we studied the effect of various reactive oxygen scavengers on the HMME-mediated SDT killing of C6 glioma cells *in vitro*. This article will report the preliminary results of *in vitro* experiments.

2. Material and Methods

2.1 Sensitizer

Hematoporphrin monomethyl ether (HMME) was provided by Meifakang Ltd. (Beijing, China). HMME is a domestic 2nd generation PDT photosensitizer and has been used in PDT protocols for treating port wine stains and various tumors.[12]

2.2 Cell Line

Rat glioma C6 cells were obtained from the Neurosurgery Institution of Harbin Medical University. Cells were maintained in BPMI-1640 medium (Hylone, USA) supplemented with 10% fetal calf serum (FCS; Invitrogen, USA) at 37°C in 5% CO_2.

2.3 Experiment Set Up

A multi-function physical therapy ultrasound device with a planar ultrasound transducer (ϕ = 4 cm) (Tianshi Technologies Ltd. Co., Beijing) was used to generate ultrasound of 1 MHz and 1 W/cm^2. For sonicating cells, the transducer was placed in a 37 °C water bath and 3 cm below the cells. The water bath was filled with degassed water and stained with black ink.

2.4 Experimental Procedures

Freshly prepared glioma cells were suspended in RPMI-1640 medium and transferred to 8 wells in the center of

96-well plates. Plates were divided randomly into 4 groups: control, ultrasound alone, HMME alone, and HMME+ ultrasound (SDT). In SDT group, cells were incubated with HMME at the final concentration of 10 μg/ml in the dark for 2 hours and then subjected to ultrasound treatment for 2 min. After SDT treatment, cells were re-incubated for 4 hours and then subjected to viability assay.

To study the effect of reactive oxygen scavengers on the HMME-mediated SDT killing of glioma cells, the above procedures were repeated in the presence of either 2 mmol/L of sodium azide (NaN_3; Sigma, USA), 200 KU/L of hydrogen peroxidase (CAT; Sigma), 1 μg/ml of superoxide dismutase (SOD; Sigma), or 1 μg/ml of mannitol ($C_6H_{14}O_6$; Sigma).

Cell viability was analyzed with the MTT assay. Killing rate was calculated as: Killing rate (%) = $(1 - OD_{\text{treatment group}} / OD_{\text{control group}}) \times 100\%$.

2.5 Statistical analysis

Statistical evaluation was performed with *t* test using the analytical tools of SPSS Software system. Data are presented as mean values±SD, P value of less than 0.05 was considered statistically significant.

3. Results

The sensitivity of rat C6 glioma cells to SDT was studied in an in vitro experiment using a low power ultrasound transducer and a photosensitizer hematoporphrin monomethyl ether. To demonstrate the effect of oxygen free radicals on SDT-induced cell killing, HMME-mediated SDT killing of glioma cells were studied in the absence or presence of various reactive oxygen scavengers (e.g. NaN3, CAT, SOD, and manioc). HMME or scavenger did not almost alter cell viability. Ultrasound alone could only slightly kill cells. However, the combination of HMME (10 μg/ml) and ultrasound (1 MHz and 1 W for 2 min) could induce 78% cell killing. The presence of scavenger during ultrasound exposure had different effects on the HMME SDT-induced cell killing. The presence of NaN3 could significantly reduce the killing rate of HMME SDT. At the level of 2 mmol/L, the singlet oxygen scavenger could cause nearly 50% reduction ($p < 0.01$). The presence of CAT or SOD could also reduce ~13% of HMME SDT killing ($p < 0.05$). Although the presence of mannitol could also reduce the killing rate slightly (6%), statistical analysis indicated there was no significance.

4. Discussion

The primary motivation of this study was to elucidate the mechanism(s) underlying the sonodynamic effect induced by porphyrin derivatives. Since the first report of sonodynamic effect, its mechanism has been controversial. Some suggest that the mechanism is mainly due to certain mechanical stress such as the augmentation of physical disruption of cellular membrane by sensitizers in the close vicinity of cells and/or cavitation bubbles.[13] Others suggest that active oxygen species play an important role in anti-tumors SDT. [8-11] There are two major standpoints in the formation of reactive oxygen species. (1) Singlet oxygen ($^1O_2$) mechanism: as the molecules of sonosensitizer accept some energy from radiation of ultrasonic wave, they can undergo a series of transmission to change into excited triplet state oxygen of long life. If oxygen molecules acquire some energy during this period, they will become into excited singlet state oxygen from ground triplet state oxygen. Singlet state oxygen is an oxidizing agent which has an ability of electrophilicity and can initiate oxidizing reaction to kill targeted tumor cells. (2) Free radical mechanism: ultrasound cavitation activates sonosensitizers in tissues where cavitation bubbles occur on the liquid-air interface. It can decompose weak link by thermolysis or react with H^+ or OH^- which is generated by thermolysis of water. Subsequently, free radicals which come from sonosensitizer react with oxygen to form hydroperoxy and alkoxyl. Those substances can enable to form a series of lipid peroxidation to kill tumor cells. Although other theories continue to be proposed,[14] it is indeed that the reactive oxygen species play significant rules in anti-tumor SDT.

To confirm the formation of various oxygen free radicals during SDT, we examined the protective effect of different reactive oxygen scavengers on SDT-induced cell killing in an in vitro model. We used NaN_3 for singlet oxygen, CAT for hydrogen peroxide, SOD for superoxide, and mannitol ($C_6H_{14}O_6$) for hydroxy radical. We demonstrated that sodium azide could significantly suppress the effect of SDT on C6 glioma cells. SOD and CAT also showed similar effect in a certain extent. But mannitol showed little effect on SDT. This suggests that the effect of SDT on glioma cells could be mainly due to the formation of singlet oxygen. On the other hand, superoxide anion and hydrogen peroxide also participate in the SDT process. But hydroxide radical might be less important. These results suggest that HMME-mediated SDT can kill rat C6 glioma cells through the generation of oxygen free radicals and especially the singlet oxygen but other oxygen free radicals, such as superoxide anion and hydrogen peroxide, are also involved.

References

1. N. Yumita, R. Nishigaki, K. Umemura and S. Umemura, *Jpn. J. Cancer Res.* **80,** 219–222 (1989).
2. D. Kessel, J. Lo, R. Jeffers, J. B. Fowlkes and C. Cain, *J. Photochem. Photobiol. B.* **28,** 219-221 (1995).
3. N. Yumita, R. Nishigaki, K. Umemura and S. Umemura, *Jpn. J. Cancer Res.* **81,** 304–308 (1990).
4. T. Uchida, K. Tachibana, S. Hisano and E. Morioka, *Anticancer Drugs* **8**, 329-335 (1997).
5. N. Yumita, R. Nishigaki and S. Umemura, *J. Cancer Res. Clin. Oncol.* **126**, 601-606 (2000).
6. Z. H. Jin, N. Miyoshi, K. Ishiguro, *J. Dermatol.* **27**, 294-306 (2000).
7. W. Yue, D. Y. Song, Z. Zhao, S. D. Cao, X. X. Meng, X. Zhang and Z. Huang, *Proc. SPIE.* **6438**, 64380I-1 (2007).
8. N. Miyoshi, T. Igarashi and P. Riesz., *Biochim. Biophys. Acta.* **1620**, 179-184 (2003).
9. V. Misik and P. Riesz P, *Antioxid. Redox Signal.* **2,** 355-362 (2000).
10. V. Misik and P. Riesz, *Ann. N. Y. Acad. Sci.* **899**, 335-348 (2000).
11. W. Hiraoka, H. Honda, L. B. Feril Jr, N. Kudo and T. Kondo, *Ultrason. Sonochem.* **13,** 535-542 (2006).
12. Z. Huang, *Photodiag. Photodyn. Ther.* **3**, 3-10 (2006).
13. T. Yu, Z Wang and T. J. Mason, *Ultrason. Sonochem*, **11**, 95–103 (2004).
14. H. Honda H, Q. L. Zhao and T. Kondo, *Cancer Sci.*, **95**, 181-185 (2004).

Topic V

Multimodal and Hybrid Biomedical Imaging

THE STUDY ON THE RECONSTRUCTION ALGORITHMS FOR WIDE-FIELD OPTICAL COHERENCE TOMOGRAPHY*

YONGYONG YANG, JIA JIA, YINQI FENG ††

Opto-Mechatronic Equipment Technology Beijing Area Major Laboratory
Beijing Institute of Petro-chemical Technology, Beijing 102617 China

In a wide-field optical coherence tomography system, the principle of 4-step and 8-step phase-shifting algorithms is illustrated. The experimental results show that 8-step method is a better vibration-insensitive image reconstruction algorithm. The interference images with a smaller value of phase difference can offer the more detail signals and enhance the quality of reconstruction images. The study will offer a reference to the deep research on the wide-field optical coherence tomography.

1. Introduction

An imaging technique with non-invasion, safety to biological bodies, low price and miniaturization is needed imminently. In order to monitor natural biological tissues, and chemistry components in biological bodies, or to check the subsurface flaw of material, optics tomographic imaging and optics imaging techniques in turbid media are actively researched in relative fields. Recently a technique called optical coherence tomography (OCT)[1] has been developed for non-invasive, in vivo, real time and 3-dimentional reconstruction imaging measurement. With a high scanning speed, a high spatial resolution and sensitivity, it combines confocal, low-coherence, optical heterodyne, scanning tomographic imaging techniques. Especially, the advantage of its non-invasive and in vivo measurement makes OCT achieve wide applications in the inspection of biological tissues such as eyeball, skin, stomach, intestines and heart blood vessel, etc, in clinic medicine field. At present, making OCT instruments and developing practical OCT techniques are the research hotspots for OCT. However, OCT techniques have been limited in laboratory and only the OCT technique applied into ophthalmic inspection has been realized commercially. It is clearly that the successful application of OCT techniques into the future inspections and clinic medicine applications mainly depend on the imaging resolution and the imaging acquiring speed.

In real time and in vivo measurements, the scanning speed of an OCT system is quicker than the moving speed of samples. In a conventional OCT system, 3-dimentional images are acquired through mechanical point-to-point and line-to-line scanning, for example, a reference mirror is first moved to-and-fro to generate longitudinal scanning and objective or sample is then moved parallelly to generate lateral scanning. The scanning speed in longitudinal and lateral direction limits the image acquiring speed in the system, which is an essential factor that can't achieve a higher imaging speed rate. In recent 16 years, researchers have been working hard and brought forward some resolving methods. At the beginning, a stepping motor was used to drive the reference mirror to generate longitudinal scanning. Because torque and response time could not reach the demand of the high-speed hardware, the scanning speed was only 1.6mm/s. In 1996, Tearney etc.[2] put forward a method using to extending fiber made by piezoelectricity crystal. Although the scanning speed was high, the great dispersion can't meet the request of imaging. At present, a popular method proposed by Tearney[3] in 1997 is that using optical delay line generated longitudinal scanning. The method improved the scanning speed of the system, but it introduced some noises into the reconstructed images. To resolve the problem, J.Szydlo, etc.[4] drove a rotating prism using an air turbine to make the scanning speed arrive at 175 m/s in 1998, but dispersion phenomena happened when light went into the prism resulting in not using wide-band light source in OCT system. In

* This work is supported by a 2007 grant from General Program of Beijing Natural Science Fund and a 2006 grant (KM200610017005) from Science Research Development Program of Beijing Education Council.
† Work partially supported by the grants from the returned scholars and students programs funding of Ministry of Education of China and Beijing Ministry of Personnel.
† Any correspondence should be addressed. Email: fengyinqi@bipt.edu.cn

1999, Delachenal, etc.[5] used two different light sources to control the scanning speed and the scanning range respectively, which made the scanning speed reached 37 m/s. However, a compensating light route had to be set up because of the greater nonlinear error. Karamata, etc.[6] used a voice coil to realize scanning in 2004, but this method was proved not to apply in clinic medicine although its structure was easy, its operation was convenient and its scanning speed was faster than that of a stepping motor.

Those scanning methods referred above have merits or demerits. Generally speaking, the common ground of those methods is that the speed of OCT system is increased through improving the longitudinal scanning speed, while, at the same time, the lateral scanning speed didn't have too much improvement. The line-to-line scanning method needs longitudinal scanning (Z-direction) and lateral scanning (X-direction) when a 2-dimensional image is rebuilt. If a 3-dimensional image is reconstructed, another lateral scanning (Y-direction) will be added. Therefore, the line-to-line scanning method is imperfect inherently on the improving the scanning speed. In 2002 Bordenave, ect.[7] proposed a wide-field optical coherence tomography (WFOCT) for the problem. Its idea is derived from the principle of coherence radar and white-light interferometry. Instead of imaging a single point as in the standard confocal configuration, an entire surface of the sample was imaged with a collimated beam emitted from low-coherence light source. Using a 2-dimensional detector (CCD) to record the interferometric image directly, the necessity of the lateral scanning (X, Y-direction) and the heterodyne detection could be avoided. The images of sample surfaces or sub-surfaces could be obtained by digital processing the images recorded by CCD. Without reducing the imaging resolution, the imaging acquiring speed could be improved much better by the WFOCT, which is also called full-field optical coherence tomography (FFOCT)[8]. With this technique, the modern microscopy with a high speed and a high resolution become possible. Its sensitivity is ~80 dB at least. It could be a supplement for a wide-used fiber OCT.

As the studies on WFOCT are taken up in smaller proportion in China at present, we established a WFOCT experimental system and investigated the reconstruction algorithms of sample images. Based on these work, the experimental results are given, the factors of impacting on the quality of reconstructed image are analyzed, and some resolving measures are advised.

2. Experimental Systems

The skeleton of our WFOCT system is described in Figure 1. Its main part is the Michelson interferometer, in which there is a sample arm and a reference arm. After passing through a microscope objective (MO), a lens (L) and polarizer (P), a parallel beam is equally separated into the two arms in Michelson interferometer by the broadband polarizing beam splitter (BS). The reflected lights from the sample (S) and the reference arm are combined to produce coherence light when their light distance difference is less than the coherence length of the light source (LS). An arbitrary light beam diameter can be achieved by changing a relative position between a microscope objective and a lens. The coherence light through an analyzer (A) is detected by charge-coupled device (CCD). CCD changes a light signal into an electricity signal which is passed into a personal computer (PC) through an image acquisition board (IMAQ). The reference mirror (REF) is mounted onto a nanometer-level micro-move stage (MIS), which is driven by a piezo-transducer (PZT). The light distance in the reference arm is modulated by the PZT. Using image reconstruction algorithms, a cross correlation signal can be separated from the direct current signal of CCD. The position of micro-move stage can be exactly controlled by a corresponding control program stored in the PC. Two-dimensional images of sample in an X-Y plane in Cartesian space can be reconstructed by a corresponding image processing program stored in the PC.

To reconstruct a three-dimensional image of a sample, the micro-move stage is mounted onto a macro-move stage (MAS) which is used to realize to scan the sample in a longitudinal direction. Besides, for a continuous adjustment of the relative intensity and interference contrast between the sample and reference beams, two quarter wave plates (Q) are placed in the two arms to change the orthogonal polarization states. All of the reflections from the two arms are steered toward the CCD through an analyzer. Otherwise, the MIS and the MAS are driven by a MIS control box (MISCB) and a MAS control box (MASCB) respectively.

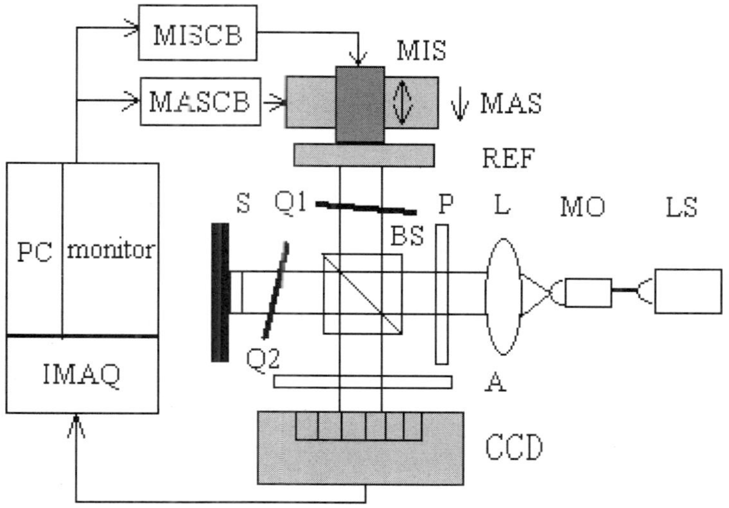

Figure 1. The skeleton of WFOCT

3. A Reconstruction Algorithm

Two-dimensional images of a sample in the system could be rebuilt by phase-shifting interferometry. The light intensity distribution I(x, y) detected by the CCD is written as:

$$I = I_O(x,y) + I_B(x,y) + I_R(x,y) + 2\sqrt{I_O(x,y)I_R(x,y)}\cos[\varphi_i + \varphi(x,y)] \tag{1}$$

where $I_o(x, y)$ is the portion of the sample reflection that is coherent with respect to the reference reflection $I_R(x, y)$ and $I_B(x, y)$ is the incoherent background. $\phi(x, y)$ is the relative phase profile between the sample and the reference beams, and ϕ_i is the phase shift introduced by shift of the micro-move stage. In a 4-step method, four image frames are acquired at quarter wavelength intervals corresponding to reference phases of ϕ_i = C, $I_{\pi/2}$, I_π, $I_{3\pi/2}$. These interference intensities are expressed as:

$$\left.\begin{aligned} I_0 &= I_O + I_B + I_R + 2\sqrt{I_O I_R}\cos\varphi \\ I_{\pi/2} &= I_O + I_B + I_R - 2\sqrt{I_O I_R}\sin\varphi \\ I_\pi &= I_O + I_B + I_R - 2\sqrt{I_O I_R}\cos\varphi \\ I_{3\pi/2} &= I_O + I_B + I_R + 2\sqrt{I_O I_R}\sin\varphi \end{aligned}\right\} \tag{2}$$

Based on trigonometric relations between the above four interference intensities in the Eq.(2), the intensity of the coherent sample rebuilt image is then given by:

$$I_O = \frac{(I_0 - I_\pi)^2 + (I_{3\pi/2} - I_{\pi/2})^2}{16 I_R} \tag{3}$$

and its correlational phase map is:

$$\varphi = \arctan\frac{I_{3\pi/2} - I_{\pi/2}}{I_0 - I_\pi} \tag{4}$$

In experiments, we found that the value of phase-shifting was affected by the vibration of our experimental platform so sensitively that the periodical fringes appear in the reconstructed images and damage the reconstructed images even if the value of the phase-shifting is just a tiny change. The effect is equal to a modulation to a high frequency carrier wave signal in reconstruction images. To solve the problem, many vibration-insensitive phase-shifting interferometries are suggested, and the method we used here is the 8-step method.

The difference between 8-step and 4-step methods is the value of the phase-shifting which is equal to $\pi/4$ in the 8-step method. Eight image frames whose intensities is similar to those in Eq. (2) are acquired at 1/8 wavelength intervals corresponding to the reference phases of $\varphi_i = I_0, I_{\pi/4}, I_{\pi/2}, I_{3\pi/4}, I_\pi, I_{5\pi/4}, I_{3\pi/2}, I_{7\pi/4}$. Based on the trigonometric relations between the intensities of the eight interference images, the amplitude of the rebuilt image of the sample is then given by:

$$I_O = \frac{(I_0 - I_\pi)^2 + (I_{\pi/4} - I_{5\pi/4})^2 + (I_{\pi/2} - I_{3\pi/2})^2 + (I_{3\pi/4} - I_{7\pi/4})^2}{32 I_R} \tag{5}$$

where I_0、$I_{\pi/4}$、$I_{\pi/2}$、$I_{3\pi/4}$、I_π、$I_{5\pi/4}$、$I_{7\pi/4}$ are the eight interference intensities of the images acquired by CCD at 1/8 wavelength intervals. And its correlational phase map is:

$$\varphi = \arctan\left\{\frac{1}{2}\left[\frac{I_{3\pi/2} - I_{\pi/2}}{I_0 - I_\pi} + \frac{(I_{3\pi/4} - I_{7\pi/4}) + (I_{\pi/4} - I_{5\pi/4})}{(I_{3\pi/4} - I_{7\pi/4}) - (I_{\pi/4} - I_{5\pi/4})}\right]\right\} \tag{6}$$

4. Experimental Results and Analyses

We establish a WFOCT experimental system based on the study of image reconstruction algorithms for WFOCT. A He-Ne laser (the wave length is 632.8 nm, the band width is 0.2 μm) is chosen as the light source. A JAI CV-M77 color CCD camera is used in the system. Its frame rate is 25 frames / s, pixels in video output are 1024(h) × 768(v), and cell size is 4.65μm. The vibration-insensitive experimental platform is adopted. A plane mirror surface and a plane mirror with a cross-hair are used as samples. According to the Eq. (3) and Eq. (5), the interference images acquired by the CCD are processed by our image processing program written in VC language, and the reconstructed images of the samples are shown by the Fig.2 and Fig.3, respectively.

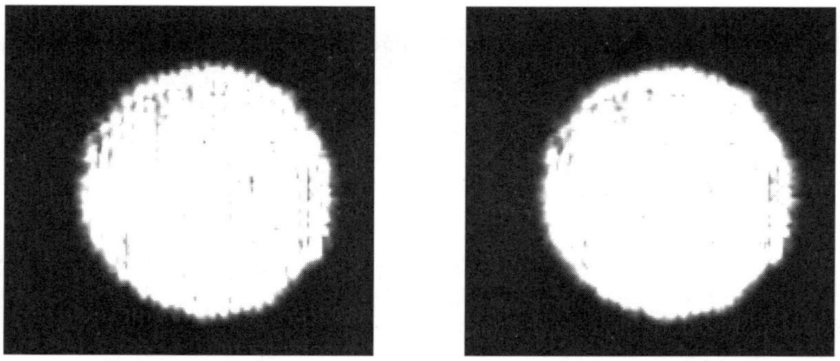

(a) 4-step method (b) 8-step method

Figure 2. The reconstructed images of a plane mirror surface in the WFOCT system

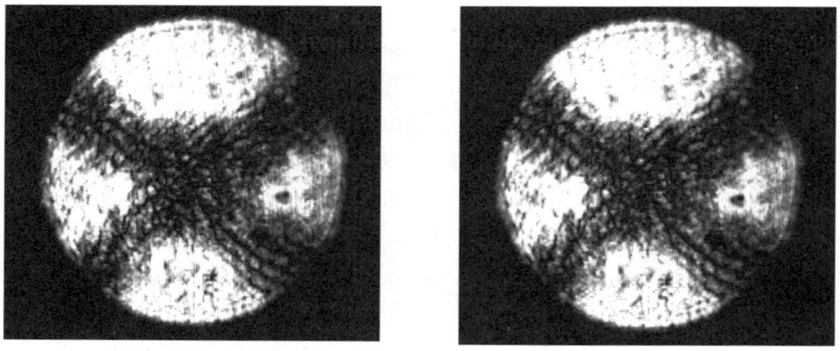

(a) 4-step method (b) 8-step method

Figure 3. The reconstructed images of a plane mirror with a cross-hair in the WFOCT system

In Figure 2, the reconstructed images are illustrated using a plane mirror surface as the test sample. And in Figure3, the reconstructed images are illustrated using a plane mirror with a cross-hair as the test sample. Because of vibrations from experimental platform, some periodical fringes appear in Figure 2(a) and Figure 3(a). The platform's vibrations mainly come from the control boxes of micro-move stage and macro-move stage, persons' walking, machines' working, various noises in our experimental environment and so on. Therefore, the vibration-insensitive optics experimental platform is chosen, the instruments which could introduce vibrations are not placed on it, and a quiet environment is also necessary. At the same time, vibration-insensitive image reconstruction algorithms are used too. In Figure 2(b) and Figure 3(b), the 8-step method is used to achieve the reconstruction images of the samples. Compared to the 4-step method, the value of phase-shifting in 8-step method is reduced by a half, so that the interference images increase to 8 frames from 4 frames in 4-step method, which offer the more abundant signal to the image reconstruction than 4-step method. With the 8-step method, the periodical fringes contrast is declined in the processing of the image reconstruction and the quality of reconstructed images is enhanced effectively. In a word, to acquire a better quality of reconstructed images, it is necessary to improve vibration-insensitive effect of experimental platform and to improve vibration-insensitive image reconstruction algorithms.

In the Figure 2 and Figure 3, some appearing spots are relative to the quality of light beam. To eliminate those spots, the intensity of light beam with well-distributed intensity must be achieved. Another factor that affects the quality of reconstruction images is the light intensity ratio between the sample arm and the reference arm. Through lots of experiments, it is shown that the quality of a reconstructed image is better when the ratio is in between 1：2 to 1:10, which are decided by reflectivity of samples.

5. Conclusion

In this article, a WFOCT experiment system is introduced, and the principle of 4-step and 8-step phase-shifting algorithms is illustrated. Experimental results are shown when a plane mirror and a plane mirror with a cross-hair are used as samples. Through analyzing and comparing, 8-step method is a better vibration-insensitive image reconstruction algorithm. It is shown that the interference images with a smaller value of phase difference can offer the more detail signals and enhance the quality of reconstruction images. In addition, It is necessary to obtain a light beam with the well-distributed intensity and to adjust the light intensity ratio between the sample arm and the reference arm in the range of 1:2～1:10.

References

1. D. Huang, E. A. Swanson, C. P. Lin, J. S. Schuman, W. G. Stinson, W. Chang, M. R. Hee, T. Flotte, K. Gregory,C. A. Puliafito, and J. G. Fujimoto, *Science*, **254**, 1178 (1991).
2. G.J. Tearney, B.E. Bouma, S.A. Boppart, B. Golubovic, E.A. Swanson, and J.G. Fujimoto, *Optics Letters*, **21**, 408 (1996).
3. G.J. Tearney, B.E. Bouma, and J.G. Fujimoto, *Optics Letters*. **22**, 1811 (1997).
4. J. Szydlo, N. Delachenal, R. Gianotti, et al., *pt. Comm.* **154**, 1 (1998).
5. N. Delachenal, R. Walti, R. Gianotti, et al., *Opt. Comm.* **162**, 195 (1999).
6. B. Karamata, P. Lambelet, M. Laubscher, et al., *Optics Letters*, **29**, 736 (2004).
7. E. Bordenave, E. Abraham, G. Jonusauskas, N. Tsurumachi, J. Oberle, C. Rulliere, P. E. Minot, M. Lassegues, and J.E.S. Bazeille, *Appl. Opt.* **41**, 2059 (2002).
8. Blandine Laude, Antonello De Martino, Bernard Drevillon, etc., *Appl. Opt.* **41**, 6637 (2002).

DECONVOLUTION RECONSTRUCTION OF LIMITED-VIEW THERMOACOUSTIC TOMOGRAPHY FOR SMALL-SCALE OBJECTS

CHI ZHANG, YUANYUAN WANG

Department of Electronic Engineering, Fudan University, Shanghai, 200433, China

Abstract: By now there doesn't exist any exact algorithm for the thermoacoustic tomography (TAT) in the limited-view case. Here an approximate algorithm, called the deconvolution reconstruction (DR), is proposed for the limited-view TAT. In this algorithm, a new function is firstly constructed from detected thermoacoustic signals. Then the electromagnetic absorption distribution of the object is reconstructed from this function based on the deconvolution method. The discrete deconvolution is the key step of this algorithm and is solvable either in the full-view or in the limited-view. With simulation studies, this algorithm is proved to have a good imaging precision for small-scale objects even in the limited-view case. It is shown that detected signals over 90 degrees in the two-dimensional case are enough to reconstruct the image of the scanned object. Simulation results of the DR algorithm are also compared with those of two popular algorithms: the time-domain reconstruction (TDR) and the filtered back projection (FBP). It is also shown that results of the DR algorithm are much better than those of the TDR and the FBP for small-scale objects in the limited-view case. Therefore, the DR algorithm is an efficient one for the limited-view TAT of small-scale objects.

Key words: Thermoacoustic tomography, limited-view, small-scale, deconvolution.

1. Introduction

The thermoacoustic tomography (TAT) is a new fast-developing technique of noninvasive medical imaging methods [1-3]. In the TAT of biological tissues, a short-pulsed electromagnetic wave (usually microwave or laser) [3-5] is used to irradiate the tissue homogeneously. The absorbed electromagnetic energy causes thermoelastic expansion in the tissue, which then radiates acoustic waves, called thermoacoustic waves (normally ultrasound waves). Thermoacoustic waves carry the electromagnetic absorption property of the tissue. A wide-band ultrasonic transducer can be employed to acquire ultrasound signals, from which the electromagnetic absorption distribution image of the tissue can be reconstructed based on certain algorithms [4-8]. The TAT is also referred to as the photoacoustic tomography (PAT), depending on the type of the irradiating electromagnetic wave. Customarily, it is called the TAT while the electromagnetic wave is the far-infrared light or microwave, or the PAT while the electromagnetic wave is the near-infrared light or visible light [9].

The TAT combines the high contrast of the pure optical imaging and the good spatial resolution of the pure ultrasound imaging [5]. Hemoglobin, common in the biological tissue, has strong absorption of the electromagnetic wave. So the TAT can be applied to detect the vascular distribution and character in the tissue [3, 10-11]. The TAT also has great value in detecting tumors [12]. In the traditional ultrasound imaging, the contrast between the tumor and the normal tissues around is very low. However, due to the abundant microvascular proliferation accompanied with the malignant tumor, the TAT can provide high contrast imaging for the tumor detection [13].

The key problem with the TAT is the imaging algorithm: how to reconstruct the electromagnetic absorption distribution from detected thermoacoustic waves. By now, there exist many exact and approximate algorithms for the TAT [2-8]. In all the exact algorithms, it is assumed that thermoacoustic signals are detected from a full view, meaning that each point of the scanned object is visible from the transducer's trajectory for 4π steradians in the three-dimensional (3-D) case or 2π radians in the two-dimensional (2-D) case [14]. So, in the typical practical application of the exact reconstruction, the transducer moves along a sphere in the 3-D case or a circle in the 2-D case. However, thermoacoustic signals can not be detected from a full view in many applications, such as the detection of the human breast. Although it can be shown theoretically that an arbitrarily small scanning arc suffices for the uniqueness of recovery [14-15], to the best of our knowledge, no exact algorithm for the limited-view case has been proposed.

In this paper, a reconstruction algorithm is proposed for the limited-view TAT based on the discrete deconvolution, which is called the deconvolution reconstruction (DR). As an approximate algorithm while the transducer moves along a sphere in the 3-D case or a circle in the 2-D case, the DR algorithm is proved to have a good precision for small-scale scanned objects. It is also shown that the DR algorithm is an efficient one in the limited-view case. Simulation studies are conducted and results confirm the theoretical prediction.

2. Methods

2.1. *The DR Algorithm*

Here the DR algorithm is derived in the 3-D case, which can be easily simplified to the 2-D case. Assuming that the tissue is irradiated by the short-pulsed electromagnetic wave homogeneously, the generated acoustic wave pressure $p(\mathbf{r}, t)$ at the position $\mathbf{r}$ can be derived from the spatial distribution of the electromagnetic absorption $A(\mathbf{r})$ as [2, 4-5]

$$p(\mathbf{r},t) = \frac{\beta}{4\pi C_p} \iiint \frac{d^3\mathbf{r}'}{|\mathbf{r}-\mathbf{r}'|} A(\mathbf{r}') I'(t - \frac{|\mathbf{r}-\mathbf{r}'|}{c}) \quad (1)$$

where t is the time, $I(t)$ is the electromagnetic pulse function, β is the coefficient of volumetric thermal expansion, c is the sound speed, C_p is the specific heat, $\mathbf{r}'$ is the integral variable and $I'(\bullet)$ is the derivative of $I(\bullet)$. The TAT algorithm faces the problem of deriving $A(\mathbf{r})$ from $p(\mathbf{r}, t)$, which is a typical inverse problem.

It is usually supposed that the electromagnetic pulse function is a Dirac delta function, *i.e.* $I(t) = \delta(t)$. The ultrasonic transducer moves along a spherical surface whose radius is r_0 and the center of sphere is the origin. So, in $p(\mathbf{r}, t)$, it satisfies that $\mathbf{r} = \mathbf{r}_0$. Then Eq. (1) can be simplified to [2, 5]

$$p(\mathbf{r}_0,t) = \frac{\beta}{4\pi C_p c} \frac{\partial}{\partial t} \oiint_{|\mathbf{r}'-\mathbf{r}_0|=ct} \frac{A(\mathbf{r}')}{t} d^2\mathbf{r}' \quad (2)$$

The DR algorithm is derived as follows.
Equation (2) can be rewritten as

$$[\int_0^t p(\mathbf{r}_0,t)dt] \cdot t = \eta \oiint_{|\mathbf{r}'-\mathbf{r}_0|=ct} A(\mathbf{r}') d^2\mathbf{r}' \quad (3)$$

where $\eta = \beta/(4\pi C_p c)$. Defining

$$S(\mathbf{r}_0,t) = [\int_0^t p(\mathbf{r}_0,t)dt] \cdot t = \eta \oiint_{|\mathbf{r}'-\mathbf{r}_0|=ct} A(\mathbf{r}') d^2\mathbf{r}' \quad (4)$$

Then a new function $B(\mathbf{r})$ is constructed as

$$B(\mathbf{r}) = S\left(\frac{\mathbf{r}}{|\mathbf{r}|} \cdot r_0, \frac{|\mathbf{r}| - r_0}{c}\right) \quad (5)$$

Substituting Eq. (4) into Eq. (5), it is obtained

$$B(\mathbf{r}) = \eta \oiint_{\left|\mathbf{r}' - \frac{\mathbf{r}}{|\mathbf{r}|} \cdot r_0\right| = |\mathbf{r}| - r_0} A(\mathbf{r}') d^2\mathbf{r}' \quad (6)$$

For the small-scale scanned tissue, it satisfies that $|\mathbf{r}'|_{max} << r_0$ in Eq. (6). To consider the integral surface in Eq. (6): $|\mathbf{r}' - \mathbf{r} \cdot r_0 / |\mathbf{r}|| = |\mathbf{r}| - r_0$, we get

$$\begin{aligned} \left|\mathbf{r}' - \frac{\mathbf{r}}{|\mathbf{r}|} \cdot r_0\right| &= \sqrt{|\mathbf{r}'|^2 + r_0^2 - 2\frac{\mathbf{r}' \cdot \mathbf{r}}{|\mathbf{r}|} \cdot r_0} \\ &= r_0 \cdot \sqrt{1 - 2\frac{\mathbf{r}' \cdot \mathbf{r}}{|\mathbf{r}| \cdot r_0} + \frac{|\mathbf{r}'|^2}{r_0^2}} \approx r_0 \cdot (1 - \frac{\mathbf{r}' \cdot \mathbf{r}}{|\mathbf{r}| \cdot r_0}) = r_0 - \frac{\mathbf{r}' \cdot \mathbf{r}}{|\mathbf{r}|} \end{aligned} \quad (7)$$

$$\begin{aligned} |\mathbf{r} + \mathbf{r}'| &= \sqrt{|\mathbf{r}|^2 + |\mathbf{r}'|^2 + 2\mathbf{r} \cdot \mathbf{r}'} = |\mathbf{r}| \cdot \sqrt{1 + 2\frac{\mathbf{r} \cdot \mathbf{r}'}{|\mathbf{r}|^2} + \frac{|\mathbf{r}'|^2}{|\mathbf{r}|^2}} \\ &\approx |\mathbf{r}| \cdot (1 + \frac{\mathbf{r} \cdot \mathbf{r}'}{|\mathbf{r}|^2}) = |\mathbf{r}| + \frac{\mathbf{r} \cdot \mathbf{r}'}{|\mathbf{r}|} \end{aligned} \quad (8)$$

Based on Eq. (7) and Eq. (8), the integral surface can be rewritten approximately as $|\mathbf{r}+\mathbf{r}'| = 2r_0$. So Eq. (6) can be rewritten as

$$B(\mathbf{r}) = \eta \oiint_{|\mathbf{r}+\mathbf{r}'|=2r_0} A(\mathbf{r}') d^2\mathbf{r}' \quad (9)$$

Defining $A'(\mathbf{r}) = A(-\mathbf{r})$ and substituting it into Eq. (9), we obtain

$$B(\mathbf{r}) = \eta \oiint_{|\mathbf{r}+\mathbf{r}'|=2r_0} A'(-\mathbf{r}') d^2\mathbf{r}' = \eta \oiint_{|\mathbf{r}-\mathbf{r}'|=2r_0} A'(\mathbf{r}') d^2\mathbf{r}' \quad (10)$$

If $A'(\mathbf{r})$ and $B(\mathbf{r})$ are considered as the input and output of a system respectively, the system is shown to be a linear and time-invariant one. Assuming

$$\begin{aligned} B_1(\mathbf{r}) &= \eta \oiint_{|\mathbf{r}-\mathbf{r}'|=2r_0} A_1'(\mathbf{r}') d^2\mathbf{r}', \\ B_2(\mathbf{r}) &= \eta \oiint_{|\mathbf{r}-\mathbf{r}'|=2r_0} A_2'(\mathbf{r}') d^2\mathbf{r}' \end{aligned} \quad (11)$$

we can get

$$\begin{aligned} & a \cdot B_1(\mathbf{r}) + b \cdot B_2(\mathbf{r}) \\ &= \eta \oiint_{|\mathbf{r}-\mathbf{r}'|=2r_0} [a \cdot A_1'(\mathbf{r}') + b \cdot A_2'(\mathbf{r}')] d^2\mathbf{r}' \end{aligned} \quad (12)$$

where a and b are arbitrary constants. So the system is linear. It can also be derived that

$$B(\mathbf{r}+\mathbf{r}_2)=\eta\oiint_{|(\mathbf{r}+\mathbf{r}_2)-(\mathbf{r}'+\mathbf{r})_2|=2r_0} A'(\mathbf{r}'+\mathbf{r}_2)d^2\mathbf{r}' \\ =\eta\oiint_{|\mathbf{r}-\mathbf{r}'|=2r_0} A'(\mathbf{r}'+\mathbf{r}_2)d^2\mathbf{r}' \tag{13}$$

which proves the system to be time-invariant. As a result, the system is linear and time-invariant. So we have

$$A'(\mathbf{r})*h(\mathbf{r})=B(\mathbf{r}) \tag{14}$$

where $h(\mathbf{r})$ is the impulse response of the system, * means the convolution. When the input of the system is the Dirac delta function, the output is $h(\mathbf{r})$. Then we can obtain

$$h(\mathbf{r})=\eta\oiint_{|\mathbf{r}-\mathbf{r}'|=2r_0}\delta(\mathbf{r}')d^2\mathbf{r}'=\begin{cases}\eta\ , & |\mathbf{r}|=2r_0\\ 0\ , & |\mathbf{r}|\neq 2r_0\end{cases} \tag{15}$$

Appling 3-D Fourier transform to $A'(\mathbf{r})$, $h(\mathbf{r})$ and $B(\mathbf{r})$, we get $\tilde{A}'(\boldsymbol{\omega})$, $\tilde{h}(\boldsymbol{\omega})$ and $\tilde{B}(\boldsymbol{\omega})$ respectively. So we have

$$\tilde{A}'(\boldsymbol{\omega})=\frac{\tilde{B}(\boldsymbol{\omega})}{\tilde{h}(\boldsymbol{\omega})} \tag{16}$$

However, $\tilde{h}(\boldsymbol{\omega})$ may be zero at some points, and Eq. (16) can not be used in this case. A solution in the common use is to calculate $\tilde{A}'(\boldsymbol{\omega})$ by [16-17]

$$\tilde{A}'(\boldsymbol{\omega})=\frac{\tilde{B}(\boldsymbol{\omega})}{\tilde{h}(\boldsymbol{\omega})\cdot(1+\lambda/|\tilde{h}(\boldsymbol{\omega})|^2)} \tag{17}$$

where λ is a constant.

To summarize, the DR algorithm consists of the following steps.

1) To construct $B(\mathbf{r})$ from $p(\mathbf{r}_0, t)$ based on Eq. (4) and Eq. (5).

2) $B(\mathbf{r})$ is 3-D Fourier transformed to yield $\tilde{B}(\boldsymbol{\omega})$.

3) To calculate $\tilde{A}'(\boldsymbol{\omega})$ based on Eq. (15) and Eq. (17).

4) $\tilde{A}'(\boldsymbol{\omega})$ is 3-D inverse Fourier transformed to yield $A'(\mathbf{r})$.

5) $A(\mathbf{r})$ equals to $A'(-\mathbf{r})$.

The precondition of the DR algorithm is that the scanned tissue is small-scale, meaning that the farthest distant $|\mathbf{r}'|_{\max}$ between the scanned tissue and the center of the scanning sphere should be far less than the scanning radius r_0. However, the simulation results show that the DR algorithm has a good precision while $|\mathbf{r}'|_{\max} < 0.3r_0$.

2.3. *The DR Algorithm Applied to the Limited-view Case*

The limited-view TAT algorithm is an incomplete data problem of deriving $A(\mathbf{r})$ from incomplete $p(\mathbf{r}, t)$. In the limited-view TAT, $\mathbf{r}$ of $p(\mathbf{r}, t)$ does not cover a whole sphere in the 3-D case or a whole circle in the 2-D case. This is much more difficult than the full-view case. However, the key step of the DR algorithm is the discrete deconvolution given by Eq. (14), which is solvable even though $p(\mathbf{r}, t)$ is incomplete.

Without loss of generality, we analyze a simple example of discrete deconvolution in the 1-D case. Assuming

$$[a_0,a_1,a_2]*[b_0,b_1,b_2,b_3]=[c_0,c_1,c_2,c_3,c_4,c_5] \tag{18}$$

Here we want to derive a from all of b and part of c. Obviously we can calculate a, if c_0~c_2 is known, by solving the following equations

$$\begin{cases}a_0\cdot b_0=c_0\\ a_0\cdot b_1+a_1\cdot b_0=c_1\\ a_0\cdot b_2+a_1\cdot b_1+a_2\cdot b_0=c_2\end{cases} \tag{19}$$

while $b_0\neq 0$.

In the DR algorithm, we construct incomplete $B(\mathbf{r})$ from limited-view $p(\mathbf{r}, t)$. The discrete deconvolution given by Eq. (14) is also solvable by listing equations like Eq. (19). So we can reconstruct the tissue image by calculating values of limited points. The only difference is that the image reconstruction is a 3-D or 2-D problem.

The number of equations, determined by the resolution of $B(\mathbf{r})$, is always much more than that of the unknowns, determined by the resolution of $A(\mathbf{r})$ (or $A'(\mathbf{r})$). However, all equations are useful, because the detected $p(\mathbf{r}, t)$ contains certain noises, so does $B(\mathbf{r})$. So, we should calculate $A'(\mathbf{r})$ by deriving the least squares solution of equations. The more are equations, the better precision can be achieved.

2.3. *Numerical Implementation*

The resolution of $B(\mathbf{r})$ and $A(\mathbf{r})$ is always quite big, which will bring the difficulty in solving equations. Here an approximate but fast method is used to calculate $A'(\mathbf{r})$. It is described as follows.

Assuming that the resolution of $A'(\mathbf{r})$ and incomplete $B(\mathbf{r})$ are $M_1 \times M_2 \times M_3$ and $N_1 \times N_2 \times N_3$ respectively, satisfying that $N_1 > M_1$, $N_2 > M_2$, $N_3 > M_3$. Obviously only $N_1 \times N_2 \times N_3$ points of $h(\mathbf{r})$ is used, which is defined as $h'(\mathbf{r})$. Filling the incomplete $B(\mathbf{r})$ with zeros to the size of $(N_1+M_1-1)\times(N_2+M_2-1)\times(N_3+M_3-1)$, defined as $B'(\mathbf{r})$. Then we get

$$A'(\mathbf{r}) * h'(\mathbf{r}) \approx B'(\mathbf{r}) \tag{20}$$

So we can calculate $A'(\mathbf{r})$ by the Fast Fourier transform (FFT) and Inverse Fast Fourier transform (IFFT) algorithms instead of solving large equations.

In the following simulation studies, we will compare the DR algorithm with two popular TAT algorithms: the time-domain reconstruction (TDR) [5] and the filtered back projection (FBP) [2, 14]. The TDR is an exact algorithm proposed in 2002. The FBP, proposed earlier, is an approximate algorithm, which is shown to work well for the limited-view case [14]. The two algorithms are originally proposed for the full view case. In the limited-view case, they both concatenate $p(\mathbf{r}, t)$ with zeros, which will cause boundaries and sharp details to become blurred.

3. Results and Discussion

We conduct the simulation in the 2-D case to reduce the computational complexity. Firstly, the model of the small-scale scanned tissue is established. Secondly, thermoacoustic signals are calculated based on Eq. (2). Finally, we use simulated thermoacoustic signals to reconstruct the tissue image, while the TDR, FBP and DR algorithms are compared.

The simulated scanned tissue and the scanning angle are shown in Fig. 1. The white circle around the tissue is the scanning track. The transducer detects signals along the circle every 2.25°.

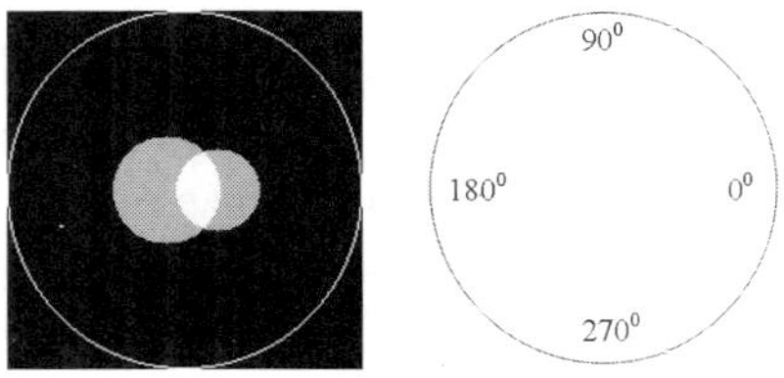

(a) (b)

Figure 1. The scanned tissue (a) and the scanning angle (b).

Fig. 2 shows the reconstruction results of the TDR, FBP and DR algorithms respectively, while the tissue is scanned from 0° to 90°. The Peak Signal to Noise Ratio (PSNR) is used to compare the DR algorithm with the TDR and the FBP. We can see that the reconstruction result of the DR algorithm is much better than those of the TDR and the FBP, and the PSNR of the DR algorithm is also much higher than those of the TDR and the FBP. Results show that the detected signals over 90° are enough to reconstruct the image of the scanned object by the DR algorithm. Moreover, PSNR is only a reference criterion in the estimation of images. It can be found that the result of the FBP is a bit better than that of the TDR, although its PSNR is lower.

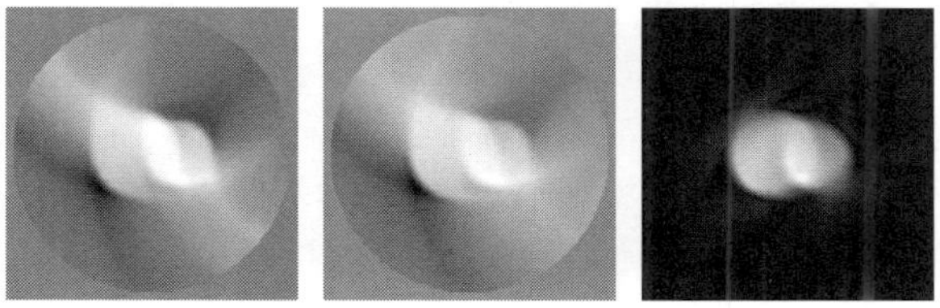

(a) (b) (c)

Figure 2. The reconstruction results (while the tissue is scanned from 0° to 90°) using (a) the TDR (PSNR = 6.65 dB), (b) the FBP (PSNR = 5.87 dB), and (c) the DR (PSNR = 23.1 dB).

Fig. 3 and Fig. 4 show the reconstruction results while the scanning angles are 135° and 252° respectively. The reconstruction results of the DR algorithm are also much better than those of the TDR and the FBP. Comparing Fig. 3 and Fig. 4 with Fig. 2, results of the TDR and the FBP are improved a lot with the increasing of scanning angle, but results of the DR algorithm are improved much less (the PSNR even become lower). The result of the DR algorithm in the 90° detection is better than those of the TDR and the FBP in the 135° detection and comparable to those of the TDR and the FBP in the 252° detection. It is because that in the DR algorithm, the detected signals over 90° are enough to reconstruct the image, but the TDR and the FBP both need full view signals.

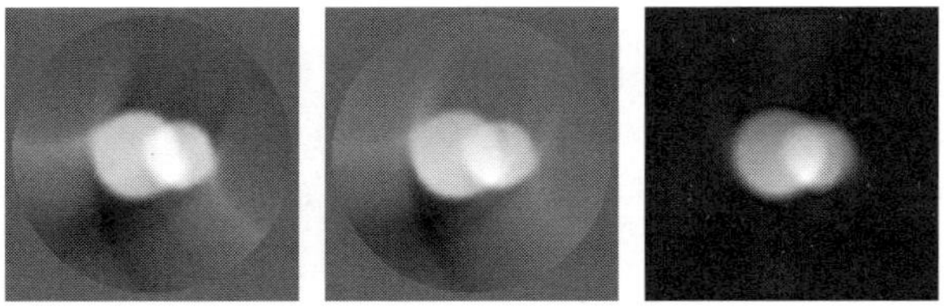

(a) (b) (c)

Figure 3. The reconstruction results (while the tissue is scanned from 0° to 135°) using (a) the TDR (PSNR = 11.4 dB), (b) the FBP (PSNR = 10.2 dB), and (c) the DR (PSNR = 22.3 dB).

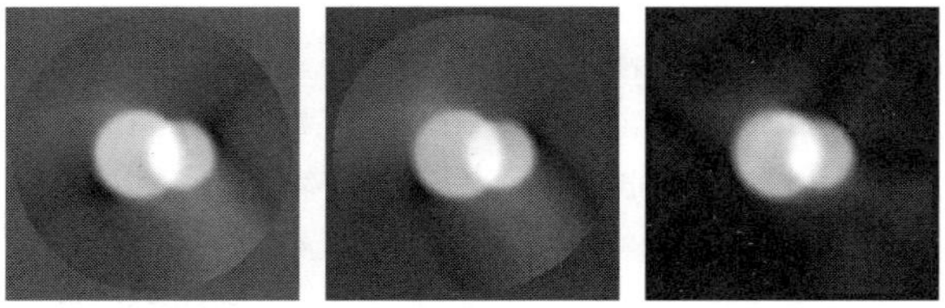

(a) (b) (c)

Figure 4. The reconstruction results (while the tissue is scanned from 0o to 252o) using the (a) the TDR (PSNR = 12.9 dB), (b) the FBP (PSNR = 15.2 dB), and (c) the DR (PSNR = 22.7 dB).

In our simulation, the size of the tissue is much smaller than the scanning radius. The DR algorithm is an approximate algorithm, which has a good precision for small-scale objects. So the DR algorithm is not suitable to be used directly in the limited-view TAT of large objects, whose size is comparable to the scanning radius. The DR algorithm for the limited-view TAT of large objects will be included in our further studies.

4. Conclusion

The DR algorithm is proposed for the limited-view TAT. It is explained theoretically that the DR algorithm is very efficient for the limited-view TAT of small-scale objects. The simulation results prove our theory and show that the result of the DR algorithm is much better than those of the TDR and the FBP, two popular algorithms, for the limited-view TAT of small-scale objects.

Acknowledgments

This work was supported by the National Basic Research Program of China (No. 2006CB705700) and Natural Science Foundation of China (No. 30570488).

References

1. K. H. Song, G. Stoica, L. V. Wang, "In vivo three-dimensional photoacoustic tomography of a whole mouse head," *Opt. Lett.*, 31(16), 2453-2455, 2006.H. Müller and B. D. Serot, *Phys. Rev.* **C52**, 2072 (1995).
2. R. A. Kruger, D. R. Reinecke, G. A. Kruger, "Thermoacoustic computed tomography-technical considerations," *Med. Phys.* **26(9)**: 1832-1837, 1999.
3. J. J. Niederhauser, M. Jaeger, R. Lemor, et al, "Combined ultrasound and optoacoustic system for real-time high-contrast vascular imaging in vivo," IEEE Trans. Med. Imaging. **24(4)**: 436-440, 2005.
4. R. A. Kruger, P. Liu, Y. Fang, et al, "Photoacoustic ultrasound (PAUS)-reconstruction tomography," Med. Phys. **22(10)**: 1605-1609, 1995.
5. M. Xu, L. V. Wang, "Time-domain reconstruction for thermoacoustic tomography in a spherical geometry," IEEE Trans. Med. Imaging. **21(7)**: 814-822, 2002.
6. Y. Xu, D. Feng, L. V. Wang, "Exact frequency-domain reconstruction for thermoacoustic tomography-I: planar geometry," IEEE Trans. Med. Imaging. **21(7)**: 823-828, 2002.
7. Y. Xu, M. Xu, L. V. Wang, "Exact frequency-domain reconstruction for thermoacoustic tomography-II: cylindrical geometry," IEEE Trans. Med. Imaging. **21(7)**: 829-833, 2002.
8. K. P. Köstli, P. C. Beard, "Two-dimensional photoacoustic imaging by use of Fourier-transform image reconstruction and a detector with an anisotropic response," Appl. Opt. **42(10)**: 1899-1908, 2003.
9. H. Yan, W. Liu, "A new technique of medical imaging-thermoacoustic imaging," Info. Med. Equip. **7**: 36-40, 2002.
10. X. D. Wang, Y. J. Pang, G. Ku, et al, "Three-dimensional laser-induced photoacoustic tomography of mouse brain with the skin and skull intact," Opt. Lett. **28(19)**: 1739-1741, 2003.
11. R. G. M. Kolkman, E. Hondebrink, W. Steenbergen, et al, "In vivo photoacoustic imaging of blood vessels using an extreme-narrow aperture sensor," IEEE J. Sel. Top. Quantum electron. **9(2)**: 343-346, 2003.
12. H. Gu, S. Yang, L. Xiang, "Photoacoustic tomography and applications in the medical clinic diagnosis," Progress Biochem. Biophys. **33(5)**: 431-437, 2006.

13. A. A. Karabutov, V. A. Andreev, B. Bell, et al, "Optoacoustic images of early cancer in forward and backward modes," In: Proc. SPIE. **4434**: 13-27, 2001.
14. Y. Xu, L. V. Wang, "Reconstructions in limited-view thermoacoustic tomography," Med. Phys. **31(4)**: 724-733, 2004.
15. M. L. Agranovsky, E. T. Quinto, "Injectivity sets for the radon transform over circles and complete systems of radial functions," J. Funct. Anal. **139(1)**: 383-414, 1996.
16. S. M. Riad, "The deconvolution problem: an overview," In: Proc. IEEE. **74(1)**: 82-85, 1986.
17. A. Bennia, S. M. Riad, "An optimization technique for iterative frequency-domain deconvolution," IEEE Trans. Instrument. Meas. **39(2)**: 358-362, 1990.

FAST PHOTOACOUSTIC IMAGING SYSTEM BASED ON A DIGITAL PHASED ARRAY TRANSDUCER

LIANGZHONG XIANG, DA XING*, DIWU YANG, SIHUA YANG, GUO HUA†
MOE Key Laboratory of Laser Life Science & Institute of Laser Life Science, South China Normal University, Guangzhou 510631, China

*Corresponding author: *xingda@scnu.edu.cn*

A fast photoacoustic imaging system consisting of a pulsed laser and a digital phased ultrasound array transducer, which was developed for visualize anatomy of the tissue structure. In this paper, theory and preliminary experimental results were presented for validating the new architecture of the fully digital photoacoustic imaging system. Advantages of this digital system over a mechanically scanned single element transducer include ability to use dynamic receive focusing (greatly increasing the depth-of-field), electronic rather than mechanical-scanning capability, and parallel data acquisition. This work demonstrates that the digital photoacoustic imaging system can potentially be used to provide fast imaging in clinic application.

1. Introduction

Photoacoustic imaging technique, which provided high ultrasonic resolution and high optical contrast tissue images, can overcome the disadvantages of pure optical imaging by measurement of laser-induced sound waves [1-4]. It has been developed as a promising tool for early detection of breast tumors or cancer [5-6]. Other applications include imaging of small animal brains [7-8], and imaging blood vessels [9].

To be accepted as a diagnostic medical imaging technique, photoacoustic imaging method should provide images in a short time. Fast imaging can not only reduce motion artifacts by respiration or heartbeat but also allow the doctor to find crucial regions of interest and to instantly diagnose the investigated region. To realize fast photoacoustic imaging, transducer array with multi-channel parallel acquisition system, and high repetition excitation sources with high-peak power are usually required. In the past decades, many researchers have focused on multi-element photoacoustic imaging. Oraevsky et al. applied a 128-channel photoacoustic imaging system for breast cancer diagnostics [10]. Niederhauser et al. used a transducer array for vascular imaging [11]. Xing et al. applied a 320-element array transducer to obtain the images of phantom and in vivo blood vessels [12-14]. The multi-element PAI has the potential to address some needs of diagnosing and monitoring the diseases, but the imaging quality need to be further improved.

This paper describes a novel fast photoacoustic imaging based on a digital phased array system. This system relies on pumping laser source to irradiate the biological tissue to produce photoacoustic signal, a linear ultrasonic transducer array is connected to a multichannel signal acquisition and real-time digital beam-formation system providing techniques of real-time dynamic receiving focus and dynamic receiving apodization to process the photoacoustic signal. Theory and preliminary experimental results were presented for validating the new architecture of the fully digital photoacoustic imaging system. This work demonstrates that the digital photoacoustic imaging system can potentially used to provide fast imaging in clinic application.

2. Materials and Methods

2.1. *Imaging Principle*

The relationship between a photoacoustic signal and the distribution of optical absorption can be expressed as

$$\oiint_{|r'-r_i|=c_0 t} A(r')dr' = -\frac{4\pi c_p}{\beta I_0 \tau} t \int_0^t p(r,t')dt' \qquad (1)$$

Where A (r') is the fractional energy-absorption per-unit-volume of soft tissue; ß is the isobaric volume expansion coefficient; C_o is the sound velocity in the target; C_p is the specific heat; P(r, t') is the PA pressure; τ is the pulse width of pulse laser, respectively.

† Work partially supported by grant 2-4570.5 of the Swiss National Science Foundation.

Because the sub-group of the linear transducer array has a characteristic impulse response, we can't directly record the photoacoustic pressure P(r, t) at the detector position, and the measured signals P'(r, t) are convolution of the PA pressure and the impulse response function h (t), so we can compute P(r, t') as

$$P'(r, t) = p(r, t) * h(t) \qquad (2)$$

Then P(r, t) can be obtained by deconvoluting the impulse response of the transducer from the measured PA signal.

$$p(r,t) \approx IFFT\left[\frac{j\omega p'(\omega)}{I(\omega)}\right]\left[1+\cos\left(\frac{\pi\omega}{\omega_c}\right)\right] \qquad (3)$$

Where p' (ω) and I (ω) were the Fourier transform of the PA signal P'(r, t) and h (t), the impulse response of the detector. ω_c is the cut-off frequency, the apodizing function $1+\cos(\pi\omega/\omega_c)$ is used to bandwidth limit the signal to ω_c . The impulse response of the linear transducer array was measured by illuminating a black rubber belt with a weak focused pulse laser. The linear transducer array is behind the black rubber belt.

2.2. *Photoacoustic Imaging System*

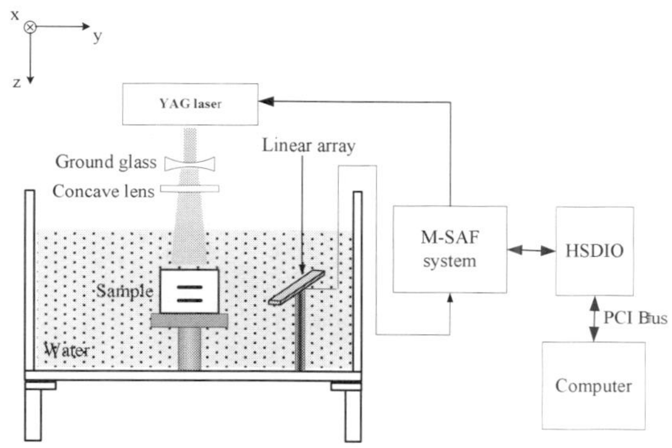

Figure 1. Experimental setup for B-scan M-SAF PA imaging system.

The schematic of the experimental setup is shown in Fig 1. A Q-switched Nd:YAG laser (LOTIS TII Ltd, Minsk, Belarus) was used to provide 532-nm laser pulses with a full-width half-maximum (FWHM) value of 6.5 ns, with a repetition rate of 10Hz. The laser beam was expanded and homogenized to provide an incident energy density of <10 mJ/cm^2 to irradiate the sample for generating photoacoustic signals. A 25MHz clock signal, provided by control circuit, was used for triggering the pulse laser and controlling the multi-element linear transducer array to acquire photoacoustic signals. The linear ultrasound transducer array (L7L38A, SIUI) for recording the photoacoustic signals is a 7.5-MHz piezoelectric array of 128 elements with greater than 70% pulse-echo -6dB fractional bandwidth. The physical dimensions of the elements are given by an elemental elevation aperture of 10-mm and a lateral pitch of 300-um. Each element of the transducer array has a thin cylinder ultrasonic lens that produces a geometric focus approximately 35mm in front of the transducer array to select 2D image plane and suppress the out-of-plane signals. The signals from the transducers, after pre-amplification and phase adjustment, are acquired with the NI PCI-6541 High-Speed Digital Wfm (50 MHz, Selectable. Voltage, 32 channels, 1 Mbit/channel). The system operation and data collection are controlled by a personal computer. The multi-element linear transducer array was driven by a step motor to circularly scan around the sample, detected the photoacoustic signals in the imaging plane at each position. The linear transducer array and the samples were both immersed in a tank of water for better coupling. The induced photoacoustic waves are recorded to transferred to a personal computer for further data processing using MATLAB (version 7.0, Mathworks), projections were calculated with the improved limited-field filtered back projection (LFBP) algorithm.

3. Results

3.1. *Photoacoustic Imaging Based on the Digital Phased Array System*

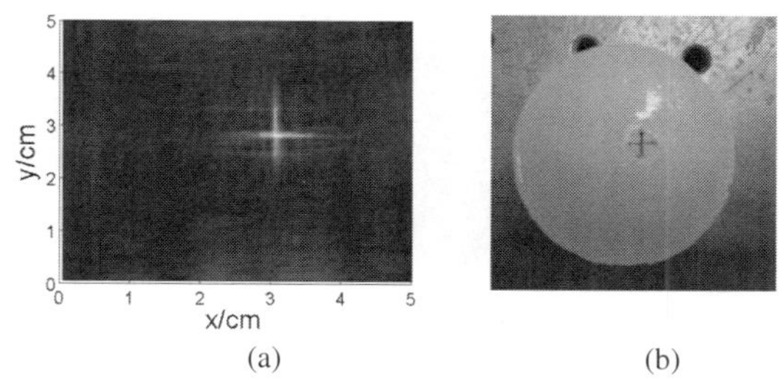

Figure 2. Photoacoustic imaging Based on a Digital Phased Array System; (a) is the reconstructed photoacoustic image of black cross of pencil trace. (b) is the photograph of the cross section of the phantom.

In our experiment, a black cross of pencil trace with diameters of about 0.1mm and a length of 10mm, which were imbedded in an intralipid-20% dilution. The targets were embedded in the cylindrical phantom at a depth of 3mm. Fig 2(a) is the reconstructed photoacoustic image of black cross of pencil trace. Fig 2(b) is the photograph of the cross section of the phantom. As we can see the reconstructed image produced by our imaging system agrees well with the original sample. The experiment demonstrates that the fast photoacoustic tomography based on a digital phased array system has the ability of reconstructing the targets inside the sample with good imaging quality.

3.2. *Photoacoustic Tomography by a Digital Phased Array System*

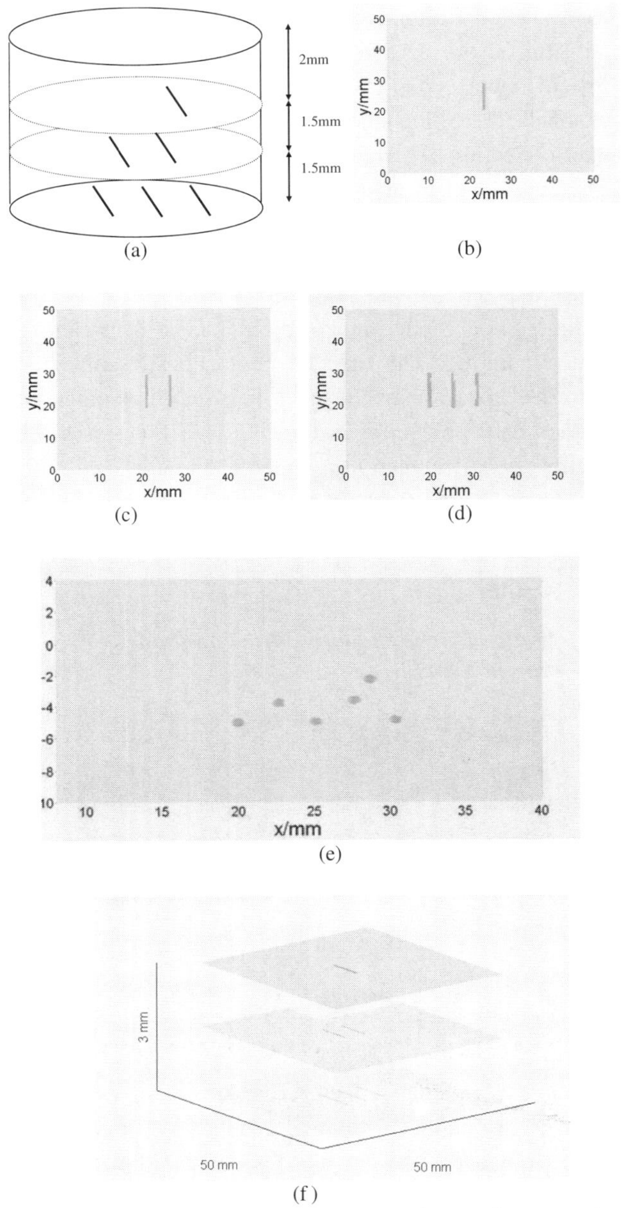

Figure 3. Series of photoacoustic tomography image of copper rods; (a) is diagram of the sample structure and the measurement; (b), (c), (d) are series of photoacoustic tomography image from different layers parallel to the phantom surface. The first layer starts from 2.0 mm beneath the phantom surface, and the layer spacing is 1.5 m; (e) is B-scan photoacoustic image in the z-x plane; (f) is thee dimensional rendering of the phantom by photoacoustic tomography images.

Each element of the transducer array has a thin cylinder ultrasonic lens that produces a geometric focus approximately 35mm in front of the transducer array to select 2D image plane and suppress the out-of-plane signals to realize the photoacoustic computed tomography. The ability of computed tomography of the photoacoustic imaging system has been tested in our experiment. The z-axis resolution of the transducer array can reach 1.4 mm.

To validate the tomographic ability of the imaging system, six copper rods (0.8 mm in diameter, 10 mm in length) buried in an intralipid-20% dilution. Fig 3(a) is the diagram of the sample structure and the measurement. Fig3.(b), (c), (d)) shows series of photoacoustic tomography image of copper rods from different layers parallel to the phantom surface. The first layer starts from 2.0 mm beneath the phantom surface, and the layer spacing is 1.5 mm. The image of the copper rod from the first layer is easily discernible (Fig 3(b)). Increasing by the depth of phantom, the sound to noise ratio (SNR) noticeably deteriorated and the copper rods became obscure, yet still recognizable (Fig 3(c), (d)). Because of the side illumination, light energy exponentially decays from the surface inward, which results in generation of much weaker signals in deep regions.

3.3. *Photoacoustic Imaging of Breast Tumor by the Digital Phased Array System*

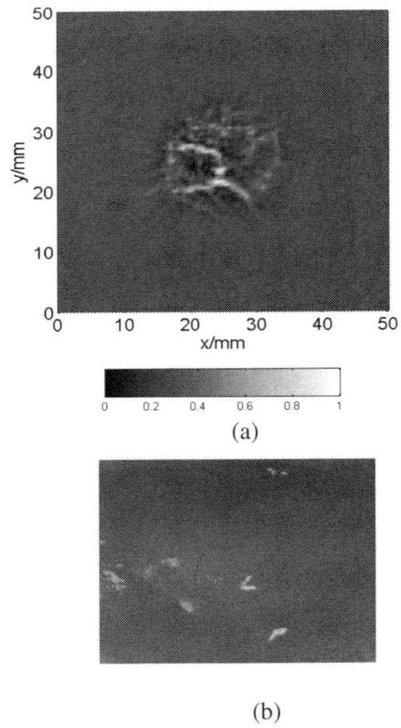

Figure 4. Photoacoustic image of early tumor detection. (a) Photoacoustic imaging of the mouse tumor; (b) Photograph of C127 breast cancer tumor in a mouse back on 10 days of cell incubation.

The digital photoacoustic imaging system was used to detect early breast tumors in a mouse tumor model. In Fig.4 (a), the tumor associated vasculature distortion can be easily recognized on the photoacoustic image. Fig.3 (b) is a photograph of the rat back tumor, an elliptical line marked the tumor area. As we can see, in the tumor area, the growing new blood vessels seriously distort the normal vascular architecture. The abnormal vessels that were owing to angiogenesis surround and feed the tumor. From the experimental results, it demonstrated that the tumoral vascular structure could be clearly and fast imaged noninvasively by our imaging system.

4. Conclusion

In this paper, we reported a fast photoaccustic imaging system based on a digital phased array. In our experiment system, an integrity multi-element synthetic aperture focusing (M-SAF) photoacousitc imaging system using real-time digital beamformer is developed. This system relies on pumping laser source to irradiate the biological tissue to produce photoacoustic signal, a linear ultrasonic transducer array is connected to a multichannel signal acquisition and real-time digital beam-formation system providing techniques of real-time dynamic receiving focus and dynamic receiving apodization to process the photoacoustic signal. This photoacoustic imaging system may have the potential for detection of tumors at the early stage, monitoring the angiogenesis around the tumors.

Acknowledgments

This research is supported by the National Natural Science Foundation of China (30470494; 30627003) and the Natural Science Foundation of Guangdong Province (7117864).

References

1. Yaguang Zeng, Da Xing and Yi Wang, *Opt. Lett.*, **29(15)**, 1760-1762 (2004).
2. Xiang Liangzhong, Xing Da, Gu Huaimin et al, *Chin. Phys. Lett.*, **24(3)**, 751 (2007).
3. X. Wang, Y. Pang, G. Ku, et al, *Nat. Biotechnol.*, **21**, 803-806 (2003).
4. L.ZH. Xiang, D. Xing, H.M Gu, et al, *Journal of Biomedical Optics,* **12**, 014001-8 (2007).
5. R. O. Esenaliev, A. A. Karabutov and A. A Oraevsky, *IEEE J. Sel. Top. Quantum. Electron.*, **5**, 981-988 (1999).
6. A. A. Oraevsky, V. A. Andreev, A .A .Karabutov, R. D. Fleming, Z. Gatalica, H. Singh and R. O. Esenaliev, *Proc. SPIE*, **3597**, 352-363 (1999).
7. Wang X, Pang Y, Ku G, Stoica G, Wang LV, *Optics Letters*, **28**: 1739-1741 (2003).
8. Sihua Yang, Da Xing, Yeqi Lao, Diwu Yang, Lvming Zeng, Liangzhong Xiang, and Wei Chen *Appl. Phys. Lett.*, **90(24)**, 24392 (2007).
9. Kolkman RGM, Hondebrink E, Steenbergen W, de Mul FFM, *IEEE Journal of Selected Topics in Quantum Electronics*, **9(2)**: 343-346 (2003).
10. A. Ermilov, A.Coniusteau, K. Mehta, R. Lacewell, P. M. Henrichs and A. A. Oraevsky, *Proc. SPIE,* **6086**: 608609 1-12 (2006).
11. J. J. Niederhauser, M. Jaeger, R. Lemor, P. Weber, and M. Frenz, *IEEE Trans. Med. Imaging*. **24**: 436 (2005).
12. W. D. Yang, D. Xing, H. M. Gu, Y. Tan and L. M. Zeng, *Appl. Phys. Lett.*, **87**: 194101-3 (2005).
13. W. D. Yang, D. Xing, Y. Tan, H. M. Gu and S. H. Yang, *Appl. Phys. Lett.*, **88**: 174101-3 (2006).
14. B. Zh. Yin, D. Xing, Y. Wang, Y. G. Zeng, Y. Tan, and Q. Chen, *Phys. Med. Biol.,* **49**: 1339-1346 (2004).

MAGNETIC RESONANCE IMAGE RECONSTRUCTION FROM PARTIAL K-SPACE USING ANALOGICAL BASIS DECONSTRUCTION THEORY

FENGQIN LIU

Department of Information Physics, School of Sciences, Zhejiang Forestry University, Zhejiang Linan, P.R. China

PENG MIAO

Department of Biomedical Engineering, Shanghai Jiao Tong University, Shanghai, P.R. China

We present a new Analogical Basis Deconstruction (ABD) theory for discrete signal processing in this paper. We also present the Asymptotic Iterative Estimate (AIE) algorithm for ABD theory. We apply this new theory to reconstructing magnetic resonance (MR) images from partial *k*-space. Our proposed theory was evaluated on both simulated data and human brain MR data. It is also compared with conventional techniques (Zero Filling, POCS and HM) for MRI image reconstruction from partial *k*-space. The results show that our new theory exhibits a great reconstruction quality. The performance of our new theory is still good even when the conventional techniques are invalid.

1. Introduction

The magnetic resonance imaging (MRI) has become the most important imaging method in many fields like medicine, neurosciences, and cognitive sciences. In a standard conventional MRI experiment, the subject needs to stay relatively long time in the MRI tunnel [1]. Although the scanning speed of MRI is faster today, there are still artifacts caused by subject's movement in the images. Recently, dynamic MRI such as cardiac imaging and functional MRI interest more and more researchers [2]. The dynamic MRI requires much shorter acquisition time than conventional MRI. The most common method for MRI is the phase-encoding method, in which the object of interest is sampled in the frequency-domain. The raw frequency-domain (*k*-space) has two frequency directions. The horizontal direction is frequency encoding, and the vertical one is phase encoding. The most common reconstruction technique for MRI is to compute the inverse discrete Fourier transform (DFT) of the *k*-space data. The magnitude of DFT output gives the magnetic resonance image for visual inspection and other applications. Based on the principles of MRI, one way to reduce the acquisition time is sampling only a part of full *k*-space which is called partial *k*-space acquisition. The partial *k*-space acquisition method needs no hardware modification [3].

Since the *k*-space data doesn't satisfy the Hermitian symmetry, we can't reduce the number of phase encoding based on the symmetry from the partial *k*-space. However, there are different reconstruction methods from partial *k*-space in the literature. These conventional methods can be classified into two catalogs. One is phase correction method [4] and the other is missing data retrieval method like *k*-space interpolation and extrapolation [5]. The phase correction method is based on the assumption that spatial phase varies slowly in the MR image. Because this assumption doesn't always match the real condition, there are always phase estimation errors. The missing data retrieval method is based on the information theory and signal estimation theory. Because the weak correlation of *k*-space data and noise influence, such reconstruction method is generally less effective than the phase correction method.

In contrast to the above two catalogs, we propose a novel reconstruction method based on Analogical Basis Deconstruction theory in this paper. The Analogical Basis Deconstruction (ABD) theory is a signal processing theory which is originally and completely presented in this paper. In the ABD theory, we define the one-dimension discrete signal as a sum of weighted basis. The same signal can be represented to different form according to different basis we choose, so there are many model for one signal. In our ABD theory, the transformation such as DFT, inverse DFT just transforms the basis to new form which is determinate and does not change the weight of each basis. In the partial *k*-space application, the partial sampling method just affects the basis in the signal model. If we can deconstruct signal (the inverse DFT of partial *k*-space) into a sum of weighted basis and estimate the weight of each basis, we can reconstruction the signal. However, the bases in partial k-space model are not orthogonal, so the weights estimation is a much difficult question in practice. We propose a new method called Asymptotic Iterative Estimate (AIE) to complete this difficult task. The ABD theory has been shown to perform better than conventional reconstruction methods in many cases.

In this paper, we first present our ABD theory from definitions to theorems with proofs. Then, the AIE algorithm for weights estimation and signal reconstruction is proposed. The ABD theory was evaluated on both simulated data and human brain MR data. It is also compared with conventional techniques (Zero Filling, HM [6] and POCS [7]) for MRI image reconstruction from partial *k*-space. The results show that the performance of ABD theory is much better than the conventional methods.

2. Analogical Basis Deconstruction (ABD) Theory

Definition 1. Suppose that the field $\mathbb{C}$ is the field of complex numbers. The linear space $\mathbb{C}^N$ is defined as the set of all *N*-dimension column vectors with components from the field $\mathbb{C}$.

Definition 2. The so-called 'one-Dimension Discrete Signal Space' is the set S which is defined as a linear subspace of $\mathbb{C}^N$. So any 'one-dimension discrete signal' $f(n)$ is one column vector of the set S where $n = 1 \ldots N$.

Definition 3. We define the inner product of any two column vectors (signals) f, $g \in S$:

$$(f,g) = \overline{g^T} f = \overline{(g,f)}. \tag{1}$$

Lemma 4. According to Linear Algebra Theory, there must be at least one standard basis for the linear space $\mathbb{C}^N$. For simplicity, we choose the basis

$$\left\{ \varepsilon_n \mid \varepsilon_n = \left[\underbrace{0\,0 \cdots \overset{n}{\hat{1}} \cdots 0\,0}_{N} \right]^T ; n = 1 \ldots N \right\}$$

as the standard basis in our ABD theory.

As a corollary we have

Theorem 5. Any vector (signal) $f \in S$ can be represented as below.

$$\begin{aligned} f &= a_1\varepsilon_1 + a_2\varepsilon_2 + \cdots + a_{N-1}\varepsilon_{N-1} + a_N\varepsilon_N \\ &= \sum_{n=1}^{N} a_n \varepsilon_n \end{aligned} . \tag{2}$$

Where $a_n = f(n) = (f, \varepsilon_n)$

Definition 6. According to Fourier Transformation theory, we define the Discrete Fourier Transformation (DFT) and Inverse Discrete Fourier Transformation (IDFT) as two linear operations in the linear space $\mathbb{C}^N$.

$$\begin{aligned} F(k) &= DFT(f) \\ &= \sum_{n=1}^{N} f(n) e^{-j2\pi(k-1)\frac{n-1}{N}} \qquad 1 \le k \le N \end{aligned} . \tag{3}$$

$$\begin{aligned} f(n) &= IDFT(F) \\ &= \frac{1}{N} \sum_{k=1}^{N} F(k) e^{j2\pi(k-1)\frac{n-1}{N}} \qquad 1 \le n \le N \end{aligned} . \tag{4}$$

Definition 7. For any two vectors (signals) f, $g \in S$, we define the '$.*$' operation as below.

$$f .* g = \sum_{n=1}^{N} f(n)\, g(n)\, \varepsilon_n . \tag{5}$$

Lemma 8. According to the definition 7, we can define the model of partial *k*-space as below.

(1) The vector $Sample \in S$ is a column vector which presents the valid data in the model of partial *k*-space.

$$Sample(n) = \begin{cases} 1 & \textit{There are data at position } n \\ 0 & \textit{There are no data at position } n \end{cases} \quad (n = 1 \ldots N)$$

(2) Assume that $F \in S$ is the *k*-space of $f \in S$, $\widetilde{F} \in S$ is the partial *k*-space of $f \in S$, then

$$F = DFT(f). \tag{6}$$

$$\widetilde{F} = F .* Sample. \tag{7}$$

Theorem 9. Suppose that $\widetilde{f} \in S$ is the Inverse Discrete Fourier Transformation of $\widetilde{F} \in S$, then

$$\tilde{f} = \sum_{n=1}^{N} a_n \tilde{\varepsilon}_n . \tag{8}$$

Where $\left\{ \tilde{\varepsilon}_n \mid \tilde{\varepsilon}_n = IDFT\left(DFT\left(\varepsilon_n\right) .* Sample \right) \right\} \quad \left(n = 1 \ldots N\right)$

Proof. The operations of DFT and $IDFT$ are two linear operations in the linear space $\mathbb{C}^N$, so we have this.

$$\begin{aligned} F = DFT(f) &= DFT\left(\sum_{n=1}^{N} a_n \varepsilon_n \right) \\ &= \sum_{n=1}^{N} a_n \left(DFT\left(\varepsilon_n\right) \right) \end{aligned} . \tag{9}$$

For the partial k-space $\tilde{F} \in S$ of $f \in S$, we can easily prove this.

$$\begin{aligned} \tilde{F} &= F .* Sample \\ &= DFT\left(\sum_{n=1}^{N} a_n \varepsilon_n \right) .* Sample \\ &= \sum_{n=1}^{N} a_n \left(DFT\left(\varepsilon_n\right) .* Sample \right) \end{aligned} . \tag{10}$$

At last, we have this.

$$\begin{aligned} \tilde{f} &= IDFT\left(\tilde{F}\right) \\ &= IDFT\left(\sum_{n=1}^{N} a_n \left(DFT\left(\varepsilon_n\right) .* Sample \right) \right) \\ &= \sum_{n=1}^{N} a_n \left(IDFT\left(DFT\left(\varepsilon_n\right) .* Sample \right) \right) \end{aligned} . \tag{11}$$

Thus

$$\tilde{f} = \sum_{n=1}^{N} a_n \tilde{\varepsilon}_n . \tag{12}$$

Where

$$\left\{ \tilde{\varepsilon}_n \mid \tilde{\varepsilon}_n = IDFT\left(DFT\left(\varepsilon_n\right) .* Sample \right) \right\} \quad \left(n = 1 \ldots N\right)$$ □

Definition 10. We define the 'Analogical Basis' of a vector (signal) $\tilde{f} \in S$ from partial k-space $\tilde{F} \in S$ as below.

$$\left\{ \tilde{\varepsilon}_n \mid \tilde{\varepsilon}_n = IDFT\left(DFT\left(\varepsilon_n\right) .* Sample \right) \right\} \quad \left(n = 1 \ldots N\right)$$

If we can deconstruct the signal $\tilde{f} \in S$ to $\tilde{f} = \sum_{n=1}^{N} a_n \tilde{\varepsilon}_n$, we can get the coefficients $\left\{ a_n \middle| n = 1 \ldots N \right\}$. Then we can reconstruct the original vector (signal) $f \in S$ using the equation below.

$$f = \sum_{n=1}^{N} a_n \varepsilon_n . \tag{13}$$

3. Asymptotic Iterative Estimate (AIE) Algorithm

Algorithm:

(1) Acquire the partial k-space data $\tilde{F}$.

(2) Compute the column vector $\tilde{f} = IDFT\left(\tilde{F}\right)$ and the complex number $Value$ as the maximum modulus of every element in the vector $\tilde{f}$. Construct the column vector $\tilde{g} = \tilde{f}$.

(3) Construct the matrix

$$E_{N \times N} = \begin{pmatrix} \tilde{\varepsilon}_1^T \\ \tilde{\varepsilon}_2^T \\ \vdots \\ \tilde{\varepsilon}_N^T \end{pmatrix} = \begin{pmatrix} IDFT\left(DFT\left(\varepsilon_1^T\right) .* Sample \right) \\ IDFT\left(DFT\left(\varepsilon_2^T\right) .* Sample \right) \\ \vdots \\ IDFT\left(DFT\left(\varepsilon_N^T\right) .* Sample \right) \end{pmatrix}$$

and the all-zeros column vector $P_{rec} = \left(\underbrace{0 0 \cdots 0 0}_{N} \right)^T$

(4) Find the maximum modulus element in the vector $\tilde{g}$. Define the integer number $P_{\max}$ as the position of that element and the real number $V_{\max}$ as the modulus of that element.

Then make $P_{rec}\left(P_{\max}\right) = 1$

(5) $\tilde{g} = \tilde{g} - \left(coef \cdot \dfrac{V_{\max}}{E\left(P_{\max}, P_{\max}\right)} \right) \tilde{\varepsilon}_{P_{\max}}$

(6) Find the maximum modulus element in the vector $\tilde{g}$. If the modulus of that element is bigger than the real number $\left(cond \cdot Value\right)$, the precision of

the iteration is reached and jump to step (7). If not, jump to step (4).

(7) Suppose that there are M ($M < N$) elements in the vector P_{rec} whose value is one and the positions of those elements are

$$\{p_m \mid m = 1 \ldots M\}.$$

Construct the Signal Reconstruction Matrix:

$$R_{N\times M} = \left(\tilde{\varepsilon}_{p_1}, \tilde{\varepsilon}_{p_2}, \ldots \tilde{\varepsilon}_{p_M}\right)$$

(8) Assume that x is the estimated signal of the original signal. Build the signal reconstruction linear system:

$$R_{N\times M} \bullet x = \tilde{f}$$

(9) Solve the linear system in step (8) using Least Squares Estimation (LS) method:

$$x = \left(R_{N\times M}^T \bullet R_{N\times M}\right)^{-1} \bullet R_{N\times M}^T \bullet \tilde{f}$$

(10) Construct all-zeros column vector

$$f = \left(\underbrace{0\,0\cdots0\,0}_{N}\right)^T$$

Then, $f(p_j) = x(j) \qquad j = 1 \ldots M$ is the reconstructed signal of the original signal.

4. Experiments and Discussions

We use the simulated Shepp-Logan image and real human brain MR data to evaluate the performance of our proposed reconstruction method. We also compare our method with the conventional method such as zero filling, HM and POCS. The reconstruction quality is assessed visually and by comparing the images reconstructed from partial and complete *k*-spaces respectively.

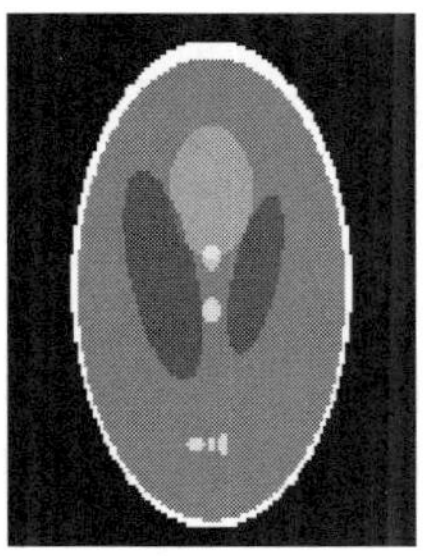

Figure 1. The reconstruction Shepp-Logan image from full k-space.

4.1. *Simulated Shepp-Logan Experiment*

The size of simulated Shepp-Logan image is 128×128. The reconstruction image from full *k*-space (128 *k*-space line data) of simulated Shepp-Logan image is shown in Fig 1. The full *k*-space (128 *k*-space line data) of simulated Shepp-Logan image is truncated which means that only the n~128 *k*-space line data is used to reconstruct the image. The results of reconstruction from different truncated *k*-space data (from left to right columns: 53~128, 65~128, 66~128, 73~128, 81~128 *k*-space line data) by different methods (form top to bottom rows: Zero filling, HM, POCS and the proposed ABD method) are shown in figure 2. The HM and POCS methods both need the phase correction from the symmetrical line data around the center of partial *k*-space (65th line data). When the beginning line of partial *k*-space is less than 65, the phase correction makes the reconstruction quality much better than Zero filling method. If the beginning line of partial *k*-space is bigger than 65, the phase correction can not work and the performances of HM and POCS methods are not good as Zero filling method. The performance of POCS method is better than HM. However, the proposed ABD method can work well in all situations. We define the

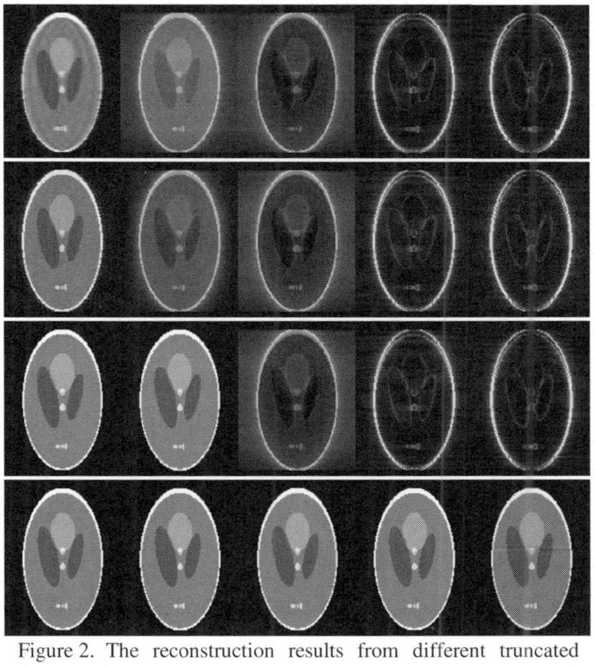

Figure 2. The reconstruction results from different truncated *k*-space Shepp-Logan data (from left to right columns: 53~128, 65~128, 66~128, 73~128, 81~128 *k*-space line data) by different methods (form top to bottom rows: Zero filling, HM, POCS and the proposed ABD method).

reconstruction error to quantify the performance of different reconstruction methods.

$$RE = \frac{1}{M \times N} \sum_{m=1}^{M} \sum_{n=1}^{N} \left| \hat{I} - I \right|. \qquad (14)$$

I is the reconstruction image ($M \times N$) from full k-space. $\hat{I}$ is the reconstruction image from partial k-space. Table 1 gives more quantitative results and comparisons between different reconstruction methods. The results also show that the proposed ABD method can give a perfect performance even when the HM and POCS method can not work well. We can reconstruct the image with high quality from 81~128 k-space line data (only 37% data of the full k-space data) by the ABD method.

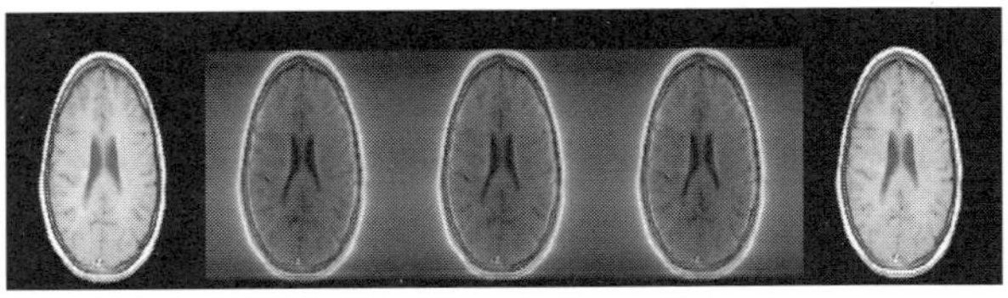

	Full K-space	Zero filling	HM	POCS	ABD
RE		23.38	23.38	23.48	0.40

Figure 3. Experiment from human brain k-space data by different reconstruction method. The first image is reconstructed from full k-space data. The others are reconstruction images from partial k-space (66~125 k-space line data) by the conventional and ABD method. The reconstruction error (RE) of each image is also presented.

Table 1. Reconstruction errors (RE) as a function of truncated bandwidth (~128).

Bandwidth	41	45	49	53	57	61	65	69	73	77	81
ZeroFilling	5.37	6.11	6.86	8.04	9.41	11.43	32.89	63.14	62.22	62.53	63.23
HM	4.25	5.06	5.73	6.74	7.44	9.49	28.23	63.14	62.22	62.53	63.23
POCS	0.34	0.37	0.41	0.47	0.54	0.64	1.79	63.98	63.52	63.67	64.03
ABD	0	0	0	0	0	0	0	0	0	0.04	0.25

4.2. *Human Brain Data Experiment*

We use the real human brain MR data to evaluate the proposed ABD method and the conventional methods. The results are shown in figure 3. The first image (128×128) in figure 3 is the reconstruction image from full k-space. Then we just use the 66~125 k-space line data (about 47% data of the full k-space data). The reconstruction error (RE) of each reconstruction image is also presented. Both the visual quality and the quantitative RE evaluation show that the performance of proposed ABD method is much better than the conventional methods.

5. Conclusion

We propose a new Analogical Basis Deconstruction (ABD) theory and apply it to reconstruct MR images from partial k-space. The results obtained with the aid of both simulated and real data demonstrate that the ABD method exhibits a great reconstruction quality. The performance of ABD method is still good even when the conventional techniques are invalid.

Acknowledgments

This work is supported by Scientific Research Development Fund of Zhejiang Forestry University under Grant No. 2351000655.

References

1. Ahmed S. Fahmy, Bassel Tawfik, and Yasser M. Kadah. *Proc. IEEE International Conference on Image Processing (ICIP)*, **Vol. 2**, 487 (2000).
2. Mark Bydder and Matthew D. Robson. *Magn Reson. Med.* **53**, 1393 (2005).
3. P. Miao, B. Hiba, J. Luo, D. Sappey-Marinier, I. Magnin, Y. Zhu. *Nucl. Instr. and Meth., A* **571,** 239 (2007).
4. G. McGibney, M.R. Smith, S.T. Nichols, A.Crawley. *Magn. Reson. Med.* **30**, 51 (1993).
5. V. Rasche, T. K¨ohler, and R. Proksa. *Proc. 15th Meeting, ESMRMB*, p. 352 (1998).
6. D. C. Noll, D. G. Nishimura and A. Macovski. *IEEE Trans. Med. Imag.*, **MI-10 (2)**, 154 (1991).
7. E. M. Haacke, E. D. Lindskog, W. Lin. *J. Magn. Reson.* **92**, 126 (1991).

STUDY OF MEDICAL X-RAY IMAGING BASED ON OPTICAL FIBER-TAPERED CCD*

CHEN SHUYUE
College of Information Science and Engineering, Jiangsu Polytechnic University, Changzhou, 213164, China

WANG ZHISHE
School of Applied Science, Taiyuan University of Science and Technology, Taiyuan, 030024, China

The paper studied a kind of high resolution and digitized medical X-ray radiographic device based on optical fiber-tapered CCD. The presented system is composed of optical fiber-tapered CCD, medical X-ray II-generation intensifier appended the optical fiber plate, image grabber, computer and image processing software. The paper emphasized on the coupling techniques between the fiber taper and CCD, the fiber taper and the fluorescent screen of the image intensifier. A formula of calculating spatial resolution of the optical fiber-tapered CCD was given. The results were discussed through the experiments. Finally, the medical applications of the hand bones diagnosis were shown.

1. Introduction

The medical X-ray imaging is widely applied in the clinic for diagnosis of disease, including the traditional medical radiography by film and image intensifier, CR(computed radiography) where a storage phosphor in a cassette is irradiated and later on read out by a laser scanner, and DR based flat panel (direct radiography)[1-3]. With the development of imaging technology, the traditional methods are challenged due to their non-digitalization or low resolution. CR and DR are up to date by means of the advanced imaging material, micro electronics and optical techniques, so their imaging quality very high and the digital image can be obtained directly. But these systems keep expensive so as to be not afforded for the small and middle-scale hospital. The paper studied a kind of high resolution and digitized medical X-ray imaging device based on an optical fiber-tapered CCD sensor to overcome above disadvantages.

The presented system is composed of optical fiber-tapered CCD, X-ray II-generation intensifier appended the optical fiber plate, image grabber, computer and image processing software. To satisfy the need of X-ray imaging with high resolution, the optical fiber taper, which consists of a bundle of optical fibers and widely used in many fields[4-8], acts as optical relay component. The lighted image is exported from the fluorescent screen of X-ray II-generation image intensifier. Then an optical fiber panel connecting with it is coupled with photosensitive surface of CCD.

We can employ two kinds of simple medical imaging methodology with CCD capturing, one is to use lens, the other is to connect output fluorescent screen and CCD with optical fiber taper. The former may cause the image sharpness down due to de-focus and optical aberration, the latter can keep up with the better image quality.

The paper emphasized on the coupling techniques between the fiber taper and CCD, the fiber taper and the fluorescent screen of the image intensifier. The formula of calculating spatial resolution of the optical fiber-tapered CCD was given. The results were discussed through the experiments. Finally, the medical applications in the hand bones diagnosis were shown.

2. Construction of the Medical X-ray Imaging and Methodology

The researched system consists of the optical fiber taper, CCD, X-ray II-generation intensifier, optical fiber plate, image grabber, computer and image processing software. Its structure is shown in Fig.1.

Where the output screen of the image intensifier is sealed with an optical fiber plate instead of the general glass, the big side of the fiber taper is connected to it. The small one is coupled with the sensitive surface of CCD. The high resolution can be achieved because the direct interface is adopted by means of the taper, in which the diameter of each fiber keeps several microns matching the pixel size of CCD.

* This work is supported by Changzhou city (China) science and technology plan project (CE2007029).

When X-ray penetrates through an object, it decays exponentially. The projection of the object forms an image on the fluorescent screen of the intensifier. Then it outputs to CCD under the conduct of the optical fiber taper. Finally, CCD captures and transforms it through an image grabber to computer. We may employ CMOS device to capture because of low power consume and wide application in lots of fields.

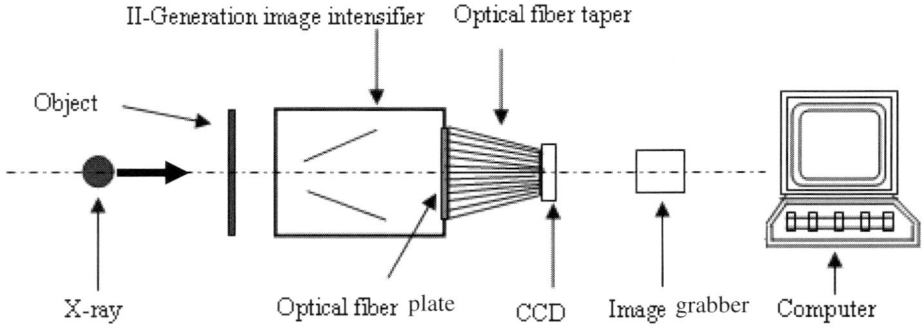

Figure 1. Construction of medical X-ray imaging based on optical fiber-tapered CCD

3. Spatial Resolution Calculation

For the sub-system of the optical fiber-tapered CCD, the input is on the big side of taper and the image output yields by CCD.

Suppose the cross section of each fiber of taper to be round and the image creation to be on the small side of taper, the maximum spatial resolution may be expressed as

$$R = \frac{1}{2\phi} \tag{1}$$

Where ϕ is the center distance between two adjacent fibers on the surface of the taper big side, generally several to decades microns. The unit of R is Line-pair/mm.

Two cases take place when the size of CCD pixel is not the same one of the fiber on the small side. Let k be ratio of cone degree of taper, L be diagonal length of CCD pixel. The spatial resolution of the sub-system in 45 degree direction may be written as

$$\eta = \begin{cases} \dfrac{R}{L/(\phi/k)} & L \geq \phi/k \\ \dfrac{R}{\phi/k / L} & L < \phi/k \end{cases} \tag{2}$$

Using Equ.(1), we re-write (2) as the Equ.(3).

For example, CCD pixel size is 3.24 μ m*3.275 μ m, fiber diameter of taper big side is 8 μ m, and ratio of cone degree of the taper is 3. We can obtain that L=4.607, ϕ /k=2.667. Hence the spatial resolution is equal to 36.18 Lp/mm by use of Equ.(3). In other word, the detectable minimum size is 13.82 μ m.

$$\eta = \begin{cases} \dfrac{1}{2kL} & L \geq \phi/k \\ \dfrac{kL}{2\phi^2} & L < \phi/k \end{cases} \tag{3}$$

4. Medical Application

We applied the presented medical X-ray system in hand bones radiography. In the system CCD pixel size is 7.58 μ m*7.58 μ m, fiber diameter of taper big side is 8 μ m, and ratio of cone degree of the taper is 3. Therefore the spatial resolution of the optical fiber-tapered CCD attains 15.55 Lp/mm according to the Equ.(3). The imaging results, including the fingers, palm, are shown in Fig.2 and Fig.3. It is feasible for doctors to diagnose disease.

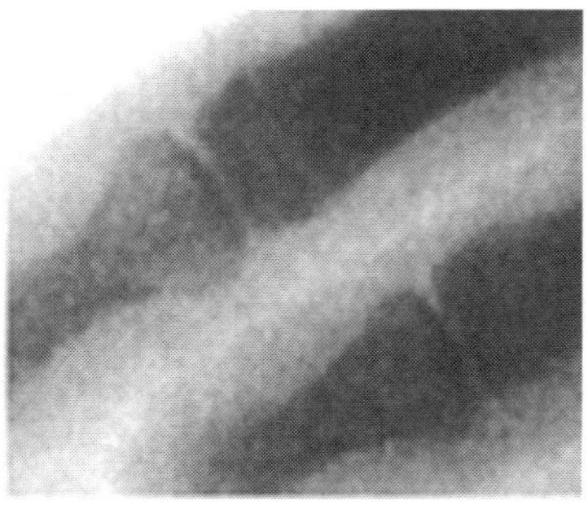

Figure 2. X-ray image of fingers

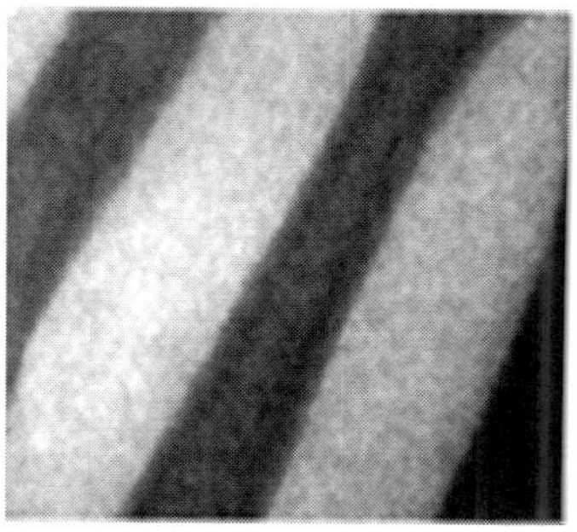

Figure 3. X-ray image of palm

Although the resolution of the optical fiber-tapered CCD is high enough, the imaging quality of the whole system is not certain to be qualified. The X-ray intensifier becomes a bottleneck of limiting the image sharpness. Therefore we may improve the system properties by use of a high-resolution intensifier.

5. Conclusions

Our work may summarize that:

(1). The methodology of the medical X-ray imaging system based on the optical fiber-tapered CCD is presented.

(2). The given system has a high image quality, which is competent for the medical application.

(3). The spatial resolution formula of CCD Coupled optical fiber taper is presented, by which the system design can be referred.

(4). The sub-system of the optical fiber-tapered CCD has a high resolution, but the X-ray intensifier affects the quality of the whole system. We should make the section of X-ray to visible light better.

References

1. Keiichi Nakagawa, Seiichirou Mizuno, Yukimasa Aoki and Kuni Ohtomo, C-MOS Flat-panel Sensor for Real-time X-ray Imaging, Radiation Medicine: **18**, No.6, 2000, Pages 349–353.
2. Wei Zhao, D.C. Hunt, Kenkichi Tanioka and J.A. Rowlands, Amorphous selenium flat panel detectors for medical applications, Nuclear Instruments and Methods in Physics Research Section A: Accelerators, Spectrometers, Detectors and Associated Equipment, **549(1-3)**, 1 September 2005, Pages 205-209.
3. Mark McEntee, Helena Frawley and Patrick C. Brennan, A comparison of low contrast performance for amorphous Silicon/caesium iodide direct radiography with a computed radiography: A contrast detail phantom study, Radiography, In Press, Corrected Proof, Available online 10 March 2006.
4. Jin-Fei Ding, Zhang, A.P., Li-Yang Shao, et al., Fiber-taper seeded long-period grating pair as a highly sensitive refractive-index sensor, Photonics Technology Letters, IEEE, **17(6)**, June 2005, Pages 1247-1249.
5. Thierry Pagnot and Christian Pierali, Fabrication of multimode fiber tapers with a high reproducibility of the subwavelength extremity size: application to scanning near-field optical microscope probes, Optics Communications, **132(1-2)**, 15 November 1996, Pages 161-169.
6. George P. Anderson, Joel P. Golden and Frances S. Ligler, A fiber optic biosensor: combination tapered fibers designed for improved signal acquisition, Biosensors and Bioelectronics, **8(5)**, 1993, Pages 249-256.
7. Yongzhi He and Frank G. Shi, A graded-index fiber taper design for laser diode to single-mode fiber coupling, Optics Communications, **260(1)**, 1 April 2006, Pages 127-130.
8. Pawel J. Wiejata, P. M. Shankar and R. Mutharasan, Fluorescent sensing using biconical tapers Sensors and Actuators B: Chemical, **96(1-2)**, 15 November 2003, Pages 315-320.

A PARALLEL MULTI-FOCUS IMAGE FUSION METHOD BASED ON CONTRAST VISION MODEL

XU XUEBIN, ZHANG XINMAN, ZHANG DEYUN

School of Electronics and Information Engineering, Xi'an Jiaotong University
Xi'an, Shaanxi,China

Abstract: Multi-focus image fusion is to combine information from multiple images of the same scene for producing a merged image. A reconstructed scheme based on contrast vision model is studied briefly exploiting segmentation techniques and blocks search. The method's computing complexity is exponentially increased with the growth of image size. After analyzing the efficiency of image fusion algorithm, the inherent complexity and the parallelism, a parallel multi–sensor image fusion method based on contrast vision model in the distributed computing environment is proposed. The experimental result shows that the efficiency of parallel computing is obviously improved.

1. Introduction

The concept of data fusion goes back to the 1950s and 1960s, with the search for practical methods of merging images from various sensors to provide a composite image which could be used to better identify natural and manmade objects.

Image fusion plays important roles in many different fields such as defense system, computer vision, weather forecast, biomedical imaging and remote sensing. Image fusion may be used to combine images from different sensors to obtain a single composite with extended information content. The aim of image fusion is to integrate complementary and redundant information from multiple images to create a composite that contains a better description of the scene than any of the individual source images [1].

Multi-focus image fusion is a hotspot research field of image fusion. Optical lenses, particularly those with long focal lengths, suffer from the problem of limited depth of field. Consequently, it is often difficult to obtain good focus for all objects in the picture. One possible solution is to take several pictures with different focus points, and then combine them together to form a single image [2, 3, 4, 5, 6].

In this paper, we propose a pixel level multi-focus image fusion method based on contrast vision model and image blocks search. The method's computing complexity is exponentially increased with the growth of image size. After analyzing the efficiency of image fusion algorithm, the inherent complexity and the parallelism, a parallel multi–sensor image fusion method based on contrast vision model in the distributed computing environment is proposed.

2. A Multi-focus Image Fusion Method Based On Contrast Vision Model

Firstly, a multi-focus image fusion method based on contrast vision model will be described [5, 6, 7].

2.1. *Contrast Vision Model*

This paper define d_k , namely uniform parameter of partition block of an image G as follows:

$$d_k = \frac{1}{m' \times n'} \sum_{(i,j)\in B_k} \frac{|G(i,j) - \mu_k|}{\mu_k} \tag{1}$$

Where $m' \times n'$ is the block size, $G(i,j)$ is the gray level of the pixel in the i th row and the j th column, μ_k is the mean of block B_k .

We adopt contrast to measure the various visibility of block signal improving the matching extent between segmentation process and vision system.

2.2. *The Basic Algorithm Procedure*

The basic algorithm procedure consists of following steps:

Step 1: Assume that there are two multi-focus image inputs *A* and *B*. The fusion algorithm decompose both of them into smaller rectangle regions i.e. $m' \times n'$ blocks. BA_i is the i th image block of image *A* and BB_i is the i th image block of image *B*.

Step 2: Calculate uniform parameters of block BA_i and BB_i . Denote dA_i and dB_i are uniform parameter of the relative blocks BA_i and BB_i .

Step 3: Compare the uniform parameters of two corresponding blocks BA_i and BB_i , then construct block BF_i as:

$$BF_i = \begin{cases} BA_i, dA_i > dB_i \\ BB_i, dA_i \le dB_i \end{cases} \quad (2)$$

Step 4: Repeat Step 2 and Step 3 on $m' \times n'$ blocks, the fused image will be constructed.

2.3. *Accurate Reconstruction*

RMSE is used to evaluate the effect of image fusion. *RMSE* is defined as:

$$RMSE = \sqrt{\frac{\sum_{i=1}^{m}\sum_{j=1}^{n}\left[G(i,j) - G'(i,j)\right]^2}{m \times n}} \quad (3)$$

Where $m \times n$ is image size, G and G' are reference image and composite image respectively, with size $m \times n$ pixels.

If *RMSE* equal to *0*, it means that implement accurate reconstruction.

It is important to search final image in terms of different block sizes. Here we can search the best block sizes based on general genetic search strategies. Spatial frequency (*SF*) is adopted as the fitness function [6, 7].

SF is defined as:

$$SF = \sqrt{RF^2 + CF^2} \quad (4)$$

Where *RF* and *CF* are the row and column frequencies and defined as:

$$RF = \sqrt{\frac{1}{m \times n}\sum_{i=1}^{m}\sum_{j=2}^{n}\left[F(i,j) - F(i,j-1)\right]^2} \quad (5)$$

$$CF = \sqrt{\frac{1}{m \times n}\sum_{j=1}^{n}\sum_{i=1}^{m}\left[F(i,j) - F(i-1,j)\right]^2} \quad (6)$$

The procedure consists of following steps:

Step 1: Set the proper reference genus numbers X by random method to produce the initial genus. The most optimized block size is likely to be chromosome number. Assume for an $M \times N$ image, the search range is $(0,M)$ and $(0,N)$, so chromosome C_i of number i is composed of two parts representing blocks' length and width code.

$$C_i = \left[\dot{n}_{i,l-1}\ \dot{n}_{i,l-2}\ \cdots\ \dot{n}_{i,0}\ \dot{n}_{i,k-1}\ \dot{m}_{i,k-2}\ \dot{m}_{i,k-3}\ \cdots\ \dot{m}_{i,1}\ \dot{m}_{i,0}\right]$$

Where l is chromosome code length of size $\log_2 N$), k is chromosome code length of size $\log_2 M$), and i belongs to a range of $[0, X-1]$.

Step 2: Computing the fitness function of each chromosome.

Step 3: Choose the most excellent chromosome copy directly to next generation, make it more chances to reproduce new chromosomes.

Step 4: Crossover and mutation operation.

Step 5: Perform genetic folding operation, judge if contents the terminate condition, end the operation, if not, go on the next step.

Step 6: Choose the optimized blocks to reach the best effect.

3. A Parallel Multi-sensor Image Fusion Method

Assume that there are two multi-focus image inputs *A* and *B* (of size *M*×*N*), the complexity of the algorithm in Section 2.2 is *M*×*N* combination. With the growth of image size *M*×*N*, the large multi-focus images cannot be fused efficiently and rapidly, especially with arrival of high resolution images (for example *512*×*512*, *1024*×*1024*). So it is important to improve on the he algorithm in Section 2 using parallel computing [8, 9].

3.1 Parallel Algorithm Design

When we begin the design of a parallel algorithm, we typically try to discover as much parallelism as possible. Partitioning is the process of dividing the data and the computation into pieces.

Partitioning can be applied to the program data (to dividing the data and operating upon the divided data at one time). It is called "data partitioning". Partitioning can also be applied to the functions of program (to dividing the program into independent functions and running the functions at one time). It is called "functional decomposition".

After analyzing the parallelism and the efficiency of image fusion algorithm, we adopt "data partitioning" scheme to implement parallel algorithm. If there are K PC nodes or processors, *M*×*N* combination will be divided into K pieces. There are $M \times N/K$ combination in a piece. Then we can search the best block sizes based on genetic search strategies in all pieces.

3.2 Parallel Algorithm Procedure

In our system, we adopt the master–slave processor model. The parallel algorithm procedure consists of following steps:

Step 1: Image prepare and image accurate registered.

Step 2: Data partitioning. $M \times N$ combination will be divided into K pieces by the master computer.

Step 3: According to genetic search strategies in section 2.3, Search the best block sizes based on in *K* pieces independently.

Step 4: The master computer is in charge of results mergence and choose the optimized blocks to obtain the best effect.

Step 5: Compute root mean square error (*RMSE*) for the fused image with a reference image.

Step 6: Output the results and shows the fused image.

3.3 Parallel Algorithm Implementation

According to the design of our parallel algorithm, we implement parallel algorithm in the Windows 2000/XP operating system with Matlab2006 distributed computing toolbox (DCT toolbox).

The Matlab2006 distributed computing engine and the distributed computing toolbox enable scientists to develop parallel computing Matlab program and execute them in a cluster of computers without leaving GUI development environment. The toolbox includes many functions based on Message Passing Interface (MPI standard) to support communication among nodes and tasks. The MPI standard is the most popular message-passing specification supporting parallel programming .Virtually every commercial parallel computer supports MPI, and free libraries meeting the MPI standard are available for "homemade" commodity clusters. We can prototype applications in Matlab and use the distributed computing toolbox functions to define independent or interdependent tasks. The Matlab2006 distributed computing engine evaluates and schedules tasks on multiple parallel Matlab sessions, reducing running time compared to execute the same algorithm program in a single Matlab session.

For algorithm implementation, we use Matlab scripts and Matlab2006 DCT toolbox to build the Parallel fusion algorithm software. We distributed the image data and program to different image-processing computers of the distributed computing environment. It is helpful to reduce the message communication of different computers to gain more efficiency.

4. Experimental Results

In this paper, we use acceleration ratio and parallel efficiency to evaluate the performance of the parallel fusion method based on contrast vision model.

We design and implement parallel Matlab programs in the hope that they will run faster than their sequential counterparts. Acceleration ratio or speedup is the ratio between sequential running time and parallel running time.

Acceleration ratio S(n) is defined as:

$$S(n) = \frac{T_S}{T_K} \tag{7}$$

Where T_S is sequential running time and T_K is parallel running time.

The efficiency of a parallel program is a measure of process utilization. Efficiency *E* is defined as:

$$E = \frac{S(n)}{n} \tag{8}$$

Where *S(n)* is acceleration ratio and n is the number of processors used.

The experiments are performed on two 256-level gray images of size *512×512* and a reference image.

The parallel algorithms were implemented using Matlab2006 distributed computing toolbox.

Table 1 shows the results of tests in the experimental environment.

Table 1. Performance of multi-focus image fusion method

Nodes Number	Test values		
	Average time(s)	S(n)	E
1	10.1242		
2	5.7727	1.7538	87.69%
4	4.1470	2.4413	61.03%
8	2.1424	4.7256	59.07%

It can be seen that the parallel computing method can obtain better effects than a stand-alone computer.

5. Conclusions

Two conclusions obtained are as follows via the research.

(1) We propose a pixel level multi-focus image fusion method based on contrast vision model and

blocks search. It is clear that this method provides better fused results.

(2) A parallel multi-focus image fusion method based on contrast vision model in the distributed computing environment is proposed and written in Matlab Scripts. The experimental result shows that the efficiency of parallel fusion method is obviously improved.

References

1. He You, Wang Guohong etc. Multi-sensor Information Fusion With Applications, *Publishing house of electronics industry*, 2001, 4.
2. Hiura S, Matsuyama T. Depth measurement by the multi-focus camera. *Proceedings IEEE Computer Society Conference On Computer Vision and Pattern*, 1998, 953-959.
3. Katto J, Ohta M. Novel algorithms for object extraction using multiple camera inputs. *IEEE International Conference On Image Processing,* 1996, **1**: 863-866.
4. Burt P J, Kolczynski R J. Enhanced image capture through fusion. *Proceedings of Fourth International Conference on Computer Vision*, Berlin, Germany, 1993, **4**, 173~182.
5. Li Shutao, Wang Yaonan, Zhang Changfan. Feature of human vision system based multi-focus image fusion. *Acta Electronica Sinica*, 2001, **29(12)**: 1699-1701.
6. Shutao Li, James T. Kwok, Yaonan Wang. Multifocus image fusion using artificial neural networks. *Pattern Recognition Letters*, 2002, 23: 985~997.
7. Zhang Xinman, Han Jiuqiang. Restoration and fusion optimization scheme of multifocus image using genetic search strategies. *Optica Applicata* , 2004, **34(3)**: 453-461.
8. Cheng Yinglei, Hu Fuyuan, Zhao Rongchun. A New Parallel Multi-sensor Image Fusion Method. *Journal of Air Force Engineering University*, 2006, **7(6)**: 40-43.
9. Chi-Chang Chen, Meng-Xiang Chen. A Generic Parallel Computing Model for the Distributed Environment. *Proceedings of the Seventh International Conference on Parallel and Distributed Computing, Applications and Technologies*. 2006, 291-294.

STUDY TO IMAGE CLASSIFICATION ON FEATURE LEVEL FUSION

XU XUEBIN, ZHANG XINMAN, ZHANG DEYUN
School of Electronics and Information Engineering, Xi'an Jiaotong University
Xi'an, Shaanxi, China

Abstract: Firstly, a basic belief assignment function is constructed, and three features of images are extracted. Following this the feature level image fusion method-based Dempster-Shafer reasoning is proposed to resolve image classification problems. The result of computer image classification simulation indicates that this method can greatly improve the ability of classification than that of single feature.

1. Introduction

To use multisource image and extract information effectively, image classification based-feature level fusion is the key issue. In various inaccurate reasoning techniques, one such technique, Dempster-Shafer evidence reasoning theory is commonly adopted in the application of image classification and recognition's fusion [1, 2]. Based on Dempster-Shafer theory the core of this method is Dempster rule of combination which its main merit is meeting the condition that is weaker than Bayesian probability theory, and has the ability of expressing directly uncertainty and ignorance. According to feature level image fusion based D-S theory, image's multi features' extraction are realized, then image classification problem is performed using the fused feature [3]. At last, the experiment results are analyzed and some conclusions are got for the feature level image fusion and image classification.

2. Algorithm Descrition

2.1. *Feature Extraction of Images*

2.1.1. *Edge feature*

This feature is based on the number of edges extracted from the image. This feature is used as the comparable complexity of images intuitively with more edges. Here, we apply the Sobel edge detector to each image sample.

$$E_g = sobel(G)$$

The total number of one's in the resultant binary image is then taken as the edge feature.

$$S = \sum_{i=1}^{m}\sum_{j=1}^{n} E_g(i, j) \tag{1}$$

2.1.2. *Variance*

Variance reflects the deviation between image pixel and gray's mean. With a larger variance, the distributing of gray level is more decentralized.

$$\sigma^2 = \frac{1}{m \times n}\sum_{i=1}^{m}\sum_{j=1}^{n}[G(i, j) - \mu_G]^2 \tag{2}$$

2.1.3. *Entropy*

Which is known to be a measure of the amount of uncertainty about the image. It is then given by

$$H = -\sum_{i=0}^{L-1} p_i \log_2 p_i \tag{3}$$

where L is the number of graylevels and p_i is the normalized graylevel histogram.

2.1.4. *Entropy the construction of BPA*

In the feature fusion of image classification based on evidence theory, the issue of basic probability assignment (BPA) is crucial to the success of the technique. An appropriate probability model is proposed according to image types and correlative coefficient.

According to the experience, the value x_j of image feature meets normal distribution $X \sim N(\mu, \sigma^2)$, viz. the function of conditional probability density is as follows:

$$p(x_j \mid A_i) = \frac{1}{(2\pi)^{1/2}\sigma_{ij}} \exp\left[-\frac{(x_j - \mu_{ij})^2}{2\sigma_{ij}^2}\right]$$

The correlative coefficient between feature j and image type A_i is defined by:

$$c_{ij} = \begin{cases} 2 - 2\int_{-\infty}^{\infty} p(x_j \mid A_i)\,dx_j & x_j \ge \mu_{ij} \\ 2\int_{-\infty}^{\infty} p(x_j \mid A_i)\,dx_j & x_j \le \mu_{ij} \end{cases}$$

The basic probability assignment m_{ij} is obtained on the basis of the following method:

$$\alpha_j = \max_{i=1}^{m}(c_{ij})$$

(the largest correlative coefficient between feature j and image type A_i)

$$\beta_j = \frac{\alpha_j}{\sum_{i=1}^{m} c_{ij}}$$

(the distributed coefficient between feature j and the respectively correlative coefficient)

$$r_j = \frac{\alpha_j \beta_j}{\sum_{j=1}^{n} \alpha_j \beta_j}$$

(the reliable coefficient of feature j)

Then the basic probability assignment that feature j endows the image type A_i is defined by:

$$m_{ij} = \frac{c_{ij}}{\sum_{i=1}^{m} c_{ij} + 2(1-r_j)(1-\alpha_j\beta_j)} \quad (4)$$

The uncertainty probability value of feature j is defined by:

$$\theta_j = \frac{2(1-r_j)(1-\alpha_j\beta_j)}{\sum_{i=1}^{m} c_{ij} + 2(1-r_j)(1-\alpha_j\beta_j)} \quad (5)$$

2.2. *The Basic Concepts of D-S Theory of Evidence*

For a finite set of mutually exclusive and exhaustive propositions Θ, it is referred to as a frame of discernment (FOD) of any subset X .

2.2.1. *Mass function*

Let Θ be a frame of discernment. A basic probability assignment (BPA) is a function m : 2Θ→[0,1] such that

1) $\sum_{A \subset \Theta} m(A) = 1$

2) $m(\phi) = 0$

2.2.2. *The belief function*

A belief measure is a function m : 2Θ→[0,1], and is drawn from the sum of probabilities that are subsets of the probabilities in question, defined by

$$Bel(A) = \sum_{B \subset A} m(B)$$

2.2.3. *The plausibility function*

The plausibility function provides the upper limit of the probability.

$$pl(A) = 1 - Bel(\overline{A})$$

Clearly $pl(A)$ represents the extent to which we fail to disbelieve A.

2.3. *Dempster's Rule of Combination*

Evidence can be continued by computing the orthogonal sum using Dempster's rule of combination where evidences A and B are used for computing a new belief function for a focal element C.

If m_1 and m_2 are the masses associated with Bel_1 and Bel_2 independent information, then the function $m = m_1 \oplus m_2$: 2Θ→[0,1], defined by:

$$m(C) = \begin{cases} 0 & C = \phi \\ \dfrac{\sum_{A \cap B = C} m_1(A) m_2(B)}{1-K} & \forall C \subset \Theta \quad C \ne \phi \end{cases} \quad (6)$$

A denominator K is defined by:

$$K = \sum_{A \cap B = \phi} m_1(A) m_2(B) < 1 \quad (7)$$

Briefly, Dempster-Shafer theory is a powerful method for combining accumulative evidence or for changing prior opinions in the light of new evidence [4, 5, 6].

3. The Application of D-S Image Fusion IN The Image Classification

We needs to carry on image classification to three types of images $A_i\,(i = 1, 2, 3)$ which are coins, leafs and doors. First different databases are constructed on the large numbers of images. Following by this 5-image test samples of each type is used in classification recognition, just like Figure 1.

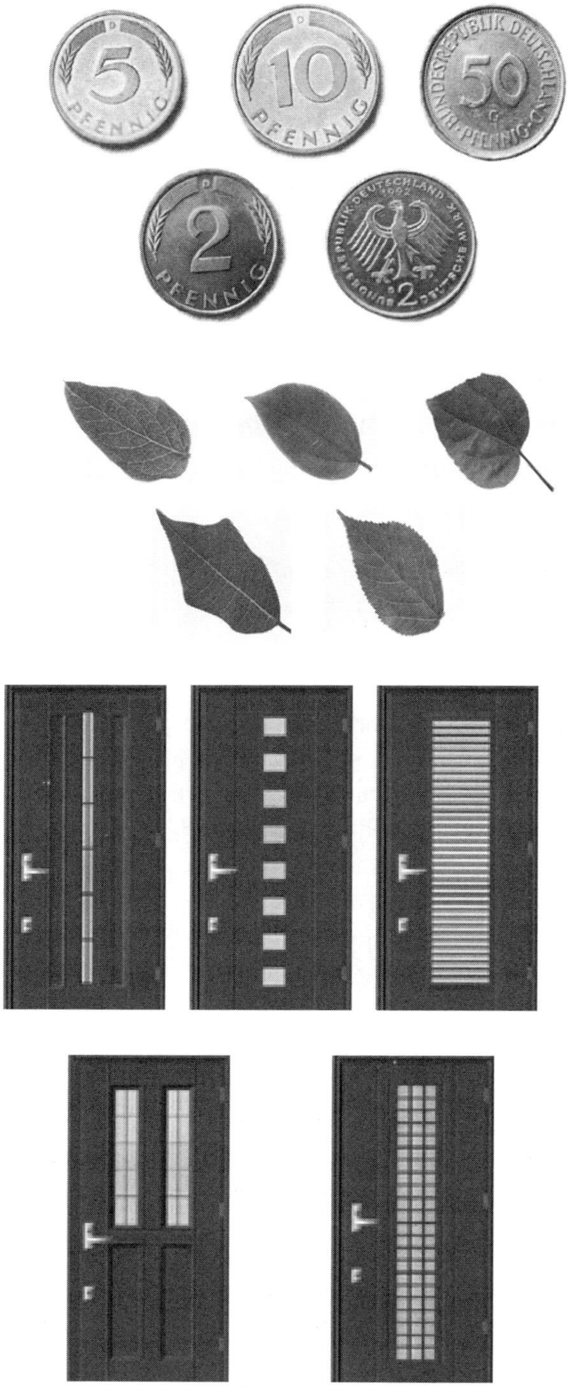

Figure 1. The test samples of Coin, Leaf and Door

One-group feature values of test sample are shown in Table 1. Table 2 shows the BPAs obtained via the above method and the feature fusion results.

Table 1. A group feature test values of different image types.

	Test values		
	1	2	3
Edge feature	1002	417	284
Variance	71.3413	66.3796	44.6473
Entropy	6.2168	4.4850	5.2741

Table 2. Feature image fusion results.

Image types	Features	Basic probability assignment（m_{ij}）			θ_i	Classification results
		Coin	Leaf	Door		
Coin	Edge feature	0.3906	0.0000	0.0000	0.6094	uncertainty
	Variance	0.2453	0.4532	0.0010	0.3005	Leaf
	Entropy	0.2602	0.0605	0.1390	0.5404	uncertainty
	Fusion	0.5283	0.2766	0.0404	0.1547	Coin
Leaf	Edge feature	0.0000	0.2754	0.5267	0.1979	Door
	Variance	0.3209	0.3360	0.0033	0.3398	uncertainty
	Entropy	0.0000	0.2883	0.0771	0.6346	uncertainty
	Fusion	0.0858	0.5376	0.2858	0.0908	Leaf
Door	Edge feature	0.0000	0.2131	0.1371	0.6498	uncertainty
	Variance	0.0527	0.0210	0.4506	0.4757	uncertainty
	Entropy	0.0159	0.3324	0.3108	0.3409	uncertainty
	Fusion	0.0038	0.3885	0.5425	0.0652	Door

Table 3. Single and multi-feature image fusion correct classification rates.

Image samples	Edge feature	Variance	Entropy	DS theory
1	1	0	0	1
2	1	0	0	1
3	0	0	1	1
4	1	0	0	1
5	0	0	0	1
6	0	0	0	1
7	0	0	0	1
8	0	1	0	1
9	0	0	0	1
10	0	0	0	1
11	0	0	0	0

12	0	0	1	1
13	0	0	1	1
14	0	0	0	1
15	0	1	0	1
Rate of classification recognition	20%	13%	20%	93%

The three type images can be carried in image classification through feature level fusion. From the results obtained it is clear that the application of D-S in image classification of feature fusion is effective and has avoided contradiction. It is optimal than that of a single feature with a higher classification recognition rate, thereby demonstrating the effectiveness and correctness of this approach.

4. Conclusions

Two conclusions obtained are as follows via the research.

(i) It is clear that the application effect is performing better based on D-S data fusion method improved the average image classification rate of the three types of images (i.e. coins, leafs and doors) and establishing favorable basis to image recognition system.

(ii)This Dempster-Shafer implementation has the powerful ability of processing uncertainty information. It is suitable for being employed in image classification system. And its algorithm is achieved simply and rapidly through computer's software programming or the algorithm of Matlab language. It has preferable practicability to meet engineering's needs in general.

References

1. C.R.Parikh, M.J.Pont, N.B.Jones, Application of Dempster-Shafer theory in condition monitoring applications: a case study, *Pattern Recognition Letters*, 2001, **22(6-7)**: 777~785.
2. He You, Wang Guohong etc, Multisensor Information Fusion With Applications, *Publishing house of electronics industry*, 2001, 4.
3. Wang Haihui, Peng Jiaxiong, Wu Wei. Application of the information fusion based on evidence theory in image classification. *Computer engineering and application,* 2003, **33**: 11~14.
4. Huber R. Scene classification of SAR images acquired from antiparallel tracks using evidential and rule-based fusion. *Image and Vision Computing*, 2001, **19(13)**: 1001~1010.
5. Lein J K. Applying evidential reasoning methods to agricultural land cover classification. *International Journal of Remote Sensing*, 2003, **24(21)**: 4161~4180.
6. Mertikas P, Zervakis M E. Exemplifying the theory of evidence in remote sensing image classification. *International Journal of Remote Sensing*, 2001, **22(6)**: 1081~1095.

APPLICATION OF RADIOIMMUNOASSAY ON THE STUDY OF CATGUT EMBEDDING IN POINTS ON OVARIECTOMIZED RATS *

CHEN GUI-ZHEN

The Laboratory of Photonic Chinese Medicine, School of Information and Optoelectronic Science and Engineering, South China Normal University, Guangzhou, 510631 China

XU YUN-XIANG†

College of Acupuncture & Massage, Guangzhou University of TCM, Guangzhou, 510405 China

Objective: To observe the varieties of reproductive endocritic hormone when female SD rats' ovaries removed, and the effect of catgut embedding in points on them. Methods: The female SD rats with ovaries removed, were adopted as the animal model of menopausal pathology, and were divided randomly into ovariectomized model (OVX model) group (A), catgut embedding in points group (B), medicine group (C).At the same time, built sham model group (D). After being treated 4 times in succession, all of the experimental animals were put to death on the 5th day after the last treatment, and collected serums and plasma, measured and observed the following indexes: the content of estradiol (E_2), testosterone (T), gonadotropin releasing hormone (Gn-RH), follicle stimulating hormone (FSH), luteinizing hormone (LH). Results: The levels of plasma Gn-RH, serum FSH and LH of group A were higher than those of group D ($P < 0.05$), while serum E_2, T were lower than those of group D ($P < 0.05$), the change range of E_2 was more significant than that of T. After treatment, the levels of E_2 and T of group B were improved and the levels of serum FSH, LH and plasma Gn-RH of groups B were reduced, the change of the content FSH between group B and group A were significant ($P < 0.05$), but E_2, T, LH and Gn-RH were no significant ($P > 0.05$); There was no significant difference between group D and group B in the level of serum E_2, FSH, LH ($P > 0.05$), while significant difference between them the level of T and Gn-RH ($P < 0.05$); There was not significant difference between group B and group C in the change of E_2, T, FSH, LH and Gn-RH ($P > 0.05$). Conclusion: Catgut embedding in points could regulate the hypothalamic-pituitary- gonad axis, and increase serum E_2 and T, and reduce plasma Gn-RH, serum FSH and LH of the ovariectomized rats.

Keywords: radioimmunoassay; catgut embedding in points /acup-mox eff; ovariectomized rats

1. Introduction

The ovary functional decline as well as the overly degraded estrogen level is generally regarded as the etiopathogenesis of menopausal syndrome in modern medicine. Consequently, the varieties of reproductive endocritic hormone are taken as the effective indexes in either clinical observations or animal studies on menopausal syndrome treated with acu-moxa therapy. This protocol aims to observe the varieties of reproductive endocritic hormone when female SD rats' ovaries were removed and the effect of catgut embedding in points on them.

2. Materials and Methods

2.1. *Main Instruments, Reductants and Materials*

2.1.1. *Main instruments*

Radio-immunity measurement instrument (SN-695B, produced by Rihuan Nuclear Instrument Factory, Shanghai); hydroextractor (DDL-5,produced by An'ting Scientific Instrument Factory, Shanghai), cryogenic box (MDF-D5410,produced by SANYO Company, Japan).

2.1.2. *Main reductants*

Radio-immunity kits for E_2, FSH, LH and T: provided by Peking Kemei Dongya Biotechnology Limited Company, Batch No: 20050425; Gn-RH kit: provided by teachers' room of neurobiology in the 2nd Military Surgeon University, Batch No.20050501

* This work was supported by grant No. 20070410242 of the Post-doctor of China Foundation and grant No. 2007166 of the Administration of Traditional Chinese Medicine of Guangdong Province

† Correspondence to: Professor Xu yun-xiang, College of Acupuncture & Massage, Guangzhou University of TCM, Guangzhou, 510405 China, Email: xuyx.1968@yahoo.com.cn

2.1.3. *Materials*

Sumianxin II parenteral solution (the Research Institute for Military Veterinary Surgeons in the Academy of Military Medical Sciences, Changchun); Catgut embedding needles the germ free needle No. 6(0.6×28 TWLB)produced by Zhejiang Oujian Medical Equipment Limited Company was selected as the cannula, and the mini-needle No.28 sized 1.5 inch manufactured by Suzhou Medical-product Factory as needle core. Cut the needle tips so that it would not stick together with the end of catgut, and the mini-needle should be kept 0.5cm longer than the cannula so as to guarantee the pulsion of catgut; Catgut the absorbable medical catgut N0.4/0 from Shanghai Pudong Jinhuan medical-product Limited Company was picked. Nilestriol Tablets (Produced by Shanghai Hualian Tragacanth Limited Corporation, 5mg per tablet, Batch No: 040101); Iodine (Produced by Shanghai Likang Sterilization High Tech Limited Company).

2.2. *Sample Size Assessment*

According to the parameter of E_2 shown in Preliminary experiment data (5 rats for each group, *pmol*/L as unit): sham model group (D) 31.84±12.42,ovariectomized model (OVX model) group (A) 15.68±4.43, catgut embedding in points group (B) 22.40±8.27, medicine group (C) 26.82±8.47;set α = 0.05 bilaterally, β= 0.2,analysis efficacy (1-β) = 0.8, g = 4. According the formula Eq. (A.1), Eq. (A.2) [1], the value of f is calculated before looking for n in the table. Subsequently, n is calculated to be 7.36, and is taken as 8, which is the minimum sample size for each group. Considering 20%in sham model group (D), and 40% in ovariectomized model (OVX model) group (A) of the maximum mortality rate, 10 samples should be recruided in sham model group (D) and 40 samples in ovariectomized model (OVX model) group (A), altogether 50 samples of rats should be included in this study.

2.3. *Experimental Animals and Modeling*

2.3.1. *Experimental animals*

50 unpregnant femail SD (Sprague-Dawley, SD) rats of reproductive age which weigh 250±20g provided by Guangdong Medicine Experimental Animal Centre. Certificate No. 0006318.

2.3.2. *Conditions for experimental animals*

The subjects are placed indoors for acclimatization for one week prior to experiment, self-cibation of animal feed (provided by our animal experiment centre) as well as water was performed in a room at 18～22^{0}C, 65% of relative humidity, and lighting alternated for day and night, namely 12 hours respectively. Environmental certificate No.0005021.

2.3.3. *Grouping*

10 samples were randomly picked among the 50 SD rats for the sham model group (SHAM, D), and bilateral ovaries removal were performed on the other 40 rats. During the operation, one rats died in sham model group (SHAM, D) and five in ovariectomized modeling. Based on the vaginal cass-off cells detection, 32 rats which were successfully modeled are selected as menopause experimental samples, and they were further randomized into 3 groups : 10 for ovariectomized model group (OVX, A), 11 for catgut embedding in points group(OVX+CEP, B) and 11 for medicine group (OVX +CEE_3, C). This met the requirement of the design, and all the conditions for feeding and living were identical among the groups as regards to lighting, ventilation, temperature, humidity, et al. No rat died in the course of treatment.

2.3.4. *Animal modeling*

Rats subjected to ovariectomization were anesthetized with Sumianxin II parenteral solution (0.8ml/kg given intramuscularly), and then fur was prepared, and the rat was laid one-sidedly on the operation table for the bilateral ovariectomized operation performed under the guidance of *Practical Manual for Medical Animal Experiment*[2]. After the **ovariectomy**, dephlogisticate powder was spread on the injured surface after interval suture done on the skin. Penicillin normal sodium was injected intramuscularly for post-operation anti-inflammation for successive three days with 1ml per day. Sham model group were dealt with in the same way except for the bilateral ovariectomized operation, which was not performed in this group.

2.3.5. *Modeling standards*

The 5th day after the female SD rats ovaries removed, examined exfoliated cell of vaginal smear continuously for five days, it indicates complete emasculation that the basal cells of all ovariectomized rats were above 50%, and the ovariectomized rats did not show the varieties of the estrous cycle which characterized the menopause. Therefore, they were adopted as the animal model of menopausal pathology. [3].

2.4. *Therapeutic Methods*

2.4.1. *Catgut embedding in points group (B)*

Point selection: *Shenshu* (BL23), *Sanyinjiao* (SP6), *Guanyuan* (CV4) (The point location on rats refers to *Atlas for Acupoints on Experimental Animals* established by research institute for acu-moxa experiments affiliated to National Acu-moxa Academy.)
Manipulation

1. The catgut No 4/0 for surgery were cut into 0.5cm of length and soaked in 95% alcohol for usage.
2. Practice began with sterilization on the acupoint location for catgut embedding.

The germ free needle No.6 (0.6×28 TWLB) were chosen as catgut embedding needles. Cut the needle tips of the mini-needle No.28 sized 1.5 inch to serve as the needle core. Pull the core out for about 1cm of the cannula, where a catgut about 0.5cm of length is inserted through the pinhead with forceps. Spread or pinch the skin with left thumb and index finger to facilitate needle insertion, and then prick the needle quickly into acupoint and manipulate lifting and thrusting, and push the needle core inwardly after the arrival of qi is obtained, and retreat the cannula outwardly simultaneously. By this way, the catgut is embedded into the acupoint deeply. Lastly, check the acupoint to make sure that no end of catgut is exposed in the air, and press the needle hole for a while to prevent bleeding before iodin disinfection.

2.4.2. *Medicine group (C)*

The groups were treated with nilestriol by intragastric administration. The dosage was referred to adult usage. Nilestriol (CEE_3) was smashed into powder and mixed with distilled water to make solution 0.2g/L of density [4]. The dosage was 5ml/kg (1 mg/kg), given once per week. CEE_3 solution was freshly made every time prior to intragastric administration.

2.4.3. *Ovariectomized model (OVX model) group (A)*

Normal saline was given by intragastric administration with 5ml/kg of dosage.

2.4.4. *Sham model group (D)*

Normal saline was given by intragastric administration with 5ml/kg of dosage.

2.4.5. *Course of treatment*

Therapeutic measures were undertaken from the 11th day after operation. Treatments were given every 5 days for each group for successive 4 times.

2.5. *Methods for Specimen Collection*

2.5.1. *Blood sample collection*

Decapitation was performed on the 31th day at 8:30～11:30 for blood sample, ① 5ml of blood was prepared beforehand for serum in test tube; ② another 3ml was reserved in prepared anticoagulation tube (trasylol 20μl/ml, 10%EDTA-Na_2 20μl/ml)for blood serum.

2.5.2. *Preparation for specimen*

When all specimen were collected, they were placed in a environment at 4°C for 2 hours before centrifugalization, the conditions of which are as follows: 4°C of temperature, 3000r/min×15min. After that, blood serum and plasma were separately contained in two 1.5ml plastic centrifuge tubes which were stored in -30°C cryogenic box for test.

2.6. *Index Assay Method*

Manipulation was carried out by professional stuff in Nuclear Medicine Lab of Guangzhou University of TCM.

Radio immunoassay was adopted for detection of E_2, T, FSH and LH, operating instruction of the kit was based on the directions enclosed.

Method for detection of Gn-RH was the same as that for E_2, T, FSH and LH.

2.7. *Statistical Method*

The assay data were expressed by mean ± standard deviation ($\bar{x} \pm s$)and statistically analyzed with Spss for window 11.5 software. Differences of measurement data between any two groups were analyzed with One-way ANOVA model, LSD Post Hoc Tests. Significance was considered at the level of 0.05.

3. Results

Table 1 Quantity of E_2 and T in Rats' Blood Serum after Treatment

Group	Cases (n)	E_2(p mol/L)	T(n mol/L)
OVX（A)	10	17.79±3.45**△	0.19±0.08*
OVX±CEP (B)	11	21.14±8.88	0.25±0.12*
OVX±CEE$_3$ (C)	11	26.98±10.79▲	0.22±0.12*
SHAM (D)	9	28.44±10.24▲▲	0.44±0.11▲△

Notes: *$P < 0.05$, ** $P < 0.01$ compared with group D; ▲$P < 0.05$, ▲▲$P < 0.01$ compared with group A; compared with group A. △$P < 0.05$ compared with group C.

Statistical analysis showed that in table 1: (1) The difference between OVX model group A and sham model group D was strongly significant for quantity of E_2, T in rats' blood serum ($P < 0.01$, $P < 0.05$); (2) Comparison between catgut embedding in points group B and OVX model group A showed uprise in E_2, T but no marked difference ($P > 0.05$). There was no significant difference between group B and A for E_2, T($P < 0.05$), regardless of the fact that group B was inferior to group A; also, no significant difference is shown between group B and C for E_2, T ($P > 0.05$).

Table 2 Variation of FSH, LH in Rats' Blood Serum after Treatment

Group	Case(n)	FSH (IU/L)	LH (IU/L)
OVX（A)	10	1.92±0.29**	2.98±0.31*
OVX±CEP (B)	11	1.60±0.23▲	2.69±0.44
OVX±CEE$_3$ (C)	11	1.75±0.29	2.91±0.52*
SHAM (D)	9	1.59±0.25▲	2.52±0.30▲△

Notes: *$P < 0.05$, ** $P < 0.01$ compared with group D; ▲$P < 0.05$, ▲▲$P < 0.01$ compared with group A; compared with group A; △$P < 0.05$ compared with group C.

Statistical analysis in table 2 showed that: (1) The difference between OVX model group A and sham model group D was strongly significant for quantity of Gn-RH in rats' blood serum ($P < 0.05$) with uprise in group A; (2) There was no siginificant difference between catgut embedding in points group B and medicine group C for Gn-RH ($P > 0.05$) which degraded in both groups. However, the difference between group B and D was significant ($P < 0.05$); (3) No remarkable difference was shown between group B and C for Gn-RH ($P > 0.05$).

Table 3 Variation of Gn-RH in Rats' Blood Serum after Treatment

Group	Cases (n)	Gn-RH (pg/ml)
OVX（A)	10	5.26±2.07*
OVX±CEP (B)	11	4.98±1.60*
OVX±CEE$_3$ (C)	11	4.56±1.56*
SHAM (D)	9	3.26±1.24▲

Notes: *$P < 0.05$, ** $P < 0.01$ compared with group D; ▲$P < 0.05$, compared with group A

Statistical analysis in table 3 showed that, (1) The difference between OVX model group A and sham model group D was strongly significant for quantity of Gn-RH in rats' blood serum ($P < 0.05$) with uprise in group A; (2) There was no significant difference between catgut embedding in points group B and medicine group C for Gn-RH ($P > 0.05$) which degraded in both groups. However, the difference between group B and D was significant ($P < 0.05$); No remarkable difference was shown between group B and C for Gn-RH ($P > 0.05$).

4. Discussion

In spite of the diverse clinical manifestations in menopause syndrome as well as the unknown mechanism, it is generally acknowledged that overly degraded estrogen level induces functional disorder of the hypothalamic-pituitary- gonad axis resulting in imbalance of neurotransmitter, hormone, and cell factor is the fundamental etiology. Massive data from clinical and experimental studies demonstrates that those suffering from menopause syndrome have higher level of FSH and LH than the asymptomatic, while the asymptomatic is superior to the menopause syndrome sufferers in the quantity of E_2. These collectively suggest that the excessive drop of estrogen is the basic pathology of menopause syndrome. Additionally, the ovary functional decline induces estrogen reduction and subsequently depression of reverse feedback, which triggers the pulsatile secretion of gonadotrophic hormone in the arcuate nuclei of hypothalamus as well as the paraventricular nuclei into the portal circulation. In this way, the secretion of FSH and LH in pituitary is enhanced. At the same time, the decrease of secretion of inhibin in ovarian follicle also contributes to the uprise of FSH and LH level. Moreover, the increase of FSH is dominant, and the highest level appears 2～3 years after menopause, which would last for 10 years until senectitude arrives[5]. In conclusion, the detection of alternation in quantity of E_2, T, Gn-RH, FSH and LH is of significant importance in the diagnosis and treatment of menopause syndrome.

As seen in the experimental results, the level of serum E_2 and T was markedly lower in OVX model group A than those in sham model group D($P < 0.05$), while group A had higher level of FSH and LH, which conforms to the endocrine alteration after menopause in human. The serum E_2 level was heightened in catgut embedding in points group B after treatment compared with OVX model group A, to the contrary, serum levels of FSH, LH and Gn-RH had dropped sharply, which may be due to the lack of samples. Other than T and Gn-RH, there was no statistical significance as for the difference between two groups for E_2, T, LH and Gn-RH ($P > 0.05$); comparison between catgut embedding in points group B and medicine group C also didn't show any statistical significance for Gn-RH,E_2, T, FSH and LH ($P > 0.05$).These collectively confirm that catgut embedding therapy and oestrogen treatment are somewhat effective on regulating the internal productive hormone in female ovariectomized rats and help to raise their hypo-estrogen level. In this study, the ovarian secretion of estrogen was cut off by ovariectomy, and the uprise of estrogen level therefore had to rely on the periphery transformation after treatment with catgut embedding, namely aromatized transformation by adrenal cortex and periphery tissue such as the skin, adipose layer, liver, et al. By contrast, the increase in estrogen in medicine group is the result of direct ectogenic supplement, which contributes little to the variation of T, FSH, LH and Gn-RH. Consequently, it is estimated that the effect of catgut embedding in points on estrogen level in the ovariectomized rats is probably mediated by the hypothalamic-pituitary- gonad axis, which activates and promotes the aromatized periphery transformation process after menopause. Another possibility is that it works to enhance the estrogen receptor level as well as the estrogen activity. As for its mechanism, further study is required.

5. Conclusions

Catgut embedding in points could regulate the hypothalamic-pituitary-gonad axis, and increase serum E_2 and T, and reduce plasma Gn-RH, serum FSH and LH of the ovariectomized rats.

Appendix

$$f = \sigma_g / \sigma \quad \text{(A.1)}$$

$$\sigma_g = \sqrt{\sum_{i=1}^{g} (m_i - m)^2 / g} \quad \text{(A.2)}$$

Note: σ_g is the standard deviation among the 4 groups of means and σ within the group; m_i denotes mean of group i, m denotes total mean.

Acknowledgments

This work was supported by the Post-doctor Foundation of China under grant No. 20070410242 and the Administration of Traditional Chinese Medicine of Guangdong Province under grant No. 2007166.

References

1. Zhao, Q.B. Xu, Y.Y. and Cao X.T. Estimation of sample size for comparing more than two treatmentgroups. J Fourth Milit Med Univ. **14(4)**, 275-280 (1993).
2. Shi, Y. Mei, S.C. Practical Manual for Medical Animal Experiment. Beijing: China Agriculture Press, 2002.
3. Xu, S.Y. Bin, R.L. and Chen, X. Experimental Pharmacology Methodology. Beijing: People's Medical Publishing House, 1991.
4. Yan, X.L. Zhu, K.L and Guan, Y. The Study of Histopathology of the Pituitary Body, Utera, Breast after Using HRT in Ovariectomized Mice. Journal of Sichuan Continuing Education College of MS. **20(1)**, 3-4 (2001).
5. Le, J. Obstetrics and Gynecology. Beijing: People's Medical Publishing House, 2001.

STABILITY OF T-PERIODIC SOLUTION ON ESP-P MODEL

JIN-GUO LIAN
Department of Mathematics and Physics, North China Electric Power University, Baoding, Hebei, 071003, China
E-mail. lianjinguo@gmail.com

The system of reaction-diffusion equations with zero-flux and periodic boundary conditions, which has stable steady-states, will induce the cell reproduction pattern by basic analysis of Turing Principle. For models with constant coefficients, their cell division patterns do not relate to the time factor. Time-related process of cell division is called aberrance of cell division, whose patterns have significant meaning to cancer pathology, especially, for the mechanism of split of cancer cell. Unlike normal cells, cancer cells do not carry on maturing once they have been made. In fact, the cells in a cancer can become even less mature over time. With all the reproducing, it is not surprising that more of the genetic information in the cell can become lost. So the cells become more and more primitive and tend to reproduce more quickly and even more haphazardly. As a result, to explain cancer cells reproduction, one has to study time-related patterns. Therefore, we extend Predator-Prey model with constant coefficients to the model with coefficients to be positive T-periodic functions (we call such a model EP-P model). The goal of this research is to study the pattern formations of EP-P model and simulate the process of tumor forming.
In this paper, by employing the powerful and effective coincidence degree method, we show the existence of T-periodic solutions of ESP-P model in $\mathcal{D}$, where $\mathcal{D}$ is a strictly positively invariant region. Furthermore, Floquet theory is provided to show that the T-periodic solution $x_0(t)$ of ESP-P model is unique in $\mathcal{D}$ and locally uniformly asymptotically stable. This establishes a solid foundation for studying the patterns of EP-P model.

Keywords: ESP-P model; Periodic solution; Existence; Uniqueness; Stability

1. Introduction

The Predator-Prey model [5, p. 72-78] is

$$\begin{cases} u_t(r,t) = d_1\Delta u(r,t) + u(r,t)(1-u(r,t)) \\ -\frac{au(r,t)v(r,t)}{u(r,t)+d}, & r \in \Lambda, \\ v_t(r,t) = d_2\Delta v(r,t) + b(t)v(r,t)(1-\frac{v(r,t)}{u(r,t)}), & r \in \Lambda \end{cases}$$

with boundary conditions

$$n(r)\cdot\nabla u(r,t) = n(r)\cdot\nabla v(r,t) = 0$$

for $r \in \partial\Lambda$, where $n(r)$ is the unit outward normal vector along the boundary of $\Lambda = [0,l]\times[0,l]$ $(l > 0)$ and a, b, d, d_1, d_2 are positive constants. The simplified Predator-Prey (SP-P) model is of form:

$$\begin{cases} \dot{u} = u(1-u) - \frac{auv}{u+d} \\ \dot{v} = bv(1-\frac{v}{u}). \end{cases}$$

If in SP-P model, we allow the coefficients a, b and d to be positive continuous T-periodic functions of t with period $T > 0$, then the corresponding model is called ESP-P model. Similarly, Predator-Prey model will be called extended Predator-Prey (EP-P) model, if we replace constants a, b and d by positive continuous T-periodic functions.

Our goal is to concentrate on the study of periodic patterns of EP-P model. This is a quite different job compared to the study of pattern formation of Predator-Prey model by using Turing's Instability. [5] In order to study the pattern formation of EP-P model, we need to find a T-periodic solution (a resource of the T-periodic pattern) for EP-P model in a certain patch, and study the stability of the T-periodic solution and its bifurcation along some parameter. The Turing's Instability method only works on an autonomous system; however, in this research, we focus on a non-autonomous system. Therefore, the study of EP-P model is a challenging work.

Periodic patterns occur throughout the natural world. For example, the behaviors or functions of living organisms, single cells, etc., could vary periodically with the environment around them; the pattern of wind-blown ripples on the surface of a sand dune periodically varies with the alternation of the four seasons in a year. Therefore, the study of EP-P model has significant meaning to life science, systems biology, and self-organizing science.

In this paper, by employing the powerful and effective coincidence degree method, we show the existence of T-periodic solutions of ESP-P model in $\mathcal{D}$, where $\mathcal{D}$ is a strictly positively invariant region. This analysis establishes a foundation for studying the periodic patterns of EP-P model. Since $x_0(t)$ is also a T-periodic solution of EP-P model, the study of stability and bifurcation of $x_0(t)$ in EP-P model will lead to find the periodic patterns in future works.

2. Existence of T-periodic Solutions in $\mathcal{D}$ for EP-P Model

Let

$$x(t) = \begin{pmatrix} u(t) \\ v(t) \end{pmatrix}$$

and

$$F(t,x(t)) = \begin{pmatrix} u(t)(1-u(t)) - \frac{a(t)u(t)v(t)}{u(t)+d(t)} \\ b(t)v(t)(1-\frac{v(t)}{u(t)}) \end{pmatrix}.$$

Then ESP-P model is defined by

$$\dot{x}(t) = F(t, x(t)) \tag{1}$$

with conditions

$$b(t) > 0, \ \frac{1}{8} < a(t) < \frac{1}{4}, \ \frac{1}{4} < d(t) < \frac{1}{2}. \tag{2}$$

Lemma 2.1. *There exists a strictly positively invariant region*

$$\mathcal{D} = \{(u,v) \in \mathbb{R}^2 \,:\, 0.501 \leq u \leq 0.999, \ \frac{1}{2} \leq v \leq 1\}$$

for ESP-P model given by (1) with conditions (2).

Proof. Clearly $\mathcal{D}$ is a closed convex subset of $\mathbb{R}^2$. We only need to check whether $n(u,v) \cdot F(t,(u,v)) < 0$ along the boundaries of $\mathcal{D}$, where $n(u,v)$ is the unit normal vector field along the boundary of $\mathcal{D}$ and $F(t,(u,v))$ is defined in (1). Notice that for any $(u,v) \in \mathcal{D}$,

$$0.501 \leq u \leq 0.999, \ \frac{1}{2} \leq v \leq 1. \tag{3}$$

Let $l_1 = \{(u,v) \in \mathbb{R}^2 \,:\, u = 0.501\}$, for any $(u,v) \in l_1 \cap \partial\mathcal{D}$, $n(u,v) = (-1,0)$, and

$$F(t,(u,v)) = (0.249999 - \frac{a(t)v(t)}{1+\frac{d(t)}{0.501}}, b(t)v(t)(1-\frac{v(t)}{0.501})),$$

by (2) and (3),

$$n(u,v) \cdot F(t,(u,v)) = -(0.249999 - \frac{a(t)v(t)}{1+\frac{d(t)}{0.501}}) < 0.$$

Let $l_2 = \{(u,v) \in \mathbb{R}^2 \,:\, u = 0.999\}$, for any $(u,v) \in l_2 \cap \partial\mathcal{D}$, $n(u,v) = (1,0)$. It follows from (2) and (3)

$$n(u,v) \cdot F(t,(u,v)) = 0.000999 - \frac{0.999a(t)v(t)}{0.999+d(t)} < 0.$$

Let $l_3 = \{(u,v) \in \mathbb{R}^2 \,:\, v = \frac{1}{2}\}$, for any $(u,v) \in l_3 \cap \partial\mathcal{D}$, $n(u,v) = (0,-1)$. Then by condition (2) and (3), we get

$$n(u,v) \cdot F(t,(u,v)) = -\frac{1}{2}b(t)(1 - \frac{1}{2u(t)}) < 0.$$

Let $l_4 = \{(u,v) \in \mathbb{R}^2 \,:\, v = 1\}$, for any $(u,v) \in l_4 \cap \partial\mathcal{D}$, $n(u,v) = (0,1)$. Then by condition (2) and (3), we get

$$n(u,v) \cdot F(t,(u,v)) = b(t)(1 - \frac{1}{u(t)}) < 0.$$

Since $n(u,v) \cdot F(t,(u,v)) < 0$ for all $(u,v) \in \partial\mathcal{D}$, $\mathcal{D}$ is a strictly positively invariant region. □

Linearizing the system (1) with respect to its a T-periodic solution $x(t) = (u(t),v(t))^T \in \mathcal{D}$ for any $t \in \mathbb{R}$ (if such a T-periodic solution exists), then we get

$$\dot{W}(t) = A(t)W(t). \tag{4}$$

Proposition 2.1. *Linear system (3) satisfies $tr(A(t)) < 0$ and $det(A(t)) > 0$ for any $t \in \mathbb{R}$.*

Proof. By (2) and (3),

$$tr(A(t)) = 1 - 2u(t) - \frac{a(t)d(t)v(t)}{(u(t)+d(t))^2} + b(t) - \frac{2b(t)v(t)}{u(t)} < 0.$$

Furthermore,

$$\det(A(t)) = [1 - 2u(t) - \frac{a(t)d(t)v(t)}{(u(t)+d(t))^2}]b(t)[1 - \frac{2v(t)}{u(t)}] + \frac{a(t)b(t)v^2(t)}{u(t)(u(t)+d(t))} > 0.$$ □

3. Preliminaries

Consider the nonlinear system

$$\dot{x}(t) = V(t,x(t)), \ x(t_0) = x_0, \ x(t) \in \mathbb{R}^2. \tag{5}$$

Lemma 3.1. *If $x^*(t)$ is an exponentially stable solution of (5), then it is also a uniformly asymptotically stable solution of (5).*

For proof, see [3, p. 178-179].

Let $\mathcal{X} = \{x \in C([0,T]) \,|\, x(0) = x(T)\}$. Clearly $\mathcal{X}$ is a Banach space with the supremum norm. Define $Lx(t) = \dot{x}(t)$ with domain

$$Dom(L) = \{x \in C^1([0,T]) \,|\, x(0) = x(T)\}.$$

It is easy to verify that $Dom(L)$ is contained in $\mathcal{X}$, the range of L is $Im(L) = \{z(t) \in \mathcal{X} \,|\, \int_0^T z(t)dt = 0\}$ and L is a Fredholm mapping of index 0. Let

$$\Theta = \{x \in Dom(L) \,|\, x(t) \in \mathcal{D}, \ \ \forall t \in [0,T]\}. \tag{6}$$

Define $\mathcal{F}_1 : \Theta \to \mathcal{X}$ by $\mathcal{F}_1(x) = F(\cdot, x(\cdot))$ and $H_1(x)(t) = \mathcal{F}_1(x)(t) - Lx(t)$.

Now, construct a homotopy family

$$H_\lambda : (Dom(L) \cap \Theta) \times [0,1] \to \mathcal{X}$$

to be of the form

$$H_\lambda(x)(t) = \mathcal{F}_\lambda(x)(t) - Lx(t), \tag{7}$$

where $\mathcal{F}_\lambda : \Theta \times [0,1] \to \mathcal{X}$ with

$$\mathcal{F}_\lambda(x)(t) = \begin{pmatrix} u(t)(1-u(t)) - \frac{\tilde{a}(t)u(t)v(t)}{u(t)+\tilde{a}(t)} \\ b(t)v(t)(1-\frac{v(t)}{u(t)}) \end{pmatrix}. \quad (8)$$

Here $\tilde{a}(t) = (1-\lambda)\frac{1}{6} + \lambda a(t)$, and $\tilde{d}(t) = (1-\lambda)\frac{1}{3} + \lambda d(t)$ with $\lambda \in [0,1]$. It is easy to verify that $\mathcal{F}_\lambda : \overline{\Theta} \times [0,1] \to \mathcal{X}$ is L-compact. For more details of degree theory, see [4, Ch. I–IV].

Lemma 3.2. *Given $\lambda \in [0,1]$, if $x_\lambda \in \overline{\Theta}$ is a T-periodic solution of the system*

$$\dot{x}(t) = \mathcal{F}_\lambda(x)(t), \quad (9)$$

then $x_\lambda \in \Theta$.

Proof. Clearly, $\tilde{a}$ and $\tilde{d}$ satisfy conditions (2). System (9) is an ESP-P model. By Lemma 2.1, $\mathcal{D}$ is still a strictly positively invariant region of system (9). None of T-periodic solutions of (9) in $\overline{\Theta}$ can touch the boundary of $\mathcal{D}$. □

Corollary 3.1. $0 \notin H_\lambda((Dom(L) \cap \partial\mathcal{D}) \times [0,1])$.

Lemma 3.3. *$D_L(H_0(x)(t), \Theta) = D_B(H_0(x)(t), \mathcal{D}) = 1$, where D_L denote Leray-Schauder degree and D_B denote Brouwer degree.*

Proof. For the system $H_0(x)(t) = 0$, there is only one steady-state $p = (u_0, v_0) = (\frac{(1-a-d)+\{(1-a-d)^2+4d\}^{\frac{1}{2}}}{2}, \frac{(1-a-d)+\{(1-a-d)^2+4d\}^{\frac{1}{2}}}{2}) = (\frac{1}{4}+\sqrt{\frac{1}{16}+\frac{1}{3}}, \frac{1}{4}+\sqrt{\frac{1}{16}+\frac{1}{3}})$ in the strictly positively invariant region $\mathcal{D}$, which is a trivial T-periodic solution. Since $H_0(x)(t) = 0$ is an autonomous system, Proposition 2.1 and Bendixson's Criteria [6, p.264] guarantee that p is only one T-periodic solution in $\mathcal{D}$.

For the system $H_0(x)(t) = 0$, Leray-Schauder degree of H_0 in $\mathcal{D}$ is in fact reduced into Brouwer degree. Therefore, by Proposition 2.1, $D_L(H_0, \Theta) = D_B(H_0, \mathcal{D}) = D_B(\mathcal{F}_0, \mathcal{D}) = sign(\det A(t)) = 1$, where $A(t)$ is defined in (4). □

Lemma 3.4. *For system (4) with conditions (2), zero is the only T-periodic solution.*

Proof. Suppose (4) has a non-trivial T-periodic solution called $W_1(t)$. By Proposition 2.1 and Floquet theory [2, p. 93-105], its orbit Γ is an orbitally asymptotically stable. For $s \in \mathbb{R}$, $sW_1(t)$ is also a T-periodic solution of (4). Then orbit of $sW_1(t)$ can not be attracted to Γ for any $s \in \mathbb{R}$. This leads a contradiction to the orbital asymptotic stability of Γ. □

Remark 3.1. For the linear system (4), if $tr(A(t))$ does not change sign in some simply connected region $E \subset \mathbb{R}^2$, then (4) has no non-trivial periodic solution in E; since the system (4) is a linearization of an non-autonomous system, Bendixson's Criteria can not be used to prove Lemma 3.4.

Lemma 3.5. *Suppose $\mathcal{F} : \mathcal{X} \to \mathcal{X}$ is a completely continuous map of a Banach space such that $\mathcal{F}(0) = 0$ and $\mathcal{F}$ is Frechet differentiable at 0 with Frechet derivative $T \in K(\mathcal{X})$, where $K(\mathcal{X})$ is a set of all compact operators defined on $\mathcal{X}$. If $I - T \in L(\mathcal{X})$ is regular (invertible), then there exists $\eta > 0$ such that, for $B = \{x \in \mathcal{X} : \|x\|_\infty < \eta\}$, we have*

$$D(\mathcal{F} - I, B) = D(T - I, B).$$

For proof, see [1, Ch. 14].

Assume system (4) is the linearization of system (1) with respect to $x_0(t)$, by Theorem 2.10 of [2, p.97], system (4) can be transformed into an autonomous system

$$\dot{Z}(t) = RZ(t), \quad (10)$$

where R is called a Monodromy matrix of $A(t)$.

Lemma 3.6. *Suppose, in $\mathcal{D}$, $x_0(t)$ is a unique T-periodic solution of system (1) with condition (2). Let $A(t)$ and $W(t)$ be defined in (4), $LW(t) = \dot{W}(t)$; $B(x_0(t), \epsilon) \subset \Theta$ denote a small neighborhood of $x_0(t)$, $B(0, \epsilon) \subset \Theta \setminus \{x_0(t)\}$ denote a small neighborhood of 0. Set*

$$Q(W(t), \lambda) = (\lambda R + (1-\lambda)A(t))W(t) - LW(t),$$

then

$$D_B(Q(\cdot, 1), B(0, \epsilon) \cap \mathbb{R}^2) = 1.$$

Proof. Clearly, $tr(R) = \rho_1 + \rho_2 = \frac{1}{T}\int_0^T tr(A(s))ds(mod\frac{2\pi i}{T}) < 0$, where ρ_1 and ρ_2 are eigenvalues of R. By Proposition 2.1, for any $\lambda \in [0,1]$,

$$tr(\lambda R + (1-\lambda)A(t)) = \lambda tr(R) + (1-\lambda)tr(A(t)) < 0,$$

and Remark 3.1 implies that $Q(W(t), \lambda) = 0$ has only trivial T-periodic solution in $B(0, \epsilon)$. By degree invariance with respect to homotopy family,

$$D_L(Q(\cdot,0),B(0,\epsilon)) = D_L(Q(\cdot,1),B(0,\epsilon)) = D_B(Q(\cdot,1),B(0,\epsilon)\cap\mathbb{R}^2).$$

Consider the Taylor expansion of $H_1(x)(t)$ at $x_0(t) \in B(x_0(t),\epsilon)$, where $H_1(x)(t)$ is defined in (7) as $\lambda = 1$. Then we have $H_1(x)(t) = H_1(x_0)(t) + M(t)(x(t)-x_0(t)) + h(t,x(t)-x_0(t))$, where $M = \mathcal{F}_1' - L$ is $(H_1)'_x$ and $h(t,x(t)-x_0(t))$ is a function of $o(\|x(t)-x_0(t)\|_\infty)$. Since $x_0(t)$ is the unique solution of $H_1(x)(t) = 0$ in $\mathcal{D}$, by excision property of the degree, $D_L(H_1(x)(t),\Theta) = D_L(H_1(x)(t),B(x_0(t),\epsilon))$ and by Lemma 3.5, $D_L(H_1(x)(t),B(x_0(t),\epsilon)) = D_L(M(t)(x(t)-x_0(t)),B(x_0(t),\epsilon))$.

Let $W(t) = x(t) - x_0(t)$, then by Lemma 3.3,

$$\begin{aligned}&D_L(H_0(x)(t),\Theta) = D_L(H_1(x)(t),\Theta)\\&= D_L(M(t)W(t),B(0,\epsilon)) = D_L(Q(\cdot,0),B(0,\epsilon))\\&= D_L(Q(\cdot,1),B(0,\epsilon)) = D_B(Q(\cdot,1),B(0,\epsilon)\cap\mathbb{R}^2)\\&= 1.\end{aligned}$$

□

4. Main Results

Theorem 4.1. *For ESP-P model, there exists only one uniformly asymptotically stable T-periodic solution $x_0(t)$ in $\mathcal{D}$.*

Proof. (**Existence**) Combine Lemma 3.2, Corollary 3.1 and Lemma 3.3, by a general existence theorem of the Leray-Schauder type, we get $D_L(H_1(x)(t),\Theta) = D_L(H_0(x)(t),\Theta) = D_B(\mathcal{F}_0(x)(t),\mathcal{D}) = 1$, which implies that there at least exists one T-periodic solution $x_0(t) = (u_0(t),v_0(t))^T$ of ESP-P model in $\mathcal{D}$. If a and d are constants, it is easy to show that there is only one trivial T-periodic solution $x_0 \in int(\mathcal{D})$, otherwise, we can easily verify that $x_0(t)$ is a nontrivial T-periodic solution of ESP-P model in $\mathcal{D}$ by substituting $x_0(t)$ into ESP-P model.

(**Uniqueness**) Define $C_T = \{x(t) \in \Theta | x(t)$ satisfies (1) with conditions (2)$\}$. Since $x_0(t) \in C_T$, C_T is not an empty set. If a and d are constants, there is only one constant solution in C_T.

If one of $a(t)$ and $d(t)$ is a non-trivial T-periodic function, then $x_0(t) \in C_T$ is a non-trivial T-periodic solution. Assume C_T is not a singleton; we pick $x_1(t) = \begin{pmatrix}u_1(t)\\v_1(t)\end{pmatrix}$, $x_2(t) = \begin{pmatrix}u_2(t)\\v_2(t)\end{pmatrix}$ in C_T and substitute them into (1) to get

$$\dot{x_i}(t) = F(t,x_i(t)), \quad i = 1,2. \tag{11}$$

Define $z(t) = x_1(t) - x_2(t)$. By the mean value theorem, we get

$$\dot{z}(t) = z(t)\int_0^1 F_x'[t,x_2(t)+\theta(x_1(t)-x_2(t))]d\theta \tag{12}$$

and

$$\int_0^1 F_x'[t,x_2(t)+\theta(x_1(t)-x_2(t))]d\theta = \begin{pmatrix}m(t) & p(t)\\ q(t) & n(t)\end{pmatrix},$$

where

$$m(t) = \int_0^1 \{1 - 2[u_2(t)+\theta(u_1(t)-u_2(t))] - \frac{a(t)d(t)[v_2(t)+\theta(v_1(t)-v_2(t))]}{[u_2(t)+\theta(u_1(t)-u_2(t))]+d(t)}\}d\theta,$$

$$n(t) = \int_0^1 b(t)\{1 - 2\frac{[v_2(t)+\theta(v_1(t)-v_2(t))]}{[u_2(t)+\theta(u_1(t)-u_2(t))]}\}d\theta.$$

Since

$$m(t) = 1 - u_1(t) - u_2(t) - a(t)d(t)\int_0^1 \frac{a(t)d(t)[v_2(t)+\theta(v_1(t)-v_2(t))]}{[u_2(t)+\theta(u_1(t)-u_2(t))]+d(t)}\}d\theta$$

and

$$n(t) = b(t)\int_0^1 \{1 - 2\frac{[v_2(t)+\theta(v_1(t)-v_2(t))]}{[u_2(t)+\theta(u_1(t)-u_2(t))]}\}d\theta,$$

By conditions (2) and (3), we have the following facts:

$$1 - u_1(t) - u_2(t) < 0,$$

$$v_2(t) + \theta(v_1(t)-v_2(t)) \geq 0, \forall\theta \in [0,1],$$

$$[u_2(t)+\theta(u_1(t)-u_2(t))] + d(t) > 0,$$

so $m(t) < 0$; also since

$$0 < u_2(t) + \theta(u_1(t)-u_2(t)) \leq 0.999, \forall\theta \in [0,1]$$

and

$$\int_0^1 2\frac{v_2(t)+\theta(v_1(t)-v_2(t))}{0.999}d\theta = \frac{v_1(t)+v_2(t)}{0.999} > 1,$$

then $n(t) < 0$.

It follows that

$$tr\left(\int_0^1 F_x'[t,x_2(t)+\theta(x_1(t)-x_2(t))]d\theta\right) < 0. \tag{13}$$

(13) implies that the zero solution is the only one T-periodic solution for (12) by Remark 3.1. Hence, $x_1(t) = x_2(t)$. C_T is a singleton.

(**Stability**) If $a(t)$ and $d(t)$ are constant functions, then $x_0 \in \mathcal{D}$ is a constant solution of system (1). By Proposition 2.1, x_0 is a locally uniformly asymptotically stable solution.

If one of $a(t)$ and $d(t)$ is a non-trivial T-periodic function, $x_0(t)$ is a non-trivial T-periodic solution of system (1). By Proposition 2.1, one Floquet exponent ρ_1 has negative real part. If $Real(\rho_2) < 0$, $x_0(t)$ is locally uniformly asymptotically stable by Theorem 2.13 of [2, p.101] and Lemma 3.1.

We now show that $Real(\rho_2) < 0$ always holds by treating two cases: (1) ρ_2 is a complex number. Notice that ρ_1 and ρ_2 are conjugate eigenvalues of R. Therefore, $Real(\rho_1) < 0$ implies that $Real(\rho_2) < 0$; (2) ρ_2 is a real number. Clearly ρ_1 is also a real number and $\rho_1 < 0$ implies that $\rho_2 \neq 0$ (otherwise, (4) must have one non-trivial T-periodic solution, which contradicts Lemma 3.4). If $\rho_2 > 0$, then $\det(R) < 0$, $sign(\det(R)) = -1 = D_B(Q(\cdot, 1), E(0, \epsilon) \cap \mathbb{R}^2)$, which contradicts Lemma 3.6. □

5. Future Works

In this paper, we proved the existence, uniqueness and stability of the periodic solution $x_0(t)$ of ESP-P model. This establishes a foundation for further studying patterns of EP-P model. Of course, the problem mentioned here is still open, the investigation of this question is currently underway.

Acknowledgment

This work was supported by Research Fund of North China Electric Power University (93509001).

References

1. R. F. Brown, *A topological introduction to nonlinear analysis*, Birkhäuser, Boston (1993).
2. J. Cronin, *Differential equation*, 2^{nd}ed., Marcel Dekker, New York (1994).
3. M. A. Krasnosel'skii, *Translation along trajectories of differential equations*, American Mathematical Society, Providence, Rhode Island, Translations of Mathematical Monographs **19** (1968).
4. J. Mawhin, *Topological degree method in nonlinear boundary value problems-Regional Conference Series in Mathematics 40*, Rhode Island, American Mathematical Society (1979).
5. James D. Murray, *Mathematical Biology*, Second Edition, New York, Springer-Verlag (1993).
6. L. Perko, *Differential equations and dynamical systems*, Third Edition, Springer, New York (2001).

CT AND MRI IMAGE FUSION IN RADIOTHERAPY FOR TRANSCRANIAL TUMOR*

XIUPING LIU, QIAN LIU

Britton Chance Center for Biomedical Photonics, Wuhan National Laboratory for Optoelectronics, Huazhong University of Science and Technology, Wuhan 430074, P. R. China

Application of the image fusion method, which employs both computed tomography (CT) and magnetic resonance (MR) images, to dimensional orientation of the transcranial tumor greatly contributes to current radiotherapy for its clear presentation of the disease focus, the volume of tumor as well as the accurate localization of the target region, thus facilitating the high-precision, high-dose, high-curative efficiency and low-damnification featured programming of modern radiation oncology. Method: The patient-specific alignment and registration parameters for MR image transformation with respect to CT images are usually explored by examination on most mutual information of grads shared by each two corresponding images. Anatomical characteristics are further used to acquire best parameters for image fusion. Result: The resulted fused images that incorporates complementary advantages between CT and MRI images lends localizing information for early diagnosis of transcranial tumor and the best reference dose for final clinic target irradiation.

1. Introduction

Modern medical imaging techniques have offered extensive information and methodological possibilities for disease diagnose and clinic therapy. Comprehensive integration and utilization of image information across modalities, which seeks to achieve advantage complementation, has been demonstrated to be necessary in clinical activities.[1] On the radiotherapy for transcranial tumor, for instance, precise delineation of the target is always considered as the most fundamental procedure. Image fusion method was developed based on the notion that the fused image, which integrate and optimize the information for a given point or area from different sources, is always more expressive and contributive than individual images since it is able to lend detailed relationship and characteristics between the tumor and surrounding normal regions as well as accurate localizing of the target.[2] It also provides guidance for early diagnosis of tumor and the best reference dose for applied target irradiation in clinic. The research on medical image fusion usually has two aspects: the registration of different images and the study on fused images.

This paper describes the fusion procedure of CT and MR images for radiology of patients suffering by transcranial tumor. The experiments of most mutual information based image registration and pixel-level image fusion on CT and MR images selected from clinic cases[3] have effectively contributed to the solution of difficulty lying in the fusion of dual-mode images.

2. Materials

2.1. *Patient Specimen*

The first patient specimen was a 47-year-old male with presenting compliant of discontinuous headache for 2 years and was also accompanied by weakness on left upper arm and twitch in the most recent 2 weeks. Medical record showed that he was conscious; capable of walking; weakness on left half of the body, but without any neuro-paralysis. The brain CT examination reports abnormal low-density area at the right frontal lobe. The visualized wall knur, edema around tumor focus, press at right and third ventricles and the shift of median line toward the left side were presented by irregular annular enhancement after Iopromide injection. An irregular area of $5.5 \times 3.5 \times 5.0$ cm^3 with vague edge are indicated in the right frontal lobe by a long farraginous signal of T1 and T2 types in MR examination. The callosum is strained downwardly. The press at the right and third ventricles and the shift of median line were also exhibited after enhancement. Pathological change of the right frontal lobe with neuroglial tumor was diagnosed when entering the hospital. Tumor excision approached along the right path was performed in April of 2007, where the examination of frozen slice reported II class astrocytoma. Patient got well convalescence and the symptoms disappeared after the operation. Glioblastoma multiforme was reported in the post-operation pathologic examination reported. Radiotherapy began at

* This work is supported by the National High-Tech Research and Development Program of China (863 program: 2006AA02Z343).

3 weeks later. X-ray with an energy of 6 MV was adopted to scan the whole cerebrum form both sides with an irradiation area of 16 × 13 cm^2. The dose for median tumor was set to 40 Gy. Then the dose was raised to 60 Gy for a target depth of 4 cm at a reduced area of 8 × 7 cm^2 from the right side. There is no recrudescence reported by periodic CT examinations after radiotherapy.

The second patient specimen was a 43-year-old female with presenting complaint of headache for more than 2 months accompanied by memory declination and character changing. The farraginous-density tumor that appears to split in leafs with a macula observed at the median line and the damaged slope and osseous tissue were shown in CT image after slight enhancement. Farraginous signal with a comparative lower T1 signal and a higher T2 signal was displayed in MR examination. Chordoma was diagnosed for X-ray Radiotherapy with dose of 32 Gy at median line. Irradiation area was limited within a radius of 15 mm. 80% dose was placed on the tumor edge and set for prescription dose to target. SRS therapy with 6 MV X-ray from 5 different planes was performed for edge of median tumor with dose of 17 Gy, which was within the central dose range of 21 Gy.

Table 1. Comparison of image parameters of Ct and MR image sets.

	Male		Female	
	CT	MR	CT	MR
Scanning interval (cm)	3	3	3	3
Scanned layers	38	40	38	40
Pixel resolution	128 × 128	512 × 512	128 × 128	512 × 512
Color bit	16 (gray)	16 (gray)	16 (gray)	16 (gray)
Image format	DICOM	DICOM	DICOM	DICOM

2.2. *Facility Requirement*

The Marconi Mx8000 Quad Multi-layer Helix Computed Tomography scanner and Marconi Eclipse 1.5 T Super-conduct Magnetic Resonance scanner were employed respectively for image acquisition. Skulls of both patient specimens were fixed before horizontal scanning in order to achieve the spatial consistence between the osseous tissue in CT images and the soft tissue in MR images.

2.3. *Image Materials*

The CT and MR images used in this study are acquired from the Medical Imaging Center in Hubei Cancer Hospital. The CT images that resulted from 38-layer successive CT scan with an interval of 3 mm have a resolution of 128 × 128 pixels corresponding to a pixel size of 2.08 mm^3. The MR images based on continuous T1W1 imaging under SE sequence have a 3 mm slice thickness and have a resolution of 256 × 256 pixels with a pixel size of 0.92 mm^3. The images, as shown in Figure 1 and 2, are stored and treated in format of DICOM 3.0. Mode parameters of the source image set are tabulated in table 1.

3. Method

Registration refers to the process of pixel alignment and geometrical transformation of the original images.[4, 5] Mutual information (MI) has always been an important factor in various approaches of image registration across modalities.[6] The MI-based method is able to automatically match the images from different equipments without manual interference. After image pretreatment,[7] the patient specific alignment and registration parameters for MRI image transform with respect to CT images are usually explored by examination on most mutual information of grads shared by each two corresponding images.

Image fusion refers to the integration of images acquired from different facilities and the generation of new image that contains the information inaccessible from individual equipment.[8, 9] Image fusion on pixel level turns to be highly anti-jamming and advantageous at detail representation within various methods for image fusion.

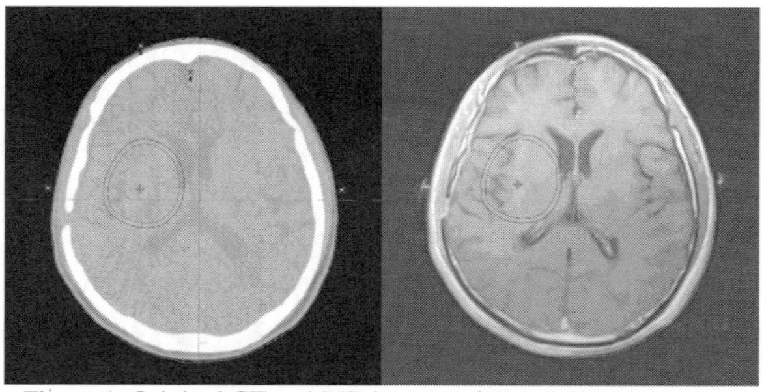

Figure 1. Original CT and MR images of the adult male patient.

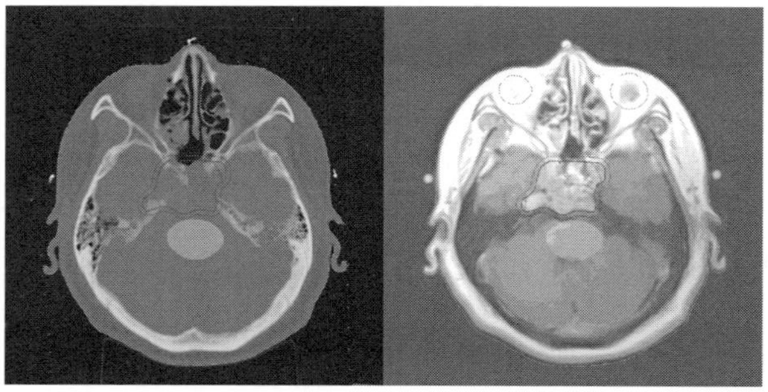

Figure 2. Original CT and MR images of the adult female patient.

3.1. *Registration of Mutual Information Based on Image Grad Characteristic*

3.1.1. Mutual information

Mutual information (MI) is a conception referring to the measurement or calibration of the degree of correlation lying in a couple of random variables in information statistics.[10] The MI of two given images (A and B) is defined as:

$$I(A,B) = \sum_{a,b} P_{A,B}(a,b)\log\frac{P_{A,B}(a,b)}{P_A(a)\cdot P_B(b)}$$

Where the $P_A(a)$ and $P_B(b)$ are the Independent Probability and the $P_{A,B}(a, b)$ is the Associated Probability. The MI reaches maximum when the spatial location of the two images are completely agreed with each other.

3.1.2. MI calculation of grad characteristic

MI is a quantity based on statistics of grad relevance, while does not utilize the inherent information of images. It has been demonstrated that direction of grads at most pixels on the contour of a certain organ or tissue always keeps consistent or opposite across modalities, although the gray scale and resolution may remain different.[11] Familiarity between images can be estimated by discrepancy of grad angle at corresponding points after transform, as shown below:

$$a_{AB} = \arccos\frac{\nabla x\cdot\nabla x'}{|\nabla x|\cdot|\nabla x'|}$$

Where the ∇x and ∇x' refer to the grad value before and after transform, respectively. The final objective function can be defined as:

$$I_{new}(A,B) = G(A,B)I(A,B)$$

The registration is considered optimized when I_{new} (A, B) reaches the maximum value.

3.1.3. Image registration

The noise and foreground within the original images could be distinguished by automatically set threshold on gray value, below which the pixel was uniformly set to zero and considered as background. Then the pixel size of CT and MR images are magnified to 2.08 mm^2. The resolution of CT images is increased to 256 × 256 pixels. In reference to the points at edge of the cerebrum, each two corresponding image sequences are able to be aligned and registered by gradual reduction of distance between characteristic points through MI based image registration approach.

3.2. *Pixel Level Image Fusion Method Based on SGNN*

3.2.1. Image pretreatment

One method that effectively reduces the noise interference as maintaining the image details can be developed based on the theory of Wavelet Transform and Self-Generating Neural Network (SGNN).[12] The Discrete Wavelet Transform for an input image f(x, y) with a size of $n_1 \times n_2$ is

$$w_\phi(j_0,m,n) = \frac{1}{\sqrt{n_1 n_2}}\sum_{x=0}^{n_1-1}\sum_{y=0}^{n_2-1} f(x,y)\phi_{j_0,m,n}(x,y)$$

$$w_\phi^i(j,m,n) = \frac{1}{\sqrt{n_1 n_2}}\sum_{x=0}^{n_1}\sum_{y=0}^{n_2} f(x,y)\phi_{j,m,n}^i(x,y)$$

under the assumption of f(x, y) = f_1(x, y) + f_2(x, y), where the f_1(x, y) is the expected signal and the f_2(x, y) is the noise signal. The Wavelet Transform is focused on the finite part of frequency space, while the coefficients for transform of noise can be any part. Systems transform are often categorized as noise transform based, where the range of wavelet coefficient is in direct proportion to the number of wavelet and should be kept for subsequent processing, or signal transform based, including the transforming result with noise, where the range of wavelet coefficient is in reverse proportion to the number of wavelet and should be eliminated. In this way, the image details are able to be preserved as reducing the noise since most of the wavelet coefficients are preserved.

3.2.2. Pixel level image fusion

Image fusion is to increase to precision and accuracy of the ambiguous quantity, also the degree of membership. Assuming that there are M images with a uniform size of $n_1 \times n_2$ pixels for fusion, the degree of membership of pixel at (m, n) with respect to j category can be calculated by

$$\mu_j(p_t(m,n)) = \max_{1 \le i \le M}\{\mu_j(p_i(m,n))\}$$

$$j = 1,\cdots,k; m = 1,2,\cdots,n_1; n = 1,2,\cdots,n_2$$

where the $p_i(m, n)$ refers the gray value at (m, n) in image i. The gray value at (m, n) in fused image F can be calculated by weighting averaged method as

$$p_f(m,n) = \frac{(\sum_{j=1}^{k} \mu_j(p_f(m,n))w_j)}{\sum_{j=1}^{k} \mu_j(p_f(m,n))}$$

$$m = 1,2,\cdots,n_1; n = 1,2,\cdots,n_2$$

The fused image can be finally acquired by serial calculation of gray value at every pixel.

3.2.3. Procedures of image fusion

The procedure of image fusion includes characteristic extraction, uncertainty estimation and final image alignment and visualization. The fusing procedure of original images, from I_1 to I_M, is illustrated in figure 3. MI based fusion method is employed after filtering, registration and characteristic transform. The resulted image needs reverse transform before final completion.

4. Fusion Experiments of CT & MR Images

4.1. *Experiment 1*

This experiment serves to examine the anti-jamming ability of image fusion method by comparison of fused images, where the Discrete Wavelet Transform (SWT) and Laplace Pyramid algorithm (LAP) and SGNN-based Image Fusion method (SIF) were adopted respectively. The original CT image of 128 × 128 pixels selected from the sample image set in this study first undergo intended contamination by two different types of noise, as shown by A_1 and B_1 in figure 3. Then DWT, LAP and SIF approaches were then employed individually to generate the fused images (d, e, f). Images A_2 and B_2 were acquired from A_1 and B_1 by Wavelet Transform for noise elimination. Another group of fused images (i, j, k) were also generated by the three approaches. Analysis of statistical errors of three methods is tabulated in table 3. MSE (Mean Square Error) and MAE (Mean Absolute Error) are defined as

$$MSE = \frac{\sum_{m=1}^{n_1}\sum_{n=1}^{n_2}(z(m,n) - o(m,n))^2}{n_1 \times n_2}$$

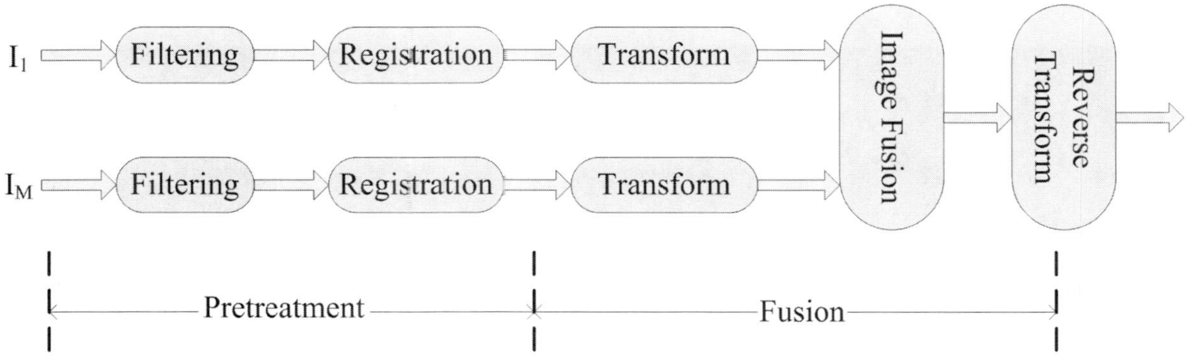

Figure 3. Procedures of image fusion.

$$MAE = \frac{\sum_{m=1}^{n_1}\sum_{n=1}^{n_2} |(z(m,n)-o(m,n))|}{n_1 \times n_2}$$

Where n1 and n2 refer to the length and width of the image, z(m, n) refers to the gray value of the comparative image. The o(m, n) refers to the gray value of original image.

It could be observed that, before the pretreatment of Wavelet Transform, the MSE and MAE levels of SIF are 68% and 44% than DWT, 66% and 46% lower than LAP, respectively,. After Wavelet Transform, they are 39% and 23% than DWT, 36% and 22% lower than LAP, respectively. SIF is always advantageous than the other two approaches. The values of both MSE and MAE have been evidently reduced by Wavelet Transform. Therefore, SIF method after Wavelet Transform should be the best choice.

Table 2 . Analysis of statistical errors for image fusion of A1 and B1

	A_1	B_1	DWT	LAP	SIF
MSE	0.019	0.016	0.027	0.009	0.028
MAE	0.111	0.100	0.141	0.078	0.136

Table 3. Analysis of statistical errors for image fusion of A_2 and B_2

	A_2	B_2	DWT	LAP	SIF
MSE	0.005	0.006	0.006	0.006	0.004
MAE	0.051	0.058	0.059	0.058	0.045

4.2. *Experiment 2*

In this experiment, a CT image of 128 × 128 pixels and a MR image of 256 × 256 pixels are fused by DWT, LAP and SIF approaches, respectively, as displayed in figure 4. The result from SIF method is obviously better than the other two.

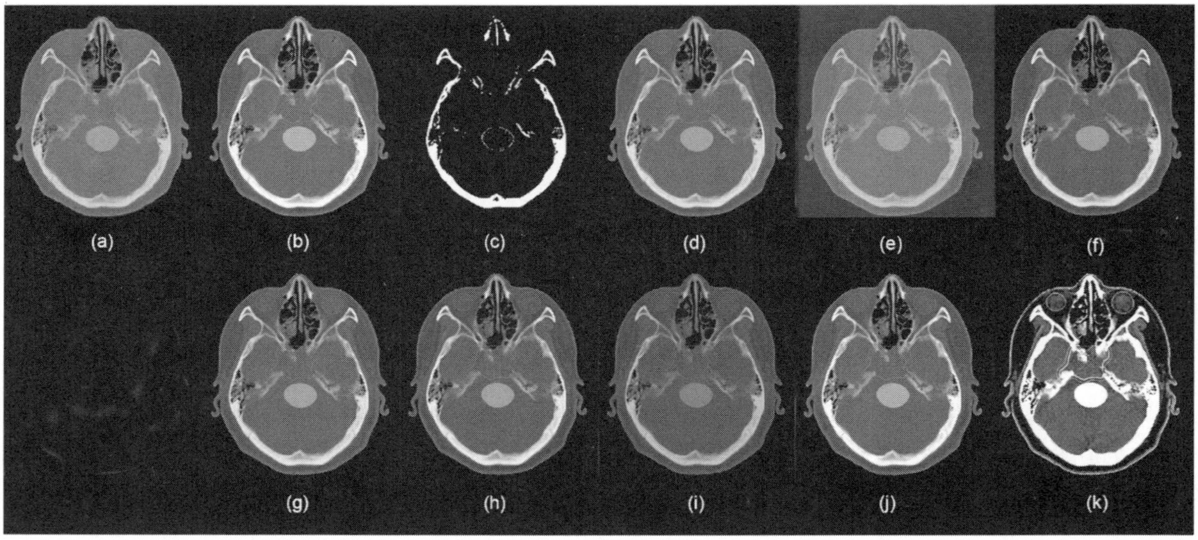

Figure 4. Original image (a), contaminated images (b, c, g, h) and fused images (e, f, j, k).

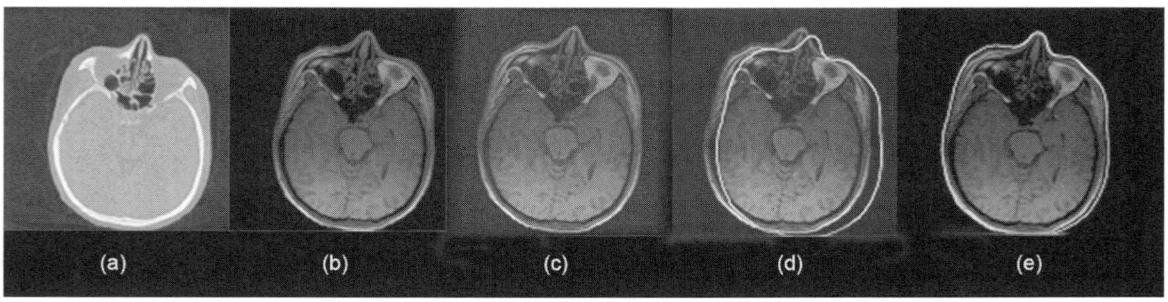

Figure 5. Original CT (a) and MR (b) images, and fused images by 3 different methods (c, d, e).

5. Results and Discussion

5.1. *Fusion of CT & MR Image*

The image fusion method as described above, which effectively integrates the detailed information and complementary advantages of CT and MR images, provides precise information of tumor localization and optimized dose reference for clinical irradiation. The original CT image, the corresponding MR image before and after registration, and the result of fused image for disease focus of the two patients is shown in figure 5 and 6, respectively. It could be observed that the fused image expresses more information on details and contours.

5.2. *Application of CT & MR Fused Image*

Medical images are always categorized as anatomical or functional.[13] The first type, such as CT images, provides detailed anatomical information while fails to reflect the functional situation of organs and tissues. The later type, such as MR images, offers irreplaceable information on metabolism but remains in low resolution level. In clinical activities, CT images are usually used for basic dose reference in precise irradiation,[14] though limited on distinguishing of the tumor edge. In contrast, MR images are able to clearly present the anatomical structure and focus edge by its high resolution on density. Therefore, it is always inconsistent for delineation and localization of tumors using single modality, especially in cerebrum. The fused image has been demonstrated to be successful in this field. Comparison of various items in clinical practice for CT, MRI and fused images is tabulated in table 3. The individual advantage and disadvantage of CT and MI image in presentation of pathological changes, such as tumor, edema, and denaturalization, anemia, intensification, acute and chronic hemorrhage, calcification, are compared in table 3(1). The significant difference of CT and MR images in displaying basic situations, including definite organ, examined focus, focus volume, localization and nature determination, functional imaging and panoramic imaging, damage and allergic reaction, are shown in table 3(2) and could be complemented by CT and MR image fusion.

5.3. *Difference with Other Approaches*

Image superposition is another way for image fusion in Medical activities. It employs the perspective principle,

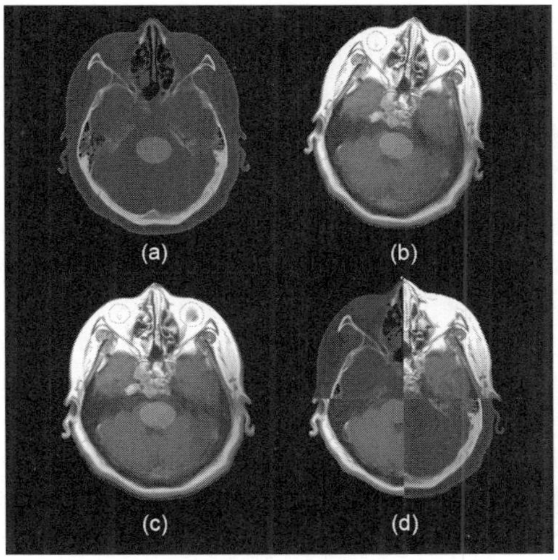

Figure 6. (a) Original CT image, (b) MR image before and (c) after registration, and (d) CT & MR fused image of the male patient.

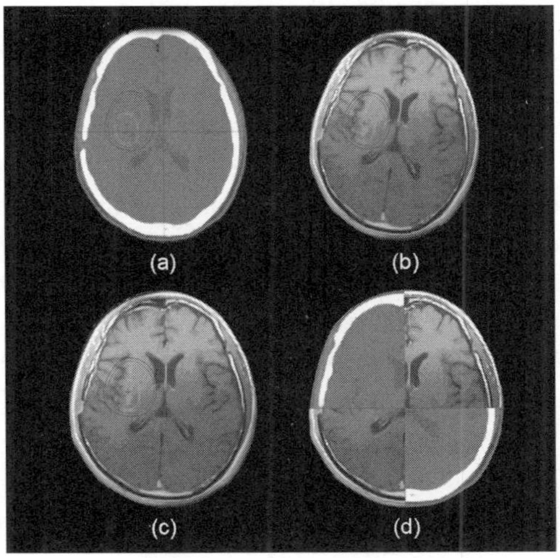

Figure 7. (a) Original CT image, (b) MR image before and (c) after registration, and (d) CT & MR fused image of the female patient.

which simply overlaps the images based on degree of transparency. The resulted image is unable to perform effective control on original information and is always difficult for three-dimensional reconstruction due to isotropic image sequences.[15] In contrast, the Wavelet Transform approach can effectively solve these problems but is usually more complex for characteristic extraction and more sensitive to registration uncertainties. The pixel level image fusion method is highly anti-jamming to registration errors since it fuses the image at a pixel level. It also has conspicuous advantage on characteristic extraction, computational mass reduction, and real-time visualization in comparison with the other two methods. The resulted image presents clear structure and detailed functional information and has a certain degree of tolerance to registration uncertainties.

5.4. *Application of Image Fusion Method in Radiology of Transcranial Tumor*

The information based on single modality is insufficient for precise pathological localization due to the structural complexity of human brain. Experience and evidence from dissection provides common reference but is still unavailable for accurate identification since there are not as plenty of time in real clinic as in cadaver study. Image fusion method has been demonstrated to be a viable alternative for clinical diagnosis.[16] The approach described in this paper has great advantage on registration, segmentation and image fusion. The test on patient specimen has demonstrated that this strategy has great potential and benefit to Medicine.

6. Conclusion

A detailed strategy for dual-mode image fusion has been illustrated by selected CT and MR image registration based on mutual information and image fusion at pixel level. The resulted image of accurate registration and nice presentation is able to provide complementary information for anatomical and pathological research as well as offer quantitative references to optimized dose in clinical irradiation.

References

1. Yu C, Petrovich Z, Apuzzo ML, *et al.* An image fusion study of metric accuracy of geometric accuracy of magnetic resonance imaging with the leksell sterotactic localization system. J. Appl. Clin. Med. Phys. 2001, 2:42.
2. David W. PET/ CT: Today and tomorrow. J. Nucl Med, 2004, 45:4.
3. Woods Roger P, Grafton Scott T, Holmes CJ, et al. Automated image registration: I. general methods and int-rasubject, intra2modality validation [J]. J. Comput. Assist. Tomogr., 1998, **22 (1)**:139-152.
4. Pluim JP, Maintz JB, Viergever MA. Image registration by maximization of combined mutual information and gradient information. IEEE Trans Med Imaging, 2000, **196(6)**:809.
5. Woods RP, Grafton ST, Watson JD, et al. Automated image registration: II. Intersubject validation of linear and nonlinear models. Journal Computer Assisted Tomography, 1998, **22(1)**:153.
6. M Frederik, Andre C. Multimodality image registration by maximization of mutual information [J]. IEEE Trans Medical Imaging, 1997, **16(2)**:187-198.
7. Zitova B, Flusser J . Image registration methods: a survey [J]. Image Vis. Comput., 2003, 24:97721000
8. Maiter H, Bloch I. Image Fusion. Vistas in Astronomy, 1997, **41(3)**:329-335.
9. Taxt T, Solberg A H S. Information Fusion in Rusion in Remote Sensing. Vistas in Astronomy, 1997, **41(3)**:337-342.
10. Collignon A, Maes F, Delaere D, et al. Automated multimodality medical image registration using information theory [A]. Proc 14th Int Conf Information Processing in Medical Imaging (IPMI'95) [C]. I1e de Berder, France: IEEE Press, 1995, 263-274.
11. Pluim J, Maintz J, Viergever M. Image registration by maximization of combined mutual information and gradient information [J]. IEEE Transactions on medical imaging, 2000, **19(8)**:809-814.
12. Li S T, Kwor J T, Wang Y N. Using the Discrete Wavelet Frame Transform to Merge Landsat TM and SPOT Panchromatic Image. Information Fusion, 2001, **(3)**:17-23.
13. Li J, Zhou Y, Li D. The research of high resolution colorful and multispectral image's fusing by means of wavelet transform. RST, 1999, **3(2)**:1-6.
14. Poter PJ, Shin AY, Detsky A. Set al. Surgery versus sterotactic radiosurgery for small, operable cerebral arteriovenous malformations: a clinical and cost comparison. Neurosurgery, 1997, **41**:757.
15. Liu Q, Gong H, Luo QM. Parallel Visualization of Visible Chinese Human with Extremely Large Datasets [C]. Proceedings of the 2005 IEEE Engineering in Medicine and Biology 27th Annual Conference, Shanghai, China, September 1-4, 2005.
16. Van den Elsen PA, Maintz JB, Pol ED, et al. Automatic regist ration of CT and MR brain images using correlation of geomet rical features [J]. IEEE Trans. Med. Imaging, 1995, **14(2)**:3842396.